Cavitation in Biomedicine

Mingxi Wan · Yi Feng · Gail ter Haar
Editors

Cavitation in Biomedicine

Principles and Techniques

Springer

Editors
Mingxi Wan
Department of Biomedical Engineering
Xi'an Jiaotong University
Xi'an
China

Yi Feng
Department of Biomedical Engineering
Xi'an Jiaotong University
Xi'an
China

Gail ter Haar
Joint Department of Physics, Division of Radiotherapy & Imaging
The Institute of Cancer Research
Sutton, Surrey
UK

ISBN 978-94-017-7254-9 ISBN 978-94-017-7255-6 (eBook)
DOI 10.1007/978-94-017-7255-6

Library of Congress Control Number: 2015944449

Springer Dordrecht Heidelberg New York London

Printed on acid-free paper

Springer Science+Business Media B.V. Dordrecht is part of Springer Science+Business Media (www.springer.com)

Foreword

Acoustic cavitation is one of those fundamental scientific concepts that play an important role in almost every application that involves nearly all liquid-based environments that support a propagating acoustic wave. Because the acoustic wave involves successive compressions and rarefactions, if the propagating environment permits that environment to fail under tension, then acoustic cavitation of one form or another is generated. It may involve the violent fracture of a liquid under enormous negative pressures or the simple oscillation of a preexisting gas bubble under very modest acoustic intensities. It can exist in every liquid that has been investigated, with a wide range of inception thresholds, and in the most delicate tissues of the body, it can have both positive and negative implications. Accordingly, it is a topic of enormous breadth and relevance and must be considered whenever such acoustic propagation is considered. This book examines acoustic cavitation principally from it potential application to the general area of medical ultrasound, and primarily with regard to therapeutic ultrasound, in which the acoustic intensities are relatively large. It is composed of nine chapters, each of which covers a topic in considerable depth and breadth. It is not exhaustive, although it examines the overwhelming majority of topics of interest to anyone interested in therapeutic ultrasound.

Chapter 1 starts with a brief review of the fundamental physics of cavitation, with an introduction of such fundamental concepts as the cavitation threshold, nucleation, and bubble dynamics in a variety of environments. Once cavitation occurs, it is necessary to determine when and where it happens. Then, Chap. 2 develops the concept of cavitation detection and mapping. Topics covered are high-speed photography, and the detection of cavitation via sonoluminescence and sonochemiluminescence. When cavitation bubbles collapse, they generate acoustic emissions and one can use a transducer to detect the existing of cavitation by a concept called passive cavitation detection (PCD). Furthermore, if the transducer is an array, one can use these emissions to map the cavitation both temporally and spatially, called passive cavitation mapping (PCM). These topics are covered in some detail in this chapter. Chapter 3 then treats the distribution of bubbles within the cavitation field and the effect of such relevant parameters as acoustic pressure, frequency, and continuous or pulse modes of operation. Chapter 4 introduces the topic of the role of cavitation in tissue, with a discussion of the concept of cavitation-enhanced thermal heating, the bioheat transfer equation, and other cavitation-enhanced thermal effects. A brief description is also given of boiling histotripsy, micro- and nanoscale cavitation heating and thermometry. Violent cavitation can induce significant mechanical effects, such as the erosion of a ship's propeller or of a blade in a water turbine. Chapter 5 examines the role of cavitation-induced mechanical effects. At low intensities, acoustic microstreaming can occur and at high intensities shock wave, sonoluminescence, and reentrant liquid jets can occur. These mechanical effects are examined in the role they play in lithotripsy, histotripsy, sonothrombolysis, and in the enhanced delivery of drugs and genetic materials. Although cavitation can be a violent and disruptive mechanism, there are also means of controlling cavitation, a topic covered in Chap. 6. The dependence of liquid temperature, static pressure, overpressure, single, and multiple frequencies, as well as pulse length and duty cycle on the cavitation threshold and cavitation intensity are examined. The topic of cavitation histotripsy is also covered, in which the control of cavitation can lead to mechanical disruption of tissue and provides a nonthermal means for inducing tissue ablation. When cavitation occurs in tissue, the gas that is associated with the event lends itself to detection by ultrasound scanners, and the emissions that are associated with the bubble oscillations are detectable by sensitive hydrophones. Chapter 7 provides some detailed information on the various imaging techniques, including both active and passive approaches, time and frequency domain conditions, and pulsed and continuous operating conditions, as well as various super-resolution reconstruction and adapted beam-forming techniques. One method of producing cavitation that can be controlled in both space and time is that of laser-induced cavitation, a topic that is covered in some detail in Chap. 8. In this approach, a focused laser beam is rapidly absorbed in the liquid, thermal heating results in localized boiling, and vapor bubbles are produced. The subsequent unstable vapor bubble then goes through an undriven series of collapses and rebounds, which can be detected by both active and passive techniques. Also examined in this chapter is the photoacoustic cavitation in which the laser beam is activated during simultaneous application of an acoustic field, thus greatly lowering the intensities required of both the acoustic and optical beams. The final chapter is a broad examination of the effects, both desirable and undesirable, of cavitation on tissues. The mechanical effects of cavitation on tissues, termed mechanobiology, are examined via the physical and chemical mechanisms that give rise to these effects. It is possible also to induce desirable bioeffects by mechanobiology, such as sonothrombolysis, sonoporation, tissue erosion, and blood–brain barrier opening. These topics are covered in some detail in this chapter.

In summary, this book is a "must-have" for any researchers involved in therapeutic ultrasound research. It covers both the fundamentals and the latest research, with over 800 references, including many that cover the latest research on the topic. The various chapters are well tied together by the principal author, Mingxi Wan.

The authors are all well-respected authorities in their respective disciplines, with contributions by well-known researchers such as Charles Church, Gail ter Haar, and Tom Matula.

Prof. Lawrence A. Crum
University of Washington

Preface

The definition of cavitation is the formation of bubbles/cavities in a liquid or other medium, and the subsequent bubble dynamics. Cavitation has been studied in several scientific areas and used in a wide variety of applications, such as acoustics, hydrodynamics, fluid machinery, biomedicine, chemistry, material science, botany, food science, and petroleum engineering. Cavitation was avoided in earlier studies of fluid machinery and hydrodynamics. However, in recent years, it is increasingly utilized in a wide range of applications, such as ultrasonic cleaning, cell and bio-material processing, ultrasound imaging and therapy, cavitation-assisted chemical reactions, dialysis, and extraction.

Cavitation is becoming increasingly important in biomedical research and applications. It has been studied in three important categories: (i) Physics where the study of the physical phenomena and biophysical mechanisms has established the foundations of cavitation in biomedicine. This has become more interesting with the introduction of several novel micro-/nano-materials which act as artificial implanted cavitation nuclei; (ii) The development of the principles and engineering techniques in cavitation detection, imaging, characterization, and control has promoted the research and development of biomedical devices and equipment; (iii) Applications of cavitation in biomedical diagnosis and therapy and appropriate of evaluation are the subjects of active research. A deeper understanding of cavitation in biomedicine should bring new opportunities for the medicine of the future.

Cavitation has been the topic of some publications. Most are interested with its physical mechanisms and bubble dynamics. A few have discussed cavitation in non-Newtonian fluids, concentrating on its applications in fluid machinery and biomedicine. Some important applications, such as erosion, laser/ultrasonic surgery, and cell surgery, have been briefly mentioned. Although cavitation has been a "hot topic" in biomedical research, no book has given a comprehensive account of this field. This is not hard to be understood because the study of cavitation in bio-medicine is interdisciplinary and covers areas of interest from physics, engineering to the biological and medical sciences. The greatly increased publications on cavitation are widely spread in journals in the fields mentioned above. Thus, the proposed book would provide, for the first time, a systematic understanding of the engineering principles and techniques of cavitation in biomedicine based on its physics and mechanism.

The fundamentals of cavitation are introduced in Chap. 1. This gives the basic definition, the process, and the threshold of cavitation and then outlines the types of cavitation nuclei and cavitation, and cavitation bubble dynamics are described.

In Chaps. 2 and 3, it describes cavitation characterization under free-field conditions and analyzes the spatial and temporal distributions of cavitation, both qualitatively and quantitatively.

Chapters 4 and 5 presents cavitation-enhanced thermal and mechanical effects and their applications. Although both effects are inseparable, they are discussed in in two chapters separately to help the reader build the complete framework for each effect. Then, the cavitation process control is introduced in Chap. 6 with some specific cavitation control-related applications being discussed.

Addressing diagnosis and therapy monitoring in tissue, cavitation imaging, including active cavitation imaging (ACI) and passive cavitation imaging (PCI), is introduced in Chap. 7.

Some interesting applications of cavitation in biomedicine which are at early stages of research are described in the last section. Based on a deep understanding in laser-induced cavitation, photoacoustic cavitation (cavitation initiated by acoustic pressure and laser energy spontaneously) is discussed briefly in Chap. 8. In Chap. 9, cavitation mechanobiology is considered in a discussion of the bioeffects of ultrasound related to cavitation and the transduction mechanisms induced, with the cascade of cellular and molecular events.

Cavitation research is under rapid development. This book is intended as a combination of reference book for graduate students, and a monograph for scientists and engineers who work with cavitation in biomedicine. It will help students gain a broad and solid foundation in the field. The aim is to create a bridge for the different disciplines involved and to promote the integration of cross-curricular interests, thus encouraging innovations in the scientific research and engineering application.

I would like to sincerely thank all the contributors of the chapters as well the reviewers for their excellent contributions toward the successful completion of this book. This book is also supported by grants from the National Natural Science Foundation of China and the National Basic Research Program 973 of the Ministry of Science and Technology of China.

Mingxi Wan
Xi'an Jiaotong University

Contents

Chapter 1
Fundamentals of Cavitation

Shanshan Xu, Yangzi Qiao, Xiaodong Liu, Charles C. Church and Mingxi Wan

Keywords Nucleation and collapse thresholds · Artificial nuclei · Tensile and energetic cavitation · Bubble dynamics in liquid · Bubble dynamics in tissue

1.1 Introduction

Cavitation is defined as the formation of one or more cavities in a liquid. The word "formation" can refer, in a general sense, to both the creation of a new cavity and the expansion of a preexisting gas pocket to a size where macroscopic effects, e.g., its shape and size, acoustic emissions, sonoluminescence, and erosive properties, can be observed. The cavity's gas content refers to the liquid's vapor, gas dissolved in the liquid, or combinations thereof. Cavitation usually occurs as a response when the pressure has been reduced sufficiently below the vapor pressure of the liquid or when the temperature has been elevated above the boiling point. In addition, chemical-, electrical-, and radiation-induced cavitation also exists. This book only

S. Xu · Y. Qiao · X. Liu · M. Wan (✉)
The Key Laboratory of Biomedical Information Engineering of Ministry of Education, Department of Biomedical Engineering, School of Life Science and Technology, Xi'an Jiaotong University, Xi'an 710049, P.R. China
e-mail: mxwan@mail.xjtu.edu.cn

S. Xu
e-mail: xushanshan@mail.xjtu.edu.cn

Y. Qiao
e-mail: qiaoyangzi@gmail.com

X. Liu
e-mail: lxdmwz123@163.com

C.C. Church
Department of Physics, University of Mississippi, National Center for Physical Acoustics, 1 Coliseum Drive, University, Oxford 38677, Mississippi
e-mail: cchurch@olemiss.edu

M. Wan et al. (eds.), *Cavitation in Biomedicine*,
DOI 10.1007/978-94-017-7255-6_1

considers the first case of cavity formation—that associated with a drop in pressure or a rise in temperature—in response to an acoustic field. This type of cavitation, called acoustic cavitation, was described as encompassing any observable activity involving a bubble or a population of bubbles stimulated into motion by an acoustic field (Apfel 1984).

In biomedicine, acoustic cavitation is the main physical mechanism by which ultrasound delivers its therapeutic applications, such as localized drug delivery, sonoporation, gene transfer, noninvasive sonothrombolysis, lithotripsy, and histotripsy (ter Haar 2007); cavitation may also enhance the thermal ablation of tumors or other diseased tissue. The bioeffects of cavitation, which stem from the high temperatures, pressures, and fluid velocities it generates include cellular erosion or lysis, molecular degradation, and the formation of free radicals. The onset of inertial cavitation can be also observed with applications of diagnostic ultrasound in vitro, suggesting that even diagnostic may present risks for structural tissue damage or cell lysis in the body (Apfel and Holland 1991; Miller et al. 1996). Thus, controlling the occurrence of inertial cavitation is important for applications that require the bioeffects of inertial cavitation to be maximized, minimized, or prevented entirely.

The cavitation thresholds including the minimum acoustic pressure necessary to initiate cavitation (called the nucleation threshold) and the minimum acoustic pressure necessary to induce transient collapse (called the collapse threshold) are the critical aspects of cavitation research. In addition, because gasses have large compressibility compared with liquids, bubbles interact very strongly with alternating acoustic pressures. Hence, the bubble dynamics in a sound field is an essential aspect of acoustic cavitation. Furthermore, in biomedical fields, several sorts of cavitation nuclei, possessing their own intrinsic functionality in various applications, have been introduced to reduce the cavitation threshold, although these nuclei may involve potential risks.

Several books have been published on the bubble dynamics of cavitation and cavitation thresholds, so in this chapter, we introduce these key concepts but omit their specific derivations. We also present a detailed review of how various cavitation nuclei are employed for biomedical applications.

1.2 Process and Thresholds of Cavitation

1.2.1 Nucleation, Growth, Oscillation, Collapse, and Dissolution

The cavitation process encompasses two basic stages: the nucleation or initial formation of the cavity and the subsequent bubble dynamics involving growth, oscillation, and collapse.

Nucleation can also be called the inception of cavitation, which can typically occur in two forms, depending on the quality of the liquid. Starting with a pure homogeneous liquid that contains no impurities or gas, cavities will be formed as the liquid is torn apart, which occurs because the acoustic pressure imposed on the liquid is greater than the tensile strength of the liquid. The process by which cavities form in pure liquid is called "homogeneous nucleation." However, in real systems or practical experiments, almost all nucleation is thought to occur not because neighboring liquid molecules are being spontaneously broken apart, but because the liquid contains some "weak" sites where cavitation can start more easily. These weak sites include boundaries between the liquid and a solid, liquid–solid–gas interfaces (such as a crevice in an imperfectly wetted container's surface or solid impurity suspended in the liquid) or even at a site in the liquid established by impinging radiation, as in the bubble chambers used in nuclear physics. This process is called heterogeneous nucleation.

Subsequently, under relatively low acoustic irradiation intensities, the nucleated cavity or preexisting bubbles initially grow larger because there is a greater surface area available for dissolved gas to diffuse into the bubbles during the expansion phase than there is for gas to diffuse out of the bubbles during the compression phase. This process, known as rectified diffusion, occurs until the bubbles reach a resonant size. They then oscillate around this resonant size during alternating ultrasonic pressures. However, if acoustic pressures are high enough, the bubbles expand rapidly during the negative half cycle, often too many times greater than their original size. Following this, on the compression half cycle, they collapse violently, often breaking up into many smaller bubbles, generating highly localized extremes of temperature and pressure, which in turn generates shock waves, free radicals, and local heat. The cavitation process is illustrated in Fig. 1.1. Many biomedical applications are based on cavitation bubble collapse.

In order for acoustic cavitation to occur, an alternating pressure field must be present. Once the pressure field is removed by turning off the acoustic transmission, the remaining bubbles undergo dissolution due to surface tension, whereby they dissolve completely or shrink to a size that can then be grown to the resonant size during the next acoustic pulse. The kinetics of bubble dissolution is presented in detail in Sect. 3.2.

1.2.2 Nucleation and Collapse Thresholds

The cavitation threshold generally refers to the minimum acoustic pressure amplitude necessary to initiate a bubble and cause it to violently collapse. However, these two events occur at different pressure amplitudes. Thus, the former is called the nucleation threshold, and the latter is called the collapse threshold, as follows:

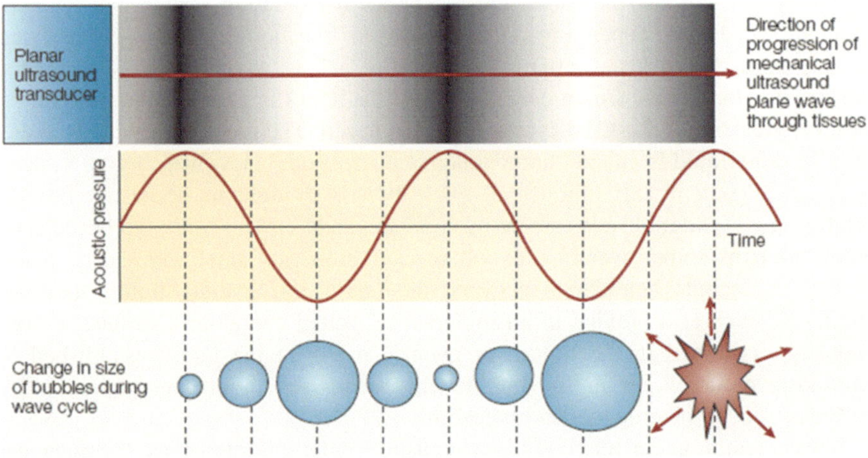

Fig. 1.1 The cavitation process. A mechanical ultrasound wave progresses through tissues (*top*), causing alternating cycles of increased and reduced pressure (compression and rarefaction, respectively, *middle*). Gas is drawn out of solution during rarefaction, creating bubbles. These bubbles can grow and shrink in a stable fashion with the changing tissue pressure, but ultimately they might collapse, causing local energy release and temperature increases at the microscopic level (*bottom*). Reproduced with permission from Kennedy (2005). © Nature Publishing Group

1. *Nucleation Threshold*. The nucleation threshold refers to the force or energy required for homogeneous and heterogeneous nucleation in a liquid, although in acoustics the threshold is more frequently given in terms of pressure.

Homogeneous nucleation in a pure liquid cannot occur until the liquid pressure has become sufficiently negative to overcome the forces of natural cohesion, and it is well known that degassed liquid can withstand very high negative pressures (Briggs et al. 1947; Harvey et al. 1947; Temperley 1947). According to the classical nucleation theory, a small bubble of vapor formed in the liquid must overcome an energy barrier. Thus, during bubble formation, the energy in the system can be altered by two counteracting factors. When a new interface is introduced, the free energy of the system increases by an amount proportional to the surface area of the vapor bubble. However, the vapor has a lower free energy per unit volume than the liquid due to the principle of thermodynamic stability, which decreases the free energy. So, in most cases, the shape of the critical nucleus is spherical because a spherical nucleus can minimize its surface area for a given volume. H.J. Maris has described the change of total free energy, ΔF, which is the sum of a positive surface term and a negative volume term (Maris 2006):

$$\Delta F = 4\pi R^2\sigma - \frac{4}{3}\pi R^3|\Delta f| \tag{1.1}$$

where R is the radius of the nucleus, σ is the surface free energy per unit area, and Δf is the bulk free energy difference per unit volume between the vapors and liquids, having a different form for different transitions. For example, for the nucleation of bubbles in a liquid in tension,

$$\Delta f = p_0 - p \tag{1.2}$$

where p is the local pressure in the liquid and p_0 is the pressure at which the liquid and vapor are in phase equilibrium when the temperature is T. Conversely, for the nucleation of liquid droplets in over-pressurized vapor,

$$\Delta f = \frac{v_G}{v_L}(p_0 - p) \tag{1.3}$$

where v_G and v_L are the molar volumes of the gas and liquid, respectively.

In Fig. 1.2, ΔF is plotted versus R for p. It is apparent that at first, the total free energy increases with increasing bubble radius before reaching a maximum value at a critical radius R_c, after which it decreases. The critical radius can be obtained from the derivative of Eq. (1.1), thus $R_c = 2\sigma/\Delta f$, and the corresponding maximum free energy is

$$\Delta F_{\max} = \frac{16\pi\sigma^3}{3\Delta f^2} \tag{1.4}$$

Furthermore, Eq. (1.1) assumes the molar volume of the liquid to be negligible compared to the molar volume of the vapor. The probability of nucleation of the vapor phase is then

$$\Gamma = \Gamma_0 \exp\left(\frac{-\Delta F_{\max}}{kT}\right) = \Gamma_0 \exp\left(-\frac{16\pi\sigma^3}{3kT\Delta f^2}\right) \tag{1.5}$$

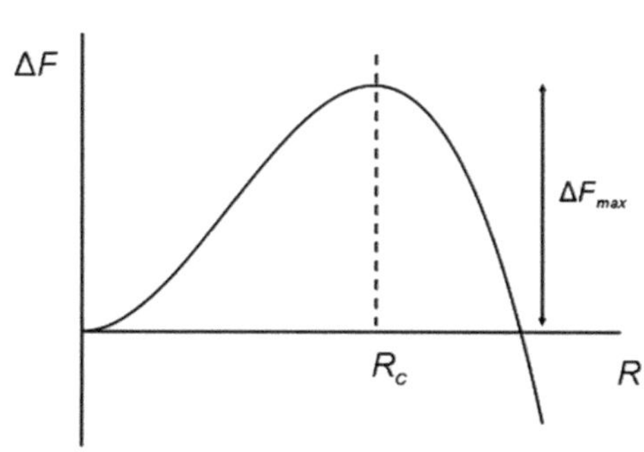

Fig. 1.2 Energy required for creating a spherical bubble of radius R. Reproduced with permission from Maris (2006). © Elsevier B.V.

where Γ_0 is a factor representing the frequency per unit time and volume.

To apply this relationship to experimental conditions, we must consider the volume V_{exp} and time τ_{exp} over which the negative pressure is applied. Nucleation is likely to occur when the negative pressure is increased to a value such that $\Gamma V_{\text{exp}}\tau_{\text{exp}} \approx 1$. Employing Eqs. (1.2)–(1.5), the pressure p_{cav} at which this condition is satisfied is

$$p_{\text{cav}} = p_0 - \left[\frac{16\pi\sigma^3}{3kT\text{In}(V_{\text{exp}}\tau_{\text{exp}}\Gamma_0)}\right]^{1/2} \tag{1.6}$$

Because the cavitation pressure involves only the logarithm of the experimental volume and time, p_{cav} depends only weakly on these quantities. Thus, a rather definite value of the cavitation pressure can be identified despite the statistical nature of the nucleation process.

On the other hand, for heterogeneous nucleation, the thermodynamic barrier is lower than that of the homogeneous nucleation because the effective surface energy that contributes to the work of critical radius formation is lower. For example, consider again a liquid at negative pressure p and take the liquid–vapor, vapor–wall, and liquid–wall surface energies to be σ_{LV}, σ_{VW}, and σ_{LW}, respectively. Now, consider a bubble of the form shown in Fig. 1.3. The volume of the bubble is

$$V = \frac{1}{3}\pi R^3(1+\cos\theta)^2(2-\cos\theta) \tag{1.7}$$

while the area of the interface between the liquid and the vapor is

$$S_{\text{LV}} = 2\pi R^2(1+\cos\theta) \tag{1.8}$$

and the area of the interface between the vapor and the wall is

$$S_{\text{VW}} = \pi R^2 \sin^2\theta \tag{1.9}$$

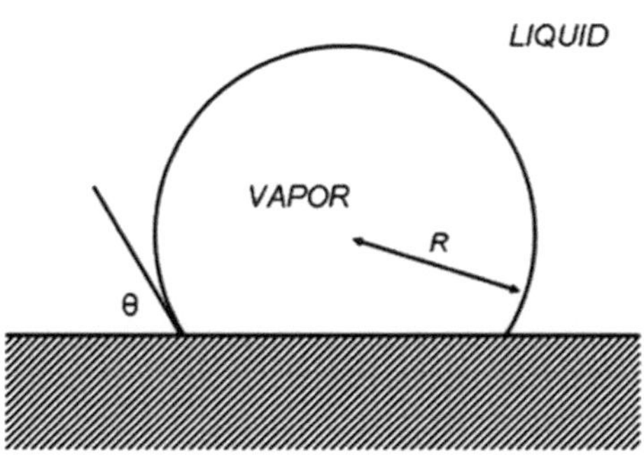

Fig. 1.3 A bubble of radius R formed at a flat solid wall with contact angle θ. Reproduced with permission from Maris (2006). © Elsevier B.V.

Thus, the energy needed to introduce a bubble like this according to the classical nucleation theory is

$$\Delta F = \sigma_{\mathrm{LV}} S_{\mathrm{LV}} + (\sigma_{\mathrm{VW}} - \sigma_{\mathrm{LW}}) S_{\mathrm{VW}} - |\Delta f| V \tag{1.10}$$

Because $\sigma_{\mathrm{VW}} = \sigma_{\mathrm{LV}} \cos\theta + \sigma_{\mathrm{LW}}$, Eq. (1.10) can be written as

$$\Delta F = \left[4\pi R^2 \sigma_{\mathrm{LV}} - \frac{4}{3}\pi R^3 |\Delta f|\right] \frac{(1+\cos\theta)^2(2-\cos\theta)}{4} \tag{1.11}$$

When the radius $R = 2\sigma_{\mathrm{LV}}/|\Delta f|$, that is, at the same radius as for nucleation in bulk, we can obtain a maximum value of ΔF, which is the energy barrier height:

$$\Delta F_{\max} = \frac{4\pi}{3}(1+\cos\theta)^2(2-\cos\theta)\frac{\sigma_{\mathrm{LV}}^3}{|\Delta f|^2} \tag{1.12}$$

Compared with Eq. (1.4), (1.12) is different by a factor of

$$\lambda = \frac{(1+\cos\theta)^2(2-\cos\theta)}{4} \tag{1.13}$$

From Eq. (1.13), we can see that if the liquid completely wets the wall, then $\theta = 0$ and so $\lambda = 1$, producing no change in the nucleation energy barrier from the homogeneous case. Alternatively, incomplete wetting leads to a reduction in the energy barrier, such that the energy barrier goes to zero as $\theta \to \pi$.

The schematic in Fig. 1.3 indicates that the surface of the solid wall is flat. However, if the surface is not flat, the energy barrier becomes more complex and needs to be further modified. In recent decades, the nucleation for different surface geometries has been calculated. For simplicity, here, we first discuss the simple geometry of a cone (Fig. 1.4). As we let the semiangle of the cone be β, we will consider two different cases, shown in Fig. 1.4a, b, respectively. First, consider the moment that the interface between the liquid and vapor looks as it does in Fig. 1.4a, with a radius R. Then the energy barrier is

$$\Delta F = \left(4\pi R^2 \sigma_{\mathrm{LV}} - \frac{4\pi}{3} R^3 |\Delta f|\right) \lambda(\theta, \beta) \tag{1.14}$$

where

$$\lambda(\theta, \beta) = \frac{1}{2} - \frac{\sin(\theta - \beta)}{2} + \frac{\cos^2(\theta - \beta)\cos\theta}{4\sin\beta} \tag{1.15}$$

Analyzing Eq. (1.14), we can easily obtain the maximum value of ΔF (that is, $\Delta F_{\max}$) at the point $R_c = 2\sigma_{\mathrm{LV}}/|\Delta f|$. The maximum value of $\Delta F_{\max}$ is

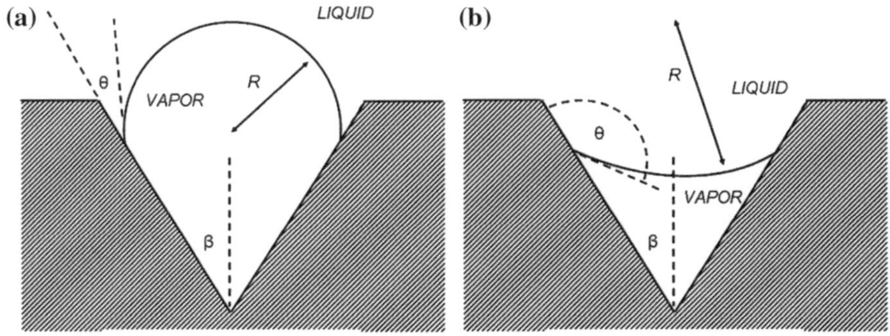

Fig. 1.4 A bubble of radius *R* formed in a cone. Reproduced with permission from Maris (2006). © Elsevier B.V.

$$\Delta F_{\max} \frac{16\pi\sigma_{\mathrm{LV}}^{3}}{3|\Delta f|^{2}} \lambda(\theta, \beta) \tag{1.16}$$

Next, if θ is greater than $\beta + \pi/2$, as is shown in Fig. 1.4b, then the value of λ (θ, β) is negative. If the pressure is positive, the energy barrier has a minimum value when $R = 2\sigma_{\mathrm{LV}}/|\Delta f|$, which is equal to the value in Eq. (1.16). But, if the pressure is sufficiently negative that theta becomes less than the advancing contact angle for the material, then the interface moves out to the end of the cone and there may or may not be any barrier against nucleation (Borkent et al. 2009).

Finally, we must also consider nucleation on the surface of a nano- or micro-particle (Zhang et al. 2014). The effective apparent contact angle of a gas bubble on a spherical surface (θ) in contact with a liquid is bigger than the intrinsic contact angle of bubbles on planar surfaces (θ_0) (Fig. 1.4a) (Zhang et al. 2014), which may influence the surface energy, the volume work, and the nucleation energy barrier of the bubble.

Bubble formation at a liquid–particle interface is sketched in Fig. 1.5, and the nucleation energy barrier is given by Eq. (1.17)

$$\Delta E = \frac{4\pi\sigma^{3}(2 + 3\cos\theta - \cos^{3}\theta)}{3P^{2}} \tag{1.17}$$

where σ is the gas–liquid surface tension (73 mN/m) and P is the sum of the vapor pressure (P_{V}), the gas pressure (P_{G}), and the acoustic pressure (P_{A}). P_{A} is at least 1 atm, which is much bigger than other partial pressures; thus, other contributions were negligible for this volume work. The symbol θ is the effective apparent contact angle of a nucleation bubble on a curved surface in contact with a liquid (Fig. 1.5a). When the particle is very small (e.g., 500 nm), θ is higher than it would be on a bigger particle surface (Fig. 1.5a, b). If the particle diameter is big enough (e.g., 50 μm), the nucleation of a 100-nm bubble on the particle surface is similar to that on a planar surface. According to Eq. (1.17), θ drastically influences the nucleation

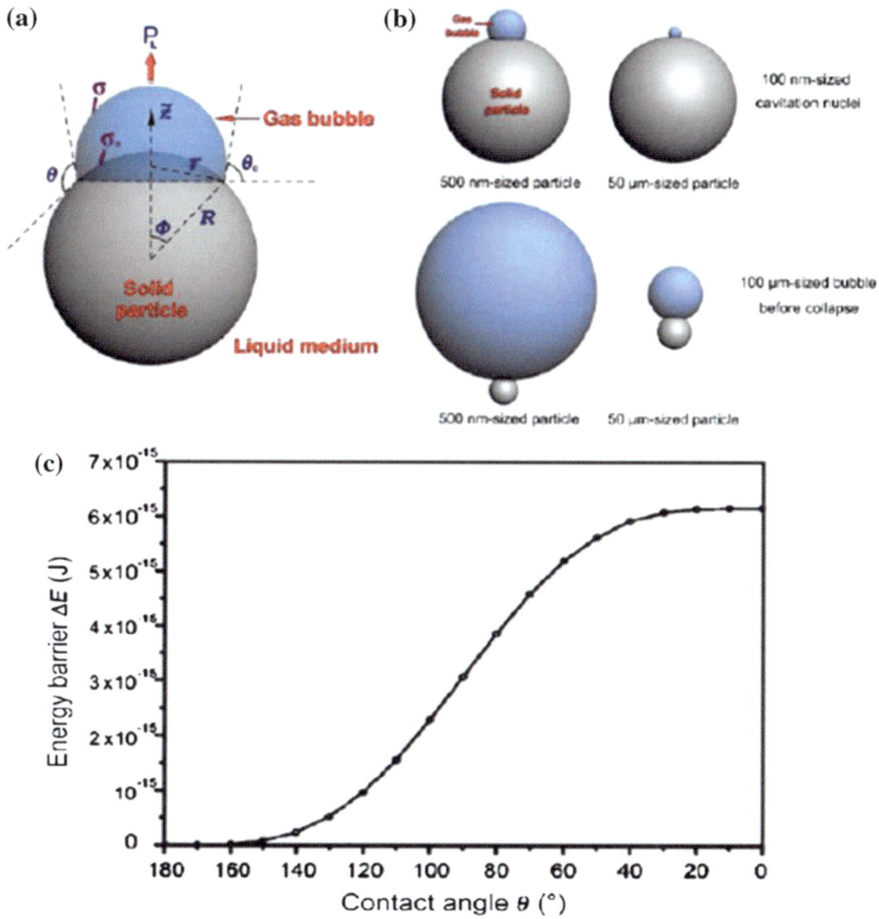

Fig. 1.5 **a** Scheme for nucleation and growth of a cavitation bubble at a particle surface; θ is the effective apparent contact angle, θ_0 is an intrinsic contact angle. **b** Scheme for cavitation nuclei (*top row*) and bubble before collapse (*bottom row*) on small (*left*) and large (*right*) particles. **c** Nucleation energy barrier as a function of contact angle θ derived from the maximum of the sum of the surface energy and the volume work. Reproduced with permission from Zhang et al. (2014). © ACS Publications

energy. Figure 1.5c shows the nucleation energy barrier as a function of contact angle, which is derived from the maximum or the sum of the surface energy and the volume work. It is thus suggested that the nucleation energy barrier increases as the contact angle decreases, indicating that bubbles preferably nucleate on submicrometer particle surfaces.

2. *Collapse Threshold.* As we have discussed, when the acoustic pressure is greater than the value that causes the liquid to go into tension during the rarefaction phase of cycle, gas bubbles in the liquid will expand rapidly, often too many

times larger than their original size. After this, on the compression half cycle, they collapse violently—they implode—and can break up into many smaller bubbles. This pressure is called the collapse threshold, and it is the minimum value necessary to initiate bubble growth and subsequent adiabatic collapse. Based on the above criterion, Apfel (1981) calculated the theoretical threshold for bubble collapse as follows:

$$R_{\mathrm{T}} = \begin{cases} \frac{0.13}{f}\left(\frac{P_0}{\rho}\right)^{1/2}\left\{\frac{p-1}{\sqrt{p}}\left[1+\frac{2}{3}(p-1)\right]^{1/3}\right\}; & p \leq 11 \\ \frac{0.3}{f}\left(\frac{P_0}{\rho}\right)^{1/2}\left[\frac{2}{3}(p-1)\right]^{1/2}; & p \geq 11 \end{cases} \quad (1.18)$$

Here, the collapse threshold, P_{T}, is expressed inversely in terms of the threshold bubble radius, R_{T}, where the collapse threshold is a function of p (= P_{T}/P_0, with P_0 being the liquid ambient pressure).

1.3 Cavitation Nuclei

If a liquid is entirely gas-free, when the acoustic pressure falls below the vapor pressure of the liquid, the cavities develop and grow very rapidly because of evaporation into the growing cavity. The process is called vaporous cavitation. However, if we consider a liquid with dissolved gas, then, depending upon the magnitude of the pressure reduction which is occurring in the system and its rate of change, gas cavities may initially be formed due to the presence of suitable nuclei (a contaminant, for example). These cavities grow slowly by the diffusion of dissolved gas from the liquid. The process is called gaseous cavitation. This type of cavitation, where cavitation originates at a suitable nucleus (called a cavitation nucleus), is much more common than cavitation occurring within the bulk of a pure liquid. In order for gaseous cavitation to occur though, gas molecules must be dissolved in the surrounding liquid. This section describes some types of cavitation nuclei according to their applications in biomedicine.

1.3.1 Microcavity

Microcavities can be divided into two subgroups, made up of either free-floating micron-sized bubbles suspended in a liquid or the cavities in solid surfaces in contact with the liquid. Numerous experiments have shown that the sizes of these nuclei range from a few micrometers to hundreds of micrometers. Thus, they can be collectively called microcavities, which refers to any bounded volume, whether it is empty or contains a gas or vapor, where at least part of the boundary is a liquid (Neppiras 1984).

Generally, free micron-sized bubbles dissolve in water within just a few seconds (Epstein and Plesset 1950), because of surface tension. As such, these bubbles are unsuitable as cavitation nuclei. Recently however, pulsed ultrasound techniques have used these microbubbles to optimize cavitation effects in industrial and ultrasonic therapeutic applications. Apparently, these pulsed ultrasound techniques work by employing the free bubbles nucleated by the previous pulse, possibly because hydrophobic ions may collect on or near the surface of free bubbles and help stabilize them (Akulichev and Rozenberg 1966).

The first model to explain why free bubbles can exist stably was the skin model, proposed in 1954 by Fox and Herzfeld. They suggested that free gas bubbles in a liquid could be stabilized by a skin of organic molecules such as fatty acids (Fox and Herzfeld 1954). Obviously, such molecules would be soluble in organic substances such as alcohol and carbon tetrachloride, so these liquids should therefore not contain nuclei and be unable to undergo cavitation. However, cavitation can occur in these liquids. Thus, this hypothesis was abandoned by Herzfeld in 1964 (Flynn 1964). Subsequently, a revised skin model by Yount (1979) assumed the skin to be formed by amphiphilic molecules (surface-active substances) that allow the gas in a bubble to be in diffusion balance with the gas molecules dissolved in the surrounding liquid; the skin allows the bubble to resist collapse. In water, the hydrophilic heads of the amphiphilic molecules form hydrogen bonds with the surrounding water molecules, while the hydrophobic tails bound the gas content of bubbles, thus forming a skin on each bubble surface that allows the gas in the void to be in diffusion balance with the gas in the water. Yount's model agrees with observations by Johnson and Cooke (1981), whereby at atmospheric pressure, gas bubbles in seawater can be stabilized by a skin.

The other type of microcavity (other than free-floating microbubbles) is cavities in solid surfaces that are in contact with liquid. This "crevice model" was first put forward by Harvey et al. (1947), who suggested that the gas pockets trapped in hydrophobic conical cracks and crevices of solid surfaces act as cavitation nuclei. Here, a gas pocket may have a negative radius of curvature, which allows it to be stable to gas diffusion, as illustrated in Fig. 1.4. Strasberg considered the implications of this conical crevice model on predictions of the cavitation threshold. In his discussion, he illustrated the importance of the liquid's gas content and of prepressurization of the liquid sample on the threshold (Strasberg 1959). Later, Apfel (1984) sought to extend his work to describe how the threshold depends on vapor pressure, temperature, and mote size. The resulting theory can qualitatively predict the static pressure required to nucleate a vapor cavity from a mote and describe how the threshold depends on temperature, vapor pressure, hydrostatic pressure, gas content, mote wettability, and mote size (Apfel 2005). The kinetics of this cavity are also discussed briefly in Sect. 1.2.2, but for more details, please refer to Chap. 2 of Leighton (1994).

Several kinds of cavitation nuclei that have been widely used in biomedical applications will be discussed below.

1.3.2 Encapsulated Bubble

Above, we discussed the two types of nuclei that arise from microcavities—free micron-sized bubbles and cavities at solid surfaces. Here, we now discuss how encapsulated bubbles can be used as nuclei. Because they are encapsulated with a coating, these bubbles have improved stability against gas dissolution and microbubble coalescence. These bubbles are constructed of a lipid, polymer, or albumin shell encapsulating a core of air or a gas with a low diffusivity, and they range in diameter from 1 to 10 μm (Correas et al. 2001). Although encapsulated bubbles have different dynamics in an ultrasonic field than spherical micron-sized bubbles, they still respond similarly to an ultrasonic pressure wave, in terms of expanding and contracting during the rarefaction and compression phases, respectively. We present the dynamics of encapsulated bubbles in Sect. 1.5.

Microbubbles exposed to a relatively low acoustic excitation undergo stable oscillation and emit nonlinear signals, which allow them to be detected in small capillaries located deep in tissue, where blood velocity is low (Chang et al. 1995; Simpson et al. 1999). Furthermore, a transmission pressure sufficiently high can cause the microbubbles to be rapidly destroyed, releasing their gas (Bouakaz et al. 2005). The rapid destruction occurs because the diameter of contrast agents sharply decreases in one transmitted pulse, and this is often accompanied by the separation of the microbubble shells into several small fragments, releasing broadband noise signals, complete diffusion of the encapsulated gas, and a large change in the received echo amplitude (Chomas et al. 2001). Many researchers have investigated microbubble fragmentation from both acoustic and optical perspectives, and experimental results suggest that microbubble fragmentation relies on many factors, such as acoustic excitation pressure, center frequency, pulse length, and resting diameter. Microbubbles tend to be fragmented with increasing acoustic excitation pressure and pulse length, but with decreasing center frequencies and resting diameters. (Chomas et al. 2001; Bouakaz et al. 2005; Porter et al. 2006; Smith et al. 2007).

However, the destruction mechanisms for lipid-coated and polymer-coated microbubbles are different (Bloch et al. 2004). For lipid-coated microbubbles, high-speed optical experiments have shown that both the shell and its contents fragment. Conversely, polymer-coated microbubbles remain largely intact and do not oscillate significantly in response to ultrasound; instead, the gas contained within the microbubble is extruded and ejected through a defect in the polymer shell, while the microbubble retains its shape. Figure 1.6 shows the respective destruction process for lipid-coated and polymer-coated microbubbles.

Intentional microbubble fragmentation has, recently, been introduced in therapeutic ultrasound applications, such as ultrasound-induced localized drug delivery, thrombolysis, and hemolysis to enhance cavitation effects. Unsurprisingly, the therapeutic effects of these ultrasound applications arise via cavitation mechanisms and largely depend on the existence of gas bubbles. Hence, with the advent of microbubbles, the bioeffects of cavitation for these therapies could be further

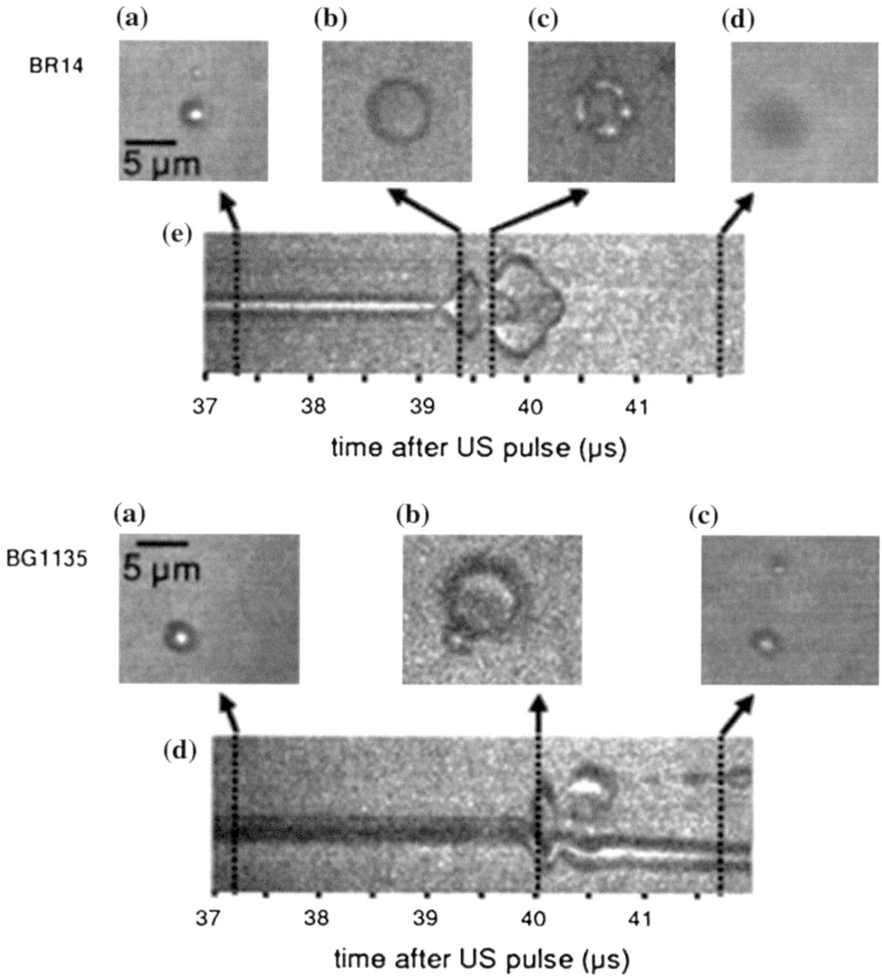

Fig. 1.6 Destruction of a lipid-coated microbubble (BR 14) and a polymer-coated microbubble (BG 1135). Still images **a–d** or **a–c** depict the agent before, during, and after exposure. The streak image **e** or **d** shows one line of sight through the agent versus time, with the acquisition times of the still images indicated. Reproduced with permission from Bloch et al. (2004). © American Institute of Physics

increased. However, using microbubbles may give rise to high velocity jets, strong shear forces, and shock waves that can damage surrounding tissues, even when ultrasound is only meant to be used at a diagnostic level.

Overall, when encapsulated microbubbles are exposed to an acoustic pressure with sufficiently high amplitude, they can be destroyed rapidly, thus initiating inertial cavitation.

1.3.3 Phase-Shift Nanodroplet

Recently, several biomedical applications have begun to employ acoustic droplet vaporization (ADV), where nano- or microdroplets are activated and then vaporized into gaseous bubbles upon ultrasound irradiation, even at low power. ADV is a physical process in which the pressure waves of ultrasound induce a phase transition that causes superheated liquid nanodroplets to form gas bubbles. Diagnostically, the droplets have the potential to leak out of blood vessels, such as those found in tumors, into the interstitial tissue (Rapoport et al. 2007). Immediately after vaporization, the newly induced microbubbles scatter diagnostic ultrasound nonlinearly and thus can be detected with contrast-specific imaging techniques, such as pulse-inversion imaging method (Reznik et al. 2011). ADV has been investigated for a wide range of therapeutic applications, such as embolotherapy (Zhang et al. 2010), therapeutic drug delivery (Fabiilli et al. 2010), and high-intensity focused ultrasound (HIFU) thermal ablation (Zhang and Porter 2010).

The droplets widely used for biomedical application are perfluorocarbons since they are non-toxic, biocompatible, and immiscible with water. Figure 1.7 illustrates the local pressure variation as ultrasound propagates in a liquid medium (Pitt et al. 2014). There are positive and negative pressure fluctuations that oscillate around atmospheric pressure. When the rarefaction cycle of the acoustic wave decreases the local pressure to values below the vapor pressure of the PFC liquid for sufficient time, it boils and forms gas bubbles. Sufficient time means that the local pressure is lower than the vapor pressure for the time needed to produce heterogeneous or homogeneous nucleation of the gas phase. The pressure inside a nanoemulsion droplet is greater than the surrounding liquid pressure because of the additional contribution from the Laplace pressure: $\Delta P_{\text{LP}} = 2\gamma/r$, where γ is the interfacial energy and r is the radius of the emulsion droplet. As the sum of the local pressure and the Laplace pressure drops below the vapor pressure of the liquid PFC, there is a subpressurization driving potential to nucleate the gas phase. The subpressurization is defined as the difference between the vapor pressure and the local pressure. If the acoustic cycle is too short or the subpressurization is inadequate, no gas phase will form.

Currently, the potential mechanisms for ADV have been proposed to be thermal absorption and inertial cavitation. It is known that although PFC_5 has a boiling temperature of 29 °C at atmospheric pressure, the heat-induced droplet-to-bubble transition is not efficient at 29 °C due to an additional pressure resulting from surface tension inside the droplets, defined as the Laplace pressure. However, researchers (Sheeran and Dayton 2012) have shown successful vaporization of droplets using ultrasonic energy at or even below their boiling point because of the decrease of ambient pressure exerted on the droplet during the rarefaction phase of ultrasound, thus implying inertial cavitation is involved in ADV. In addition, most studies have shown that increasing the ultrasonic pulse duration at clinically relevant ultrasound frequencies results in no significant decrease in the vaporization threshold, unless the duration is on the order of 1 ms or greater. Elongation of the

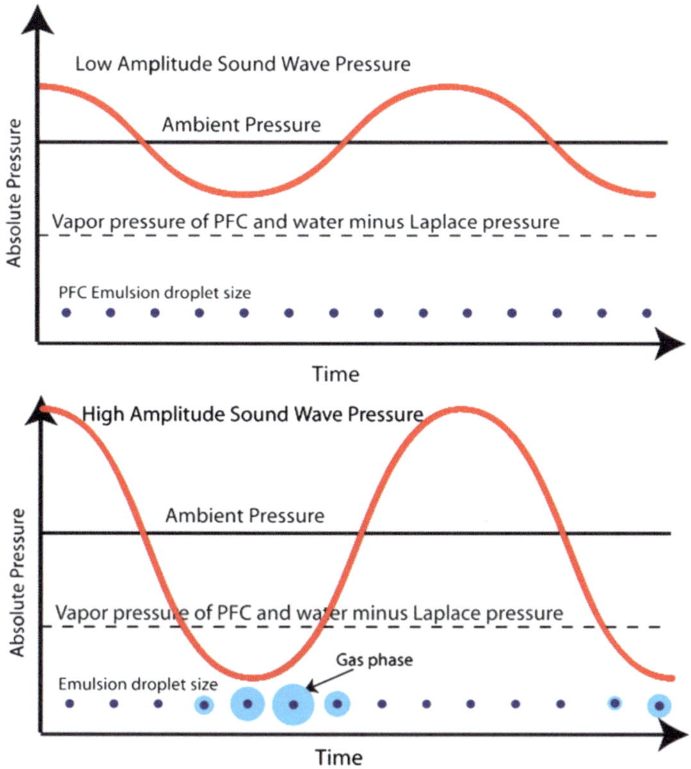

Fig. 1.7 Schematic of an ultrasonic acoustic wave in a perfluorocarbon emulsion. The *upper figure* illustrates an acoustic wave whose lowest pressure remains above the vapor pressure of the PFC minus the Laplace pressure. The *lower figure* illustrates how a gas phase forms around the emulsion droplet when the local wave pressure drops below the vapor pressure of the PFC minus the Laplace pressure. Reproduced with permission from Pitt et al. (2014). © Elsevier B.V

ultrasound pulse length to longer than 1 ms decreases the vaporization threshold, possibly because energy is converted from the incident ultrasound wave to heat absorbed by the droplets, which elevates the temperature of droplet. Figure 1.8 illustrates the ADV process for PFC_5 (Sheeran and Dayton 2012).

Overall, thermal absorption may be the dominant mechanism by which ADV occurs when the pulse is in the millisecond range, while inertial cavitation may be the primary mechanism for ADV when the pulse is in the microsecond range.

Recently, a different mechanism for this process has been proposed in which ADV is initiated by a combination of two phenomena: highly nonlinear distortion of the acoustic wave before it hits the droplet and focusing of the distorted wave by the droplet itself (Shpak et al. 2014).

At high excitation pressures, nonlinear distortion causes significant superharmonics with wavelengths of the order of the droplet size. These superharmonics strongly contribute to the focusing effect (Fig. 1.9). The presented mechanism

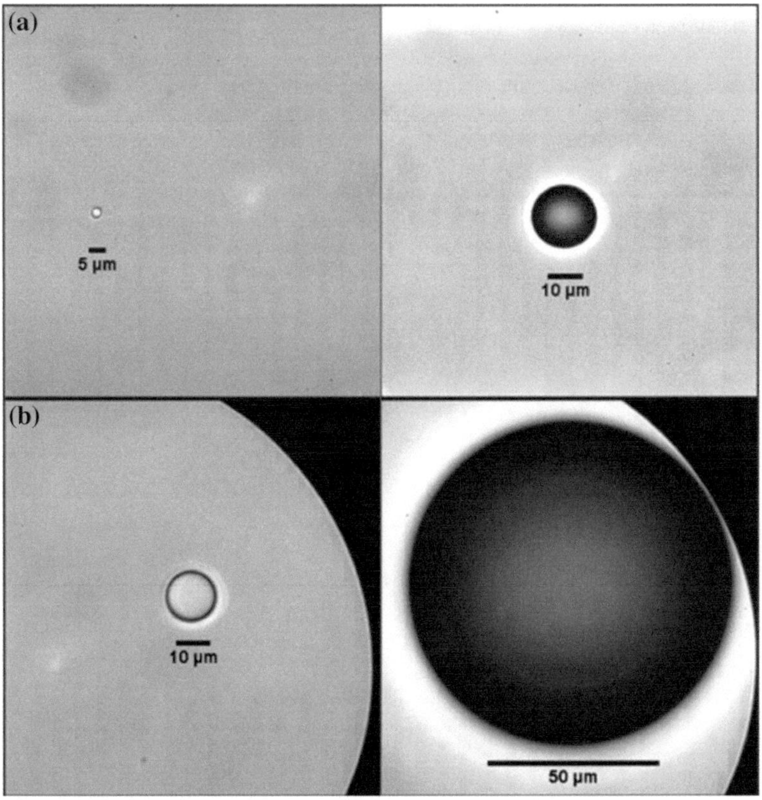

Fig. 1.8 Vaporization of PFC_5 droplets as a result of exposure to ultrasonic pulse at 37 °C. **a** The original droplets with diameters of 5 and 10 µm; **b** gas bubble induced by vaporization. The resulting increase in size after vaporization was close to that predicted by ideal gas laws (approximately 5–6 times the original droplet diameter). Reproduced with permission from Sheeran and Dayton (2012). © ACS Publications

explains the hitherto counterintuitive dependence of the nucleation threshold on the ultrasound frequency. Further, the physical insight in this presented theory allows for the optimization of ADV for therapeutic applications, in particular with respect to the acoustic pressures required for activation. This includes the design of droplets by mixing different liquids of different physical properties which allows to vary the acoustic impedance by a change of the density and speed of sound. Using dual or multiple frequency transducers, the amplitudes and phases of the transmit waves can be optimized to have maximal constructive interference within the droplets to maximize the focusing strength at any given acoustic input pressure.

1.3.4 Other Micro-/Nanoscale Particles

Gold nanoparticles, which are minute optical absorbers, are another type of cavitation nucleus. For example, cavitation can occur when vapor bubbles surrounding gold nanoparticles collapse due to pulsed laser light exposure. The gold nanoparticles absorb the laser energy, causing them to rapidly heat, leading to vaporization of the surrounding medium and the formation of a transient vapor cavity. Cavitation around laser-heated nanoparticles produces a strong and localized mechanical disruption and is effective in producing localized cell death in nanoparticle-targeted cells (Lukianova-Hleb et al. 2010).

One of the principle limitations in applications involving nanoparticle-targeted photonic cavitation is the high optical fluence required to nucleate and grow the bubble. Bubble formation begins when the superheated liquid surrounding the nanoparticle approaches the spinodal temperature of water ($\sim$550 K), and additional optical energy is required for subsequent bubble growth (Kotaidis et al. 2006). The cavitation threshold fluence can be reduced by reducing the local pressure at the nanoparticle position during the time of illumination using an applied ultrasound field; a process referred to as photoacoustic cavitation (Farny et al. 2005). When nanoparticles are subjected to negative pressure, such as is encountered during the rarefaction phase of the applied ultrasound field, they have a drastically reduced cavitation threshold fluence. If a cavity that is larger than the Blake radius has formed during the rarefaction phase of the focused ultrasound exposure, cavitation ensues. The size of the cavity is determined primarily by the

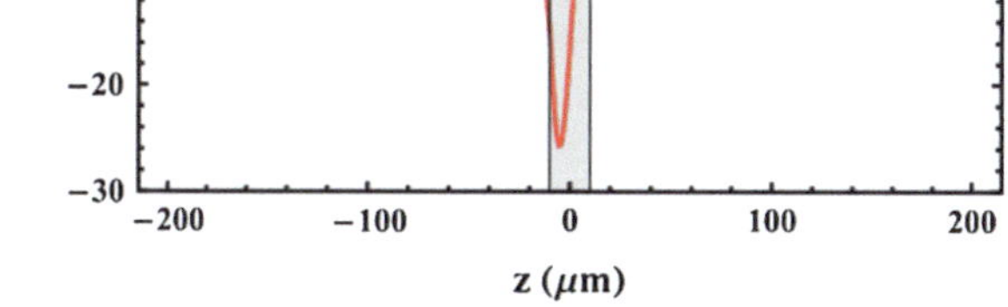

Fig. 1.9 Snapshot of the superharmonic focusing effect within a spherical droplet. The *black line* represents the acoustic pressure waveform on the axis of symmetry as a function of the z coordinate in the absence of a droplet. The *red solid line* is the focused pressure in the presence of the droplet. The snapshot is taken right at the moment of minimum focused pressure. The *horizontal axis* displays one full wavelength in the medium outside the droplet. The *gray shaded region* depicts the position of the droplet, $R = 10$ μm. The focusing spot lies around $z = -0.4R$, and the pressure is amplified 5.8 times compared with the incident acoustic pressure PNP = 4.5 MPa. Reproduced with permission from Shpak et al. (2014). © National Academy of Sciences

size of the particle, and the arrival of the laser pulse determines the timing of nucleation relative to the acoustic cycle. With this technique, it may be possible to generate optimally sized nuclei within optically accessible regions. Such nuclei can be created essentially "on demand" and with minimal collateral damage, as the absorption coefficient of the particles is much greater than that of the surrounding tissue.

Magnetic nanoparticles have also been used in the clinic as magnetic resonance (MR) contrast agents. Recently, by harnessing well-developed surface engineering, a wide range of functionalities have been loaded onto the particle surface. These include biovectors such as peptides, antibodies, or aptamers which are able to guide the migration of nanoparticles in a living subject and to accumulate them preferentially in areas of interest. On top of that, the nanoparticles can be further loaded with imaging motifs or therapeutic agents and, as a consequence, be upgraded to multifunctional "nanogadgets" with either multimodal imaging capabilities or theranostic features for both diagnostic and therapy applications.

Magnetic nanoparticles can also enhance the degree of inertial cavitation induced by HIFU by decreasing the cavitation nucleation threshold. As such, when used with HIFU, magnetic nanoparticles have been shown to enhance the rate of destruction of tumor spheroids (Ho et al. 2011). Furthermore, the superparamagnetic properties of these particles offer the possibility of magnetic targeting to the disease site.

We have long known that the presence of particles within a liquid can provide nucleation sites for cavitating bubbles (Crum 1979), thereby decreasing the cavitation threshold. Therefore, in addition to employing each particle's unique intrinsic properties, these particles they can also be used as cavitation nuclei to increase cavitation activity, making them multifunctional and thus highly useful particles.

1.4 Tensile and Energetic Cavitation

In biomedical fields, the generation of cavitation is classified into two categories according to the physical mechanisms that cause cavitation bubbles to be nucleated, either by lowering the pressure in the liquid (tensile cavitation) or by depositing energy (energetic cavitation). The classification of tensile and energetic cavitation is shown in Fig. 1.10.

While the inception of acoustic cavitation might be expected to occur only at an acoustic pressure amplitude that generates a level of tension within the liquid that is equal to or greater than the liquid's tensile strength, acoustic cavitation can also occur in the form of energetic cavitation, arising from the heat absorbed after acoustic exposure. For example, a number of ultrasound-induced therapeutic applications have been used with an acoustic pulse transmitted at several hundred cycles or more, e.g., HIFU-induced tumor ablation (Kennedy 2005) and ADV (Sheeran and Dayton 2012).

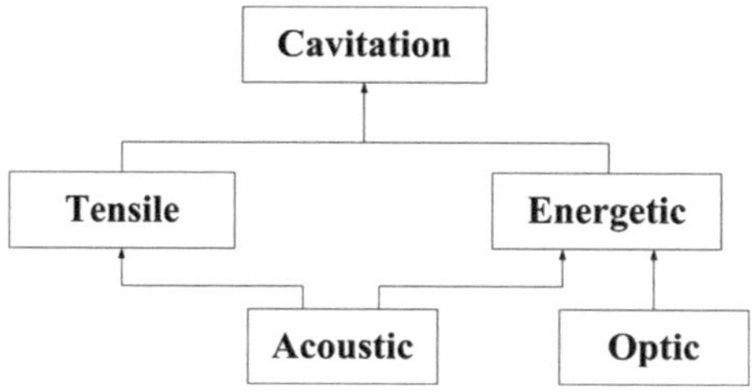

Fig. 1.10 Classification scheme for the different kinds of cavitation

Shortly after the laser was invented, cavitation phenomena were also observed in high-intensity light fields (Askar'yan and Moroz 1963; Brewer and Rieckhoff 1964). This is called optic cavitation (Lauterborn and Bolle 1975; Lauterborn 1976) and can be considered another form of energetic cavitation. When high-power laser pulses are focused into a transparent medium, the medium suddenly becomes opaque to the laser radiation as soon as a certain irradiance threshold is surpassed. The sudden rise in the absorption coefficient is due to the formation of a dense, optically absorbing plasma. Plasma formation, also known as laser-induced breakdown, has been observed in gases (Raĭzer 1966; Smith and Meyerand (1976), solids (Ready 1971; Bloembergen 1974), and liquids (Kennedy et al. 1997; Sacchi 1991). It leads to rapid heating of the material in the focal volume, followed by explosive expansion and the emission of a shock wave. The expansion of the heated volume further results in the formation of a cavity, if it occurs in a solid, or of a cavitation bubble, if it takes place in a liquid (Vogel et al. 1996). The process, called photoacoustic cavitation (Farny et al. 2005), can also be classified as energetic cavitation.

1.5 Cavitation Bubble Dynamics

1.5.1 Free Gas Bubble Dynamics in a Free Field

Because gases are much more compressible than liquids, bubbles react very strongly to alternating acoustic pressures. Hence, bubble dynamics in a sound field are an essential aspect of acoustic cavitation.

The most often used model to describe bubble oscillation in a free field is the Rayleigh–Plesset equation (Plesset 1949):

$$\ddot{R}R + \frac{3}{2}\dot{R}^2 = \frac{1}{\rho}[P - P_0 - P_{\text{driv}}(t)] \tag{1.19}$$

where the overdots indicate derivatives with respect to time. Here, it is assumed that a bubble with an initial radius of R_0 oscillates spherically symmetrically in an infinite liquid, and the gas inside the bubble is homogenous. The density of the surrounding liquid ρ is much larger than the density of the gas within the bubble, and the vapor pressure within the bubble, P_v, remains constant during its oscillation. Also, the radius of the bubble is smaller than the wavelength ($R < \lambda$), and the velocity of the bubble wall is much smaller than the speed of sound within the liquid. The bulk viscosity of the material in the bubble and the compressibility of the liquid are not considered. The above equation was extended by Noltingk, Neppiras, and Poritsky to take into account the gas within the bubble, vapor pressure, surface tension, and viscosity of the liquid (Leighton 1994):

$$R\ddot{R} + \frac{3\dot{R}^2}{2} = \frac{1}{\rho}\left\{\left(P_0 + \frac{2\sigma}{R_0} - P_v\right)\left(\frac{R_0}{R}\right)^{3\kappa} + P_v - \frac{2\sigma}{R} - P_0 - P_{\text{driv}}(t) - \frac{4\mu\dot{R}}{R}\right\} \tag{1.20}$$

These kinds of equations are called Rayleigh–Plesset, or RPNNP equations. They make the critical assumption that the liquid around the bubble is incompressible. However, this assumption becomes untrue when the amplitude of oscillation is relative large. This leads to a class of first-order equations, one of which is the Herring–Trilling equation (Leighton 1994; Flynn 1964):

$$R\ddot{R}\left(1 - 2\frac{\dot{R}}{c}\right) + \frac{3}{2}\dot{R}^2\left(1 - \frac{4}{3}\frac{\dot{R}}{c}\right) = \frac{1}{\rho}\left[P - P_0 - P_{\text{driv}}(t) - \frac{R}{c}\dot{P}\left(1 - \frac{\dot{R}}{c}\right)\right] \tag{1.21}$$

The velocity of the bubble wall cannot be neglected, so a new term $\dot{R}/c$ defined as M (Mach-number) appears here. The derivation also includes the energy stored during the liquid compression. While bubble dynamics are governed in most cases by the Rayleigh–Plesset equation, this equation ignores the acoustic attenuation due to acoustic radiation. The Herring–Trilling equation somewhat addresses this issue, but this equation fails when $M \geq 0.5$, meaning that the compressibility of the liquid is overcompensated, resulting in a negative inertial term and a numerically unstable equation. Even so, the bubble wall has a high speed only in a short compression period during the whole cycle, so even though it is inappropriate for this transient time, the equation is still useful.

Based on these equations, the modified Herring–Trilling model was developed (Hoff 2001):

$$R\ddot{R} + \frac{3}{2}\dot{R}^2 = \frac{1}{\rho}\left[P - P_0 - P_{\text{driv}}(t) + \frac{R}{c}\dot{P}\right] \tag{1.22}$$

It introduces the radiation attenuation and neglects the correction term of the first-order mach-number, which prevents numerical instability.

Another equation, developed by Gilmore, considers the nonlinear oscillation of a bubble with a high amplitude (Leighton 1994). When a bubble oscillates violently, the accompanying phenomenon of a shock wave may induce sonoluminescence.

Under these circumstances, the excitation pressure and the amplitude of bubble oscillation are high. The velocity of the bubble wall may approach the sound speed. The surrounding liquid, which is considered to be homogeneous, has high compressibility and can store a large amount of energy, and the sound velocity changes with the pressure. The bubble dynamics are related to the liquid enthalpy, as follows (Leighton 1994):

$$
\begin{aligned}
& R\ddot{R}\left(1-\tfrac{\dot{R}}{c}\right)+\tfrac{3}{2}\dot{R}^2\left(1-\tfrac{1}{3}\tfrac{\dot{R}}{c}\right)=H\left(1+\tfrac{\dot{R}}{c}\right)+\tfrac{\dot{R}}{c}\dot{H}\left(1-\tfrac{\dot{R}}{c}\right) \\
& H=\tfrac{1}{\rho}\tfrac{n}{n+1}(P+B)\left[\left(\tfrac{P+B}{P_\infty+B}\right)^{\frac{n-1}{n}}-1\right] \\
& c=c_\infty\left(\tfrac{P+B}{P_\infty+B}\right)^{\frac{n-1}{2n}}
\end{aligned}
\tag{1.23}
$$

where B and n are the constants in the equation of state; H is the liquid enthalpy; and c is the sound speed in the liquid. Both H and c are related to the excitation pressure and the state constants.

Then, to account for the effects of acoustic radiation, the viscosity of the liquid, the surface tension of bubble, and the incident acoustic waves, Keller and Miksis deduced a different first-order equation. This equation reflects the large, radical oscillations of a bubble trapped in a sound field (Keller and Miksis 1980):

$$
\begin{aligned}
\left(1-\frac{\dot{R}}{c}\right)R\ddot{R}+\frac{3}{2}\dot{R}^2\left(1-\frac{\dot{R}}{3c}\right)=\left(1+\frac{\dot{R}}{c}\right)\frac{1}{\rho_l}\left[P_B(R,t)-P_A\left(t+\frac{R}{c}\right)-P_\infty\right] \\
+\frac{RdP_B(R,t)}{\rho_l c dt}
\end{aligned}
\tag{1.24}
$$

where $P_B(R,\ t)$ is the pressure on the liquid side of the bubble interface, and $P_A(t + R/c)$ is the time-delayed driving pressure. Many modeling studies on bubble dynamics in HIFU are based on this equation.

1.5.2 Encapsulated Bubble Dynamics in a Free Field

The above equations are all for free gas bubbles, and they make up the foundation of bubble dynamics in complicated conditions. However, most of the bubbles that arise in biomedical applications are encapsulated by shell materials to make them more stable. The shell materials influence the bubble oscillation and scattering in two main ways (Hoff et al. 2000): First, encapsulated bubbles are stiffer than free gas bubbles of the same size, giving them higher resonant frequencies and fewer nonlinear effects. Second, shell materials make the bubbles more viscid, leading to more acoustic energy transformed into heat. As such, the ratio of scattering and attenuation decreases. For most encapsulated bubbles, the most important parameters that determine their acoustic response are those of the shell material. There are several models for encapsulated bubbles, such as the Morgan model, the Nico de Jong model, and the Newtonian rheological model, to name a few.

The Morgan model (Morgan et al. 2000) is based on the modified Herring–Trilling model and is expressed as follows:

$$\rho R\ddot{R} + \frac{3}{2}\rho\dot{R}^2 = \left(P_0 + \frac{2\sigma}{R_0} + \frac{2\chi}{R_0}\right)\left(\frac{R_0}{R}\right)^{3\gamma}\left(1 - \frac{3\gamma}{c}\dot{R}\right) - \frac{4\mu\dot{R}}{R} - \frac{2\sigma}{R}\left(1 - \frac{\dot{R}}{c}\right)$$
$$- \frac{2\chi}{R}\left(\frac{R_0}{R}\right)^2\left(1 - \frac{3}{c}\dot{R}\right) - 12\mu_{\mathrm{sh}}\varepsilon\frac{\dot{R}}{R(R-\varepsilon)} - (P_0 + P_{\mathrm{driv}}(t)) \tag{1.25}$$

where $2\sigma/R_0$ and $4\mu\dot{R}/\mathrm{R}$ are the effects of surface tension at that interface and the viscosity of the liquid, respectively. The term $(2\chi/R)(R_0/R)^2$ accounts for the dilational elasticity of the shell (Glazman 1983), and $12\mu_{\mathrm{sh}}\varepsilon\frac{\dot{R}}{R(R-\varepsilon)}$ is due to the shell's viscous damping, where ε and μ_{sh} are the thickness and the viscosity coefficient of the shell, respectively. Because this model can only be applied to small-amplitude oscillations, it is always assumed that the velocity in the shell is not very high. Also, both the viscosity and elasticity coefficients are included, and the inner gas is regarded to follow the ideal gas law and can be compressed infinitely. This assumption is satisfied when bubble oscillation is mild, but in reality, the gas cannot be compressed infinitely. When the applied pressure increases significantly and as the distance between the gas molecules decreases to a certain degree, the gas molecules will strongly repel.

Another model for encapsulated bubbles was proposed by de Jong et al. (1992)

$$R\ddot{R} + \frac{3\dot{R}^2}{2} = \frac{1}{\rho}\left\{P_{g0}\left(\frac{R_0}{R}\right)^{3\gamma} + P_{\mathrm{v}} - \frac{2\sigma}{R} - P_0 - P(t) - S_p\left(\frac{1}{R_0} - \frac{1}{R}\right) - \delta_t\omega\rho R\dot{R}\right\}$$
$$\delta_t = \frac{4\mu}{\omega\rho_0 R^2} + \frac{\omega R_0}{c}\cdot\frac{1}{1 + (\omega R_0/c)^2}\cdot\left(\frac{\omega_0}{\omega}\right)^2 + \frac{S_p}{4\pi R^3\rho\omega}$$
$$S_p = \frac{Et}{1-\mu} \tag{1.26}$$

They introduced two parameters, the coefficient of stiffness (S_{p}) and the damping term (δ_t), to illustrate the influence of the shell. The term $S_{\mathrm{p}}(1/R_0 - 1/R)$ reinforces the influence of the surface tension. This value becomes negative for surface-active agents, zero for air-filled gas bubbles, and positive for encapsulated bubbles. During bubble oscillation, the friction between the shell molecules generates additional viscosity damping, which is combined with acoustic damping and radiation damping, and is written as $\delta_t = \delta_f + \delta_{vis} + \delta_{th} + \delta_{ra}$. Furthermore, the process is considered to be heat adiabatic. The thickness of the shell is not specified in this model because changes in the shell thickness are reflected by changes in the mechanical index and the damping term. The damping term is obtained under linear conditions without taking into account the applied acoustic pressure.

Another model, first proposed by Chatterjee and Sarkar (2003), is the Newtonian rheological model that focuses on the interfaces of encapsulated bubbles. Here, the typical shell material is assumed to comprise several layers, and the thickness of each layer is considered to be infinitely small. Based on the rheological characteristics of the interface and the Newtonian rheological characteristics of the surrounding liquid, the bubble dynamics can be written as follows:

$$\rho\left(R\ddot{R}+\frac{3}{2}\dot{R}^2\right)=P_{g0}\left(\frac{R_0}{R}\right)^{3\kappa}-4\mu\frac{\dot{R}}{R}-\frac{2\gamma}{R}-\frac{4\kappa^s\dot{R}}{R^2}-P_0+P_{\text{driv}}(t) \tag{1.27}$$

The expansion elasticity modulus, which changes due to surface tension, is also introduced and defined as follows:

$$E^s=\left(\frac{\partial\gamma}{\partial\alpha}\right)_{\beta=0} \tag{1.28}$$

$$\begin{aligned}\gamma&=\gamma_0+E^s\beta\\ \beta&=\frac{\delta A}{A}=\left[(R/R_E)^2-1\right]\end{aligned} \tag{1.29}$$

where R_E is the bubble radius without strain. The final model is

$$\rho\left(R\ddot{R}+\frac{3}{2}\dot{R}^2\right)=P_{g0}\left(\frac{R_0}{R}\right)^{3\kappa}-4\mu\frac{\dot{R}}{R}-\frac{4\kappa^s\dot{R}}{R^2}-\frac{2\gamma_0}{R}-\frac{2E^s}{R}\left[\left(\frac{R}{R_E}\right)^2-1\right]-P_0+P_{\text{driv}}(t) \tag{1.30}$$

In all of the above models, the thickness of the shell is not given. However, this would significantly influence the scatter and other characters of the bubbles, and for encapsulated bubbles, shell thicknesses might be larger for drug delivery and gene therapy applications. Church (1995) suggested that the shell is a continuous, incompressible elastic solid material and separates the inner gas and the surrounding Newtonian liquids. When the bubble radius is less than 10 μm, the shell viscosity plays a more important role than the liquid viscosity, and the model for the thickness of the bubble shell is as follows:

$$R_1 R\left[1+\left(\frac{\rho_L-\rho_s}{\rho_s}\right)\frac{R_1}{R_2}\right]+\dot{R}_1^2\left[\frac{3}{2}+\left(\frac{\rho_L-\rho_s}{\rho_s}\right)\left(\frac{4R_2^3-R_1^3}{2R_2^3}\right)\frac{R_1}{R_2}\right]$$
$$=\frac{1}{\rho_L}\left[P_{g0}\left(\frac{R_{01}}{R_1}\right)^{3\gamma}-P(t)-\frac{2\sigma_1}{R_1}-\frac{2\sigma_2}{R_2}-4\frac{\dot{R}_1}{R_1}\left(\frac{V_s\mu_s+R_1^3\mu_L}{R_2^3}\right)-4\frac{V_sG_s}{R_2^3}\left(1-\frac{R_{e1}}{R_1}\right)\right] \tag{1.31}$$

$$V_s=R_{02}^3-R_{01}^3$$
$$R_{e1}=R_{01}(1+Z)=R_{01}\left\{1+\left[\frac{2\sigma_1}{R_{01}}+\frac{2\sigma_2}{R_{02}}\right]\left[\frac{R_{02}^3}{V_s}\right](4G_s)^{-1}\right\} \tag{1.32}$$

where ρ_s and ρ_L indicate the density of the solid shell and the surrounding liquid, respectively. R_{01} and R_{02} are the inside and the outside radius, and V_s is the volume of the bubble shell. R_{e1} is the balanced radius without strain, which is different from the static inner radius. The terms σ_1 and σ_2 are the inside and outside surface tensions, while μ_s and μ_L are the shear viscosity coefficients of the shell and liquid. G_s is the Lamé constant, which is called the rigidity modulus; the resonant frequency of a thick-shelled bubble increases with the rigidity modulus. When the density of the liquid is smaller than that of the shell, the acceleration of bubble wall will decrease, and so do its nonlinear characteristics. When the density of the liquid is larger than that of the shell, the acceleration of bubble wall and the nonlinear characteristics will increase. This model is more appropriate for small-amplitude oscillations.

Based on this model and the Morgan model, Allen et al. (2002) developed the following model:

$$R_1\ddot{R}_1\left[1+\left(\frac{\rho_1-\rho_s}{\rho_s}\right)\frac{R_1}{R_2}\right]+R_1^2\left[\frac{3}{2}+\frac{\rho_1-\rho_s}{\rho_s}\frac{4R_2^3-R_1^3}{2R_2^3}\frac{R_1}{R_2}\right]$$
$$=\frac{1}{\rho_1}\left[\rho_{g0}\left(\frac{R_{10}}{R_1}\right)^{3k}-p(t)-\frac{2\sigma_1}{R_1}-\frac{2\sigma_2}{R_2}-4\mu_s\dot{R}_1\left(\frac{R_2^3-R_1^3}{R_2^3R_1}\right)-4\mu_l\dot{R}_1\left(\frac{R_1^2\dot{R}_1}{R_2^3}\right)\right] \tag{1.33}$$

The shell is considered to comprise two liquid layers, each of which has a thickness of hundreds of nanometers. During oscillation, the inner and outer layers differ considerably. The shell is an incompressible viscoelastic material, and the volume and thickness of the shell remain constant during oscillation.

1.5.3 Bubble Dynamics Near an Interface

When bubble growth is constrained in such media, the bubble dynamics are much more complicated than in a free field. All the characteristics of the media—the distances from boundaries, the asymmetric oscillation of bubbles, and so on—should be taken into account. Here, we mainly discuss two types of constrained media: media near an interface and media in constrained vessels. We focus on interfaces because bubble–boundary interactions will generate high-speed liquid jets, which are very important in ultrasonic cleaning and therapeutic applications. Also, we focus on vessels because they are the most commonly encountered targets during focused ultrasound therapy.

Bubble collapse near an interface is very important because it can generate inertial cavitation and mechanical damage. Bubble collapse occurs once the bubble's acceleration, i.e., $\ddot{R}$, is high enough, and the bubble may become highly unstable. Many experiments have captured interactions at bubble boundaries, and they show that oscillation of bubbles is asymmetric and that liquid jets form during the collapse phases (Oguz and Prosperetti 1990; Robinson et al. 2001; Sankin and Zhong 2006). Some theoretical studies have tried to explain these observations, especially the liquid jet formation process and the energy it contains. Plesset and Chapman presented a theoretical model for an initially spherical vapor cavity in the neighborhood of a solid (Labouret and Kurz 2010), in which they proved that jets were directed toward the solid boundary; their results are consistent with the experimental observations of Lauterborn and Bolle (1975). Later, Brujan et al. (2005) studied jet formation and shock wave emission during the collapse of ultrasound-induced cavitation, and they calculated certain factors—the variation of the bubble wall with time, the maximum velocity, and the kinetic energy of the liquid jet—to illustrate how HIFU achieves its therapeutic action. They calculated the cavity pressure P_c using the following equation:

$$P_c = P_v + P_0 \left(\frac{V_0}{V} \right)^{\kappa} \tag{1.34}$$

where P_v is the vapor pressure, and P_0 is the pressure of the non-condensable content near the time of cavity inception. Before the liquid jet is formed, the governing equation can be written in terms of an integral equation:

$$\frac{1}{2}\phi(x') = \int_{\partial\Omega} \left(\frac{\partial\phi(x)}{\partial n} G(x,x') - \phi(x) \frac{\partial G(x,x')}{\partial n} \right) \mathrm{d}S \tag{1.35}$$

where S is the bubble surface and $G(x, x')$ is a Green's function. However, when the bubble forms a doubly connected region with the boundary, Eq. (1.33) fails to describe the process. After modification, the integral formula becomes the following:

$$\frac{1}{2}\phi(x') = \int_{\partial\Omega} \left(\frac{\partial\phi(x)}{\partial n} G(x,x') - \phi(x) \frac{\partial G(x,x')}{\partial n} \right) \mathrm{d}S - \Delta\phi \int_{c} \frac{\partial G(x,x')}{\partial n_+} \mathrm{d}S \tag{1.36}$$

The value of $\Delta\phi$ is obtained from the difference between ϕ at the tip of the liquid jet and ϕ at the opposite cavity surface. The energy of the liquid jet is calculated by

$$E_k = \frac{\rho}{2}\int_{S_j} \phi \frac{\partial \phi}{\partial n} \mathrm{d}S \tag{1.37}$$

where S_j is the surface of the liquid jet. In the above study, another very important parameter is γ, which is the ratio of the distance from the center of the bubble at its maximum expansion to the maximum bubble radius.

Figure 1.11 shows the computed bubble profiles when $\gamma = 1.2$, $R_0 = 8.9$ μm, and $R_{max} = 200$ μm. It is assumed that the pressure gradient at the bubble wall close to the boundary and the pressure at the bubble wall far from the boundary cause the center of the bubble to move toward the boundary, generating a high-pressure liquid jet, which can reach a velocity of ~30 m/s. Later, the bubble shape can even become toroidal. Importantly, the theoretical results are highly consistent with experiment results.

As mentioned above, γ is the normalized distance to the wall and is defined as $\gamma = d/R_{max}$. When γ is large enough so that the bubble just touches the boundary at maximum expansion, a counterjet can be observed. Figure 1.12 shows images of bubble collapse (taken by high-speed photography), whereby both jet formation and counterjet formation have occurred (Labouret and Kurz 2010). The fluid's velocity potential is obtained from the Laplace equation (Eq. 1.38), and the evolution of the bubble boundary can be calculated based on Eq. (1.39)

$$v(x) = -\nabla\phi(x) \tag{1.38}$$

$$\frac{\mathrm{d}y}{\mathrm{d}t} = vy, \quad \frac{\mathrm{d}\phi}{\mathrm{d}t} = -\frac{P}{\rho} + \frac{1}{2}|v|^2 \tag{1.39}$$

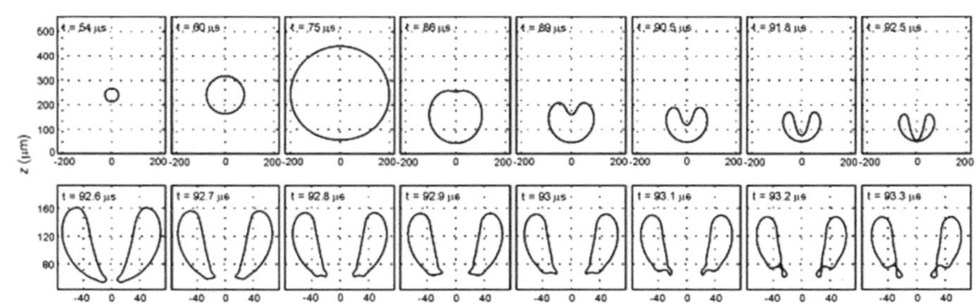

Fig. 1.11 Computed bubble profiles for $\gamma = 1.2$, $R_0 = 8.9$ μm, $R_{max} = 200$ μm. Reproduced with permission from Brujan et al. (2005). © 2005 IOP Publishing Ltd.

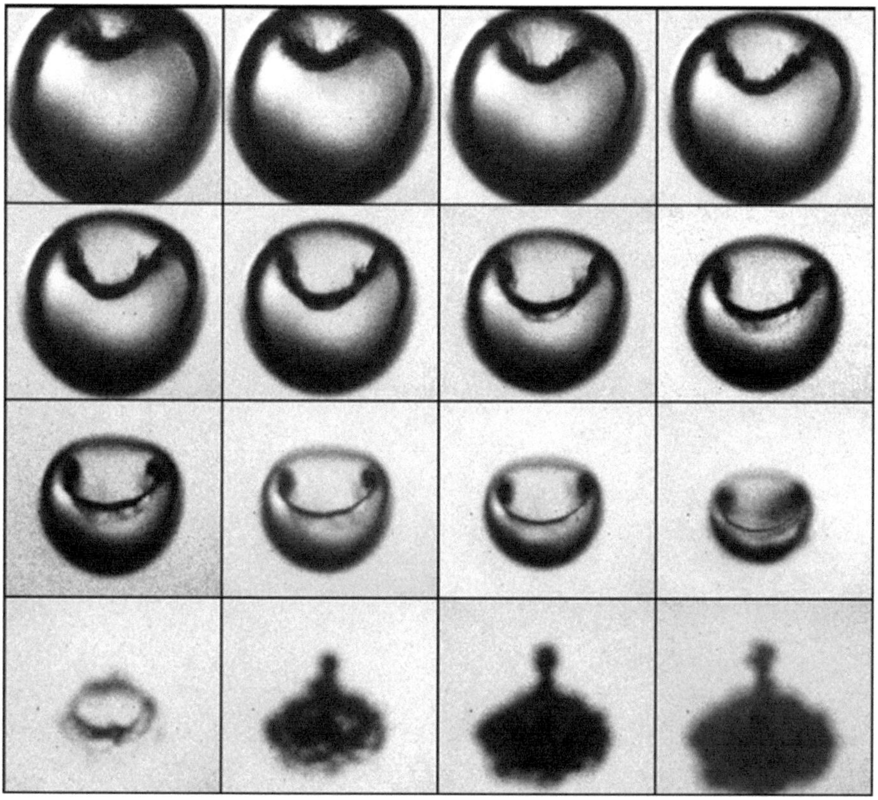

Fig. 1.12 Bubble collapse with jet formation, toroidal bubble, and counterjet. Maximum bubble size $R_{\max} = 1.5$ mm, $\gamma = 2.6$. Reproduced with permission from Lauterborn and Kurz (2010). © 2010 IOP Publishing Ltd.

Lee et al. (2007) have extended the boundary integral method to the rebound phase. In the oscillating system, the total energy is calculated according to the following equation:

$$E = -\int_{V_0}^{V} P_{g,0}\left({}^{V_0}\!/_{V}\right)^{\lambda}\mathrm{d}V + \int_{V_0}^{V} \Delta P\mathrm{d}V + \frac{1}{2}\rho\int_{\Omega} u^2\mathrm{d}\Omega \\ = \text{constant for each cylce} \tag{1.40}$$

where $\Delta P = P_\infty - P_v$ and $P_\infty = P_{\mathrm{ATM}} + \rho gh$. As bubbles oscillate less and less violently, the energy loss upon bubble collapse should be included. To do this, the

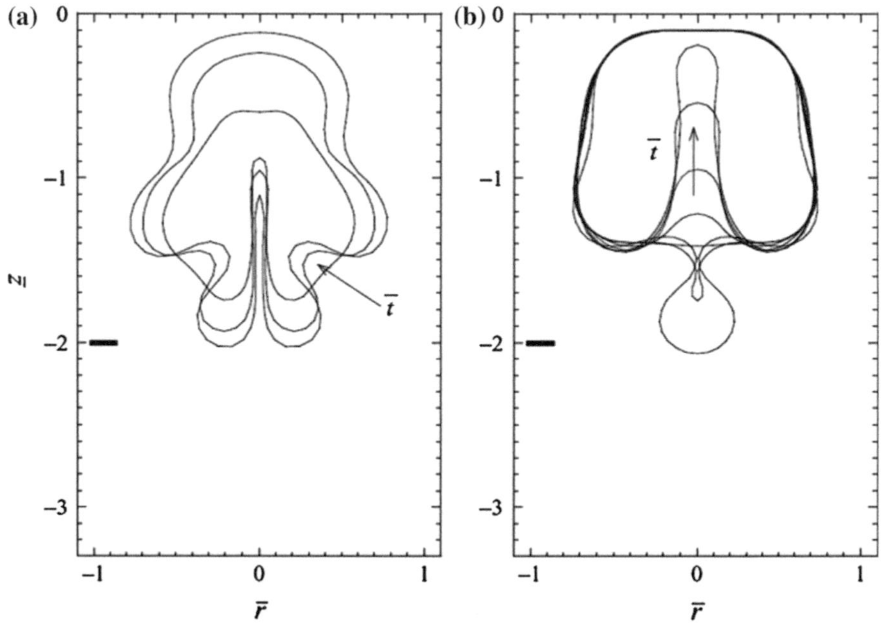

Fig. 1.13 Calculated bubble shapes in the **a** rebound and **b** collapse phase with an energy loss upon a first collapse of 70 % and a normalized bubble distance γ = 2.0. Reproduced with permission from Lee et al. (2007). © 2007 Cambridge University Press

loss of energy is considered to occur in the vicinity of each minimum volume, and the short period of emission of energy is neglected. The bubble energy in each cycle can be estimated as follows:

$$\frac{E_{n+1}}{E_n} = \left(\frac{R_{m,n+1}}{R_{m,n}}\right)^3 \left(\frac{P_\infty - P_{v,n+1}}{P_\infty - P_{v,n}}\right) \tag{1.41}$$

where E_n indicates the fraction of the energy that remains for the nth oscillation period. Figure 1.13 shows the calculated bubble evolution with an energy loss upon a first collapse of 70 % and a normalized bubble distance γ = 2.0. During bubble collapse, a counterjet develops very rapidly in the rebounding stage in conjunction with a necking process.

Bubble oscillation can occur near rigid boundaries, as discussed, but it can also occur near liquid–air surface boundaries and compliant surface boundaries. Early in 1981, Blake and Gibson (1981) first investigated the motion of a vapor cavity near a free surface, and many related studies followed (Oguz and Prosperetti 1990; Robinson et al. 2001). At a free surface boundary, a reentrant jet can be observed directed away from the boundary as a bulge of liquid develops at the surface, which may develop into a jet directed into the air. Conversely, Gibson and Blake (1982) investigated bubble collapse near a deformable surface, and they also identified a jet directed away from the boundary. Subsequently, Duncan and Zhang (1991) developed a mass-spring system to describe compliant surface boundaries.

1.5.4 Bubble Dynamics in Constrained Vessel

Because microbubbles have great potential for biomedical applications, such as drug delivery and medical imaging, many studies have modeled bubble dynamics in vessels. Within a vessel that is much larger than the bubble's diameter, the Rayleigh–Plesset equation describing the pulsation of bubbles in a free field can be used as a reference. However, when the bubble is constrained in a small vessel, in which its diameter is on the same order as the bubble's, bubble oscillation will depend on the vessel characteristics and on the interaction between the bubble and the vessel wall. In this case, bubble oscillation is confined by the vessel wall, a condition which has been studied both theoretically and experimentally. For instance, Oguz and Prosperetti (1998) studied free gas bubbles within different rigid vessels, while Miao et al. (2008) used a coupled finite element method (FME) to stimulate bubble oscillation inside a deformable tube. Recently, Hong et al. (2011) captured the translation of microbubbles in microvessels, finding that their translation is always accompanied with the displacement of the vessel wall. Hay et al. (2013) developed a model that can accurately predict the pulsating and translating of a gas bubble between two soft, thin elastic layers, and it can also approximate wall deformation.

The geometry and coordinate system used by Hay et al. (2013) is presented in Fig. 1.14. The layers, with shear moduli μ_1 and μ_2 and finite thicknesses h_1 and h_2, were immersed in an incompressible liquid. They used Lagrange's equation to describe the dynamics of the system. The equation they deduced to describe the effective radius of the bubble is as follows:

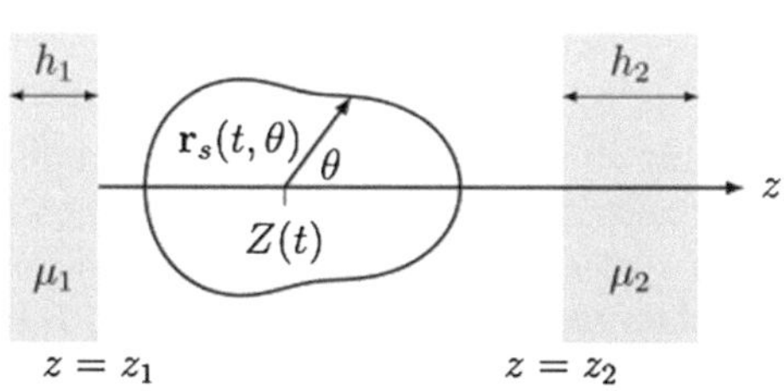

Fig. 1.14 Geometry and coordinate system for a spherical bubble with shape perturbations. Reproduced with permission from Hay et al. (2013). © 2013 Acoustical Society of America

$$
\begin{aligned}
R_e\ddot{R}_e + \frac{3}{2}\dot{R}_e^2 = &\frac{P_1}{\rho} + \frac{1}{4}U^2 + \frac{1}{2}\left(\frac{s_2^2}{3} + \frac{s_3^2}{4}\right) + \frac{s_2\ddot{s}_2}{15} + \frac{s_3\ddot{s}_3}{14} + \frac{\ddot{R}_e}{R_e}\left(\frac{2}{15}s_2^2 + \frac{1}{7}s_3^2\right)\\
&+ \frac{\dot{R}_e^2}{R_e^2}\left(\frac{s_2^2}{15} + \frac{s_3^2}{14}\right) + \frac{2\dot{R}_e}{R_e}\left(\frac{2}{15}s_2\dot{s}_2 + \frac{s_3\dot{s}_3}{7}\right) - \frac{11}{70}\frac{s_2 s_3}{R_e}\dot{U}\\
&- \frac{U}{R_e}\left(\frac{1}{5}s_2\dot{s}_3 + \frac{3}{14}\dot{s}_2 s_3\right) - \frac{U^2}{10R_e}\left(s_2 + \frac{17}{98}\frac{s_3^2}{R_e} - \frac{19}{70}\frac{s_2^2}{R_e}\right)\\
&- \sum_{j=1}^{2}\frac{\beta_j}{2}\left[\frac{R_e^3}{(Z - z_j)^4} - \frac{4R_0^3}{(Z_0 - z_j)(Z - z_j)(Z_0 + Z - 2z_j)^2}\right]
\end{aligned} \tag{1.42}
$$

The translation force exerted on the bubble can be calculated by the following equation:

$$
\begin{aligned}
\mathrm{F} = 2\pi\rho\Bigg\{ &-6\nu R_e U + \frac{R_e^2}{5}\left(s_2\dot{U} + \dot{s}_2 U + \frac{3}{14}s_2\ddot{s}_3 + \frac{2}{7}s_3\ddot{s}_2 + \frac{1}{2}\dot{s}_2\dot{s}_3\right)\\
&- \frac{\dot{U}R_e}{35}\left(\frac{19}{5}s_2^2 + \frac{17}{7}s_3^2\right) + \frac{2UR_e}{5}\left(s_2\dot{R}_e + \frac{17}{49}s_3\dot{s}_3 - \frac{19}{35}s_2\dot{s}_2\right)\\
&- \frac{\dot{R}_e U}{35}\left(\frac{19}{5}s_2^2 - \frac{17}{7}s_3^2\right) - \frac{11}{35}s_2 s_3 R_e\left(\ddot{R}_e + \frac{\dot{R}_e^2}{R_e}\right)\\
&- \frac{\dot{R}_e R_e}{5}\left(\frac{8}{7}s_2\dot{s}_3 + \dot{s}_2 s_3\right)\\
&+ \frac{2R_e^3}{3}\sum_{j=1}^{2}\frac{\beta_j}{(Z - z_j)^2}\left[\frac{R_e^3}{(Z - z_j)^3} - \frac{2R_0^3(Z_0 + 3Z - 4z_j)}{(Z_0 - z_j)(Z_0 + Z - 2z_j)^3}\right]\Bigg\}
\end{aligned} \tag{1.43}
$$

The layer displacement was estimated through integrating the velocity at the interface between the liquid and the layers in response to the bubble motion:

$$
\begin{aligned}
\upsilon_i = (\nabla\phi)_i = &\frac{R_e^2\dot{R}_e}{r_j^3}\left[x_i - (Z - z_j)\delta_{iz}\right]\\
&- \frac{R_e^3 U}{2r_j^3}\left\{\delta_{iz} - \frac{3}{r_j^2}\left[x_i - (Z - z_j)\delta_{iz}\right]\left[z - (Z - z_j)\right]\right\}
\end{aligned} \tag{1.44}
$$

Figures 1.15 and 1.16 are simulation results of a 1.5-μm-diameter bubble whose initial distances to the left and right layers were 1.5 R_0 and 3.5 R_0, respectively. The bubble was driven by 2.5 cycles of an external sinusoidal acoustic pressure source with an amplitude of 330 kPa and a frequency of 2.9 MHz. These theoretical results fit well with the experimental observations made by Chen et al. (2011), and they provide insight into understanding the interaction between bubbles and microvessels, which is important for the diagnostic and therapeutic applications of bubbles.

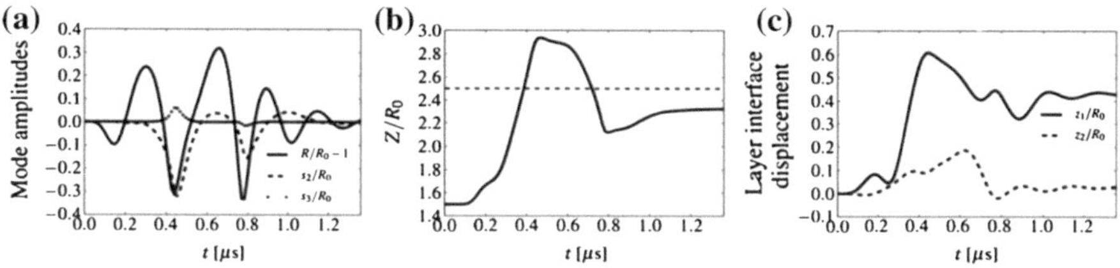

Fig. 1.15 Mode of a bubble's **a** amplitude and **b** position, as well as **c** the displacement of the adjacent elastic layers. Reproduced with permission from Hay et al. (2013). © 2013 Acoustical Society of America

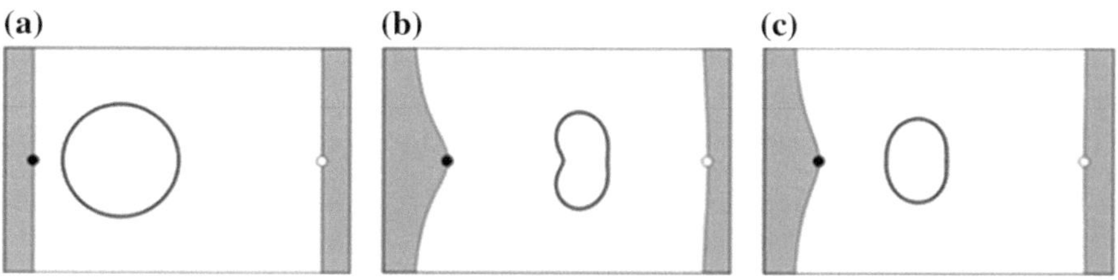

Fig. 1.16 System state at times equal to **a** 0, **b** 0.45, and **c** 0.79 μs. Reproduced with permission from Hay et al. (2013). © 2013 Acoustical Society of America

Rather than free gas bubbles, encapsulated microbubbles are more often used in both diagnostic and therapeutic applications within microvessels. As the vessel wall is susceptible to intense microjetting and shockwaves and can easily suffer from mechanical damage, it is important to understand the process of encapsulated microbubble oscillation and the mechanical effects involved. To do this, Hosseinkhah and Hynynen (2012) developed a three-dimensional numerical model of an encapsulated microbubble inside a microvessel, which accounts for the effects of the shell material, fluid viscosity, and curvature of vessel wall. Because the fluid's characteristics influence bubble oscillation and fluid motion itself can cause vessel wall deformation, a model has been introduced that reflects a two-way coupling between the bubble wall and the surrounding fluid, as well as a two-way coupling between the fluid and the vessel wall. Figure 1.17 is a schematic illustration of the three-dimensional geometry of a bubble (2 μm), blood, and a microvessel (5 μm in radius, 2 μm in thickness, and 204 μm in length).

The Navier–Stokes equations were used to describe the motion of the blood in the vessel:

$$\rho\left(\frac{\partial v}{\partial t} + v\nabla v\right) = -\nabla p + \mu\nabla^2 v \tag{1.45}$$

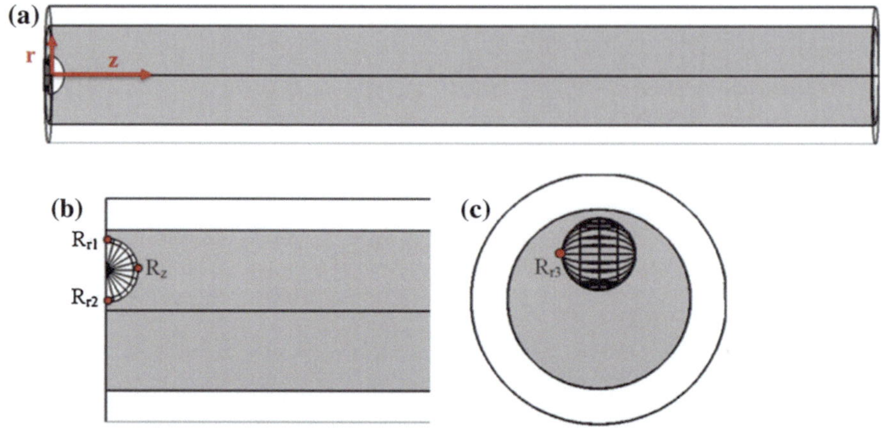

Fig. 1.17 A schematic illustration of the three-dimensional geometry of a bubble, blood, and a microvessel. **a** Bubble in the *middle* of the vessel, **b** an *off-center* microbubble, and **c** *side view* of an off-center bubble. Reproduced with permission from Hosseinkhah and Hynynen (2012). © 2012 Institute of Physics and Engineering in Medicine

Next, a bubble's oscillation should satisfy the following equation:

$$P_{\mathrm{L}} = \left(P_0 + \frac{2\sigma}{R_0} - P_{\mathrm{v}}\right)\left(\frac{R_0}{R}\right)^{3\kappa} + P_{\mathrm{v}} - \frac{2\sigma}{R} - \frac{4\eta\dot{R}}{R} - \frac{4\kappa_s\dot{R}}{R^2} = 4\chi\left(\frac{1}{R_0} - \frac{1}{R}\right) - P(t) \tag{1.46}$$

where P_{L} indicates the fluid pressure at the bubble wall. The governing differential equation of the blood vessel is the following viscoelastic model:

$$P_{\mathrm{i}} + \frac{\eta_1}{E_1}\dot{P}_{\mathrm{i}} = E_2 e + \eta_1\left(1 + \frac{E_2}{E_1}\right)\dot{e} \tag{1.47}$$

where P_{i} indicates the intravascular pressure and e is the diameter strain. E_1 and E_2 are the elastic coefficients with dimensions of pressure, and η_1 represents a viscous coefficient with dimensions of pressure times time. Figure 1.18 is a comparison of experimental and theoretical results. Figure 1.18a also gives the solution of the Rayleigh–Plesset equation, which is indicated by the dotted line. From this figure, we can see that the numerical simulation can much better predict the bubble oscillation than the Rayleigh–Plesset equation. Figure 1.18b, which represents bubble oscillation inside a 200-μm-diameter capillary vessel, also matches closely.

The oscillation of encapsulated bubbles in tubes has also been investigated for phase-shift nanodroplets, which are usually composed of a perfluorocarbon (PFC) gas surrounded by a coating material. PFCs with low boiling temperatures (~29 °C) are liquid at room temperature and are stable after being coated by a shell

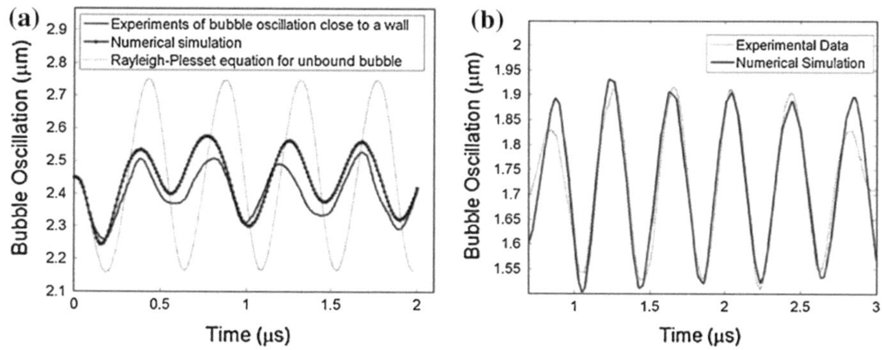

Fig. 1.18 Validation of experimental data. **a** Validation with the optical tweezer experiment for bubble oscillation close to a wall. $P_a = 200$ kPa, $f = 2.25$ MHz, $R_0 = 2.45$ μm. **b** Validation with a bubble inside a 200-μm-diameter capillary vessel. $P_a = 58$ kPa, $f = 2.5$ MHz, $R_0 = 1.7$ μm. Reproduced with permission from Hosseinkhah and Hynynen (2012). © 2012 Institute of Physics and Engineering in Medicine

material. These nanodroplets can undergo a phase shift and change into a gas bubble under intense acoustic exposure, which is termed ADV.

Ye and Bull (2006) investigated a model of the ADV process within a flexible tube and assessed the effects of vaporized bubbles on the surrounding vessel. Figure 1.19 is a schematic of a bubble in a flexible tube. The ADV process is modeled by a single bubble whose initial size ranges from 10 to 90 % of the tube diameter. The characteristic parameters of the tube include L_t (length), D (inner diameter), K_{sp} (radial stiffness), and σ_w (longitudinal tension).

The ADV process is assumed to be isothermal, and any mass diffusion that occurs across the bubble interface after the droplet vaporizes is ignored because the time period is so short. According to the ideal gas law, the governing equation used to describe the gas phase is

$$P_v^* V_v^* = \text{cons tan t} \tag{1.48}$$

Fig. 1.19 Schematic of a bubble in a flexible tube. Reproduced with permission from Ye and Bull (2006). © 2006 by ASME

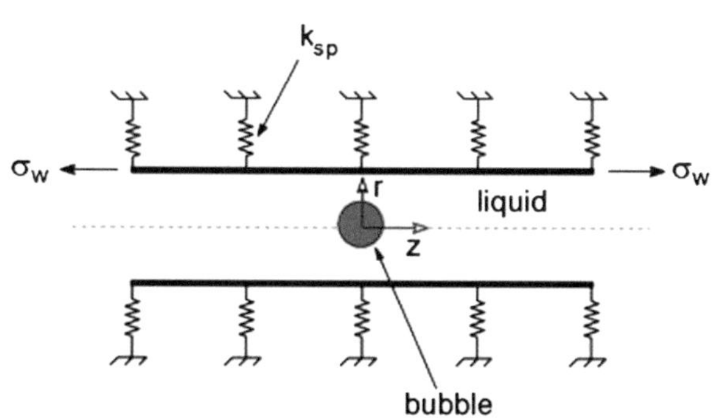

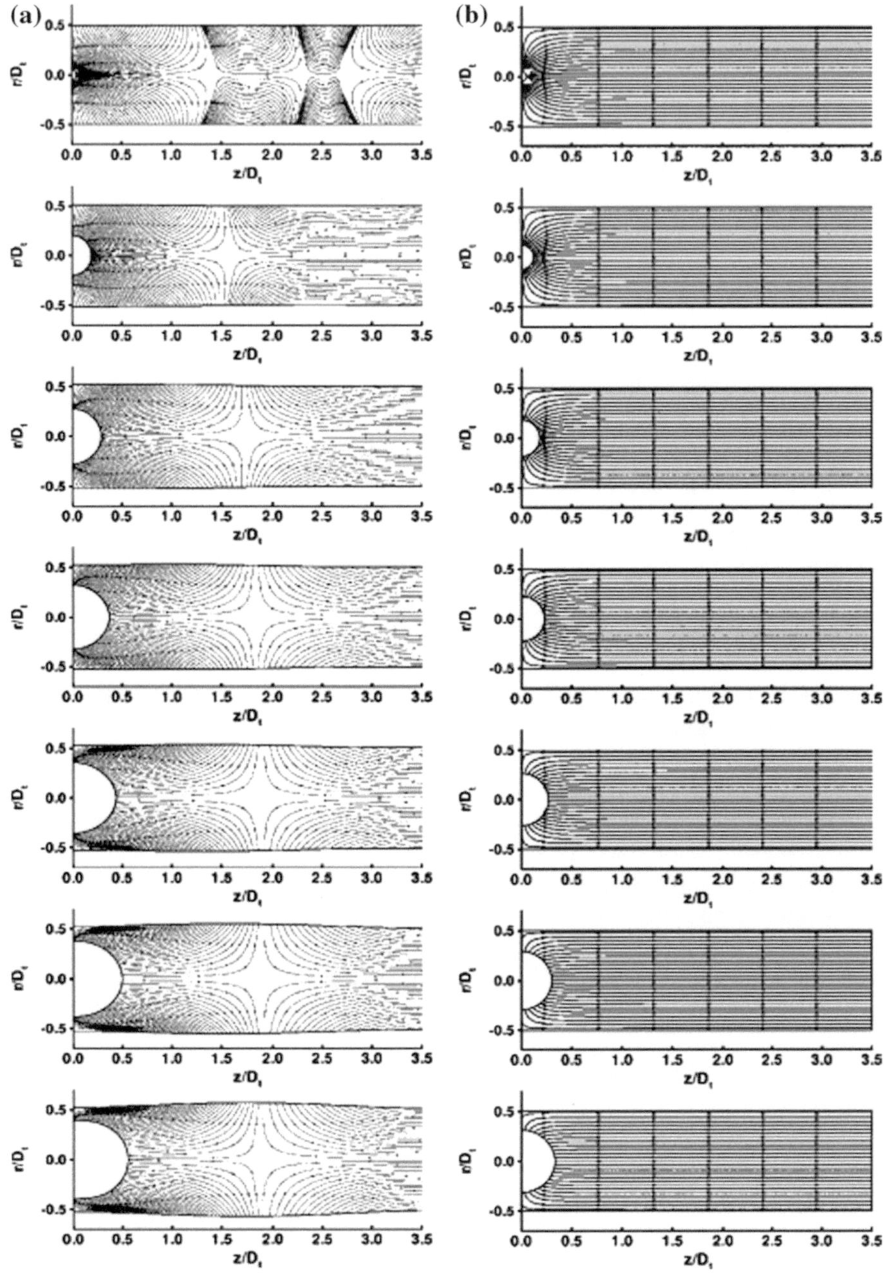

Fig. 1.20 Bubble shapes, wall shapes, and streamlines within **a** a flexible tube and **b** a rigid tube. $R_e = 427.59$, $W_e = 6.93$, $S_t = 10.47$, $\Omega_s = 5 \times 10^{-9}$, $\Omega_t = 0.05$, $P_h = 176.96$, $d_i = 0.1$. Reproduced with permission from Ye and Bull (2006). © 2006 by ASME

where P_v^* is the bubble pressure and V_v^* is the bubble volume. The liquid is governed by the Navier–Stokes equations. For the boundary and interfacial conditions, the bubble interface velocity and the liquid velocity are considered to be the same:

$$\frac{dY^*}{dt^*} = \left(u_l^* \cdot \hat{n}\right)\hat{n} \tag{1.49}$$

The conditions of the tube wall can be described according to the following equation:

$$p_l - \mu_l \left(\frac{\partial u_n^*}{\partial n}\right)_l + \sigma_w \kappa_w^* - \kappa_{sp} \Delta r^* = 0 \tag{1.50}$$

where Y^* is the interface position and $\hat{n}$ is the unit normal vector to the interface, σ_w is the wall tension, κ_w^* is the curvature of the tube wall, κ_{sp} is the spring constant, and Δr^* is the displacement in the radial direction. Figure 1.20 shows the bubble, the wall shape, and the streamlines within a flexible and rigid tube at different times. The initial bubble diameter is 0.1 μm. Bubble expansion causes the flexible tube to deform and pushes the fluid away toward the end of the tube. Pressure stagnation can be observed in the middle of the tube, which increases as the bubble volume increases. On the other hand, the situation in the rigid tube is much simpler. In this study, the influence of the initial bubble diameter and flexibility of the vessel on the wall stress was also assessed, and results showed that smaller droplets within large and more flexible vessels were less likely to damage vessels.

1.5.5 Bubble Dynamics in Soft Tissue

Along with liquids, cavitation can also occur in tissue. However, cavitation in tissues can disrupt the ablation lesion, making it unpredictable during high-intensity focused ultrasound (HIFU). In order to control HIFU during therapy, we must understand cavitation activity in tissue better. The challenges involved in studying cavitation in tissue relate to the properties of the surrounding medium, but some models have been developed to represent soft tissue, such as the Maxwell model, the Kelvin–Voigt model, and the Zener viscoelastic model. Here, we show three specific examples of bubble dynamics in soft tissue or viscoelastic fluids.

Allen and Roy studied bubble dynamics in viscoelastic fluids with linear and nonlinear viscoelasticity (Allen and Roy 2000a, b). For the linear viscoelastic medium, a system of coupled nonlinear ordinary differential equations has been derived, and it describes the radial bubble dynamics. The small-amplitude bubble oscillations are governed by the Rayleigh–Plesset equation, and the linear Jeffreys model is used to describe the viscoelastic medium:

$$\tau_{rr} + \lambda_1 \frac{\partial \tau_{rr}}{\partial t} = -2\eta_0 \left(\dot{\varsigma} + \lambda_2 \frac{\partial \dot{\varsigma}}{\partial t} \right)$$
$$\dot{\varsigma} = \frac{\partial u}{\partial r} = -\frac{2\dot{R}R^2}{r^3} \tag{1.51}$$
$$u = \frac{\dot{R}R^2}{r^2}$$

where λ_1 is the relaxation time, λ_2 is the retardation time, $\dot{\varsigma}$ is the rate of strain or deformation tensor, and η_0 is the zero shear-rate viscosity. Coupling the linear Jeffreys model with the initial bubble equation, the following bubble dynamics can be obtained:

$$\overline{R\ddot{R}} + \frac{3}{2}\overline{\dot{R}^2} = \left(1 \middle/ \overline{R^{3\kappa}} \right) - w\left(1 \middle/ \overline{R} \right) - (1 - w)(1 - \bar{a}\cos(\Omega \bar{t})) - \frac{3}{\mathrm{Re}} \int_{R}^{\infty} \frac{\overline{\tau_{rr}}(\bar{r}, \bar{t})}{\bar{r}} \tag{1.52}$$

The parameters can be defined as follows:

$$w = \frac{2\hat{\sigma}}{p_{go}R_0}; \quad \Omega = R_0 \sqrt{\frac{\rho}{p_{go}}}\omega; \quad R_e = \eta_0 \Big/ (p_{go}\rho)^{1/2} R_0; \bar{a} = \frac{p_a}{p_0}$$

where R_e indicates the Reynolds number, w is the dimensionless Laplace pressure, and Ω is the ratio of the acoustic forcing frequency to the natural resonance frequency of the bubble; the overbar indicates normalized (i.e., dimensionless) variables. This relationship can be extended to studies where the role of large bubble deformation is not critical.

The nonlinear oscillation of a spherical, acoustically forced bubble has also been investigated in a nonlinear viscoelastic medium. Here, bubble oscillation is also described by the Rayleigh–Plesset equation, while the upper-convected Maxwell (UCM) model, which is suitable for studying large-amplitude excursions of bubbles, was used to simplify the nonlinear viscoelastic medium. The equation can be written as follows:

$$\tau + \lambda_1 \tau_{(1)} = -\eta_0 \dot{\gamma} \tag{1.53}$$

The definition of $\tau_{(1)}$ is

$$\tau_{(1)} = \frac{D}{Dt}\tau - \left((\nabla u)^{\perp} \cdot \tau + \tau \cdot (\nabla u) \right)$$

The subscript (1) means the first contravariant convective time derivative. The superscript $^{\perp}$ indicates the transpose, and *D/Dt* is the material derivative. The rewritten Rayleigh–Plesset equation for nonlinear oscillation of a bubble in a nonlinear viscoelastic medium is as follows:

$$R\ddot{R}+\frac{3}{2}\dot{R}^2=\frac{p_0}{\rho\omega^2R_0^2}\left[\left(1+\frac{2\sigma}{p_0R_0}\right)\left(\frac{1}{R}\right)^{3\kappa}-\left(\frac{2\sigma}{p_0R_0}\right)\left(\frac{1}{R}\right)-(1+a\sin(t))\right]$$
$$-\frac{2}{R_e}\frac{1}{\omega R_0}\sqrt{\frac{p_0}{\rho}}\int_R^\infty\left(\frac{T_{11}-T_{\theta\theta}}{r}\right)\mathrm{d}r \tag{1.54}$$

After results from the linear and nonlinear viscoelastic models were compared, they were both found to agree perfectly for small-amplitude forcing, shown in Fig. 1.21. However, when the applied pressure increased considerably as in Fig. 1.22, the equations show significant divergences after the first cycle: The UCM model predicts stable oscillation, while the Maxwell model predicts explosive growth.

In another study on bubble oscillation in tissue, Yang and Church (2005) modeled the gas bubble dynamics in soft tissue based on the Keller–Miksis equation, which can account for large-amplitude bubble oscillation. Here, they chose the Voigt model to describe the viscoelasticity of soft tissue. The boundary and initial conditions are given as Eq. (1.55).

$$\begin{aligned}&P=P_g-\frac{2\sigma}{R}+\tau_{rr}\quad \text{at } r=R\\&P=P_\infty\quad \text{at } r=\infty\\&R=R_0,\ \dot{R}=0\quad \text{at } t=0\end{aligned}\tag{1.55}$$

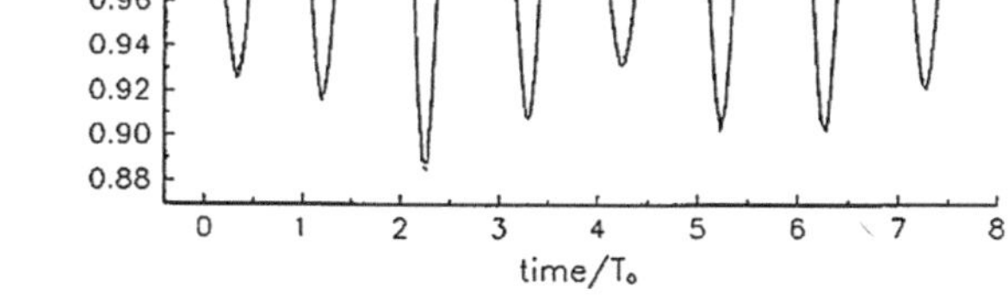

Fig. 1.21 A comparison of the linear and nonlinear model predictions for small-amplitude forcing. $P_A = 0.05$ MPa, $f = 3.0$ MHz, $R_0 = 1.0$ μm, De = 3.0, and Re = 2.5. Good agreement exists between the UCM (*dashed line*) and the linear Maxwell (*solid line*) models. Reproduced with permission from Allen and Roy (2000b). © 2000 Acoustical Society of America

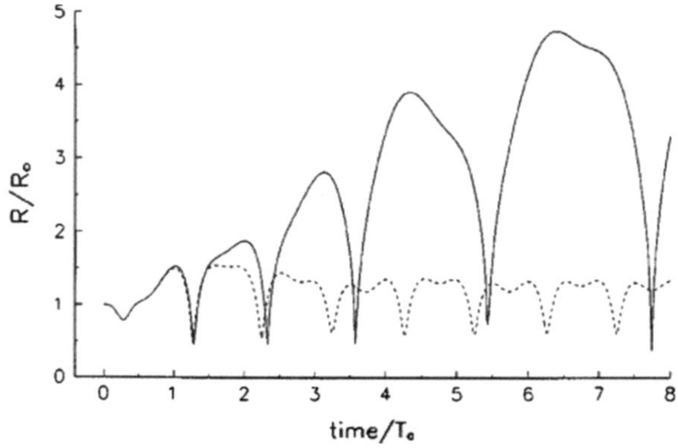

Fig. 1.22 Radius versus time plot for the same conditions as in Fig. 1.21, except the pressure amplitude has been increased to 0.2 MPa. Reproduced with permission from Allen and Roy (2000b). © 2000 Acoustical Society of America

where P_g is the gas pressure inside the bubble, R is the position of the gas–tissue interface, R_0 is the bubble equilibrium radius, and σ is the surface tension. Next, the asymptotic solution was calculated in both the near and far fields, and these solutions matched those in the internal and external zones. The following equation takes into account the compressibility of the surrounding medium:

$$\left(1-\frac{\dot{R}}{c}\right)R\ddot{R}+\frac{3}{2}\left(1-\frac{\dot{R}}{3c}\right)\dot{R}^2=\left(1+\frac{\dot{R}}{c}\right)\frac{P_a-P_l}{\rho}+\frac{R}{\rho c}\frac{\mathrm{d}}{\mathrm{d}t}[P_a-P_l] \tag{1.56}$$

where $(2\rho/c)\psi_2''=P_A g(t)$ is the driving pressure,

$$P_a=P_g-\frac{2\sigma}{R}+\tau_{rr}(R,t)$$

and

$$P_a-P_l=P_g-\frac{2\sigma}{R}-P_0+\frac{2\rho}{c}\psi_2''+3\int_R^\infty\frac{\tau_{rr}}{r}\mathrm{d}r$$

To model the viscoelasticity of soft tissue, the linear Voigt model was used, as it is appropriate for low-megahertz frequencies. This led to the following equations:

$$P_a-P_l=P_g-\frac{2\sigma}{R}-P_0+P_A g(t)-\left[\frac{4G}{3R^3}\left(R^3-R_0^3\right)+\frac{4\mu\dot{R}}{R}\right] \tag{1.57}$$

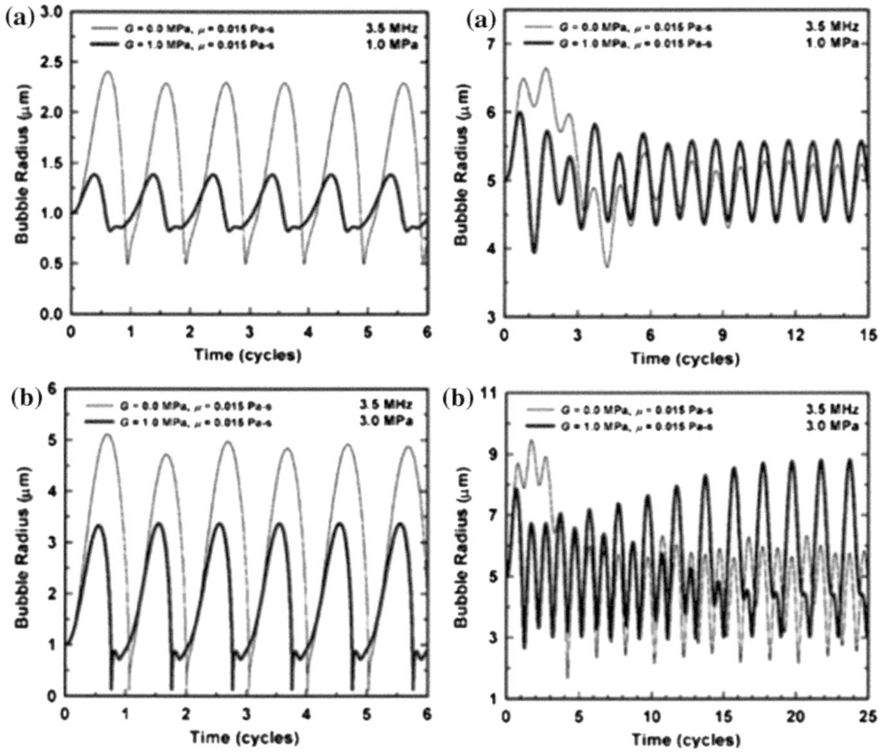

Fig. 1.23 A comparison of radial responses for 1-μm-radius bubbles (*left*) and 5-μm-radius bubbles (*right*) driven by a 3.5-MHz pulse at **a** 1 MPa and **b** 3 MPa for $G = 0$ (*thin line*), and $G = 1.0$ MPa (*thick line*). Reproduced with permission from Yang and Church (2005). © 2005 Acoustical Society of America

$$
\begin{aligned}
\frac{\mathrm{d}}{\mathrm{d}t}(P_\mathrm{a} - P_\mathrm{l}) &= \frac{\mathrm{d}}{\mathrm{d}t}\left(P_\mathrm{g} - \frac{2\sigma}{R} - P_0 + P_\mathrm{A} g(t)\right) + \frac{\mathrm{d}}{\mathrm{d}t}\left[3\int_R^\infty \frac{\tau_{rr}}{r}\mathrm{d}r\right] \\
&= \frac{\mathrm{d}P_\mathrm{g}}{\mathrm{d}t} + \frac{2\sigma\dot{R}}{R^2} + P_\mathrm{A}\frac{\mathrm{d}g(t)}{\mathrm{d}t} - 4G\frac{R_0^3\dot{R}}{R^4} - 4\mu\left(-\frac{\dot{R}^2}{R^2} + \frac{\ddot{R}}{R}\right)
\end{aligned}
\tag{1.58}
$$

Based on these equations, Yang and Church (2005) studied the effect of tissue viscoelasticity on thermal and elastic damping. Figure 1.23 shows the radial responses of 1-μm-radius bubbles and 5-μm-radius bubbles driven by a 3.5-MHz pulse at 1 and 3 MPa for $G = 0$ and $G = 1.0$ MPa when the viscosity was set at 0.015 Pa s. For the smaller bubbles (radius of 1 μm), both the amplitude and the nonlinearity of bubble oscillation were reduced. However, for the larger bubbles, the presence of elasticity increased the amplitude of the bubble oscillation and increased the nonlinearity. Because the cavitation threshold is larger in soft tissue than in free fields, it is important to study cavitation-related effects in vivo.

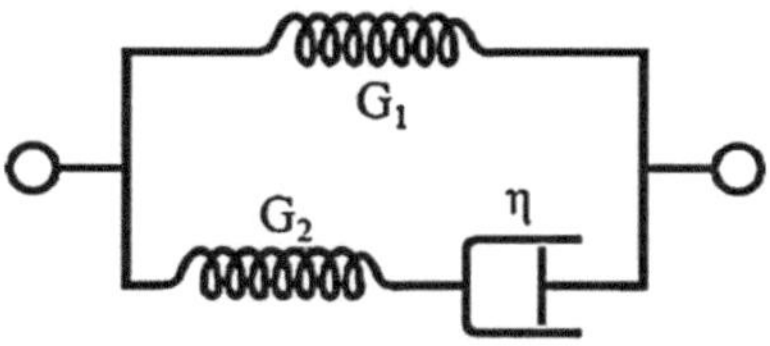

Fig. 1.24 Spring-dashpot schematic for the Zener model. Reproduced with permission from Hua and Johnsen (2013). © 2013 AIP Publishing LLC

In a different study, Hua and Johnsen (2013) used a Zener viscoelastic model to describe the relaxation, viscosity, and elasticity of soft tissue. They also used the Keller–Miksis equation to investigate large-amplitude oscillation. To be more accurate, a third-order expansion of the radius was used to represent the dynamics. As the bubble collapse is governed by the pressure difference between the surroundings and the interior, the pressure on both sides of the interface is calculated:

$$
\begin{aligned}
p_{\mathrm{B}} &= p_{\mathrm{i}} - \frac{2\sigma}{R} + 3\int_{R}^{\infty} \frac{\tau_{rr}}{r}\,\mathrm{d}r \\
p_i &= p_{\mathrm{G}_0}\left(\frac{R_0}{R}\right)^{3\gamma} \\
p_{\mathrm{G}_0} &= p_\infty + \frac{2\sigma}{R_0}
\end{aligned}
\tag{1.59}
$$

where p_{B} is the pressure on the liquid side of the interface and p_{i} is the internal pressure on the other side. The term p_{G_0} is the initial partial pressure of a non-condensable gas (air) in the bubble. The linear Zener model including a tissue viscosity parameter μ, shear modulus G, and relaxation parameter λ represents the tissue for simplicity:

$$
\lambda \frac{\partial \tau_{rr}}{\partial t} + \tau_{rr} = 2G\gamma_{rr} + 2\mu \frac{\partial \gamma_{rr}}{\partial t}
\tag{1.60}
$$

The Zener model can transform into the models like Kelvin–Voigt, Maxwell, and Oldroyd under certain definitions. Figure 1.24 is the spring-dashpot schematic for the Zener model, where springs are denoted by the spring constants G_{i} and the dashpot by the damping coefficient η.

After coupling the constitutive relations to the bubble dynamics, the modified Keller–Miksis equation can be written in a dimensionless form:

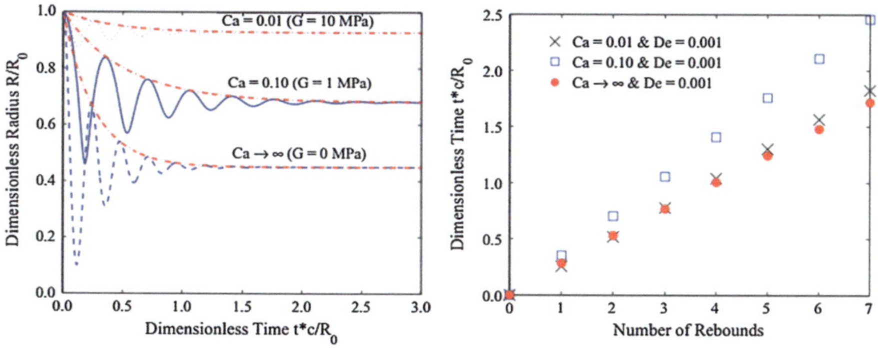

Fig. 1.25 Numerical solutions showing the dependence of bubble oscillations on elasticity (De = 0.001, PR = 34, C = 160, and Re = 3.0). (*Left*) History of the bubble radius: Ca → ∞ (Maxwell: *dashed*), Ca = 0.10 (*solid*), and Ca = 0.01 (*dotted*); decay curves (*dashed-dotted*). (*Right*) Time to achieve a given rebound number: Ca → ∞ (Maxwell: *dots*), Ca = 0.10 (*circles*), and Ca = 0.01 (*crosses*). Reproduced with permission from Hua and Johnsen (2013). © 2013 AIP Publishing LLC

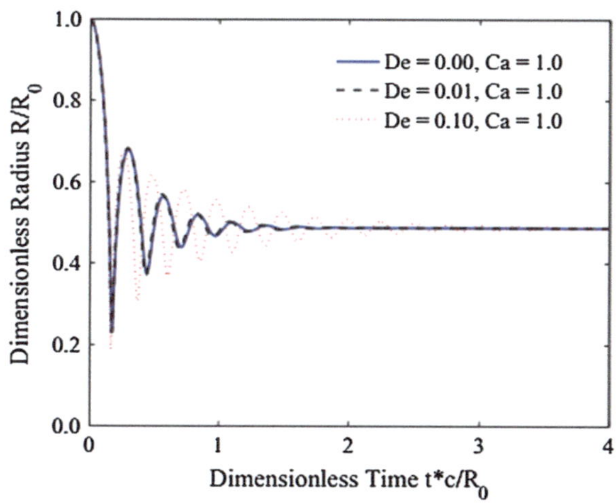

Fig. 1.26 Numerical solution of the bubble radius for different relaxation times (Ca = 1.0, P_R = 34, C = 160, *Re* = 3.0): De = 0.00 (Kelvin–Voigt: solid), De = 0.01 (*dashed*), and De = 0.1 (*dotted*). Reproduced with permission from Hua and Johnsen (2013). © 2013 AIP Publishing LLC

$$\left(1-\frac{\dot{R}}{C}\right)R\ddot{R}+\frac{3}{2}\left(1-\frac{1}{3}\frac{\dot{R}}{C}\right)\dot{R}^2=\left(1+\frac{\dot{R}}{C}\right)\left[\left(1+\frac{2}{We}\right)\frac{1}{R^{3\gamma}}-\frac{2}{WeR}+3\varsigma-1-p_R\right]$$
$$-\frac{\dot{R}}{C}\left[\left(1+\frac{2}{We}\right)\frac{3\gamma}{R^{3\gamma}}-\frac{2}{WeR}\right]+3\frac{R}{C}\dot{\varsigma} \tag{1.61}$$

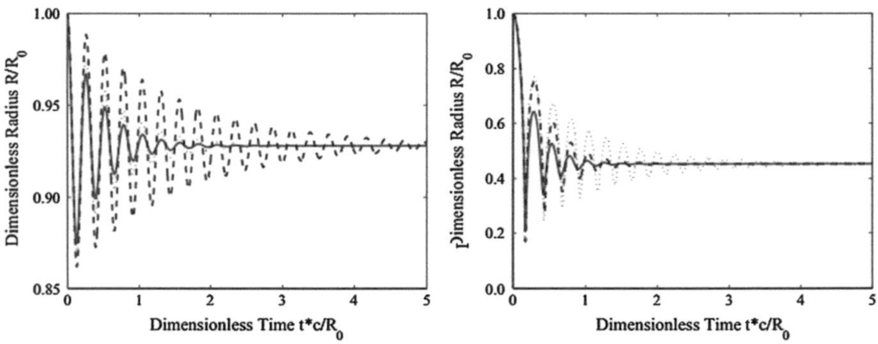

Fig. 1.27 Numerical solution of the compressibility and viscosity at a fixed Ca (PR = 34 and De = 0). (*Left*) History of the bubble radius at Ca = 0.01. (*Right*) History of bubble radius at Ca = 10. Numerical results with $C = 160$ and $Re = 3.0$ (compressible and viscous: solid), $C \to \infty$ and $Re = 3.0$ (incompressible and viscous: *dashed*), and C = 160 and $Re \to \infty$ (compressible and inviscid: *dotted*). Reproduced with permission from Hua and Johnsen (2013). © 2013 AIP Publishing LLC

with constitutive relations:

$$De\dot{\tau}_m + \tau_m = 4\frac{De}{Ca}\frac{\dot{R}}{R^4} - \frac{4}{Re}\frac{\dot{R}}{R}$$

$$De\dot{\varsigma}_m + \varsigma_m + De\frac{\dot{R}}{R}\tau_m = \frac{4}{3}\left(\frac{De}{Ca} - \frac{1}{Re}\right)\frac{\dot{R}}{R}$$

The definitions of the various parameters are given in Hua and Johnsen (2013). Figures 1.25, 1.26, and 1.27 show the dependence of bubble dynamics on elasticity, relaxation, and compressibility and viscosity, respectively. Elasticity governs the final equilibrium radius. Thus, increasing elasticity causes smaller oscillations and amplifies the damping. On the other hand, the oscillation period and damping decrease with an increasing Deborah number, which describe the relaxation properties. As for the influence of compressibility and viscosity, they compete to regulate damping. At high elasticity, compressibility has a significant effect, but at low elasticity, viscosity becomes more important.

1.6 Summary

In this chapter, we first covered fundamental information about acoustic cavitation, including the basic cavitation process, nucleation and collapse thresholds, cavitation nuclei, and cavitation types. Then, we provided an overview of bubble dynamics in various environments: free fields, near boundaries, in vessels, and in tissues.

In the biomedical field, various micro- and nanoscale particles have long been used in diagnostic applications, but they are now being incorporated into ultrasonic-based therapeutic applications as new types of cavitation nuclei, thus being developed as "theranostic" agents with both therapeutic and diagnostic functions. Therefore, the classical physical theories on cavitation might need to be updated for these new uses. As such, here we break away from the traditional literature by classifying cavitation as either tensile or energetic cavitation, which reflects the mechanism of cavitation induction as either mechanical or thermal, respectively. For biomedical applications, acoustic cavitation can be categorized as either solely tensile or as a combination of tensile and energetic, depending on the duration and amplitude of the acoustic exposure. In addition, biomedical applications of cavitation are different from other applications, given that the cavitation may occur in non-Newtonian fluids or even in soft tissue. Therefore, we also discussed how the medium affects cavitation dynamics.

Overall, the effects of the medium, the multi-imposing energy source, and multifunctional cavitation nuclei were considered in this chapter. Due to these new considerations that have arisen with the use of acoustic cavitation for biomedical applications, we must also develop new physical theories and engineering technologies to detect and further control the spatial and temporal activity of cavitation. Some of these recent developments will be discussed in the following chapters.

References

Akulichev V, Rozenberg L. Certain relations in a cavitation region (Energy dissipation of acoustic waves propagating in liquid medium and producing cavitation). Sov Phys Acoust. 1966;11:246–51.

Allen JS, Roy RA. Dynamics of gas bubbles in viscoelastic fluids. I. Linear viscoelasticity. J Acoust Soc Am. 2000a;107(6):3167–78.

Allen JS, Roy RA. Dynamics of gas bubbles in viscoelastic fluids. II. Nonlinear viscoelasticity. J Acoust Soc Am. 2000b;108(4):1640–50.

Allen JS, May DJ, Ferrara KW. Dynamics of therapeutic ultrasound contrast agents. Ultrasound Med Biol. 2002;28(6):805–16.

Apfel RE. Acoustic cavitation prediction. J Acoust Soc Am. 1981;69:1624–33.

Apfel RE. Acoustic cavitation inception. Ultrasonics. 1984;22(4):167–73.

Apfel RE. The role of impurities in cavitation-threshold determination. J Acoust Soc Am. 2005;48 (5B):1179–86.

Apfel RE, Holland CK. Gauging the likelihood of cavitation from short-pulse, low-duty cycle diagnostic ultrasound. Ultrasound Med Biol. 1991;17(2):179–85.

Askar'yan G, Moroz E. Pressure on evaporation of matter in a radiation beam. Sov J Exp Theor Phys. 1963;16:1638–9.

Blake JR, Gibson D. Growth and collapse of a vapour cavity near a free surface. J Fluid Mech. 1981;111:123–40.

Bloch SH, Wan M, Dayton PA, Ferrara KW. Optical observation of lipid-and polymer-shelled ultrasound microbubble contrast agents. Appl Phys Lett. 2004;84(4):631–3.

Bloembergen N. Laser-induced electric breakdown in solids. IEEE J Quantum Electron. 1974;10 (3):375–86.

Borkent BM, Gekle S, Prosperetti A, Lohse D. Nucleation threshold and deactivation mechanisms of nanoscopic cavitation nuclei. Phys Fluids. 2009;21(10):102003.

Bouakaz A, Versluis M, de Jong N. High-speed optical observations of contrast agent destruction. Ultrasound Med Biol. 2005;31(3):391–9.

Brewer RG, Rieckhoff KE. Stimulated Brillouin scattering in liquids. Phys Rev Lett. 1964;13 (11):334.

Briggs H, Johnson J, Mason W. Properties of liquids at high sound pressure. J Acoust Soc Am. 1947;19:664–77.

Brujan EA, Ikeda T, Matsumoto Y. Jet formation and shock wave emission during collapse of ultrasound-induced cavitation bubbles and their role in the therapeutic applications of high-intensity focused ultrasound. Phys Med Biol. 2005;50(20):4797–809.

Chang PH, Shun K, Wu S-J, Levene HB. Second harmonic imaging and harmonic Doppler measurements with Albunex. IEEE Trans Ultrason Ferroelectr Freq Control. 1995;42(6):1020–7.

Chatterjee D, Sarkar K. A Newtonian rheological model for the interface of microbubble contrast agents. Ultrasound Med Biol. 2003;29(12):1749–57.

Chen H, Kreider W, Brayman AA, Bailey MR, Matula TJ. Blood vessel deformations on microsecond time scales by ultrasonic cavitation. Phys Rev Lett. 2011;106(3):034301.

Chomas JE, Dayton P, Allen J, Morgan K, Ferrara KW. Mechanisms of contrast agent destruction. IEEE Trans Ultrason Ferroelectr Freq Control. 2001;48(1):232–48.

Church CC. The effects of an elastic solid-surface layer on the radial pulsations of gas-bubbles. J Acoust Soc Am. 1995;97(3):1510–21.

Correas J-M, Bridal L, Lesavre A, Méjean A, Claudon M, Hélénon O. Ultrasound contrast agents: properties, principles of action, tolerance, and artifacts. Eur Radiol. 2001;11(8):1316–28.

Crum LA. Tensile strength of water. Nature. 1979;278(5700):148–9.

de Jong N, Hoff L, Skotland T, Bom N. Absorption and scatter of encapsulated gas filled microspheres: theoretical consideration and some measurements. Ultrasonics. 1992;30 (2):95–103.

Duncan J, Zhang S. On the interaction of a collapsing cavity and a complaint wall. J Fluid Mech. 1991;226:401–23.

Epstein P, Plesset M. On the stability of gas bubbles in liquid-gas solutions. J Chem Phys. 1950;18:1505–9.

Fabiilli ML, Haworth KJ, Sebastian IE, Kripfgans OD, Carson PL, Fowlkes JB. Delivery of chlorambucil using an acoustically-triggered perfluoropentane emulsion. Ultrasound Med Biol. 2010;36(8):1364–75.

Farny CH, Wu T, Holt RG, Murray TW, Roy RA. Nucleating cavitation from laser-illuminated nano-particles. Acoust Res Lett Online. 2005;6(3):138–43.

Flynn H. Physics of acoustic cavitation in liquids. Physical Acoustics. 1964;1(Part B):57–172.

Fox FE, Herzfeld KF. Gas bubbles with organic skin as cavitation nuclei. J Acoust Soc Am. 1954;26:984–9.

Gibson D, Blake JR. The growth and collapse of bubbles near deformable surfaces. Appl Sci Res. 1982;38(1):215–24.

Glazman RE. Effects of adsorbed films on gas bubble radial oscillations. J Acoust Soc Am. 1983;74(3):980–6.

Harvey EN, McElroy WD, Whiteley AH. On cavity formation in water. J Appl Phys. 1947;18 (2):162–72.

Hay TA, Ilinskii YA, Zabolotskaya EA, Hamilton MF. Model for the dynamics of a spherical bubble undergoing small shape oscillations between parallel soft elastic layers. J Acoust Soc Am. 2013;134(2):1454–62.

Ho VH, Smith MJ, Slater NK. Effect of magnetite nanoparticle agglomerates on the destruction of tumor spheroids using high intensity focused ultrasound. Ultrasound Med Biol. 2011;37 (1):169–75.

Hoff L. Acoustic characterization of contrast agents for medical ultrasound imaging. Berlin: Springer; 2001.

Hoff L, Sontum PC, Hovem JM. Oscillations of polymeric microbubbles: Effect of the encapsulating shell. J Acoust Soc Am. 2000;107(4):2272–80.

Hong C, Brayman AA, Kreider W, Bailey MR, Matula TJ. Observations of translation and jetting of ultrasound-activated microbubbles in mesenteric microvessels. Ultrasound Med Biol. 2011;37(12):2139–48.

Hosseinkhah N, Hynynen K. A three-dimensional model of an ultrasound contrast agent gas bubble and its mechanical effects on microvessels. Phys Med Biol. 2012;57(3):785–808.

Hua C, Johnsen E. Nonlinear oscillations following the Rayleigh collapse of a gas bubble in a linear viscoelastic (tissue-like) medium. Phys Fluids. 2013;25(8):083101.

Johnson BD, Cooke RC. Generation of stabilized microbubbles in seawater. Science. 1981;213 (4504):209–11.

Keller JB, Miksis M. Bubble oscillations of large amplitude. J Acoust Soc Am. 1980;68 (2):628–33.

Kennedy JE. High-intensity focused ultrasound in the treatment of solid tumours. Nat Rev Cancer. 2005;5(4):321–7.

Kennedy PK, Hammer DX, Rockwell BA. Laser-induced breakdown in aqueous media. Prog Quantum Electron. 1997;21(3):155–248.

Kotaidis V, Dahmen C, von Plessen G, Springer F, Plech A. Excitation of nanoscale vapor bubbles at the surface of gold nanoparticles in water. J Chem Phys. 2006;124(18):184702.

Labouret W, Kurz T. Physics of bubble oscillations. Rep Prog Phys. 2010;73(10):106501.

Lauterborn W. Optische kavitation. Phys Blätter. 1976;32(12):553–63.

Lauterborn W, Bolle H. Experimental investigations of cavitation-bubble collapse in the neighbourhood of a solid boundary. J Fluid Mech. 1975;72(02):391–3.

Lee M, Klaseboer E, Khoo B. On the boundary integral method for the rebounding bubble. J Fluid Mech. 2007;570:407–29.

Leighton T. The acoustic bubble. London: Academic Press; 1994.

Lukianova-Hleb EY, Santiago C, Wagner DS, Hafner JH, Lapotko DO. Generation and detection of plasmonic nanobubbles in zebrafish. Nanotechnology. 2010;21(22):225102.

Maris HJ. Introduction to the physics of nucleation. CR Phys. 2006;7(9):946–58.

Miao H, Gracewski SM, Dalecki D. Ultrasonic excitation of a bubble inside a deformable tube: Implications for ultrasonically induced hemorrhage. J Acoust Soc Am. 2008;124(4):2374–84.

Miller MW, Miller DL, Brayman AA. A review of in vitro bioeffects of inertial ultrasonic cavitation from a mechanistic perspective. Ultrasound Med Biol. 1996;22(9):1131–54.

Morgan KE, Allen JS, Dayton PA, Chomas JE, Klibaov A, Ferrara KW. Experimental and theoretical evaluation of microbubble behavior: effect of transmitted phase and bubble size. IEEE Trans Ultrason Ferroelectr Freq Control. 2000;47(6):1494–509.

Neppiras E. Acoustic cavitation series: part one: Acoustic cavitation: an introduction. Ultrasonics. 1984;22(1):25–8.

Oguz HN, Prosperetti A. Bubble oscillations in the vicinity of a near plane free-surface. J Acoust Soc Am. 1990;87(5):2085–92.

Oguz HN, Prosperetti A. The natural frequency of oscillation of gas bubbles in tubes. J Acoust Soc Am. 1998;103(6):3301–8.

Pitt WG, Singh RN, Perez KX, Husseini GA, Jack DR. Phase transitions of perfluorocarbon nanoemulsion induced with ultrasound: a mathematical model. Ultrason Sonochem. 2014;21 (2):879–91.

Plesset M. The dynamics of cavitation bubbles. J Appl Mech. 1949;16:277.

Porter TM, Smith DA, Holland CK. Acoustic techniques for assessing the optison destruction threshold. J Ultrasound Med. 2006;25(12):1519–29.

Raĭzer YP. Breakdown and heating of gases under the influence of a laser beam. Sov Phys Usp. 1966;8(5):650–73.

Rapoport N, Gao Z, Kennedy A. Multifunctional nanoparticles for combining ultrasonic tumor imaging and targeted chemotherapy. J Natl Cancer Inst. 2007;99(14):1095–106.

Ready J. Effects of high-power laser radiation. New York: Academic Press; 1971.

Reznik N, Williams R, Burns PN. Investigation of vaporized submicron perfluorocarbon droplets as an ultrasound contrast agent. Ultrasound Med Biol. 2011;37(8):1271–9.

Robinson PB, Blake JR, Kodama T, Shima A, Tomita Y. Interaction of cavitation bubbles with a free surface. J Appl Phys. 2001;89(12):8225–37.

Sacchi C. Laser-induced electric breakdown in water. J Opt Soc Am B. 1991;8(2):337–45.

Sankin GN, Zhong P. Interaction between shock wave and single inertial bubbles near an elastic boundary. Phys Rev E. 2006;74(4):046304.

Sheeran PS, Dayton PA. Phase-change contrast agents for imaging and therapy. Curr Pharm Des. 2012;18(15):2152–65.

Shpak O, Verweij M, Vos HJ, de Jong N, Lohse D, Versluis M. Acoustic droplet vaporization is initiated by superharmonic focusing. Proc Natl Acad Sci. 2014;111(5):1697–702.

Simpson DH, Chin CT, Burns PN. Pulse inversion Doppler: a new method for detecting nonlinear echoes from microbubble contrast agents. IEEE Trans Ultrason Ferroelectr Freq Control. 1999;46(2):372–82.

Smith D, Meyerand R Jr. Laser radiation induced gas breakdown. New York: Wiley; 1976.

Smith DA, Porter TM, Martinez J, Huang S, MacDonald RC, McPherson DD, Holland CK. Destruction thresholds of echogenic liposomes with clinical diagnostic ultrasound. Ultrasound Med Biol. 2007;33(5):797–809.

Strasberg M. Onset of Ultrasonic Cavitation in Tap Water. J Acoust Soc Am. 1959;31(2):163–76.

Temperley H. The behaviour of water under hydrostatic tension: III. Proc Phys Soc. 1947;59 (2):199.

ter Haar G. Therapeutic applications of ultrasound. Prog Biophys Mol Biol. 2007;93(1–3):111–29.

Vogel A, Busch S, Parlitz U. Shock wave emission and cavitation bubble generation by picosecond and nanosecond optical breakdown in water. J Acoust Soc Am. 1996;100 (1):148–65.

Yang XM, Church CC. A model for the dynamics of gas bubbles in soft tissue. J Acoust Soc Am. 2005;118(6):3595–606.

Ye T, Bull JL. Microbubble expansion in a flexible tube. J Biomech Eng-Trans Asme. 2006;128 (4):554–63.

Yount DE. Skins of varying permeability: a stabilization mechanism for gas cavitation nuclei. J Acoust Soc Am. 1979;65:1429.

Zhang P, Porter T. An in vitro study of a phase-shift nanoemulsion: a potential nucleation agent for bubble-enhanced HIFU tumor ablation. Ultrasound Med Biol. 2010;36(11):1856–66.

Zhang M, Fabiilli ML, Haworth KJ, Fowlkes JB, Kripfgans OD, Roberts W, Ives K, Carson PL. Initial investigation of acoustic droplet vaporization for occlusion in canine kidney. Ultrasound Med Biol. 2010;36(10):1691–703.

Zhang L, Belova V, Wang H, Dong W, Möhwald H. Controlled cavitation at nano/microparticle surfaces. Chem Mater. 2014;26(7):2244–8.

Chapter 2
Cavitation Mapping

Ting Ding, Hui Yin, Hong Hu, Chen Bai and Mingxi Wan

Keywords High-speed photography · Sonoluminescence · Sonochemiluminescence · Active cavitation mapping · Passive cavitation mapping · Precise spatial–temporal mapping · Ultrafast mapping · Doppler method

2.1 Introduction

When the acoustic pressure of ultrasound propagating in a medium is higher than a threshold value, one or more gas pockets (bubbles) can form Apfel (1984). During ultrasound exposure, these gas bubbles may grow, oscillate, and collapse, during which they are called "active" cavitation bubbles. When ultrasound exposure is turned off, some active cavitation bubbles tend to dissolve, leaving only the remaining "residual" cavitation bubbles. Particularly with pulsed ultrasound, residual cavitation bubbles between successive ultrasound exposures may act as cavitation nuclei to enhance cavitation effects. As cavitation plays important roles in almost all ultrasound applications—such as tumor ablation (Kennedy 2005), lithotripsy (Coleman et al. 1987), and hemostasis (Vaezy et al. 1999)—a better

T. Ding · H. Yin · H. Hu · C. Bai · M. Wan (✉)
The Key Laboratory of Biomedical Information Engineering of Ministry of Education, Department of Biomedical Engineering, School of Life Science and Technology, Xi'an Jiaotong University, 710049 Xi'an, People's Republic of China
e-mail: mxwan@mail.xjtu.edu.cn

T. Ding
e-mail: dtingb@gmail.com

H. Yin
e-mail: yhlovegxm@gmail.com

H. Hu
e-mail: huqianqin18@163.com

C. Bai
e-mail: baichen1986@gmail.com

M. Wan et al. (eds.), *Cavitation in Biomedicine*,
DOI 10.1007/978-94-017-7255-6_2

understanding of the spatial–temporal distribution of cavitation bubbles is essential to optimize these applications. For decades, techniques for detecting and mapping cavitation have been developed, and these techniques mainly include optical and acoustic methods (Barnett 1998).

Optical methods, such as high-speed photography, sonoluminescence (SL), and sonochemiluminescence (SCL), can directly visualize the spatial distribution of cavitation bubbles with time, but they can only be used in transparent media. High-speed photography can be employed to capture images of bubble dynamics and distribution when ultrasound exposure is on, and this powerful tool has a sufficiently high frame rate. Its time resolution can reach the nanosecond level, and its spatial resolution can be improved by combining it with microscopes, for example, an inverted microscope or a long-distance microscope. However, compared to high-speed photography, SL and SCL can provide more information about the spatial distribution of chemically active cavitation bubbles.

The acoustic methods commonly used to detect cavitation include active cavitation detection (ACD), passive cavitation detection (PCD), and the Doppler method. In general, ACD is a pulse-echo system based on the scattered acoustic signals from bubbles, and it is sufficient to monitor the existence and dissolution of bubbles. ACD using a focused transducer with a high center frequency is more sensitive to single bubbles in the micron size range. Nevertheless, ACD can only be used when ultrasound exposure is turned off otherwise it might interfere with the interrogating pulse. Conversely, PCD has been used extensively to monitor cavitation activity during ultrasound exposure, where it passively records acoustic emissions from cavitating bubbles. Compared with ACD, PCD is generally more sensitive to either resonance-sized bubbles or clouds of bubbles. ACD and PCD are the classic methods for detecting acoustic cavitation. These have been applied not only in water, but also in tissue. However, both of these acoustic techniques are one-dimensional detection methods, and thus, they cannot reflect the spatial distribution of cavitation. The Doppler method, however, can obtain a much higher frequency shift range that caused by translation motion of a solid particle, so it can effectively distinguish a cavitation event.

Recently, ACD and PCD have been adapted to cavitation mapping techniques, specifically active cavitation mapping (ACM) and passive cavitation mapping (PCM). For these techniques, a linear array is connected to the ultrasound system to increase the spatial coverage of cavitation detection. In this chapter, we introduce two ACM methods: precise active cavitation mapping (PACM) that uses a modified ultrasound line-by-line scanning and ultrafast active cavitation mapping (UACM) with a plane wave beam. PACM can give detailed information with high sensitivity and spatial–temporal resolution, but it is time consuming and strict on the medium with repeatable cavitation distribution. UACM with a plane wave beam has a much higher frame rate than that of conventional B-mode ultrasound imaging, and thus, it can map the transient behaviors of bubbles over the entire region at once. In addition, three-dimensional (3D) plane-by-plane cavitation mapping can be achieved by sequentially measuring a series of such planes where the linear array at each unit is positioned perpendicular to the axis of the cavitation source.

Because UACM uses a plane wave beam rather than a focused pulse, the signal-to-noise ratio (SNR) and lateral resolution should be improved. One way to do this is by using minimum variance (MV) beamforming and coherence factor (CF) weighting in combination with UACM. This MVCF-based UACM can image cavitation bubbles with a relatively high SNR and good spatial–temporal resolution. However, because the primary cavitation source signal as well as the interrogating pulse may interfere with acoustic emissions from bubbles, ACM is presently performed when ultrasound exposure is off. Unlike ACM though, cavitation maps obtained by PCM are based on the emission signals from bubbles when ultrasound irradiation is on, potentially allowing for real-time spatial mapping of cavitation activity.

In this chapter, we introduce methods of cavitation detection and mapping and we describe these techniques when they are applied in water; cavitation imaging in tissues is described in Chap. 7.

2.2 Cavitation Mapping by High-Speed Photography

High-speed photography refers to technology specifically for recording very fast phenomena. In 1948, the Society of Motion Picture and Television Engineers (SMPTE) defined high-speed photography as any set of photographs captured by a camera with 128 frames per second (fps) or greater. At present, ultrahigh-speed photography can achieve as high as 200 million fps in an exposure time of 5 ns. In common applications, high-speed photography may refer to either one or both types of meanings: High-speed photography can refer to a single frame that is taken to freeze motion, especially to reduce motion blur, or it can refer to a series of photographs that are taken at a high sampling frequency. By the first meaning, high-speed photography requires a sensor with good sensitivity and either a very good shuttering system or a very fast strobe light. In the second case, it requires some means of capturing successive frames, either with a mechanical device or by loading data from electronic sensors very quickly.

The first practical application of high-speed photography was Eadweard Muybridge's 1878 investigation into whether horses' feet were actually all off the ground at once during a gallop. Later, in 1886, the first photograph of a supersonic flying bullet was taken in Rijeka by the Austrian physicist Peter Salcher, which afterward was used by Ernst Mach in his studies of supersonic motion. Many decades later, high-speed photography was employed to capture instantaneous phenomenon in many research fields, and it is now also used to view ultrasound- and laser-induced cavitation bubbles. To better understand the effects cavitation bubbles produce, considerable effort has been applied to study the behavior of cavitation bubbles, such as their oscillations, translator motion, collapse, disintegration, and coalescence, under various conditions. For a long time, high-speed photography has been the only method available to perform these investigations, and in the following sections, we discuss how it has been applied.

2.2.1 *Single-Bubble Cavitation*

2.2.1.1 Free Field

In 1989, Leighton et al. (1989) presented two image series showing the unstable growth and collapse of bubbles much smaller than resonance size, where images were captured by a Hadland Hyspeed camera. The camera had a maximum frame rate of about 8000 fps for full-frame photography, taking about 1 s of film in real time. With these images, they verified the predictions of the earlier numerical solutions. Bubbles just smaller than resonance size (which represents a radius 0.3 mm in a 10-kHz sound field) were seen to undergo stable cavitation; much smaller bubbles grow and collapse unstably.

Lauterborn and Ohl (1997) employed an image converter camera to take a photographic series showing the collapse of a laser-generated spherical bubble in water. Its frame rate reached 20.8 million fps. With this high frame rate, the shock wave radiated upon bubble collapse was easily captured. And in 2001, a high-speed camera (Imacon 700, Hadland Photonics) with 22,700 fps was employed by the same group to take 18 consecutive frames, as shown in Fig. 2.1, in which bubble expansion, contraction, and collapse were presented (Akhatov et al. 2001).

2.2.1.2 Fluid–Rigid Boundary

When a bubble is collapsing in a non-spherically symmetric environment, the situation changes considerably. A flat surface nearby causes the bubble to involute from the distal end and to develop a high-speed liquid jet toward this solid surface. However, the distance between the bubble's proximal end and the solid surface may result in different jet shapes.

In the 1982 Ultrasonics Symposium, famous high-speed photograph of a single bubble was presented (Fig. 2.2) by Prof. L.A. Crum. This image shows the liquid jet produced during collapse of a cavitation bubble with a diameter of 1 mm on a solid surface. The width of the bubble was about 1 mm (Crum 1982).

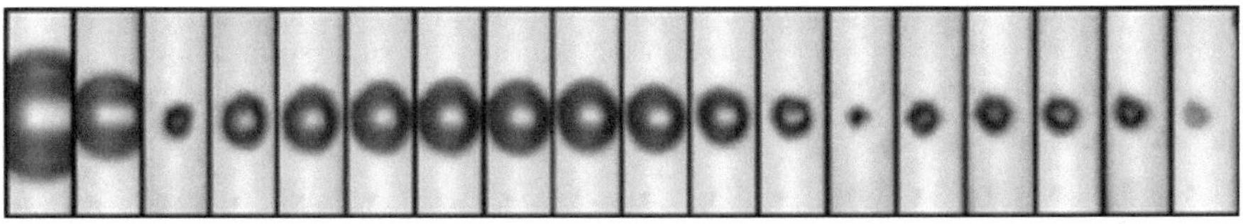

Fig. 2.1 High-speed series of bubble dynamics shortly before and after the first collapse. The interframe time and exposure time were 4.4 and 1 μs, respectively. Reproduced with permission from Akhatov et al. © American Institute of Physics

Fig. 2.2 Cavitation jet developed in a single bubble. Reproduced with permission from Crum et al. © IEEE

As the distance between a bubble and solid surface increases, the length of the liquid jet increases as well. W. Lauterborn's group presented a series of bubble collapsing photographs taken at 75,000 fps with a rotating mirror camera (Lauterborn and Ohl 1997). The jet could be visualized when the bubble was in the first rebound phase. As shown in Fig. 2.3, the dark line inside the bright central spot of the bubble is the same as that shown in Fig. 2.2. However, the difference is that the funnel-shaped downward protrusion shown in Fig. 2.3 is the elongated bubble wall containing the jet, which is driving the elongation.

In this study, each growth and collapse was suggested to take about 50 μs for the smallest bubbles (radius <0.5 μm) and 3–4× longer for larger bubbles (radius 0.15 ± 0.05 mm). Furthermore, the group predicted that the bubble sizes reached during stable cavitation were similar to earlier predictions. However, an exact correlation was not possible since the initial bubble radii could only be estimated from the high-speed photographs, and stable cavitation is highly nonlinear and very sensitive to the initial bubble size.

2.2.1.3 Free Fluid Surface

Another case that has received extensive attention is pulsating bubbles below a free surface (Cole 1965; Holt 1977). Robinson and Blake et al. (2001) presented the experimental effects of nonlinear interactions on the motion of a single bubble and a free surface using high-speed photography (Fig. 2.4).

A striking free-surface–bubble interaction is shown in Fig. 2.5 for a bubble that was generated at just more than one-half of its maximum radius from the free surface (Robinson et al. 2001). The bubbles were generated by an electrical discharge, and the images clearly show the development of both the free-surface spike and the liquid jet.

To observe single-bubble cavitation dynamics, high-speed photography is always used in conjunction with a microscope, such as an inverted microscope or a long-distance microscope, but the specific use depends highly on the cavitation situation.

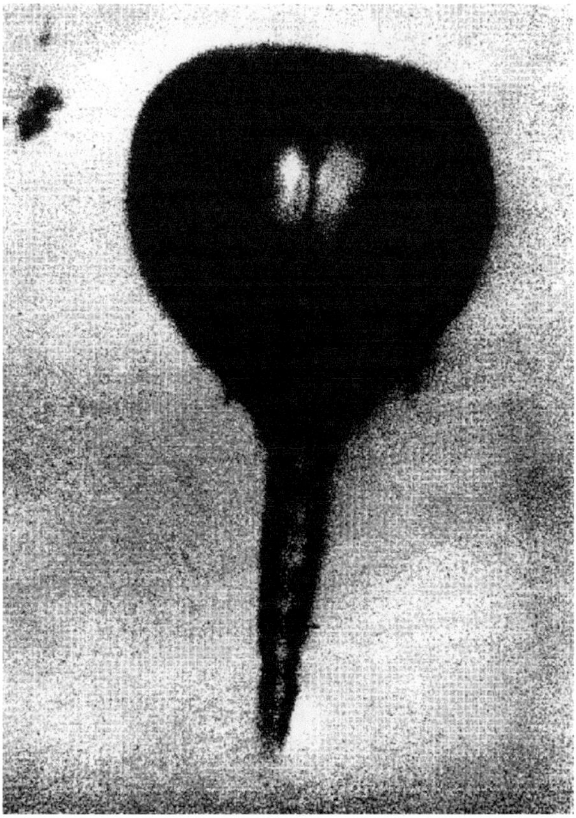

Fig. 2.3 Enlargement of a bubble with a jet and its protrusion pointing to the solid boundary. Reproduced with permission from Lauterborn et al. © Elsevier

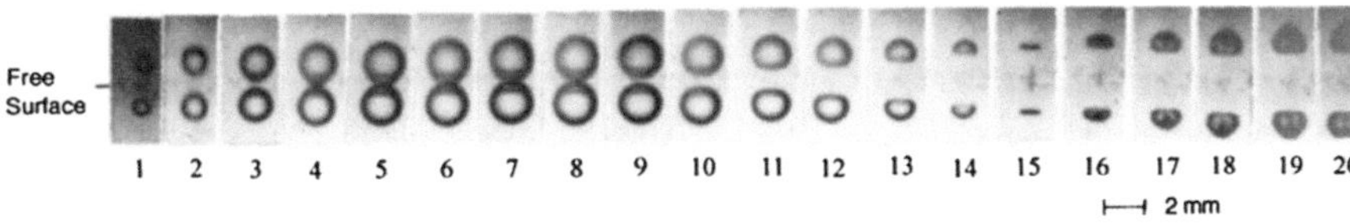

Fig. 2.4 High-speed photographs of a single bubble generated at a dimensionless distance beneath a free surface. Frame interval, 10 μs; exposure time, 2 μs. Reproduced with permission from Robinson et al. © American Institute of Physics

2.2.2 *Multiple-Bubble Cavitation*

Turning now to studying the inception of cavitation and the large-scale distribution of cavitation bubbles, high-speed photography has also been used to visualize multiple-bubble cavitation, especially in focused ultrasound fields. Focused ultrasound is characterized by its ability to noninvasively penetrate deep into the human

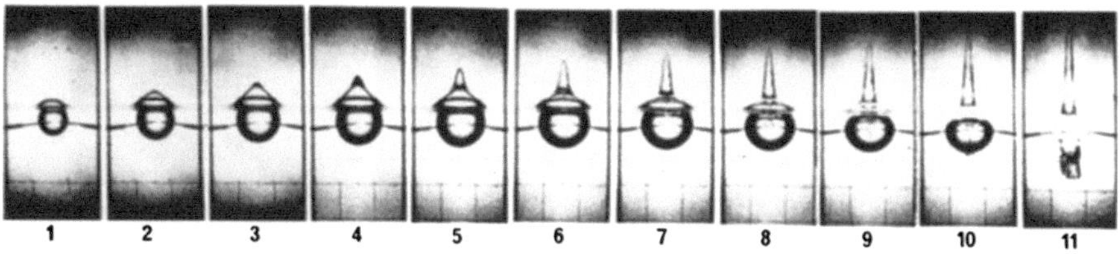

Fig. 2.5 High-speed photographs of a single bubble generated by an underwater electrical discharge at a dimensionless distance beneath a free surface. Reproduced with permission from Robinson et al. © American Institute of Physics

body and precisely deliver energy to a specific site (Vaezy et al. 2001a). Because of its high energy concentration, focused ultrasound represents a cavity source for multiple bubbles. In various applications of ultrasound in water, cavitation bubble clouds organize into different structures, and high-speed photography has been very useful for investigating the dynamics of these bubbles.

Luther et al. (2001) observed branchlike patterns of cavitation bubbles induced by an acoustic resonator in the 20 kHz frequency range, and they used a 2250-fps high-speed camera to observe the bubbles. In another study, Moussatov et al. (2003) discovered a conelike bubble structure in the vicinity of a transducer radiating surface with a frequency of 20.7 kHz, and they used high-speed images to reveal the process through which this structure was built.

Also studying multiple-bubble cavitation, Chen et al. (2006a, b) documented the temporal evolution of a cavitation bubble cloud in a 1.2-MHz focused ultrasound field and investigated the emergence of the cloud in the focal region, as shown in Figs. 2.6 and 2.7, respectively.

After analyzing these images, the authors gave two possibilities for why cavitation initiated in the prefocal region. First, in the prefocal region, some secondary pressure peaks exist, especially in the area near the focal region. Meanwhile, cavitation bubbles in the focal region may reflect ultrasound energy into the prefocal region. For example, cavitation bubbles generated at the transducer surface are driven by radiation forces to the focal region. When they move upward, the high pressure causes them to expand and fragment to small bubbles which, in turn, grow larger, resulting in the augmentation of the bubble clouds. At the same time, they are attracted to each other and form bubble clusters by the secondary Bjerknes forces. Gradually, the bubble cloud accumulates to a visible size in the prefocal region.

Chen et al. (2006a) did additional studies focused on the inception of cavitation bubble clouds induced by high-intensity focused ultrasound (HIFU). As shown in Fig. 2.8, they found that cavitation bubbles may be visible in the prefocal regions because the cavitation bubbles near the focus might shield the upper bubbles (post-focal regions) from oscillating to detectable size.

The inertial 145 μs of bubble inception in HIFU focus was also investigated by the authors (Chen et al. 2006a). They found that the whole bubble cloud was formed at 85 μs, and it then remained stable for at least 60 μs. The radiation forces

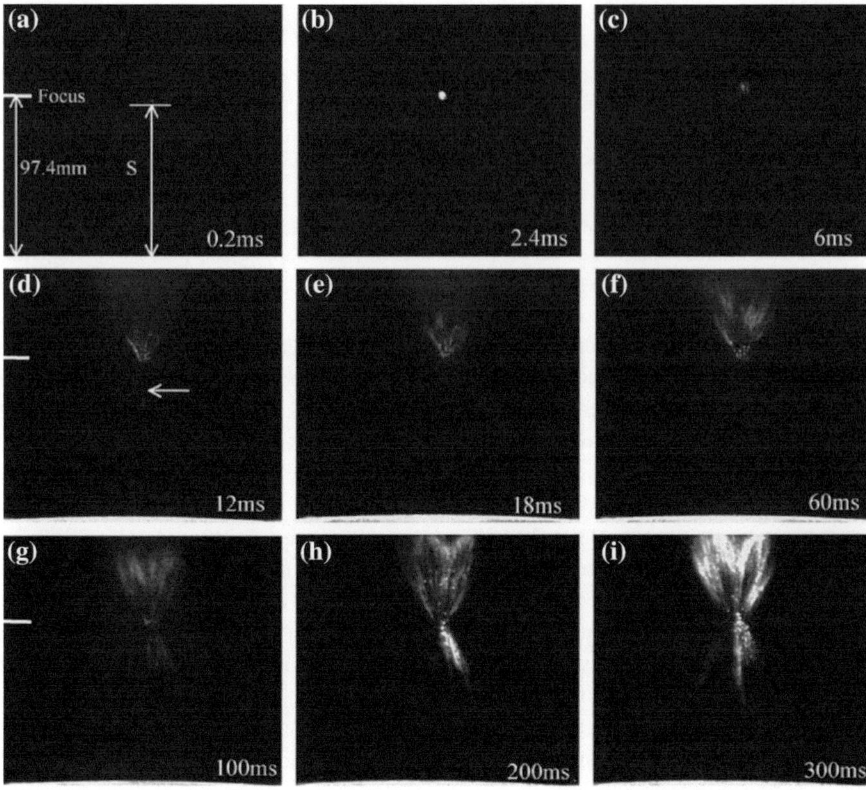

Fig. 2.6 High-speed images of the formation of a cavitation bubble cloud captured at different delay times (indicated in the *right* corner of each image). Reproduced with permission from Chen et al. © Elsevier

produced only a slight upward movement, which is visible on the streak image shown in Fig. 2.9. The axial maximum displacement was about 0.9 mm achieved by the bubble nearest to the focus (as pointed out by the arrowhead). Therefore, the bubbles seemed to be in a stable dynamic equilibrium.

Another application that employs multiple-bubble cavitation produced by focused ultrasound that has been investigated in recent years is pulsed HIFU-induced tissue erosion. The temporal evolution and spatial distribution of cavitation bubble clouds in the focal region have been studied by several groups. In 2008, Xu et al. (2008) studied the evolution of bubble clouds generated by a histotripsy pulse (10 and 14 acoustic cycles) with the help of an intensified charge-coupled device (ICCD). In the next year, Chen et al. (2009) employed ultrahigh-speed photography to capture the inception of a cavitation bubble cloud in the focal region of a tissue's surface, and they obtained two structures of bubble clouds, and the interchanging between these two bubble cloud structures are shown in Fig. 2.10.

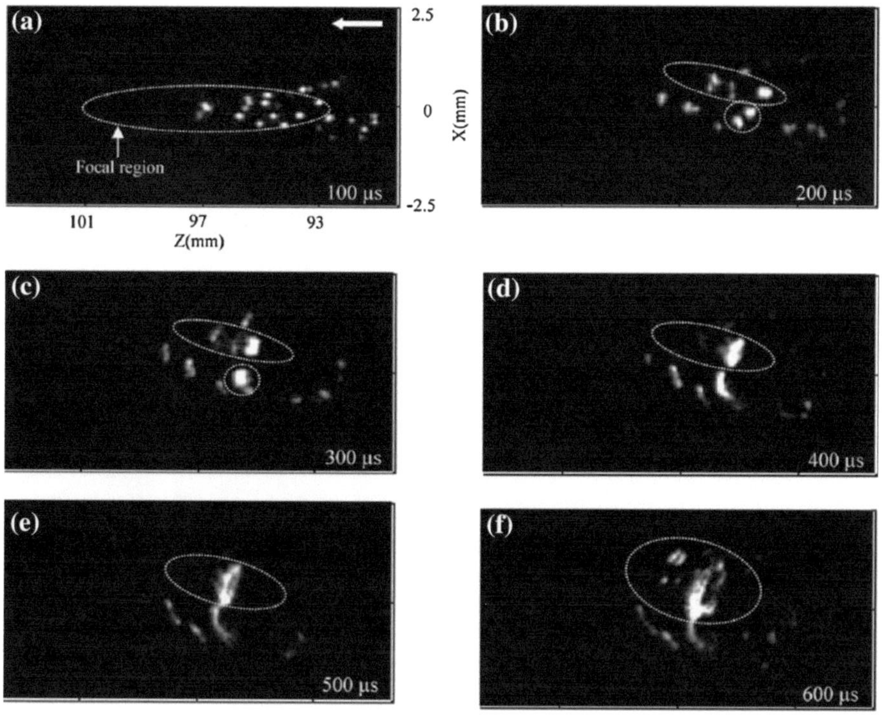

Fig. 2.7 High-speed images (exposure time, 10 μs; captured in 100-μs steps) showing the generation, mergence, and collapse of cavitation bubbles. Reproduced with permission from Chen et al. © Elsevier

2.2.3 Cavitation in Microvessel

Turning to the biomedical field, cavitation in microvessels must be understood well because ultrasonic contrast agents are usually delivered in microvessels. To study this, Chomas et al. (2001) investigated the behavior of lipid microbubbles (MP1950) in cavities. Figure 2.11 is a set of seven frames and a streak image of a microbubble undergoing Rayleigh–Taylor instability and its resultant fragmentation. Figure 2.11a shows the microbubble before insonation, with a resting diameter of 3 μm. The ultrasonic pulse is a two-cycle sinusoid, with a peak negative transmission pressure of 1.2 MPa. The microbubble is shown at the peak expansion of the first cycle in Fig. 2.11b, with a diameter of 10 μm. Figure 2.11c, d shows the microbubble during compression after the first expansion, while it is still intact. Due to the high velocity experienced by the wall during the end of compression, Fig. 2.11d appears blurred in both the frame and the streak. The microbubble fragments near the peak compression, and the result of this process is shown as a group of six fragments in Fig. 2.11e, f. Figure 2.11g shows the microbubble after the end of insonation, after three of the fragments observed in Fig. 2.11f dissolved.

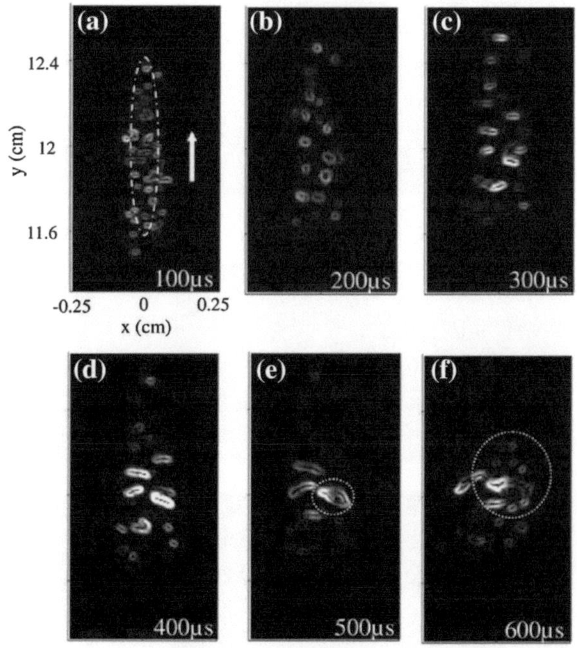

Fig. 2.8 High-speed images showing the dynamics of cavitation bubbles within the initial 600 μs. Reproduced with permission from Chen et al. © Elsevier

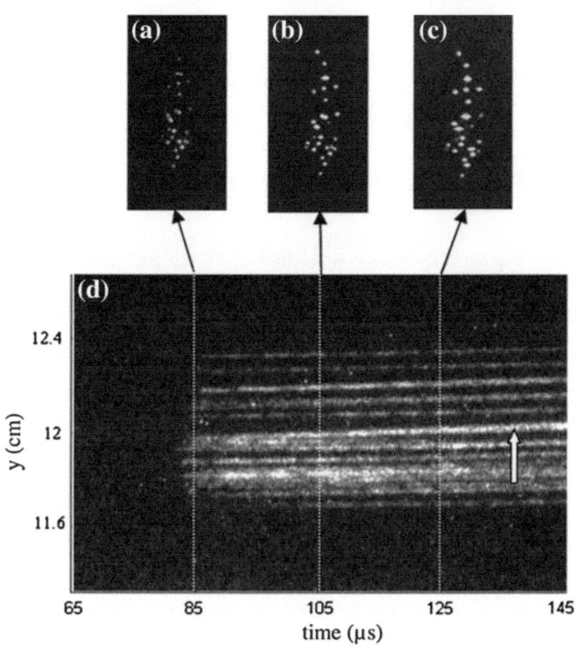

Fig. 2.9 Still images photographed at different times after the onset of ultrasound transmission: **a** 85 μs, **b** 105 μs and **c** 125 μs. Reproduced with permission from Chen et al. © Elsevier

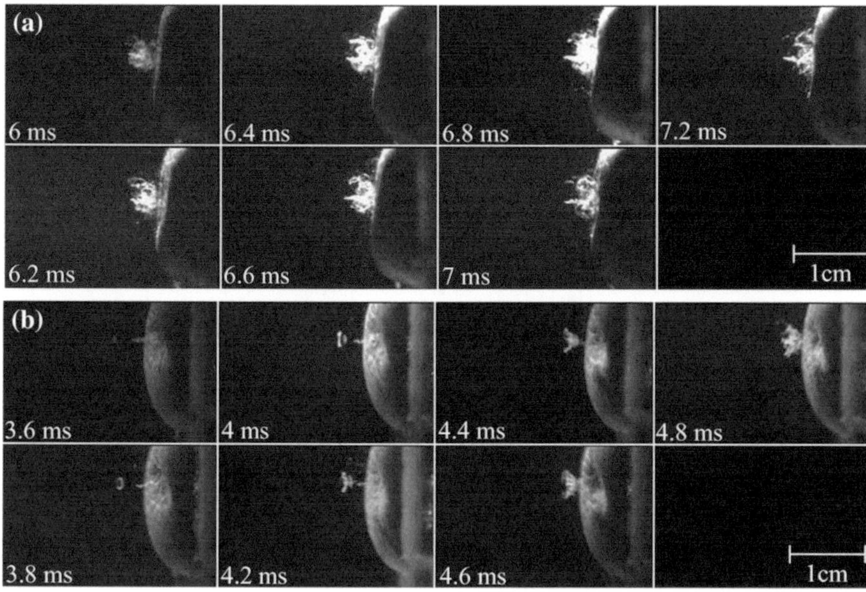

Fig. 2.10 Bubble cloud structures on tissue surfaces. **a** A crown-shaped bubble cloud developed into a cone-shaped bubble cloud. **b** A cone-shaped bubble cloud changed to a crown-shaped bubble cloud. Reproduced with permission from Chen et al. © Elsevier

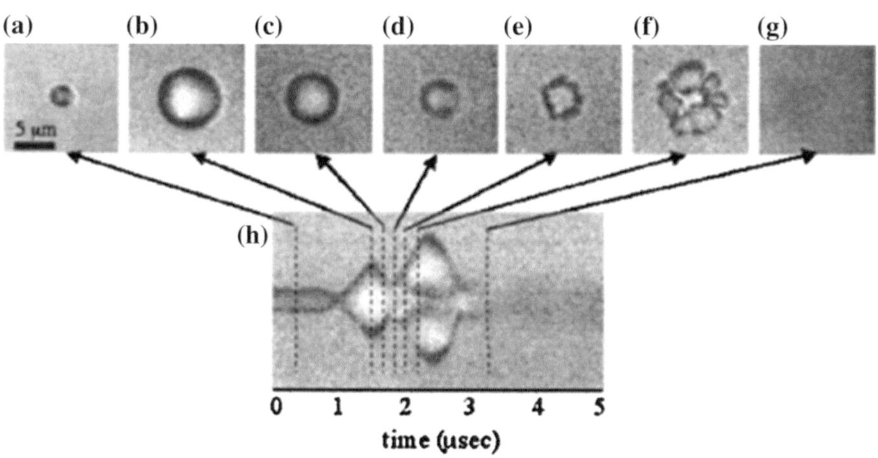

Fig. 2.11 Optical frame images and a streak image corresponding to the oscillation and fragmentation of a contrast agent microbubble, where fragmentation occurred during compression. The microbubble has a resting diameter of 3 μm shown in (**a**). The streak image in (**h**) shows the diameter of the microbubble as a function of time, and the lines indicate the times at which the two-dimensional (2D) frame images in (**a**–**g**) were acquired relative to the streak image. Reproduced with permission from Chomas et al. © Society of Photo-Optical Instrumentation Engineers

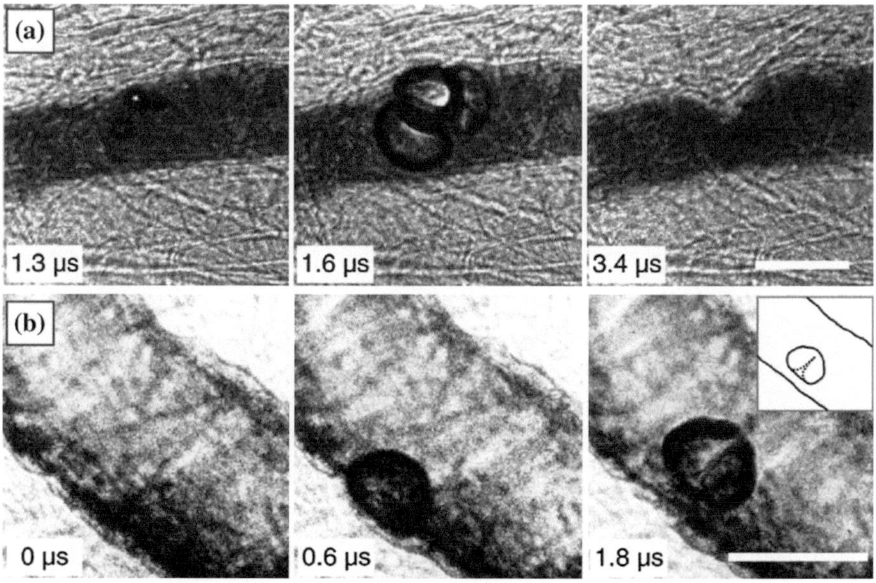

Fig. 2.12 Characteristics of observed bubble–vessel interactions. Reproduced with permission from Chen et al. © American Physical Society

Chen et al. (2011) also made two primary observations about how acoustically excited bubbles interact with microvessels in ex vivo tissue. They found that vessel deformation favored invagination over distention and that the liquid jets formed during collapse were directed away from the nearest vessel wall. As shown in Fig. 2.12, the qualitative characteristics of such bubble–boundary interactions are expected to be sensitive to the mechanical properties of the boundary material. Because the mechanical properties of tissue remain incompletely understood (especially for strains and strain rates relevant to medical ultrasound), these observations identify how bubbles are likely to interact with real vessels.

2.3 Cavitation Mapping by Sonoluminescence and Sonochemiluminescence

As a powerful tool to visualize the spatial distribution and dynamics of cavitation bubbles (Chen et al. 2006b), high-speed photography can capture all cavitation bubbles in a cloud. However, optical devices that detect bubbles' SL and SCL emissions can provide more information about the spatial distribution of chemically active cavitation bubbles (Cao et al. 2012; Suslick 1990).

SL, a phenomenon generally attributed to the high temperatures generated during the collapse of cavitation bubbles, can be observed as short light pulses occurring once every acoustic period (Gaitan et al. 1992). When observed through a microscope, these emissions can be seen to originate at the geometric center of the bubble, and they likely occur simultaneously with bubble collapse. SL in single bubbles (single-bubble sonoluminescence SBSL) has also been investigated using acoustically trapped and periodically driven gas bubbles (Gaitan et al. 1992). In these studies, when the bubble collapses strongly, the energy released at collapse leads to light emission. However, the light intensity is quite weak and thus difficult to detect. As a result, various photosensitizers must be used to enhance the light emission without changing the cavitation mechanisms.

2.3.1 *Mechanisms of SL and SCL*

Adiabatic collapse of an active cavitation bubble generates extremely high local temperatures and pressures. The extreme conditions within active cavitation bubbles can lead to SL light emission, as discussed previously. However, when luminol is added to an alkaline solution, it can significantly enhance the light intensity. This phenomenon is referred to as SCL, and it is identical to chemiluminescence that occurs in bulk liquid regions when induced by the chemical reaction of luminol molecules with $HO^{\cdot}$ radicals. When ultrasonic irradiation is applied to an alkaline luminol solution, the sonochemical effects of molecular activation and dissociation take place within the active cavitation bubble or at the solution layer near the bubble.

2.3.1.1 Sonochemical Reactions in Air-Saturated Water (McMurray and Wilson 1999)

In air-saturated water, the hemolytic cleavage of H_2O and dissolved O_2 occurs:

$$H_2O \rightarrow H^{\cdot} + HO^{\cdot}$$
$$O_2 \rightarrow O + O$$

The cleavage products may then recombine or react further as below:

$$HO^{\cdot} + HO^{\cdot} \rightarrow H_2O_2$$
$$O + H_2O \rightarrow HO^{\cdot} + HO^{\cdot}$$
$$H^{\cdot} + O_2 \rightarrow HO_2^{\cdot}$$
$$HO_2^{\cdot} + HO_2^{\cdot} \rightarrow H_2O_2 + O_2$$

Both $HO_2^{\cdot}$ and H_2O_2 are weak acids, so in conditions of neutral and higher pH values, the dissociation processes are as follows:

$$HO_2^{\cdot} \rightleftharpoons H^+ + O_2^{\cdot-}$$
$$H_2O_2 \rightleftharpoons H^+ + HO_2^-$$

The hydroxyl radicals ($HO^{\cdot}$) have the potential to oxidize both HO_2^- and H_2O_2. Thus, in the presence of significant concentrations of H_2O_2 in a sonicated solution, the following two reactions become dominant:

$$HO_2^{\cdot} + H_2O_2 \rightarrow O_2^{\cdot-} + H_3O^+$$

$$HO^{\cdot} + HO_2^- \rightarrow O_2^{\cdot-} + H_2O$$

2.3.1.2 Luminol SCL Mechanism (McMurray and Wilson 1999)

Luminol, with the structural formula as shown in Fig. 2.13, is a weak dibasic acid with first and second pKa values of 6.3 and ~13, respectively. In an alkaline solution, the luminol molecule will dissociate into a luminol monoanion (I) through the following two processes.

Then, the luminol monoanion can be oxidized by hydroxyl radicals ($HO^{\cdot}$) to produce the diazaquinone radical anion (II), as shown in Fig. 2.14 step (i). In step (ii), compound II reacts with the $O_2^{\cdot-}$ radical anion to form the hydroperoxide addition product (III), which is a weak acid with a pKa of 10.4. The monoanion form of compound III can decompose through step (iii) to give the exited state of the aminophthalate monoanion (IV), while the neutral form of III decomposes via a dark reaction (iv) to generate the starting material (I) and O_2. The excited aminophthalate monoanion IV molecule can then relax into the ground state by emitting light at 430 nm.

Even though the mechanism of SL still remains unclear, many investigations during the last few decades have tried to uncover it. Since 1934, when Frenzel and Schultes (1934) first documented the phenomenon of multi-bubble

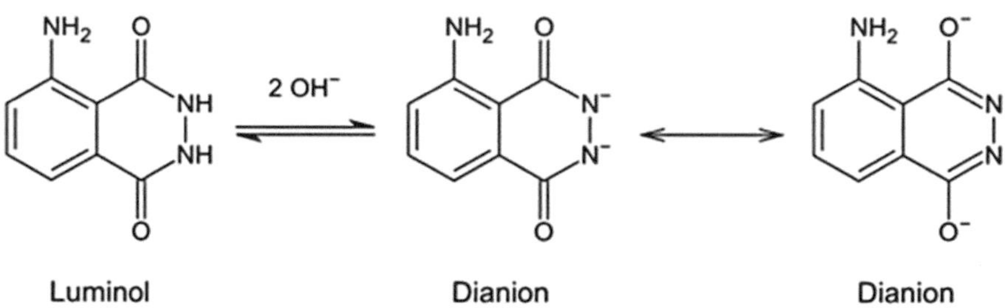

Fig. 2.13 Structural formula and dissociation of luminol

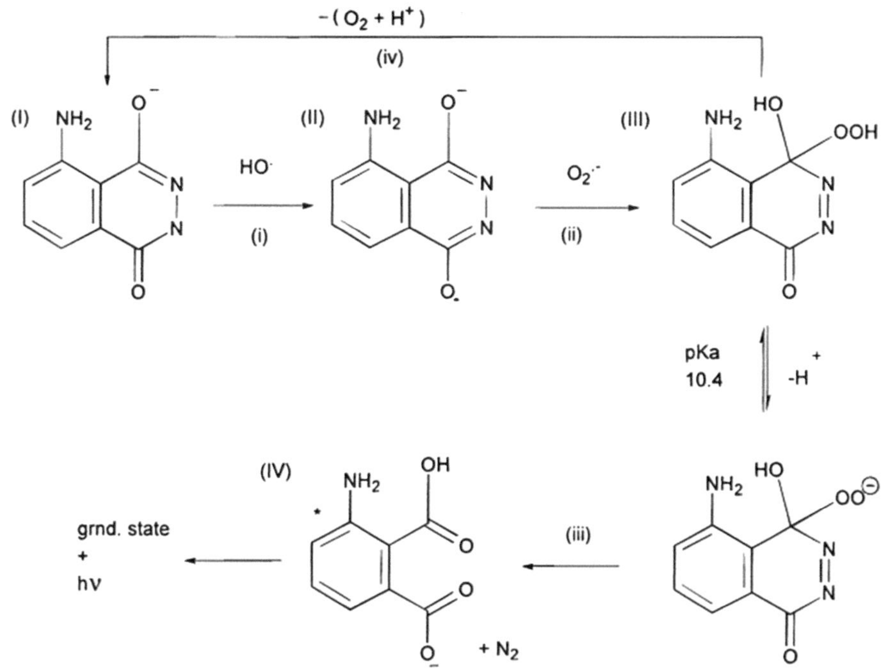

Fig. 2.14 Schematic showing the principal mechanism of sonochemiluminescence

sonoluminescence (MBSL) in ultrasonic fields, various hypotheses have been put forward to explain the origin of the light emission. For example, Chambers (1936) put forward a triboluminescence theory as early as 1936. However, with the advent of SBSL in 1989, studies on the mechanisms of SL have improved, causing many of the early theories to fall out of favor and even become abandoned.

In the next four sections, we describe the most popular and most recognized theories for the mechanisms of SL.

2.3.1.3 Hot Spot Theory

In "hot spot" theories, the energy for light emission is thought to originate from the thermal energy released in the adiabatic process of bubble collapse. These theories rely on the fact that light emission preponderantly depends on the maximum temperatures reached within the cavitation bubbles, because the recombination of dissociated molecules that can accompany light emission and the characteristic molecular radiation of light emission can occur at extremely high temperatures (Brenner et al. 2002). Thus, these hot spot theories can be supported by the findings presented by Noltingk and Neppiras (1950), and Neppiras and Noltingk (1951) who used the Rayleigh–Plesset equation of bubble dynamics to deduce that adiabatic

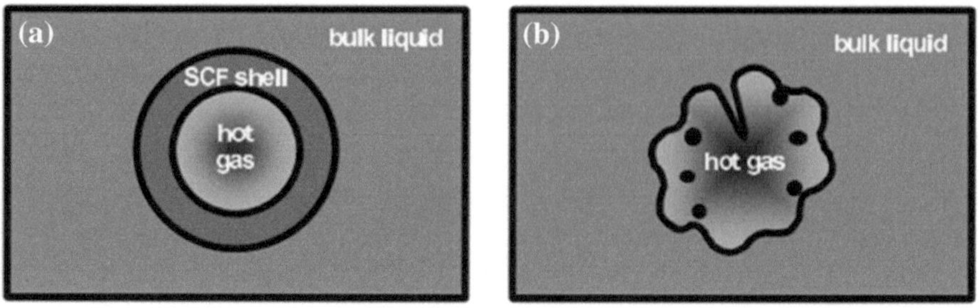

Fig. 2.15 The two-site models of sonochemical reactions. **a** The thermal diffusion shell model. **b** The surface-wave distortion and jet model. Reproduced with permission from Suslick et al. © Wiley Online Library

compression of a bubble's contents can produce temperatures of 5000 to 10,000 K. More recently in the past few decades, Kenneth Suslick and colleagues have amassed impressive evidence in favor of thermal molecular emission luminescence (Brenner et al. 2002). For example, Suslick (1986, 1990) deduced the thermal origin of MBSL from the SL spectrum, in which molecular bands can be clearly identified while lines associated with discharges do not exist. He has also proposed that there are in fact two sonochemical reaction sites shown in Fig. 2.15. The first (the dominant site) is the bubble's interior gas phase, whereby a hot liquid shell is formed around the surface of the gas bubble. The second reaction site is initially liquid and corresponds either to the heating of a shell of liquid around the collapsing bubble or to the injection of liquid droplets into the hot spot by surface-wave distortions or jets of the collapsing bubble (Suslick et al. 1999).

While hot spot theories are supported by compelling evidence, there is no clear definition of a "hot spot" in physics, so the origin of the light emission in SL cannot be completely explained by such theories.

2.3.1.4 Blackbody Theory

Another theory for the mechanism of SL was introduced by Jarman (2005), who has described that luminescence occurs from microshocks that are thermal in origin and have a spectral distribution that closely resembles that of a blackbody. In addition, Vazquez et al. (2001) measured the spectral irradiance, Mie scattering, and flash width of SL occurring from bubbles of hydrogen gas and noble gasses, and they suggested that the light emission was actually the radiation from a blackbody with temperatures ranging from 6000 K (H_2) to 20,000 K (He). For a mixture of hydrogen and water, as shown in Fig. 2.16, the radius of a bubble surface for light emission is about 0.2 μm, and the SL spectrum can match that of a blackbody at 6230 K.

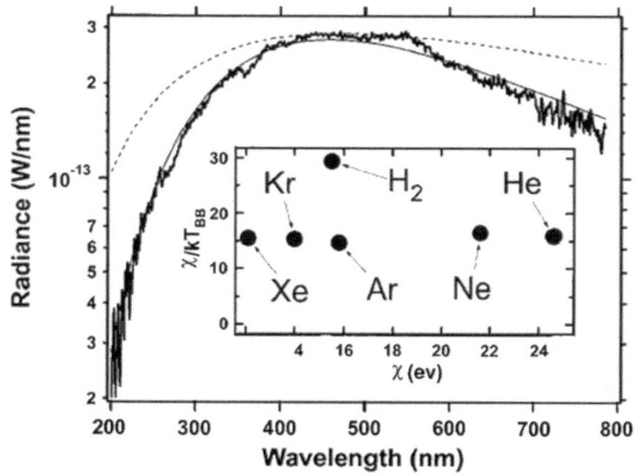

Fig. 2.16 Spectrum (resolution, 24 nm FWHM) of 33-kHz SL from a bubble formed in water (23 °C) into which H_2 was dissolved at a partial pressure of 5 Torr. The solid curve is fit to a blackbody at 6230 K. The dashed curve represents a bremsstrahlung fit with a temperature of 15,000 K. The inset shows a plot of χ/TBB as a function of χ, the ionization potential of the gas used to produce the SL. Reproduced with permission from Vazquez et al. © Optical Society of America

2.3.1.5 Bremsstrahlung

Bremsstrahlung is another theory that can be used to explain the mechanism of SL. Bremsstrahlung is the radiation produced when an electrically charged particle, especially an electron, is slowed down by the electric field of an atomic nucleus or an atomic ion. In accordance with Young's book (2004), there are three kinds of bremsstrahlung. (1) Electron ion bremsstrahlung is the light emission from an electron slowed down in the Coulomb field of a positive ion. (2) Electron atom bremsstrahlung is the bremsstrahlung that occurs when an electron passes through the field of a neutral atom. The field of a neutral atom decreases rapidly with distance, and therefore, the electron must pass very close to the atom to ensure the emission of light. (3) Finally, polarization bremsstrahlung is similar to the electron atom bremsstrahlung, but the main difference is that the neutral atom is polarized in the Coulomb field.

In 1998, Frommhold (1998) suggested that electron-neutral atom bremsstrahlung may be a principal mechanism for the SL from rare gas bubbles in a weakly ionized environment. Supporting this, Yasui (1999) later showed that SL mainly occurs via radiative recombination of electrons and ions, and electron atom bremsstrahlung.

In a different study that investigated an argon gas bubble, Xu et al. (1998, 1999, 2000) found that atoms inside a very thin layer of plasma around the bubble's origin were strongly ionized, and the SL occurred nearly simultaneously. They concluded that multiple ionization and recombination played a dramatically important role in SL.

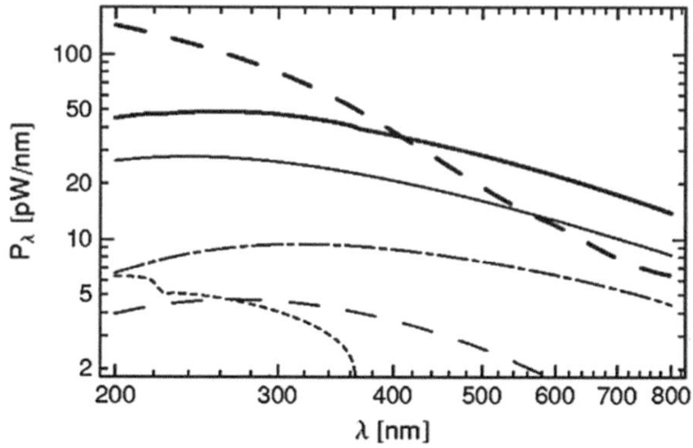

Fig. 2.17 An SBSL spectrum and its composition for argon bubbles in water at freezing temperatures. Shown in the total spectrum (*thick solid line*), electron-neutral argon bremsstrahlung (*thin solid line*), O^-radiation (*dotted line*), H^- radiation (*dashed line*), and e-Ar^+ bremsstrahlung (*dash dotted line*). The measured spectrum is indicated by the *thick dashed line*. Reproduced with permission from Hammer et al. © American Physical Society

Hammer and Frommhold (2002) have made steady contributions, and they have conclude that in conventional single-bubble sonoluminescence, electron ion bremsstrahlung amounts to but a small percentage of the overall emission and that electron-neutral bremsstrahlung is likely to be the principal radiative mechanism. Figure 2.17 shows a spectrum computed from Hammer and Frommold's model for argon bubbles in freezing water. The predominance of electron-neutral bremsstrahlung is shown.

2.3.1.6 Shock Wave Theory

Finally, shock wave theory has also been used to explain SL. This theory assumes that extreme temperatures can be attained by spherical microshocks propagating within the imploding bubble as they converge toward the center of the bubble (Taylor and Jarman 1970), and the extreme temperatures can accelerate the dissociation process of molecules, generating high-speed free electrons and ions. Then, these high-speed free electrons and ions can collide, producing the light emission seen in SL.

To deduce that a shock forms during bubble collapse, Greenspan and Nadim (1993) used the governing equations of gas dynamics inside a bubble. They also found that this converging spherical shock grows in strength as it propagates toward the center.

Later, Moss et al. (1994) performed numerical hydrodynamic simulations of the growth and collapse of a 10-μm air bubble in water using a plasma model composed of a thermal conductive component and a partially ionized component. They calculated the light pulse duration time and the spectrum character, and they

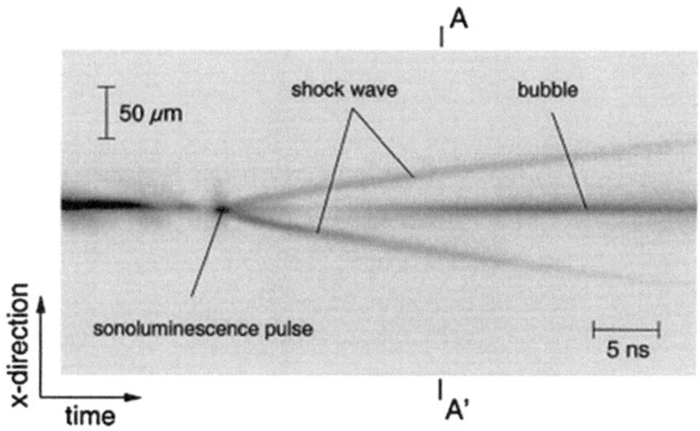

Fig. 2.18 Streak image of the SBSL bubble collapse. The time and space axis are marked by arrows, and the dark parts of the image correspond to high light intensity. Reproduced with permission from Pecha et al. © American Physical Society

concluded that a nuclear fusion reaction temperature can be achieved by the addition of the converging shock wave and an extra pressure pulse.

Further supporting the shock wave theory, Pecha and Gompf (2000) used a streak camera with high spatial and temporal resolution to image the dynamics of the violent collapse in SBSL. Figure 2.18 shows their recorded streak image of the end phase of the bubble collapse, in which the shock wave was formed.

Putterman and Weninger (2000) summarized that light comes out only while the plasma exists. The light emission is thermal bremsstrahlung from the accelerating free electrons. These electrons will accelerate and radiate light as they collide with the ions.

2.3.2 *Free Field*

SL responses have also been studied in free fields. Cao et al. (2012) reported results of SL by an ICCD camera with a 30-s exposure time in a 1.2-MHz focused ultrasound field in degassed water. They varied the electrical power level from 30 W to 50 W to 70 W as shown in Fig. 2.19, from which we can see that the SL emissions were mainly located in the post-focal region, where the density of chemically active cavitation bubbles was high.

Because the active SL bubbles grow during the expansion phase, the temperature required for SL emissions is higher, while the temperature for SCL emissions remains relatively low. Therefore, the sonochemically active region and time period are larger in comparison with SL, and the intensity of SCL is 10–100× higher than that produced by SL (Hatanaka et al. 2000).

Thus, Cao et al. (2012) also acquired SCL images under the same experimental conditions as the SL experiments shown in Fig. 2.19, as these SCL results are shown in Fig. 2.20. Compared with Fig. 2.19, the SCL region is more extensive and has higher light intensity. Also, the SCL images showed branchlike shapes

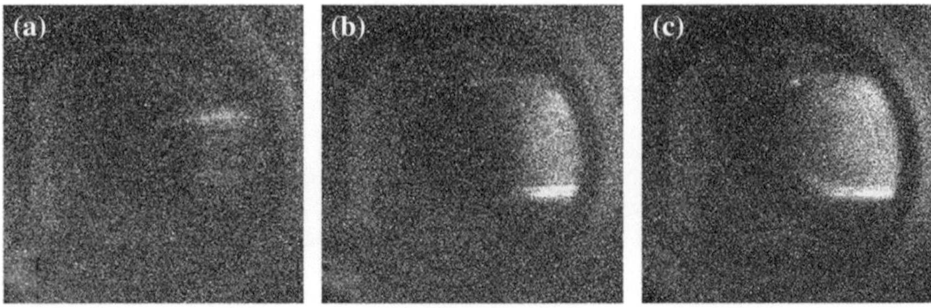

Fig. 2.19 SL images acquired by an ICCD camera at the same exposure time (30 s) but with different electrical powers: **a** 30 W, **b** 50 W, and **c** 70 W. The ultrasound wave was transmitted from the *left* to the *right* side. Reproduced with permission from Cao et al. © Elsevier

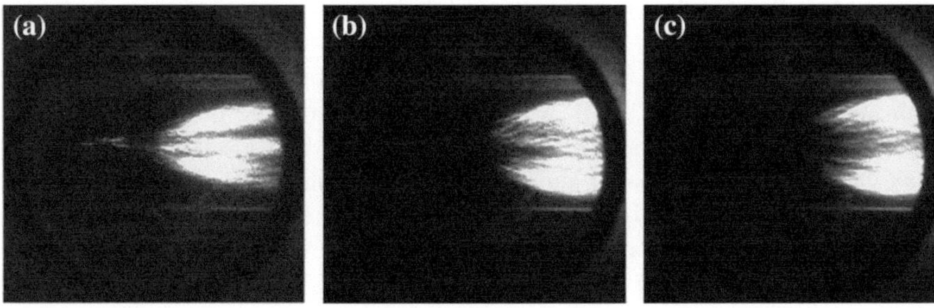

Fig. 2.20 SCL images at the same exposure time (30 s) with different electrical powers: **a** 30, **b** 50 W, and **c** 70 W. Reproduced with permission from Cao et al. © Elsevier

corresponding to the distribution of cavitation bubbles obtained by the high-speed camera (H. Chen et al. 2006b).

In a different SCL experiment, Hallez et al. (2010) presented SCL images from a digital camera with a exposure time 10 s, in which cavitation bubbles were localized in a 3-MHz focused ultrasound field. As can be noted in Fig. 2.21a, there were no obvious SCL emissions in the vicinity of the transducer and downstream of the acoustic propagation axis, and this corresponds to the fact that no active cavitation bubbles existed in those regions, shown in the tomographic image in Fig. 2.21b. Using an optic fiber to count the photons emitted, Hallez et al. quantitatively verified that the intensity of SCL emissions increased with the irradiation power, as shown in Fig. 2.22. And also, an important concept "oversaturation" was raised by L. Hallez.

2.3.3 *Tissue–Fluid Interface*

Cavitation is considered to be the most predominant mechanism for tissue erosion during HIFU treatments (Cao et al. 2012). Therefore, studies aimed at investigating

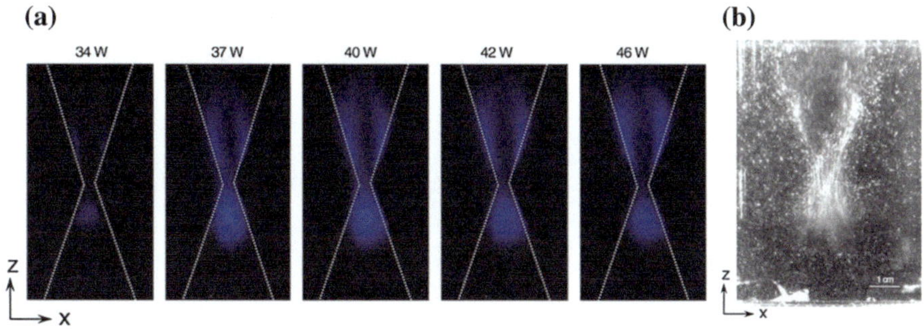

Fig. 2.21 **a** SCL for different powers. **b** Dynamics of cavitation bubbles for high-power irradiation (40 W). The transducer is located at the bottom of the picture. Reproduced with permission from Hallez et al. © Elsevier

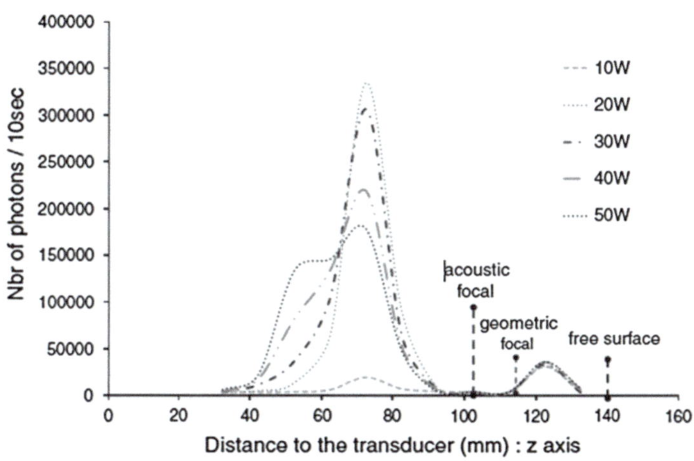

Fig. 2.22 Photon counting with an optic fiber. Reproduced with permission from Hallez et al. © Elsevier

the distribution of cavitation bubbles close to a tissue boundary are important. However, traditional acoustic detection methods and high-speed photography cannot necessarily capture active cavitation. To circumvent this issue, Cao et al. (2013) performed cavitation mapping close to a tissue–fluid interface by using SCL. Figure 2.23a shows a photograph of the tissue–fluid interface, while Fig. 2.23b–f shows the SCL images captured by an electron multiplying charge-coupled device (EMCCD) camera with a 0.5-s exposure time under various HIFU pulse lengths with a constant duty cycle of 1:10.

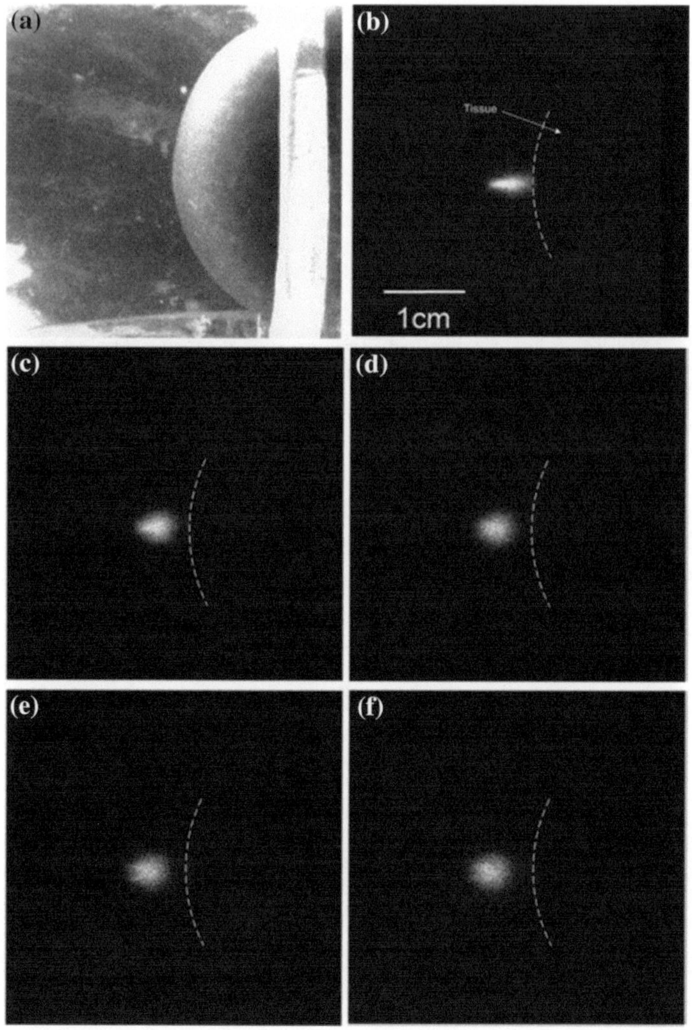

Fig. 2.23 **a** Photograph of the tissue–fluid interface. **b–f** SCL images at a tissue–fluid interface with different pulse lengths: **b** 10 μs, **c** 20 μμs, **d** 40 μs, **e** 60 μs, and **f** 80 μs. The ultrasound wave was transmitted from the *left* to *right* side in the images. Reproduced with permission from Cao et al. © Elsevier

2.3.4 Bone–Fluid Interface

Compared with soft tissues, the acoustic properties of hard tissues—bones, for example—are quite different. Hence, investigations on the cavitation bubble distribution near fluid boundaries with bones are also needed. Zhang et al. (2013) gave an account of the spatial distributions of SCL near the front surface of a bovine

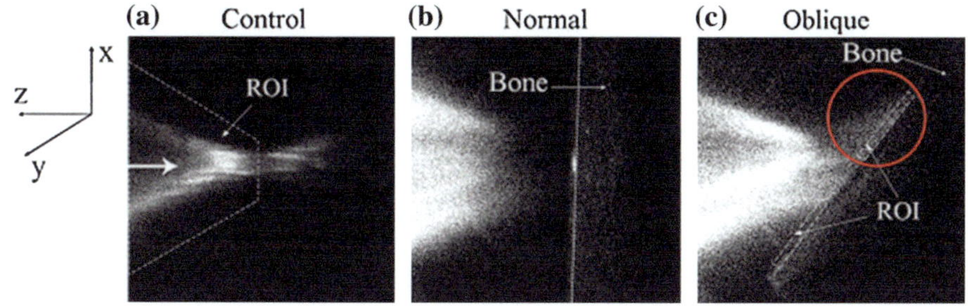

Fig. 2.24 **a** Control group: SCL emissions without the bone. **b** SCL emissions near the front surface of the bone with a normal incidence. **c** SCL emissions with an oblique incidence. Reproduced with permission from Zhang et al. © Acoustical Society of America

femur exposed to HIFU radiation with a normal and an oblique incident direction. The SCL images in Fig. 2.24 were captured by an EMCCD camera with an exposure time of 15 s. By counting the photons in the trapezoid region of interest (ROI) in the prefocal area, as shown in Fig. 2.24a, we can see that the SCL intensity in front of the bone is greater than that in the control group. This can be explained by the superposition of the reflected wave on the incident wave in the axial direction. When the incident direction was adjusted from being normal to being oblique, as seen in Fig. 2.24c, the SCL emissions became stronger and directionally extended along the straight surface of the bone (indicated with the red circle in Fig. 2.24) in the opposite direction of the HIFU field propagation.

2.3.5 *Microvessel*

Ohl et al. (2010) developed a novel microfluidic device (Fig. 2.25) for achieving intense acoustic cavitation, and then, they reported the SCL distribution using that setup. This device has one outlet, but two inlets that form a *T*-junction to create gas–liquid interfaces, as shown in the figure inset. A specially designed piezo-electric transducer is glued onto a microscope slide next to the microfluidic device.

The authors obtained the SCL image in the microfluidic using an EMCCD camera. As noted in Fig. 2.26a, the SCL emissions only occur within the liquid phase close to the gas–liquid interfaces, where cavitation bubbles are nucleated and driven into large oscillations. They also recorded the oscillations of cavitation bubbles close to the gas–liquid interface, as depicted in Fig. 2.26b. During the maximum expansion phase (third frame of Fig. 2.26), the bubbles are pancake shaped, whereas severe deformations occur during collapse (last frame of Fig. 2.26b).

Cavitation mapping has been also performed in microreactors, by means of SL and SCL visualization. Fernandez Rivas et al. (2012) described such a setup, which

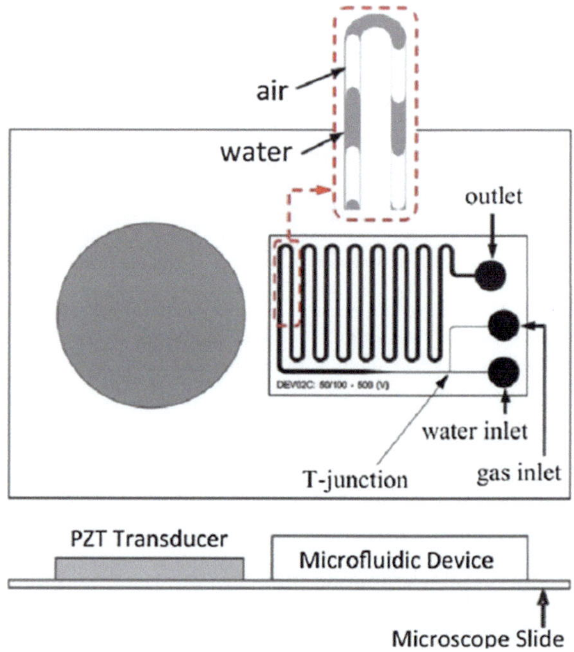

Fig. 2.25 Schematic of the microfluidic *T*-junction device and the piezoelectric transducer attached to a microscope slide. Reproduced with permission from Ohl et al. © Royal Society of Chemistry

is shown in Fig. 2.27. Here, the SL and SCL emissions are collected by the photomultiplier tube (PMT). The micromachined pits on the substrate are for nucleating and stabilizing microbubbles in a liquid.

2.3.6 *Bioreactor*

To study the bioeffects of ultrasound, normally experiments involve applying low-intensity ultrasound to carcinoma cells to determine how ultrasound interacts with the cells in vitro. There are several ways to position the cell holder in such experiments, but normally it is placed above the ultrasound propagation direction, as shown in Fig. 2.28. Whatever the experimental system though, it is necessary to decide the acoustic field and distribution before the cell irradiation.

The most common method to decide the acoustic field and distribution is by directly measuring them with a hydrophone. These measurements can map the acoustic pressure in the interested region. However, when using a bioreactor for cell–ultrasound interaction experiments, the cell holder is always sealed off from the surrounding degassed water, so it might be impossible to put the hydrophone in the cell holder in order to measure the acoustic pressure distribution in that area. To avoid this, published studies always give the acoustic pressure and distribution at the plane where cell holder is located (but when the cell holder is absent) instead of

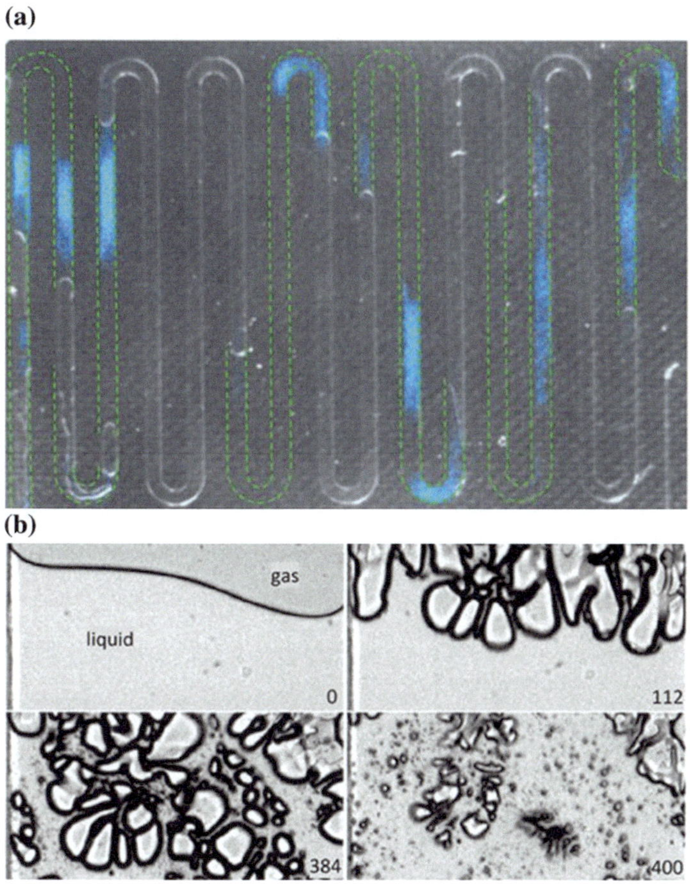

Fig. 2.26 **a** Luminol chemiluminescence from cavitation bubbles. **b** Sequence of images taken from a high-speed recording of the bubble distribution near a gas–liquid interface. Reproduced with permission from Ohl et al. © the American Association for the Advancement of Science

measuring the true acoustic pressure in the experiment. While studies do describe that the acoustic pressure and distribution might be modified by the cell holder, some suggest that these influences can be ignored, while others think the modified acoustic condition can crucially affect the results. Moreover, simulations have also been introduced to mimic the pressure distribution in the cell holder. From these studies, it is obvious that existing methods of measuring acoustic pressure and distribution do not show the true interactions between cells and ultrasound. Thus, we must develop novel methods to map the acoustic pressure in bioreactors.

Cavitation is one of the main factors resulting from low-intensity ultrasound that affects cells. Thus, cavitation mapping in the bioreactor is important for describing the true cavitation environment around cells and it can be used to further analyze the effects of cavitation on cells during ultrasound irradiation. Using SCL in a

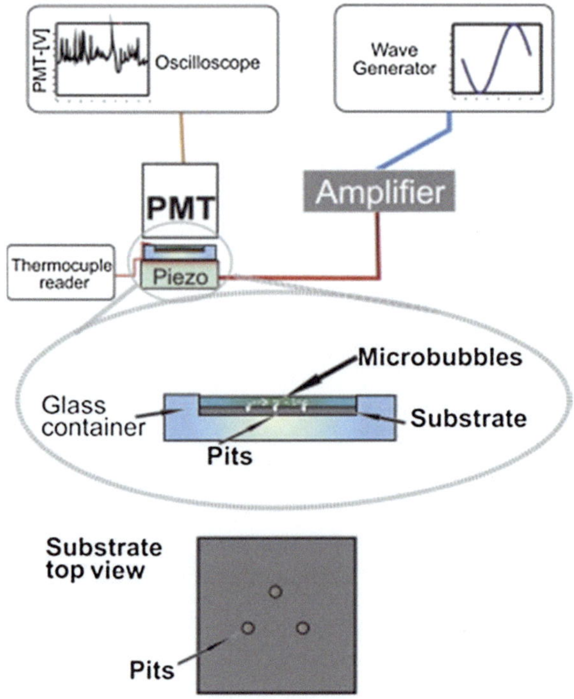

Fig. 2.27 Experimental setup (not to scale) showing the different components. The zoomed inset shows the microreactor and a *top* view of a three-pit substrate. Reproduced with permission from Rivas et al. © Elsevier

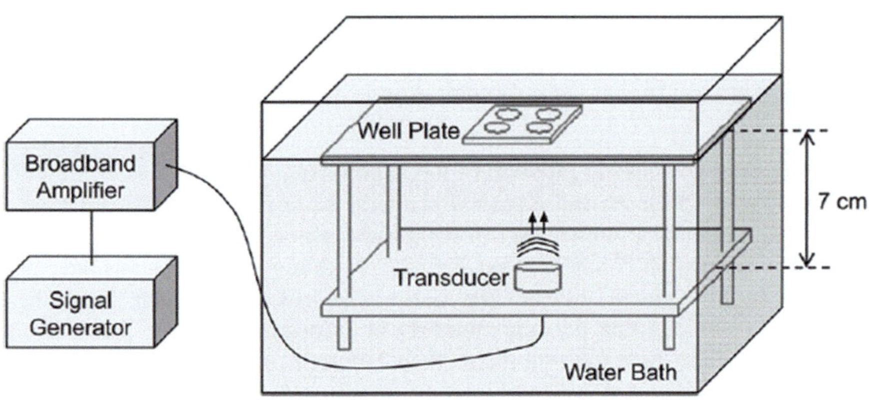

Fig. 2.28 Schematic showing an experimental ultrasound irradiation system

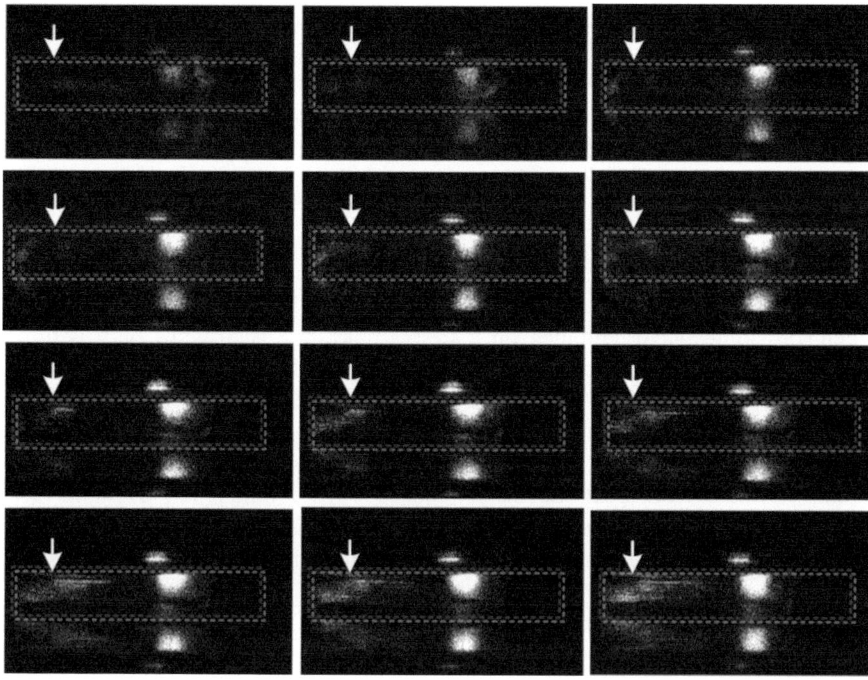

Fig. 2.29 The SCL image of cavitation in the bioreactor

bioreactor represents a new method for mapping the cavitation intensity and distribution. For a bioreactor containing cells, ultrasound irradiation, and microbubbles, studies have shown that microbubbles increase the effect of cavitation on cells. Thus, to illuminate a cavitation map, 10 μL/10 mL lipid-shelled microbubbles were added to a bioreactor system with luminol. As shown in Fig. 2.29, the SCL image sequence was acquired at acoustic power of 15 W. In the dotted rectangle which indicates the edge of bioreactor, the standing wave was developing as time went on. It became prominent and more widely distributed when the lipid-shelled microbubbles were added. This may have occurred because of more cavitation nuclei in the system or a stronger vibration of cavitation.

2.4 ACD, PCD, and Doppler Methods

When a bubble oscillates, acoustic waves are emitted. If the ultrasound power is low, then the bubble simply pulsates in an approximately linear manner emitting signals at the insonation frequency (Leighton 1994). In more powerful acoustic fields, cavitation bubbles may be generated, and the emissions contain both broadband signals and line-spectrum components at specific frequencies.

The line-spectrum components represent the harmonics, ultraharmonics, and subharmonics of the insonation frequency (McLaughlan et al. 2007). Based on the emissions from acoustically driven bubbles, that acoustic methods that can be used to detect cavitation include acoustic scattering, acoustic emission, and Doppler method.

2.4.1 Acoustic Scattering and ACD

When ultrasonic waves impinge on a free-bubble or a gas-filled cavity within a structure, much of the energy is reradiated, i.e., scattered, into the surrounding medium, especially under resonance conditions. Based on this, a system capable of sending and receiving pulses of focused 30-MHz ultrasound (Roy et al. 1990) has been used to detect short-lived cavitation events that result from in vivo exposure of rat lung tissue to four-cycle bursts of focused 4-MHz ultrasound (Holland et al. 1992). This pulser–receiver detection system is known as ACD, and the experimental apparatus is shown in Fig. 2.30. The ACD system can be divided into pulse-echo and bistatic pulse–receiver detections. Transient cavitation can be detected with the acoustic backscattering technique.

ACD can afford sufficient sensitivity and temporal resolution to noninvasively detect the presence of a single 1-μm bubble that lasts only a microsecond (Roy et al. 1990). However, bubble growth can also be produced by therapeutic ultrasound in tissues. Thus, longer interrogating tone bursts have been achieved through bistatic pulser–receiver detection systems. Such systems have been used to investigate shell disruption and fragmentation of biosphere™ and Optison™ ultrasound contrast microbubbles caused by high-amplitude tone bursts (Chen et al. 2002). In another study, Shi et al. (2000, 2001) utilized ACD to investigate the destruction of

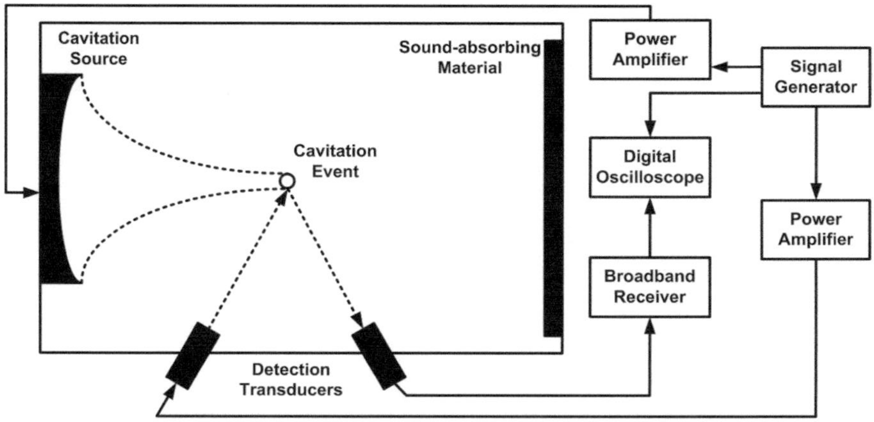

Fig. 2.30 The experimental apparatus for ACD

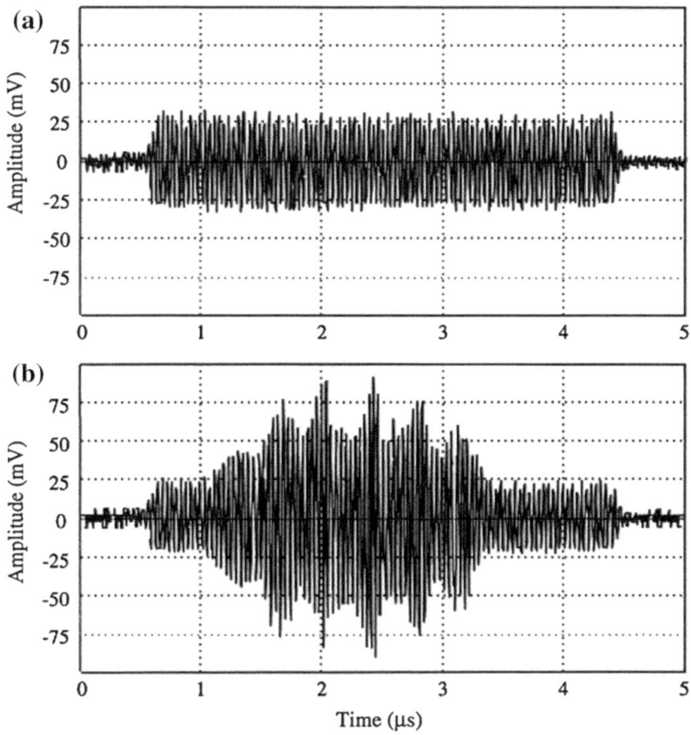

Fig. 2.31 Scattered signals from two single Sonazoid® microbubbles: **a** without exposure to imaging pulses and **b** with insonation by a 2.5-MHz acoustic pulse of 6 cycles and 0.34 MPa. Reproduced with permission from Shi et al. © World Federation for Ultrasound in Medicine and Biology

insonified Sonazoid microbubbles and how this destruction was association with inertial cavitation in vitro. Figure 2.31 shows the results of experimental ACD system, which used real scattered signals from two single Sonazoid microbubbles, without or with exposure to imaging pulses.

In general, ACD can be performed by using a single-element transducer including an unfocused or a focused probe as the pulse-echo detector to actively receive the acoustic backscatter signal generated by cavitation. When an unfocused ACD transducer is employed, the sensitivity may be less. Thus, most experimental studies use focused ACD transducers. Figure 2.32 shows the backscattered signals received by a focused ACD. Since these focused transducers only have a fixed focus, the ACD could only detect cavitation in the focal region, and it could not distinguish between different sources in the region.

Recently, ACM has been developed on the basis of ACD by using a linear array connected to the ultrasound system to increase the spatial coverage for cavitation detection. As one of ACM, conventional B-mode ultrasound imaging has been used to study bubble activity generated by therapeutic ultrasound, lithotripsy, and

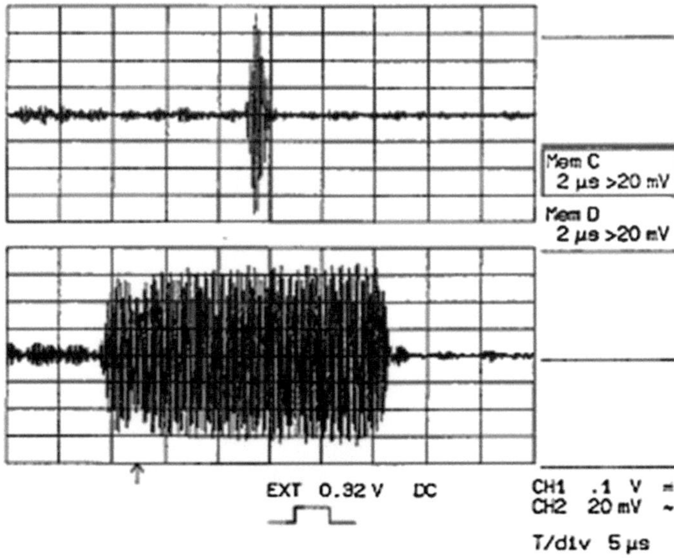

Fig. 2.32 Backscattered signals received by ACD. The *top* trace indicates the presence of a short-lived cavitation event. The bottom trace indicates the presence of an aggregate of polystyrene spheres that persisted in the focal region of the detector for the entire pulse duration (10 μs). The time base for the two traces is 2 μs/div (Crum et al. 1992). Reproduced with permission from Crum et al. © Acoustical Society of America

histotripsy. It is believed that the appearance of a hyperechoic region in ultrasound images results from bubble formation. However, due to a time difference existing between the acquisitions of two scan lines in one frame, strictly simultaneous ACM cannot be achieved. A modified ultrasound line-by-line scanning method for spatial–temporal ACM was proposed by Ding et al. (2013). The details of this method are discussed in Sect. 2.5, and the maps obtained offer high sensitivity and SNR. In addition, because there is a long enough time for the liquid to return to its original state after the acquisition of each scan line, the interactions between the interrogating pulse and bubble behavior are reduced. Therefore, this technique may eventually become standard in cavitation mapping with both high spatial and temporal resolution, similar to exact acoustic field measurements. In another study, Gateau et al. (2011a) reported ultrafast active cavitation imaging using plane wave transmission along the entire aperture of the array at a high frame rate. Figure 2.33 shows the results of ultrafast active cavitation imaging, and they show that cavitation events can be monitored during focused ultrasound treatment. However, because the ultrasound beams are unfocused, the lateral resolution of the ultrafast images may need to be improved. Hu et al. (2015) developed another UACM technique with high spatial–temporal resolution by combining plane wave transmission and adaptive beamforming; the details of this method are given in Sect. 2.6.

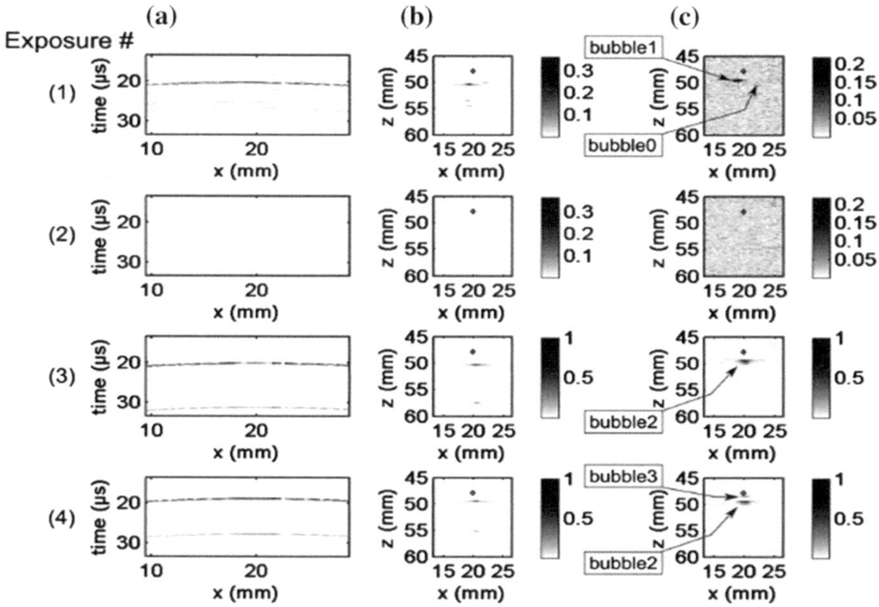

Fig. 2.33 Cavitation imaging using **a** high-pass filtered passive radio frequency data. **b** The corresponding passive beamformed image. **c** The second active change detection image for each shot. Reproduced with permission from Gateau et al. © IEEE

2.4.2 Acoustic Emission and PCD

As we have discussed, cavitation is generally classified into two types: stable cavitation and inertial cavitation. When the acoustic pressure amplitude is low, certain features of the bubble response may be absent under linear conditions. As the acoustic pressure increases, harmonic or subharmonic and ultraharmonic frequency components can be used as indicators of nonlinear bubble motion (Barnett 1998). In general, the resulting emissions at subharmonics and ultraharmonics of the main excitation frequency can be used to reflect stable cavitation. In inertial cavitation, broadband noise emissions appear above the background noise level. These emissions are generated when bubbles undergo large radial oscillations and collapse violently. Atchley et al. (1988) employed an acoustic transducer to passively listen for sound scattered by cavitation microbubbles. This technique, called PCD, relies upon the scattering of the irradiation field by the bubble clouds that arise from cavitation events. Figure 2.34 shows the PCD system, which can be employed to passively detect and monitor cavitation activity by detecting and analyzing the acoustic emissions generated from cavitation events.

Acoustic emissions collected by PCD are sampled digitally and post-processed using one of three signal processing methods, shown in Fig. 2.35. These methods include time-domain, frequency-domain, and time–frequency analysis. The

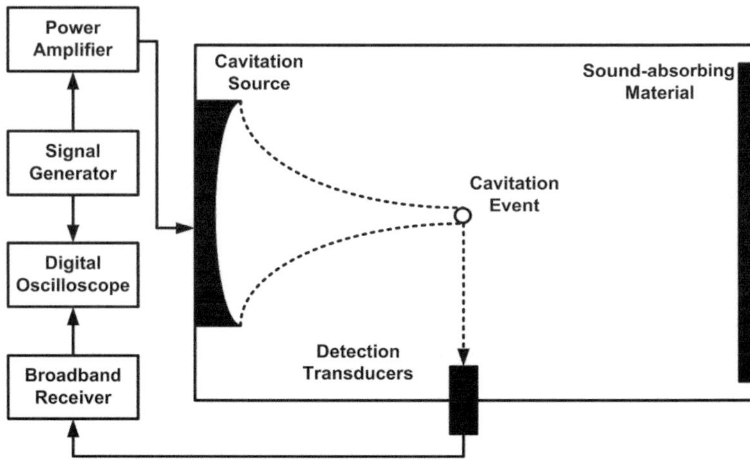

Fig. 2.34 Block diagram of the experimental apparatus for PCD

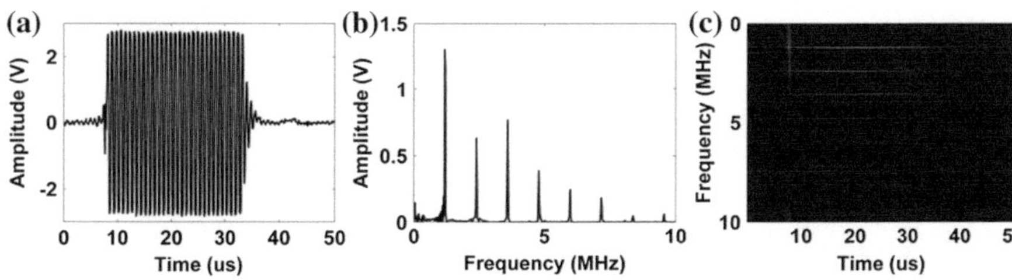

Fig. 2.35 The signal processing methods for PCD signals. **a** Time domain, **b** frequency domain, **c** time–frequency analysis

time-domain analysis is based on the signal waveform, for example, the peak values. The frequency-domain method refers to a spectral analysis of PCD signals performed by fast Fourier transform (FFT). The time–frequency analysis can be achieved by short-time Fourier transformation (STFT). Applying this technique, Ammi et al. (2006) used a time–frequency spectrogram to evaluate the rupture of shelled ultrasonic contrast agents. In another study, time–frequency plots for PCD spectra were generated at three exposure conditions in order to simultaneously compare different signal components (Mast et al. 2007). In such experiments, the changes in the frequency spectra over time can be observed directly from the time–frequency image. However, at present, the frequency-domain method is the most common for analyzing acoustic emissions from PCD signals.

During sonication, the signals generated by cavitation can be grouped into three categories (Leighton 1994; Barnett 1998): (1) The first group consists of the insonation fundamental frequency and harmonics that are scattered by particles in the tissue as well as by the stable cavitation and vapor bubbles formed during

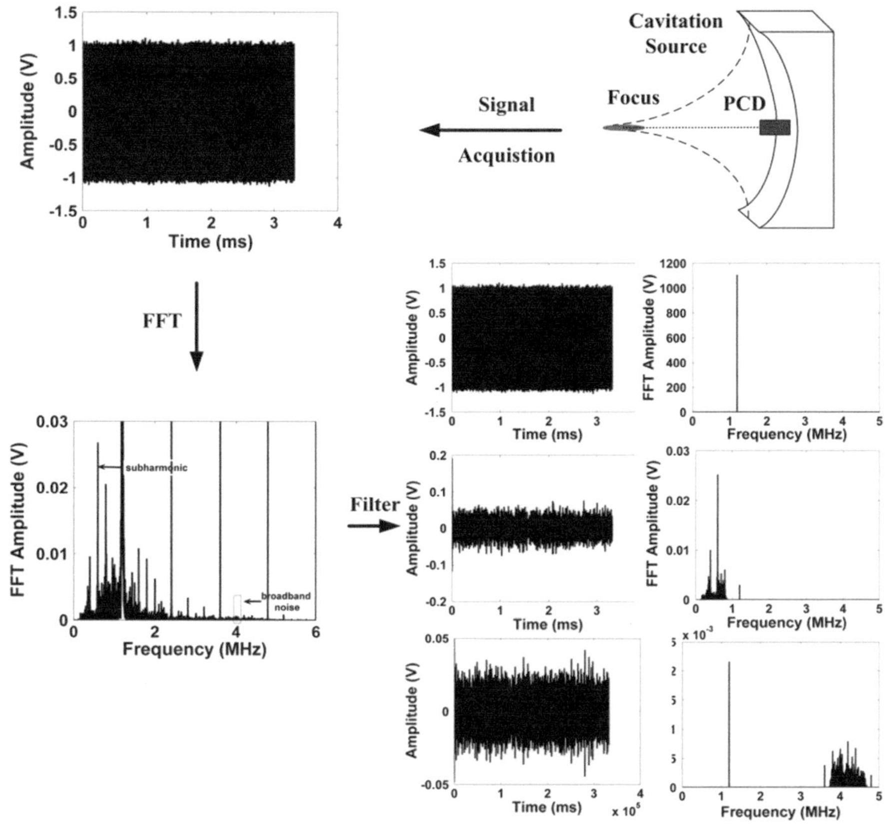

Fig. 2.36 Schematic diagram of PCD signal acquisition and frequency-domain analysis

sonication. Additionally, these signals might include reverberations off other structures present in the tissue, as well as direct propagation of the ultrasound signal. (2) The second group is made up of the subharmonic and ultraharmonic emissions generated from stable cavitation. (3) Finally, the last group contains the broadband acoustic emissions generated by inertial bubble collapses. Figure 2.36 shows the PCD signal acquisition and the frequency-domain analysis. In general, a FFT is performed, in which the signal consists of energy from the scattered sonication signal (fundamental and higher order harmonics), from stable cavitation signals (subharmonics and ultraharmonics), and from the radiated pressure arising from inertial bubble collapses (broadband noise). To isolate the contribution from cavitation, we must remove the fundamental and harmonics by digitally filtering the radio frequency (RF) signals. As such, Farny et al. applied a comb filter around each of the harmonic and ultraharmonic frequencies, so that signal energy only came from the unfiltered frequency gaps between the filtered harmonic and sub-/ultraharmonic peaks, which represented just the energy from inertial cavitation (Farny et al. 2009).

Quantifying the broadband noise in either the time or the frequency domain can allow us to calculate a value called the inertial cavitation dose (ICD), which gives information about the intensity of inertial cavitation (Tu et al. 2006; Chen et al. 2003a). Similarly, by quantifying the subharmonics and ultraharmonics associated with stable cavitation, we can calculate the stable cavitation dose (SCD). Figure 2.37 shows this method of inertial and stable cavitation data acquisition and calculation of the ICD and SCD.

McLaughlan and colleagues (2006, 2007, 2010) identified specific frequencies or frequency ranges of interest after performing an FFT calculation with a Hanning window function. In these studies, the half-harmonic as well as the fourth, fifth, and sixth harmonics plus the broadband frequency was investigated. The fourth harmonic was chosen because it is the first harmonic outside the high-pass frequency band and provided a greater signal strength than the higher harmonics after being transmitted through an attenuating medium, such as ex vivo tissue. In addition to the specific frequencies separated from PCD signals, these studies (McLaughlan et al. 2006, 2007, 2010) also recorded the electrical drive power derived from the measured HIFU drive voltage and audible emissions, and the results are shown in Fig. 2.38.

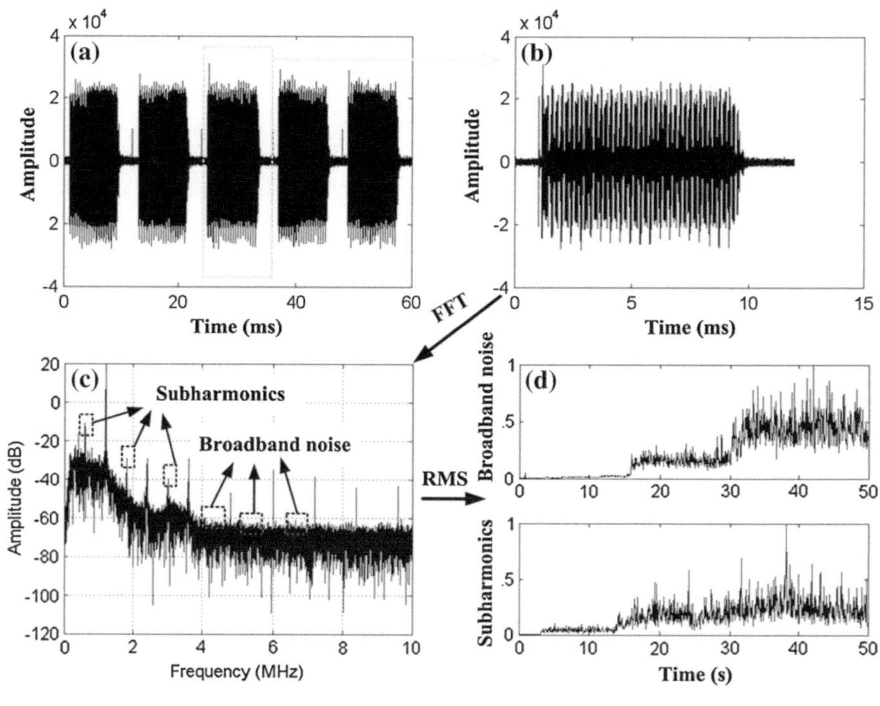

Fig. 2.37 Calculation of the ICD and SCD. **a** Detected cavitation signal from **b** each pulse is converted to **c** its frequency spectrum. The root-mean-square (RMS) amplitude in the broadband noise or in the subharmonic and ultraharmonic regions is measured and plotted in (**d**). Calculation of the ICD and SCD is defined as the integral of the area under the time versus amplitude curve over the 50-s recording

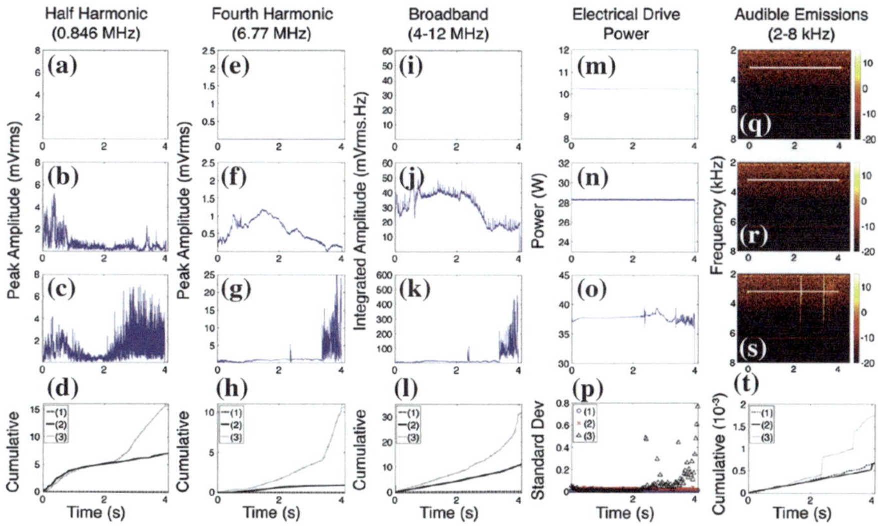

Fig. 2.38 Noise-corrected signals (as labeled) from 4-s HIFU exposures in each of the three cavitation regimes: (*top row*) 314 W/cm^2 (1.52 MPa); (*second row*) 786 W/cm^2 (1.77 MPa); and (*third row*) 1100 W/cm^2 (1.86 MPa). (*Bottom row*) The noise-corrected cumulative signals and standard deviation for the drive power for these three exposures. Reproduced with permission from McLaughlan et al. © World Federation for Ultrasound in Medicine and Biology

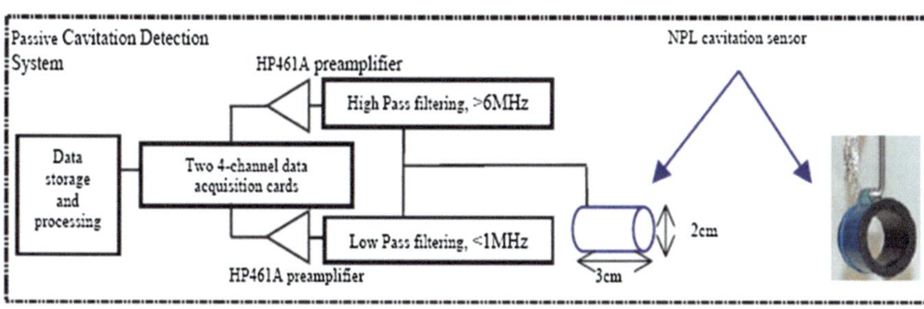

Fig. 2.39 **A** PCD system, which includes a broadband cavitation sensor. Reproduced with permission from McLaughlan et al. © American Institute of Physics

In order to study bubble activity more effectively, a dual-frequency PCD system, shown in Fig. 2.39, was designed to detect broadband high-frequency emissions (6–12 MHz) generated by acoustic cavitation during HIFU exposures (McLaughlan et al. 2006, 2007). In this system, a high-pass filter and a low-pass filter are used to reduce RF pickup from the drive and acoustic interference. This dual-frequency PCD system allows for the simultaneous investigation of multiple acoustic emissions, particularly the ultraharmonics, the half-harmonic, and the high-frequency broadband emissions. Furthermore, ACD and PCD can be combined to monitor cavitation dynamics (Everbach and Francis 2000; Gateau et al. 2011a, b). With a combined ACD-PCD system, we can obtain the return echoes of the probe signal by the emitting transducer, and these echoes can be taken as evidence of the presence of microbubbles acting as scatterers. Simultaneously, emissions of broadband noise emanating from the cavitation events are received by a second transducer, and these emissions can be taken as evidence of inertial cavitation. Recently, passive mapping of acoustic cavitation emissions using an ultrasound array has been developed, and it has potential application in real-time monitoring of ultrasound ablation. This will be described in detail in Sect. 2.8.

Unlike the ACD, PCD may not be sensitive enough to detect the occurrence of a weak cavitation event. Thus, the cavitation events detected by PCD are likely to be associated with bubble clouds rather than single bubbles. Generally, PCD is more sensitive to either resonance-sized bubbles or clouds of bubbles. Conversely, ACD using a focused transducer with a high center frequency is more sensitive to single bubbles in the micron size range. Also, the PCD technique yields threshold values lower than those measured via light production, yet higher than theoretical predictions based on free-bubble cavitation models (Roy et al. 1990). Therefore, inertial cavitation may be isolated in real time by using a PCD transducer with a sufficiently high center frequency relative to the insonation frequency, and by applying a high-pass filter to remove the energy at the first few insonation harmonics.

2.4.3 Frequency Shift Signal and Doppler

Rapid bubble growth and collapse, as occurs in cavitation, can generate a frequency shift signal that exhibits the Doppler effect (Deng et al. 1996). Thus, Doppler can be used to detect cavitation using a system such as that shown in Fig. 2.40. Using this system, cavitation was detected by a modified version of an active cavitation detector, which employed a highly focused transducer with center frequency of 30 MHz. An oscilloscope was used to monitor and record the post-processed signal received by the detection transducer. Also, a trigger level was set so that only signals of substantially higher voltage than the general background noise would be recorded and recognized as a cavitation event.

In this setup, the backscattered signal is displayed and recorded on the oscilloscope as time traces. These traces, which are the scattered signals from cavitation bubbles, are modulated both in amplitude and in frequency due to the movement of the bubble wall. Using a double balanced RF/IF mixer and a low-pass filter, frequency demodulation is achieved. The output signal from the demodulation scheme

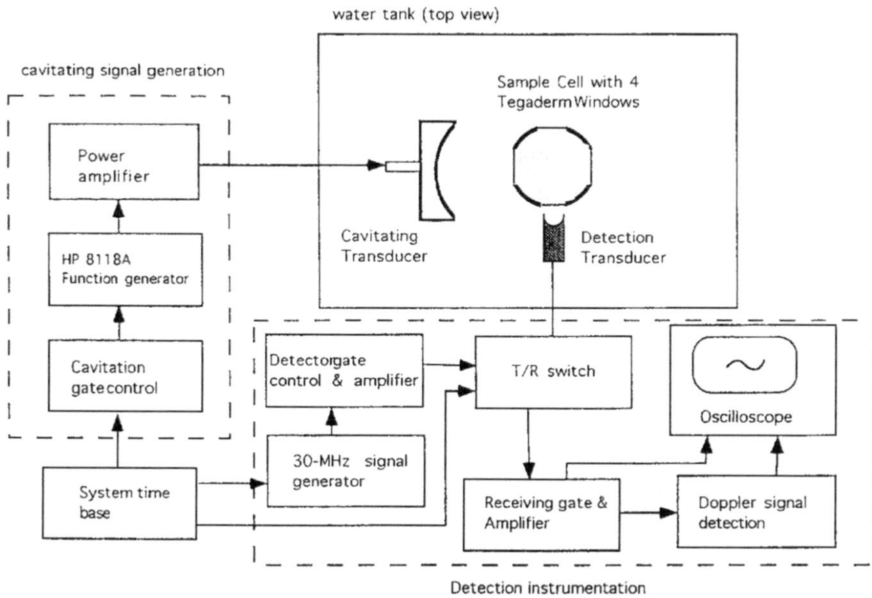

Fig. 2.40 Typical cavitation events recorded in blood showing a broadened bubble movement. *Bottom* trace is in the time trace, and the *top* trace is the Doppler frequency shift signal. Reproduced with permission from Deng et al. © World Federation for Ultrasound in Medicine and Biology

represents the frequency shift of the scattered signal from the original probing signal. The frequency shift signal obtained in this way is also displayed along with the time trace on the oscilloscope. A typical recording of a cavitation event in plasma is shown in Fig. 2.41. The Doppler frequency shift range is up to 1.5 MHz, which is much higher than that caused by the translation motion of a solid particle.

These cavitation events were also recorded in blood, and the results are shown in Fig. 2.42. The Doppler frequency shift signal showed broadened bubble movement. Based on this, a cavitation event could be distinguished from a solid particle. The backscattered signals from cavitation in these blood samples were different from those in water or plasma. Compared to the multiple growth and collapse events that occurred within one pulse in water or plasma (Fig. 2.41), the signals in blood showed that only one or two cycles of growth and collapse occurred, and the cycles were somewhat broader, as demonstrated in Fig. 2.42.

The frequency detection scheme described above has been tested using simulated signals and is advantageous in situations when noise or a large reflection is present. If the time trace shows a large reflection from a solid particle in the host medium, any scattering from a cavitation bubble may be obscured by the large reflection. Therefore, the time trace will not demonstrate the characteristic rapid

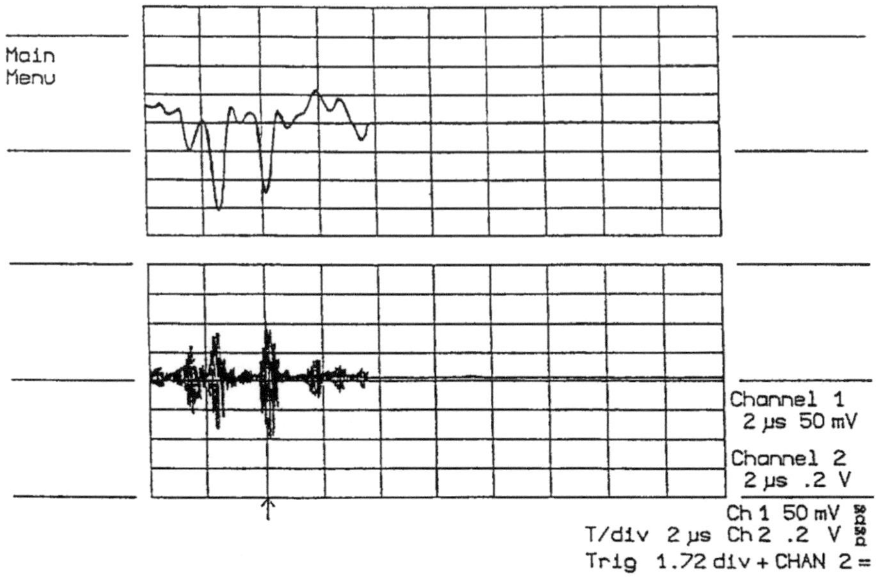

Fig. 2.41 Typical cavitation events recorded in plasma showing a distinguishable bubble and collapse. *Bottom* trace is in the time trace, and the *top* trace is the Doppler frequency shift signal. Reproduced with permission from Deng et al. © World Federation for Ultrasound in Medicine and Biology

oscillation behavior of a cavitation bubble. However, the frequency shift signal will exhibit the Doppler effect of rapid bubble growth and collapse, from which a cavitation event can be distinguished from a solid particle.

Recently, a new method for detecting cavitation has been developed, and it is based on the "twinkling" artifact (TA) of Doppler imaging (Khokhlova et al. 2013). TA that can be explained by irregular scattering of the Doppler ensemble pulses from the fluctuating microbubbles is used to detect cavitation in tissue and in polyacrylamide gel phantoms during pulsed 1-MHz HIFU exposures with different peak negative pressures (PNPs) (1.5–11 MPa). Besides, Li et al. proposed a bubble Doppler method based on a fusion of the three Doppler techniques—color Doppler, pulse-inversion Doppler, and interleaving Doppler (Li et al. 2014). These techniques were found to be not only more sensitive than conventional PCD or high-speed photography, but also complementary to each other, providing different information about the bubble distribution. With bubble Doppler, a cavitation event is considered to have occurred if color was displayed on top of the B-mode image. The details of this method are given in Sect. 7.5.

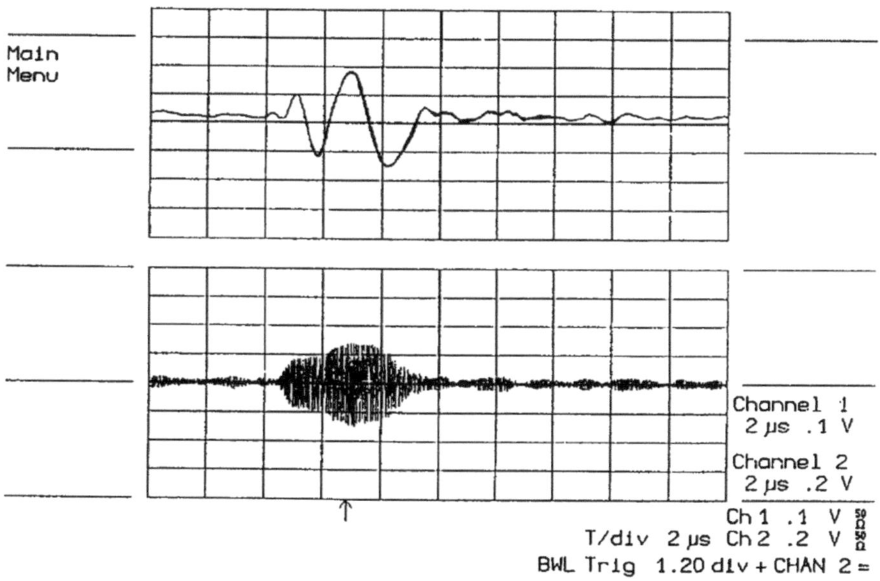

Fig. 2.42 Typical cavitation events recorded in blood showing broadened bubble movement. *Bottom* trace is in the time trace, and the *top* trace is the Doppler frequency shift signal. Reproduced with permission from Deng et al. © World Federation for Ultrasound in Medicine and Biology

2.5 Precise Active Acoustic Mapping of Cavitation Bubbles

2.5.1 Ultrasound Line-by-Line Scanning Method

As cavitation is highly complex and exhibits transient characteristics, it is essential to develop a standard cavitation mapping technique similar to exact acoustic field measurements, with both high spatial and temporal resolution. A precise active acoustic mapping of cavitation bubbles by a modified ultrasound line-by-line scanning method is proposed by Ding et al. (2013). Scattered signals from cavitation bubbles are obtained in a scan line immediately after one HIFU exposure, and then, there is a waiting between the next exposure to allow the liquid to return to its original state. As this pattern is extended, an image can be built up by sequentially measuring a series of such lines.

A linear array driven by an ultrasound scanner (Sonix-RP) was used to perform precise ACI. The input signal configuration was a line trigger. The synchronization of the HIFU exposure and RF data acquisition by the Sonix-RP scanner are illustrated in Fig. 2.43. The HIFU exposure time (T) and time delay (D) between the Sonix-RP triggering pulse and the moment when the HIFU is turned off varied from

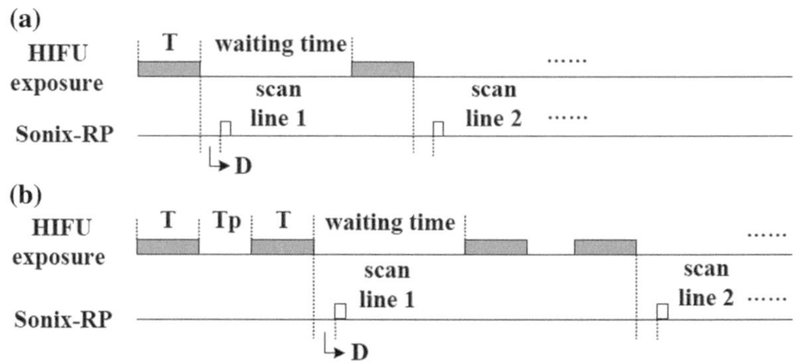

Fig. 2.43 Timing sequences for the modified ultrasound line-by-line scanning method. **a** Timing sequence used to investigate the temporal evolution of the cavitation bubble cloud and the dissolving process of corresponding bubbles, which was done by varying the parameters *T* and *D*, respectively. **b** Timing sequence used to investigate the spatial distribution affected by the previously generated cavitation, which was done by changing the parameters *T* or T_p. Reproduced with permission from Ding et al. © Elsevier

microseconds to seconds. Evolution of cavitation bubble cloud and its dissolution are acquired through varying the parameters *T* and *D* in Fig. 2.43a, respectively. In Fig. 2.43b, T_p is the interval between two HIFU exposures called pulse pairs, which consisted of the previous pulse and the next pulse. By changing the parameters *T* or T_p, as in Fig. 2.43b, the spatial distribution affected by the previously generated cavitation was investigated.

2.5.2 *Repeatability and Feasibility*

The line-by-line scanning method would strictly only be valid when the cavitation distribution is repeatable, such as when all conditions, including HIFU output and the physical properties of the surrounding medium, are the same. Because the generation of a cavitation event always accompanies with randomness, a cavitation cloud is unlikely to be perfectly reproduced. However, on the broad scale of mapping bubble clouds, building up a cavitation map using the line-by-line method may be valid, and we show the results of reproducibility tests, in which specific test cases were remeasured. For each pf the specific test cases, 20 measurements were taken. The time interval between measurements was about 5 min, and no insonation occurred in the time interval. Figure 2.44 shows the results of this remeasuring on three different timescales. Here, we can see that the initial results for the first scan line could be almost reproduced at the end. Also, the two cavitation maps obtained using the line-by-line scanning method in tap water appeared very similar. These results may prove the broadscale repeatability of this method for cavitation mapping.

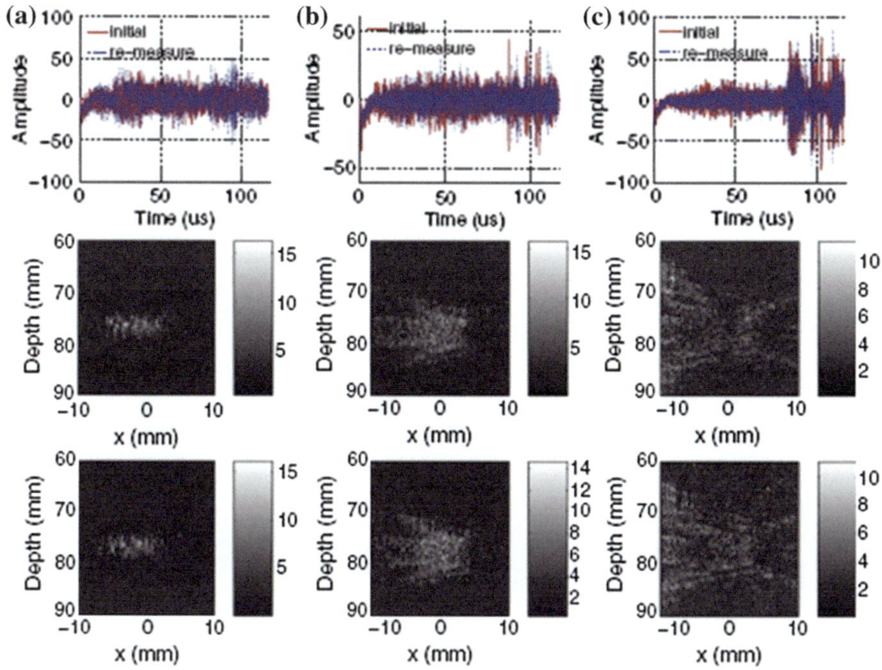

Fig. 2.44 Results of remeasuring specific test conditions on three different timescales. The exposure times of the FUS in (**a**–**c**) are 50 μs, 5 ms, and 30 ms, respectively. The first row shows the RF signal of the first scan line for both the initial and the remeasured experiments. The second and third rows illustrate two cavitation maps in tap water obtained by the ultrasound line-by-line scanning method. Reproduced with permission from Ding et al. © Elsevier

2.5.3 Spatial and Temporal Cavitation Distributions

The temporal evolution of a cavitation bubble cloud in tap water is shown in Fig. 2.45. Due to the highest acoustic pressure, the bubble cloud first emerged in the focal region. Later, the cavitation bubbles expanded symmetrically because inertial cavitation generated more bubbles, nucleating additional cavitation activity. The cavitation bubble cloud subsequently grew forward into the post-focal region, moving along a specific path and thus forming "streamers" like a tree crown, which are determined by the distribution of acoustic pressure. Bubbles in the post-focal and focal region can reflect ultrasound energy, causing cavitation bubbles to appear in the prefocal region with a separate branch structure. These results are consistent with those of previous studies (Chen et al. 2006b). When T was increased to 200 ms, the streaming due to acoustic radiation became bigger and the bubbles were pushed farther upward. Also, the prefocal region had the secondary maximum acoustic intensity, especially where close to focal region. Thus, in the prefocal region, bubbles (pointer in Fig. 2.45l) came to dynamic equilibrium. These results

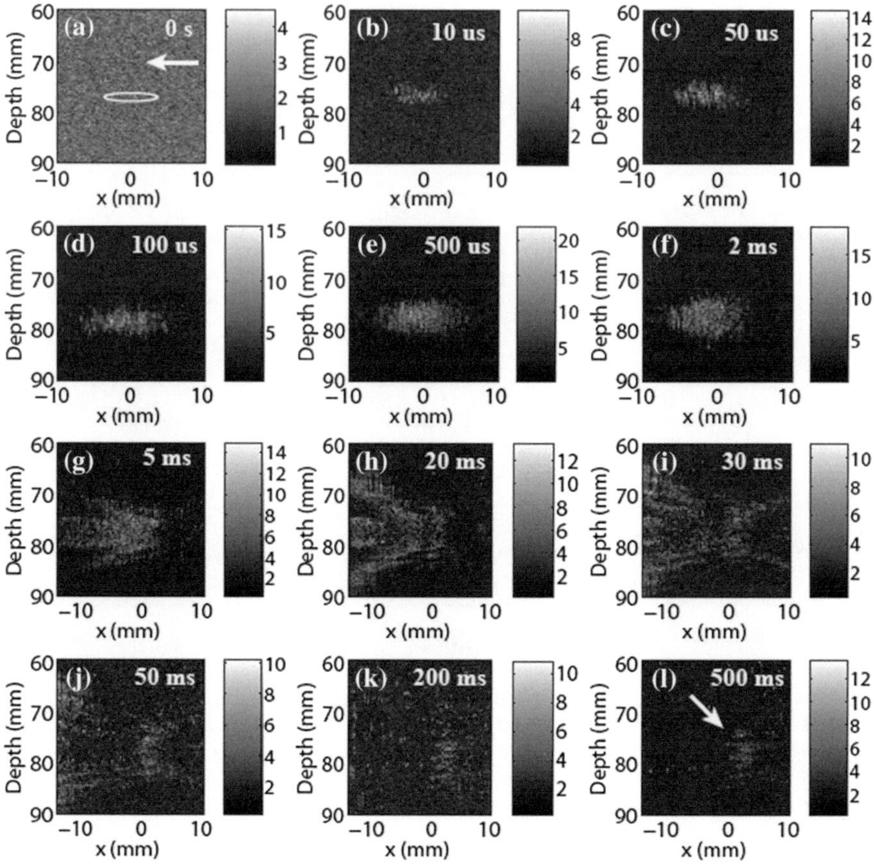

Fig. 2.45 Temporal evolution of the cavitation bubble cloud in tap water. The pointer and the *solid line* in **a** show the transmission direction of HIFU and the position of the focal region, respectively. Bubbles generated in the prefocal region near the focus are pointed out in (**l**). *D* is set to 1.000 ms for the travel time of the sound wave from the HIFU source to the focal region. The various *T* values are indicated in the *right* corner of each image. The zero of the *x*-axis is set to the center element of the array. Reproduced with permission from Ding et al. © Elsevier

suggest that a stable bubble cloud may be created when the parameters of ultrasound exposure are appropriately chosen.

Figure 2.46 shows the temporal evolution of a cavitation bubble cloud in a tap-water-filled spherical cavity in an gelatin–agar phantom. When no HIFU exposure was performed, the map acquired consisted only of the background (Fig. 2.46a). The results of cavitation mapping in the gelatin–agar phantom were obtained by subtracting the background from the cavitation maps. As shown in Fig. 2.46, the cavitation bubble cloud first emerged in the focal region with a high population density. Then, the cloud grew backward into the prefocal region, forming a triangular shape containing the accumulated bubbles. The reason for the

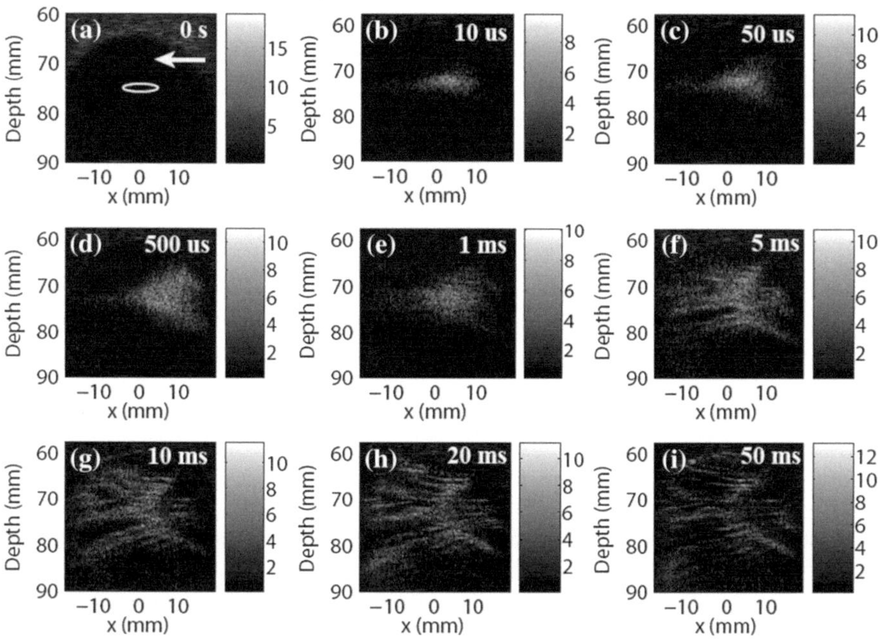

Fig. 2.46 Temporal evolution of the cavitation bubble cloud in a spherical cavity filled with tap water in a gelatin–agar phantom. The pointer and the *solid line* in **a** show the transmission direction of HIFU and the position of the focal region, respectively. *D* was set to 1.000 ms for the travel time of the sound wave from the HIFU source to the focal region. The various *T* values are indicated in the *right* corner of each image. The zero of the *x*-axis was set to the center element of the array. Reproduced with permission from Ding et al. © Elsevier

triangle shape formed is that dense bubbles in the focal region may affect the incident sound field by shielding. When *T* increased to 5 ms, the longer exposure to acoustic radiation caused the bubbles to travel to the post-focal region in a ribbonlike structure. Then, the ribbonlike structure enlarged, and bubbles were generated in almost the entire area of the cavity.

Compared with Fig. 2.45, the formation of the cloud structure in Fig. 2.46 occurred differently. Since the liquid in the cavity was also tap water, the cavitation thresholds were almost the same. However, in the phantom cavity, gas within the bubbles dissolved into the surrounding tap water, increasing the number of cavitation nuclei. Therefore, although the attenuation of the phantom existed, more bubbles were generated in the cavity. In addition, the cloud structure was relatively stable because acoustic radiation forces in the cavity may have been less than those in tap water alone.

The requirement for the cavitation distribution to be repeatable when all the experimental conditions are the same will limit the application of the line-by-line scanning technique to certain media whereby nuclei can be refreshed through movement. Also, the length of time needed to acquire a frame using this method is quite long (minimum of 128 s). However, because the method uses a focused beam, these maps (as opposed to those produced using plane wave techniques) can give good details with a high lateral resolution of 0.5 mm and an axial resolution of 0.29 mm. Furthermore, using hardware controls, the temporal resolution can reach the submicrosecond level. Also, because there is a long wait time after each scan line is acquired, the interactions between the interrogating pulse and bubble behavior are reduced. Therefore, this technique can be a useful tool to spatially and temporally study cavitation in some cases, providing a reference for clinical HIFU therapy. With more improvement, this technique may develop as a standard tool for cavitation mapping, as it offers both high spatial and temporal resolution.

2.6 Ultrafast Active Cavitation Mapping with a Plane Wave Beam

As described in Sects. 2.2, 2.3, 2.4, and 2.5, cavitation detection methods commonly include high-speed photography (Lauterborn and Ohl 1997; Chen et al. 2006b), SL/SCL (Gaitan et al. 1992; Brenner et al. 2002; Cao et al. 2012) and acoustic methods (Roy et al. 1990; Chen et al. 2003b; Vaezy et al. 2001b; Rabkin et al. 2005; Ding et al. 2013). High-speed photography (Lauterborn and Ohl 1997) can provide the distribution of cavitation bubbles intuitively with high temporal resolution ($\sim$nanosecond), but it demands transparency of the medium. SL and SCL (Cao et al. 2012) can map "active" cavitation bubbles, but these methods also require a transparent medium and their temporal resolution is low because of the weak luminous intensity.

Compared to high-seed photography and SL or SCL, acoustic methods are more suitable for in situ studies. PCD (Atchley et al. 1988; Chen et al. 2003a) and ACD (Roy et al. 1990; Chen et al. 2003b) are two common one-dimensional acoustic detection methods with high sensitivity, but they cannot reflect the spatial distribution of cavitation. ACM (Vaezy et al. 2005; Rabkin et al. 2005) has been developed on the basis of ACD and mainly refers to B-mode imaging. This technique achieves quasi-real-time mapping determined by the line-by-line scan process, and its imaging frame rate is limited, so it cannot capture transient cavitation bubble behavior. Ding et al. (2013) proposed a modified line-by-line scanning ACM (Sect. 2.5) to map cavitation bubbles with high sensitivity and high spatial–temporal resolution, but it is also time consuming and requires repeatable cavitation distributions within a medium.

However, cavitation mapping should be achieved with a high frame rate, preferably at one time over the entire region, because of the transient behaviors of the bubbles. Gateau et al. (2011a, b) have proposed a plane wave-based UACM, and they combined it with passive detection to detect and locate the cavitation events both in vitro and in vivo and performed a tilted plane wave compounding to improve the imaging quality and detection sensitivity. As reported in another study,

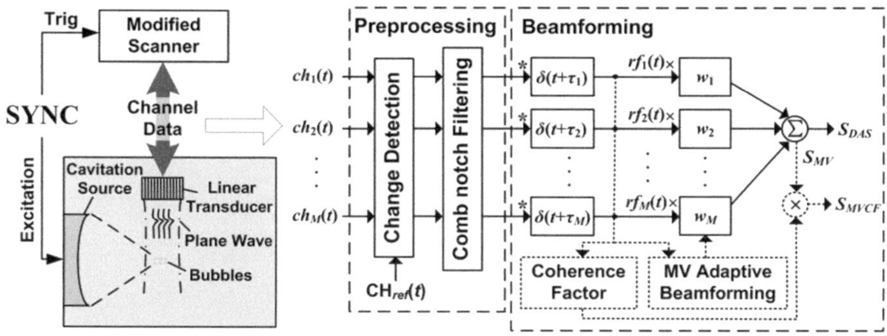

Fig. 2.47 Schematic representation of UACM

however, Hu et al. (2015) developed a UACM with high spatial–temporal resolution by combining plane wave transmission and adaptive beamforming to monitor the evolution of residual cavitation bubbles in histotripsy.

In this technique, the UACM with a plane wave beam is performed on a modified ultrasound scanner, which is accessible to channel-domain radio frequency (RF) data. A schematic of the method is shown in Fig. 2.47. The received channel-domain raw RF data are first preprocessed as follows: (1) A change detection is performed to suppress the background backscatter signal and extract the cavitation bubble signal; (2) a comb notch filter is applied to filter out the fundamental and harmonic HIFU frequencies to eliminate the strong interference from the HIFU field. After these preprocessing steps, a beamforming algorithm is then applied to form the final cavitation bubble images. According to the different beamforming algorithms, the plane wave UACM can be divided into plane wave delay-and-sum (DAS)-based UACM and plane wave adaptive beamforming-based UACM (Hu et al. 2015).

2.6.1 *Plane Wave DAS-Based UACM*

DAS is a conventional time-domain beamformer that uses a fixed set of time delays and aperture weightings to combine the channel-domain signals from the elements in the array. Time delays, adjusted for every depth on image formation to achieve dynamic receive focusing, are calculated as the propagation time from the transmitting element to the focus point and back to the receiving element, as shown in Fig. 2.48. Assuming an active aperture of M elements for a given point $\vec{r}_p$, the time delay for the mth element signal is given by:

$$\tau_m(\vec{r}_p) = \frac{\left\|\vec{r}^{\,\text{xmt}} - \vec{r}_p\right\| + \left\|\vec{r}^{\,\text{rcv}} - \vec{r}_p\right\|}{c} = \left(z + \sqrt{z^2 + (x - x_m)^2}\right)/c \tag{2.1}$$

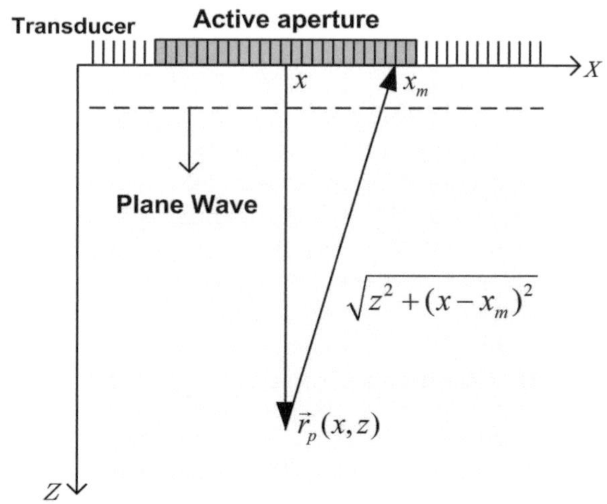

Fig. 2.48 Time delays for a plane wave insonation

for $m = 0, 1, \ldots, M-1$, where c is the speed of sound, and $\vec{r}^{\text{xmt}}$ and $\vec{r}^{\text{rcv}}$ are the spatial positions of the transmitting and receiving elements, respectively.

Then, the mth delayed element signal is given by:

$$\text{rf}_m(t) = \text{ch}_m(t + \tau_m(\vec{r}_p)) \tag{2.2}$$

where $\text{ch}_m(t)$ is the received channel-domain raw RF signal on the mth element.

Given a fixed aperture weighting, the delayed element signals are then weighted and summed as follows:

$$S_{\text{DAS}}(t) = \sum_{m=1}^{M} w_m(t)\text{rf}_m(t) = \mathbf{w}(t)^{\text{H}}\mathbf{RF}(t) \tag{2.3}$$

where $w_m(t)$ is the aperture weighting for element m, $\mathbf{w}(t) = [w_1(t)w_2(t)\cdots w_M(t)]^{\text{T}}$, $\mathbf{RF}(t) = [\text{rf}_1(t)\text{rf}_2(t)\cdots\text{rf}_M(t)]^{\text{T}}$, and $(\cdot)^{\text{H}}$ and $(\cdot)^{\text{T}}$ denote the conjugate transpose and transpose, respectively. The beamformed RF data are then enveloped and scaled to form UACM images.

2.6.2 *Plane Wave Adaptive Beamforming-Based UACM*

Adaptive beamforming calculates element weightings adaptively to the input signals to reduce the main lobe and depress the side lobes. This technique differs from conventional DAS beamforming, which uses fixed weightings. The widely used adaptive beamforming algorithm is the MV beamforming proposed by Capon (1969), of which the basic idea is to make noise as well as the signal from non-interference directions contributes minimally to the signal power while keeping the signal power from source direction constant. MV beamforming was initially used in radar, radio communications, and other far-field, non-correlated narrowband signal areas. In recent years, it has begun to be applied to near-field, highly correlated broadband ultrasound imaging (Wang et al. 2005; Asl and Mahloojifar 2009; Synnevag et al. 2007). Also, spatial smoothing and diagonal loading have been introduced to remove the strong correlation of ultrasonic signals and improve the robustness of the algorithm (Synnevag et al. 2007). CF weighting (Asl and Mahloojifar 2009) is another adaptive side lobe reduction method that works through weighting the image pixel by the corresponding CF (Asl and Mahloojifar 2009), in which the CF (Hollman et al. 1999) is initially used as a focusing index to indicate the focusing quality. Hu et al. (2015) combined plane wave transmission, MV beamforming, and CF weighting to achieve a plane wave MVCF-based UACM, as a representative of adaptive beamforming-based UACM, to image cavitation bubbles with a high SNR and high spatial–temporal resolution.

In plane wave DAS-based UACM, the aperture weightings are predefined and fixed, whereas MV beamforming uses the recorded signal to compute the weights adaptively by minimizing the variance of the beamformed output while maintaining a distortionless response of the focal point. This optimization problem can be formulated as follows (Cox et al. 1987):

$$\min_{\mathbf{w}(t)} \mathbf{w}(t)^{\mathrm{H}}\mathbf{R}(t)\mathbf{w}(t), \quad \text{subject to } \mathbf{w}(t)^{\mathrm{H}}\mathbf{a} = 1 \tag{2.4}$$

where $\mathbf{R}(t)$ is the covariance matrix of $\mathbf{RF}(t)$ and $\mathbf{a}$ is the steering vector. Due to time delaying, $\mathbf{a}$ is simply a vector of ones. The solution to Eq. (2.4) is given by Cox (1987):

$$\mathbf{w}_{\mathrm{opt}}(t) = \frac{\mathbf{R}(t)^{-1}\mathbf{a}}{\mathbf{a}^{\mathrm{H}}\mathbf{R}(t)^{-1}\mathbf{a}} \tag{2.5}$$

Spatial smoothing (Synnevag et al. 2007) is used to decorrelate the coherent received signals by dividing the array into overlapping subarrays followed by averaging the covariance matrices for all subarrays. With spatial smoothing, the covariance matrix is estimated as follows (Synnevag et al. 2007):

$$\widehat{\mathbf{R}}(t) = \frac{1}{M-L+1}\sum_{l=1}^{M-L+1} \mathbf{RF}_l(t)\mathbf{RF}_l(t)^{\mathrm{H}} \tag{2.6}$$

where L is the subarray length, and $\mathbf{RF}_l(t) = [\mathrm{rf}_l(t)\mathrm{rf}_{l+1}(t)\cdots\mathrm{rf}_{l+L-1}(t)]^{\mathrm{T}}$ is the delayed element vector for the lth subarray.

Diagonal loading (Synnevag et al. 2007) has also been introduced to improve the robustness of MV beamforming by adding a constant to the diagonal of the covariance matrix, replacing $\widehat{\mathbf{R}}(t)$ with $\widehat{\mathbf{R}}(t) + \varepsilon\mathbf{I}$. Here, ε is set to Δ times of the power in the received signals, **I** is the diagonal identity matrix (Synnevag et al. 2007):

$$\varepsilon = \Delta \cdot \mathrm{trace}\left\{\widehat{\mathbf{R}}(t)\right\} \tag{2.7}$$

The MV beamformed output is then given by:

$$S_{MV}(t) = \frac{1}{M - L + 1} \sum_{l=1}^{M-L+1} \mathbf{w}_{\mathrm{opt}}(t)^{\mathrm{H}} \mathbf{RF}_l(t) \tag{2.8}$$

CF weighting is defined as the ratio between the energy of the coherent sum to the total incoherent energy (Hollman et al. 1999):

$$\mathrm{CF}(t) = \frac{\left|\sum_{m=1}^{M} \mathbf{RF}(t)\right|^2}{M \sum_{m=1}^{M} |\mathbf{RF}(t)|^2} \tag{2.9}$$

The final output of MVCF-based UACM is given by:

$$S_{\mathrm{MVCF}}(t) = S_{\mathrm{MV}}(t) \cdot \mathrm{CF}(t) = \frac{\mathrm{CF}(t)}{M - L + 1} \sum_{l=1}^{M-L+1} \mathbf{w}_{\mathrm{opt}}(t)^{\mathrm{H}} \mathbf{RF}_l(t) \tag{2.10}$$

The beamformed RF data are then enveloped and scaled to form UACM images.

2.6.3 *Performance of UACM in a Free Field*

A focused single-element concave transducer (Imasonic, Besancon, France) driven by a double-channel arbitrary waveform generator (AWG420, Tektronix) with sine wave through a power amplifier (AG1017, T&G Power Conversion, Inc., Rochester, NY) was used as the energy source to generate cavitation in free field. A SonixTOUCH ultrasound scanner (Ultrasonix Medical Corp., Richmond, BC, Canada), equipped with a 128-element linear transducer array (L14-5/38 probe, center frequency of 7.5 MHz, 65 % bandwidth) and a plug-in module for parallel data acquisition of the raw ultrasound data (SonixDAQ, Ultrasonix Medical Corp., Richmond, BC, Canada), was used to perform UACM by programmable plane wave transmission (5 MHz, 1 cycle). The 128-element linear transducer array was aligned with the HIFU focus confocally. The time delay between cavitation excitation and UACM was 1 ms, and the sampling rate of UACM was 40 MHz.

The performances of UACMs were evaluated using the level of the SNR and the resolution. Figure 2.49 illustrates the comparison among conventional B-mode

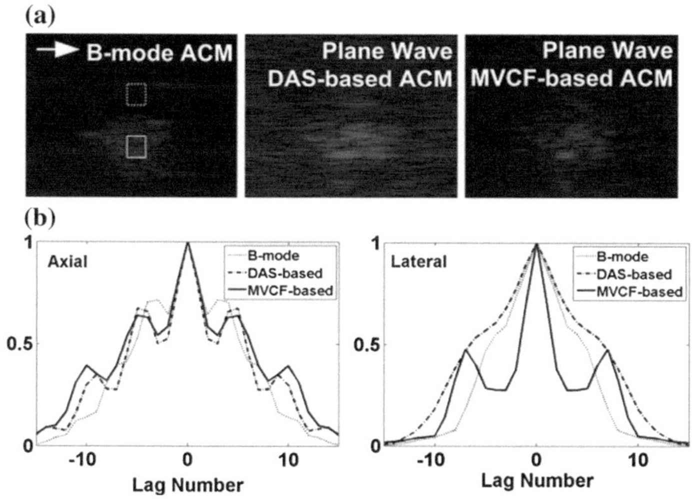

Fig. 2.49 Comparing conventional B-mode ACM, plane wave DAS-based UACM, and plane wave MVCF-based UACM in a free field. The dynamic range was 40 dB for all images (15 mm × 15 mm in dimension), and the rectangular boxes were used to calculate the SNR. Reproduced with permission from Hu et al. © Acoustical Society of America

ACM, plane wave DAS-based UACM, and plane wave MVCF-based UACM. The SNRs (defined as $10\lg(P_{\text{signal}}/P_{\text{noise}})$, where P_{signal} and P_{noise} are the power of signal and the noise, respectively) were calculated to be 19.7 ± 1.8, 11.4 ± 2.1, and 20.6 ± 2.9 dB for B-mode ACM, plane wave DAS-based UACM, and plane wave MVCF-based UACM, respectively, in five repeated experiments. From these results, we see that the SNR of the plane wave MVCF-based UACM was 9.2 ± 0.8 dB larger than that of plane wave DAS-based UACM, while it was almost the same as that of B-mode ACM. To quantify the resolution, the −6 dB autocorrelation lengths of the enveloped RF data corresponding to the images were calculated (Abeyratne et al. 1995). The axial resolution of plane wave MVCF-based UACM was 1.6 ± 0.6 times higher than that of DAS-based UACM, and it was 1.3 ± 0.7 times higher than B-mode ACM. Similarly, the lateral resolution of plane wave MVCF-based UACM was 2.4 ± 1.4 times larger than DAS-based UACM and 2.3 ± 1.3 times larger than B-mode ACM (Hu et al. 2015).

The effects of plane wave MVCF-based UACM on bubble behavior have also been studied by detecting the dissipation of cavitation bubbles in water under different UACM parameters. The UACM image sequences of bubble dissipation were quantified to integrate bubble intensities changing with time. The average results for three repeated experiments are illustrated in Fig. 2.50, from which we can see that the dissipation rate decreases as the PNP decreases, and it increases as the frame rate increases. The effect of UACM on bubble behavior is small when the imaging PNP is low. The averaged SNRs of the bubble images under different PNPs were also calculated to be 20.3 ± 1.9, 17.6 ± 0.8, and 13.6 ± 0.9 dB corresponding to PNPs of 240, 160, and 80 kPa, respectively (Hu et al. 2015).

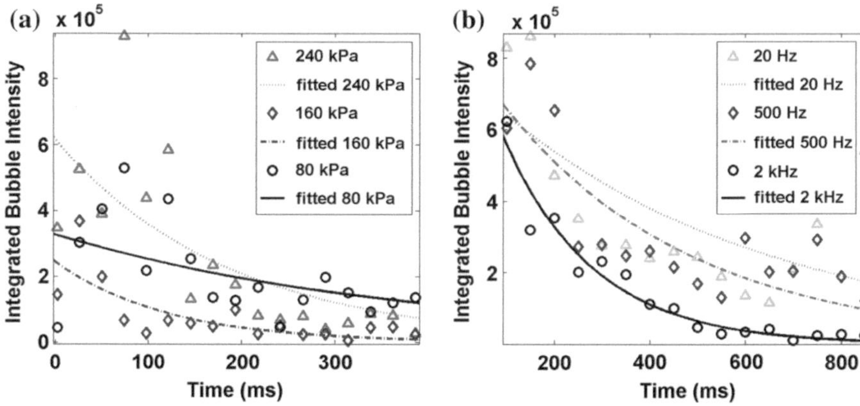

Fig. 2.50 The dissipation curves of cavitation bubbles under different UACM parameters, where exponential decay fitting was used. **a** The PNPs of UACM are 240, 180, and 80 kPa, with the same frame rate of 500 Hz. **b** The frame rates of UACM are 20, 500, and 2 kHz, while the PNP is 240 kPa. Reproduced with permission from Hu et al. © Acoustical Society of America

By combining plane wave transmission with MV beamforming and CF weighting, the frame rate of plane wave MVCF-based UACM can be significantly improved. In addition, the SNR of this technique is almost the same as in B-mode ACM, while the resolution is even higher than produced by B-mode ACM. Moreover, at a low imaging PNP, the effects of UACM on bubble behavior are small and can be neglected (Hu et al. 2015). So, plane wave MVCF-based UACM is feasible for imaging cavitation bubbles with relatively high SNR and good spatial–temporal resolution, as well as minor effects on bubble behavior.

2.6.3.1 2D UACM in a Free Fields

UACM can be employed to map both spatial–temporal evolution and dissolution of a cavitation bubble cloud that is induced by HIFU in water. This was done in the following experiments, where the electrical power was adjusted to 100 W, corresponding to a calculated total in situ spatial average intensity of about 2200 W/cm^2 (defined as I_{SAL}).

The spatial distributions of cavitation bubbles generated by a single HIFU pulse at different pulse durations are shown in Fig. 2.51. The cavitation bubbles first emerged in the focal region at a pulse duration of 50 μs, and then when the pulse duration increased, they grew and expanded forward into the post-focal region, finally appearing in the prefocal region. These results are consistent with previous studies that employed high-speed photography (Chen et al. 2006b) and a modified line-by-line scanning method (Sect. 2.5) (Ding et al. 2013).

Images of the evolution of a cavitation bubble cloud under successive HIFU pulses in a free field are shown in Fig. 2.52. Here, the duty cycle was 0.5 % and the

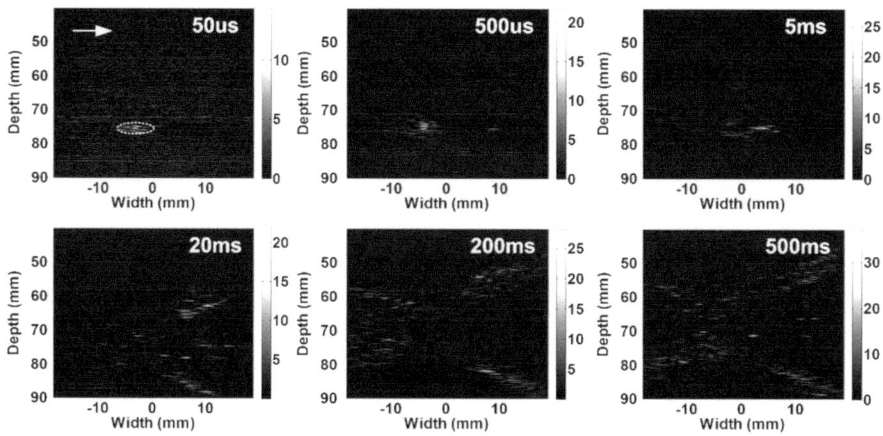

Fig. 2.51 Spatial distribution of cavitation bubbles generated by a single HIFU pulse at pulse durations varying from 50 μs to 500 ms in a free field. The *arrow* in the first image indicates the HIFU transmission direction

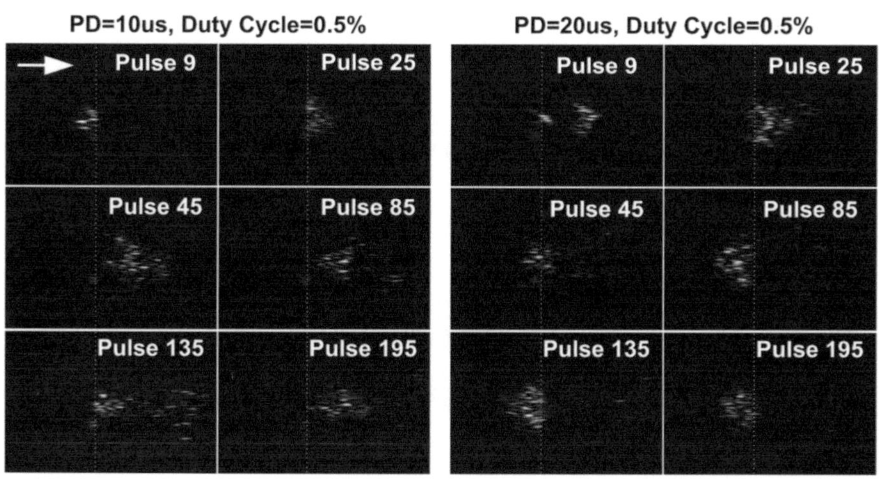

Fig. 2.52 UACM growth images of cavitation bubbles generated by successive HIFU pulses at a duty cycle of 0.5 % and pulse durations of 10 μs (*left*) and 20 μs (*right*) in a free field. The *arrow* in the first image indicates the HIFU transmission direction, and the *dotted line* indicates the center of the HIFU transducer's focal region

pulse duration was 10 and 20 μs. The bubble clouds were observed to grow and expand in the opposite direction for different pulse durations. For shorter pulses, the bubbles expanded away from the HIFU transducer because of the acoustic radiation force from HIFU. For longer pulses however, the bubbles expanded toward the HIFU transducer because enough bubbles were generated to cause acoustic "shadowing," which blocks the transmission of HIFU.

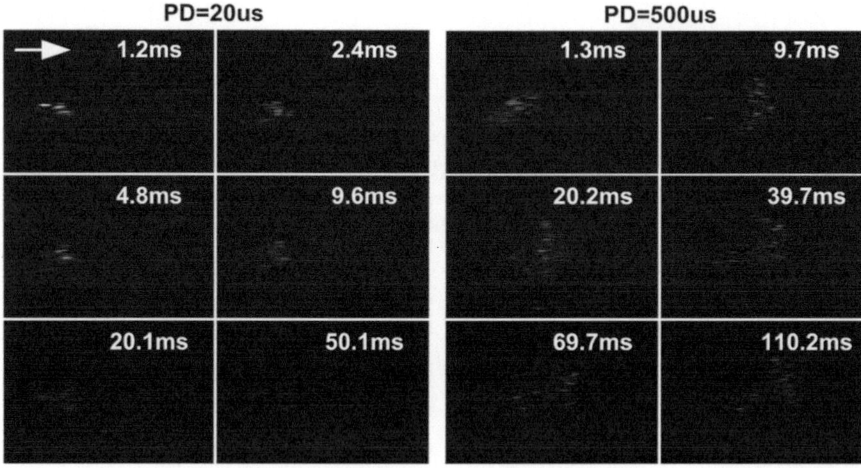

Fig. 2.53 UACM dissipation images of cavitation bubbles generated by a single HIFU pulse at pulse durations of 20 μs (*left*) and 500 μs (*right*) in a free field. The *arrow* in the first image indicates the HIFU transmission direction

Images showing the dissipation of cavitation bubbles can also be observed with UACM, as shown in Fig. 2.53, where cavitation bubbles were generated by a single HIFU pulse at different pulse durations in a free field. The results show that the overall size as well as the dissipation time of the cavitation bubble cloud increases with increasing pulse duration. The bubble cloud was almost disappeared after 50.1 ms for a pulse duration of 20 μs, while for a pulse duration of 500 μs, residual bubbles still remained after 110.2 ms.

2.6.3.2 3D Plane-by-Plane Cavitation Mapping

While UACM and PCM have been rapidly developed for cavitation detection by incorporating an array transducer to expand their spatial detection coverage for cavitation, these techniques only provide 2D cavitation maps, which give information about cavitation activity in only one plane. However, cavitation bubbles are produced in a 3D spatial area, which means that a 3D cavitation map is necessary for providing more complete information about cavitation activity.

A 3D plane-by-plane cavitation mapping technique based on the linear array of a current ultrasound diagnosis system has been proposed by Ding et al. (2013). With this technique, a cavitation event is mapped by a linear array of the ultrasound imaging system in plane wave transmitting and receiving mode, and then to extend this, a series of planes are sequentially measured, whereby these planes consist of the linear array positioned at different unit positions perpendicular to the axis of the HIFU transducer. Between two adjacent unit positions, there is a waiting time to allow the cavitation nuclei to distribute within the liquid such that it can return to its original state. With this spatial series of channel-domain raw RF data, a 3D cavitation image can be formed that encompasses the entire cavitation volume detected at one time. To do this, the acquisition of the raw RF signals for each unit position is synchronized with the HIFU exposure. The duration of HIFU exposure, as well as the delay of the interrogating pulse relative to the moment when HIFU is turned off, can vary from microseconds to seconds. In addition, while experiments by Ding et al. (2013) produced a temporal resolution of several microseconds, this can be adjusted through hardware controls to reach submicrosecond-level resolution.

To map cavitation bubbles with a high spatial–temporal resolution and a high SNR, MV adaptive beamforming and CF weighting are used to process the raw RF data. In order to retain the quality of adaptive beamforming but to reduce the amount of data needed for calculations, MV beamforming processing is performed only on random resampling points from a certain bandwidth rather than on every sampling point. Next, the total information in the frequency domain is reconstructed by compressive sensing (CS), which effectively reduces the volume of the data involved in the beamforming calculation and significantly increases the rendering speed. After acquiring each output of beamformed RF data in 3D space, the 3D cavitation map can be reconstructed using the 3D reconstruction algorithm.

This ultrasound plane-by-plane 3D mapping is only valid when the cavitation distribution is repeatable when all conditions, including the HIFU output and the physical properties of the surrounding medium, are same. However, cavitation clouds will not likely be perfectly reproduced. Even so, Ding et al. (2013) showed that when specific test cases were examined multiple times, the method is repeatable on a broad scale of cavitation mapping. In their studies, a 5-MHz linear array (L14-5/38, Ultrasonix, Richmond, Canada) aligned parallel to the axis of an HIFU transducer (1.2 MHz, focal width 1.6 mm, focal length 8 mm, and aperture diameter 15 cm; Imasonic, Besancon, France) was used to detect the cavitation bubble population immediately after one HIFU exposure in plane wave transmitting and receiving mode (Sonix DAQ, Ultrasonix, Richmond, Canada). Using a three-axis stepper motor controller (Zolix SC300-3B, Beijing, China), the linear array was

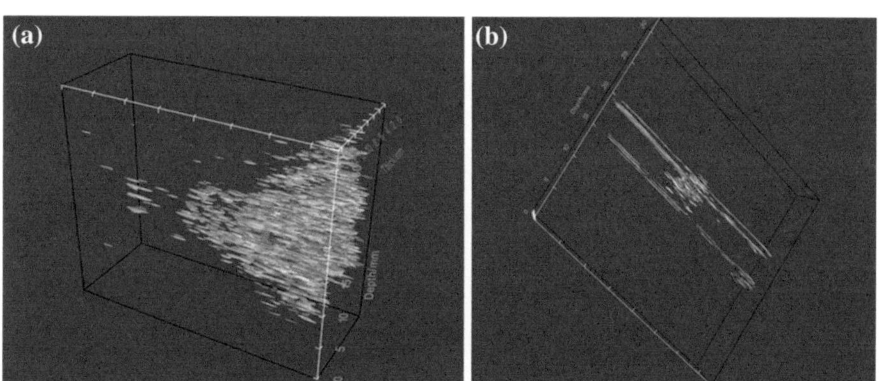

Fig. 2.54 Three-dimensional (3D) cavitation mapping **a** in tap water with a pulse duration of 20 ms and **b** in a phantom vessel with pulsatile flow and a pulse duration of 20 μs

positioned at each unit position perpendicular to the axis of HIFU transducer. This ultrasound plane-by-plane 3D cavitation mapping method was used to acquire the 3D spatial–temporal cavitation distribution in tap water and in a phantom vessel with pulsatile flow. As seen from the results in Fig. 2.54, the cavitation activity in 3D space was observed. Furthermore, the SNR of plane wave MVCF-based UACM was 20.45 ± 2.33 dB, which was almost same as that achieved in B-mode ACM. To quantify the improvement in resolution due to the beamforming, the −6 dB autocorrelation lengths of the enveloped RF data corresponding to the images were calculated. The resolution gain was defined as the ratio of the −6 dB autocorrelation lengths of the RF data before beamforming to those after beamforming. The lateral and axial resolution of plane wave MVCF-based UACM turned out to be $1.69 \pm 0.12\times$ and $2.33 \pm 0.69\times$ higher, respectively, than those resolutions in B-mode ACM.

2.7 Passive Cavitation Mapping

High-speed photography and the use of PCD have substantially improved our understanding of dynamic cavitation and its affiliated biological effects. Although PCD can resolve spatial sensitivity when receiving scattered signals, initial PCD strategies cannot distinguish individual bubbles in the cavitation bubble cloud because they rely on a single-element transducer. However, using a diagnostic ultrasound system with a linear array to map cavitation can allow us to identify cavitation along the transverse dire better axial resolution (Gyöngy and Coussios 2010a). The use of multi-element arrays as passive receivers can potentially provide substantial spatial information, thereby improving spatial and temporal resolution (Salgaonkar et al. 2009a). Therefore, PCM has been recently applied for this purpose, and it demonstrates very promising results for a wide range of applications, such for predicting and suppressing HIFU-induced vessel rupture (Hoerig et al. 2014), monitoring drug delivery (Nguyen et al. 2014), and transcranial mapping of diverging pressure waves (Arvanitis et al. 2014), to name a few.

In order to visualize the distribution of a cavitation bubble cloud produced after insonation with HIFU, post-beamforming of the spatial information is necessary. Several groups have approached this by performing passive beamforming in the time domain, which directly utilizes the cavitation recording acquired by the data acquisition board (Gyöngy and Coussios 2010a). In addition, cavitational features can also be characterized by transforming cavitation signals from the time domain to the Fourier domain (Salgaonkar et al. 2009a). In the following sections, we examine cavitation mapping from the perspectives of the time domain and the Fourier domain.

2.7.1 Time-Domain PCM

A signal of vibration or collapse from a single bubble can be considered as an emission from a cavitating point source. Therefore, a source field term s can be added that varies with position r and time t (Gyöngy et al. 2008):

$$\nabla^2 \mathrm{p} - \frac{1}{c}\frac{\partial^2 p}{\partial t^2} = -s(r,t) \tag{2.11}$$

where $s(r,t)$ represents the acoustic emissions arising from cavitation. This relationship represents the acoustic energy converted from the primary HIFU signal to broadband emissions, which is an indicator of inertial cavitation and gives the rate of the change in mass flow out of a unit volume. Integrating this source field over a volume V will yield the rate of change of mass flow out of that volume. According to the wave propagation, acoustic pressure can be depicted as follows:

$$p(r,t) = \int_{v} \frac{1}{4\pi|r-r'|} s(r', t - |r-r'|/c)dr' \tag{2.12}$$

One goal for cavitation maps is that they must directly correspond to the distribution of tissue damage that can be achieved by beamforming (DAS, for instance) the signals of array elements and then calculating the corresponding energy in time. Assuming that a linear array with elements at $(x_i, 0, 0)$ and focused in an $x - z$ plane, the source intensity map $S(x,z)$ for a time interval $[t_0, t_0 + T]$ should be given as follows (Kennedy 2005):

$$S(x,z) = \frac{1}{T}\int_{t_o}^{t_0+T} \left(\sum_i H_i(x,z,t)\right)^2 \mathrm{d}t \tag{2.13}$$

where $H_i(x,z,t)$ is the back-propagated signal:

$$H_i(x,z,t) = \alpha(d_i(x,z))p_i(t + d_i(x,z)/c) \tag{2.14}$$

Here, $p_i(t)$ is the pressure sensed by array element i at time t, c is the speed of sound, and d_i is the propagation distance from point (x,z) to array element i:

$$d_i(x,z) = \sqrt{z^2 + (x_i - x)^2} \tag{2.15}$$

$\alpha(d_i)$ is the receiver spatial sensitive compensation term. Considering 2D spreading where the linear array is focused in a plane, $\alpha(d_i)$ can be chosen as $\alpha(d_i) = \sqrt{d_i}$.

Interestingly, B-mode imaging with the use of traditional receive-only arrays and the dynamic beamforming method has been shown to be similar to passive seismic imaging achieved by time-exposure acoustics (TEA) (Kennedy 2005).The difference between them is that TEA effectively integrates the beamformed signal power over a time T, while B-mode imaging can only afford the envelope of the echo at a given time. Another distinguishing feature of TEA is that the axial resolution depends heavily on the receiving aperture, which is the same situation for passive beamforming. By contrast, the axial resolution of active B-mode ultrasound images is largely dependent on the transmitted pulse. Consequently, the receiving aperture may be varied with depth, which is called dynamic receive apodization.

Suppose a cavitation source $s(x, z, t)$ will generate the pressure filed $p(x, z, t)$. In order to estimate the source intensity S at position (x, z), we should calculate the back-propagating RF data from the signal on each receiving element to obtain the source intensity:

$$S(x, z) = \frac{1}{T}\int_{0}^{\mathrm{T}} \left\{ \left[\sum_{i=1}^{128} F_i(\tau) \right]^2 - \sum_{i=1}^{128} F_i(\tau)^2 \right\} \mathrm{d}t \tag{2.16}$$

where $F_i(\tau) = l_i * p_i(t - d_i/c + \tau)$ depicts the back-propagating signal. The term l_i represents the distance from point (x, z) to the array elements as follows:

$$l_i = \sqrt{z_s^2 + (x - x_s)^2} \tag{2.17}$$

Also, noting that:

$$\left(\sum_{i=1}^{128} k_i \right)^2 - \sum_{i=1}^{128} k_i^2 = 2 \sum_{\substack{i<j \\ i,j=1}}^{128} k_i k_j \tag{2.18}$$

we can change the integral order to obtain the cavitation source intensity:

$$S(x, z) = \sum_{\substack{i<j \\ i,j=1}}^{128} \left[\frac{2}{T} \int_0^T F_i(\tau) F_j(\tau) d\tau \right] \tag{2.19}$$

In addition, the axial resolution of passive cavitation maps created by a receive-only array would be degraded gradually as the distribution of the initial cavitation is extended along the axis of the HIFU transducer.

PCM, as mentioned above, is often produced by applying a passive beamforming algorithm for each location of interest. The recordings are appropriately delayed and summed together, with the mean square of the resulting signal providing a measure of the power emitted (Gyöngy and Coussios 2010b). However, passive beamfoming algorithms are unable to overcome the limit of Rayleigh resolution (Gyöngy and Coviello 2011), and the quality of passive maps degrades when coherent sources interfere with each other (Gyöngy 2010).

Thus, producing high-resolution cavitation maps for coherent sources may rely on adaptive beamforming methods. For example, parametric techniques that are based on the separation of subspaces between the signal and the noise are one important class of approaches for improving the resolution of cavitation maps. However, these parametric techniques require that bubble sources must be considered as points, so they would not be useful in cases where the number of cavitation sources is unknown (Gyöngy and Coviello 2011).

Gyöngy and Coussios (2010a) proposed an effective method of mapping cavitation by using a linear array to receive emissions from cavitation bubbles. PCM can conveniently provide quantitative analysis of the source power and the change rate of source intensity. An appropriate system to study cavitation allows transmitters and receivers to be positioned at any angle, and the axis of HIFU transducer should be perpendicular to the receiver of diagnostic device. Figure 2.55 shows the resulting passive cavitation maps that describe three cavitation clusters during 1.06-MHz HIFU exposures. The change of source strength with time is also shown in this figure. The source power and the source intensity at a peak rarefactional focal pressure (PRFP) of 1.6 MPa are shown on the left and right, respectively. The right part of the figure illustrates the PCD signals of a single element, which shows that the bubble clusters near the HIFU source underwent cavitation every 3 cycles, while those farther away underwent cavitation every 2 cycles. Due to the signal attenuation and the signal bandwidth, the measurements of acoustic intensity and power were lower than their actual values. Also, the low lateral resolution of the mapping system may have been caused by a collection of massive microbubbles, and more detailed behavior of the bubbles should have been deduced by other information, such as periodic collapses.

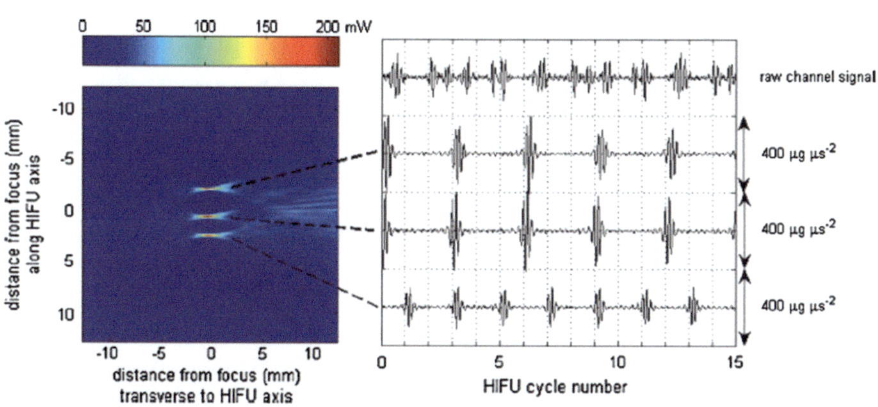

Fig. 2.55 Passive source power map (*left*) and source strengths (*right*) generated from the first frame recorded after HIFU was turned on, at a setting of 1.6 MPa PRFP. Reproduced with permission from Coussions et al. © Acoustical Society of America

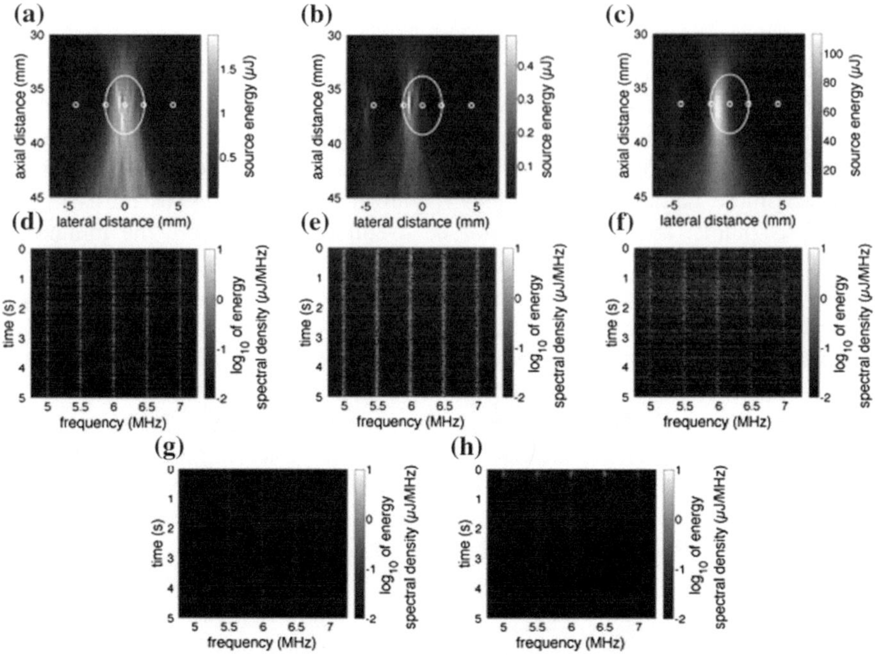

Fig. 2.56 PCI of single pulses and the accumulated total of all pulses, and the joint time–frequency-domain analyses of several points of interest labeled with little annulus in (**a**–**c**). Microbubbles were situated in 450 kPa PRPF, 50 Hz PRF, and 10 mm/s flow velocity. Large annulus in (**a**–**c**) indicates the focus region of focused ultrasound. Reproduced with permission from J.J. Choi et al. © Acoustical Society of America

Choi and Coussions (2012) also proposed a flow model to investigate spatial–temporal cavitation activity under disparate parameters by using passive acoustic imaging method in the time domain. They also combined a joint time–frequency-domain analysis with PCI to give information about the location of cavitation. Overall, this method was used to clarify the dynamic processes of focused ultrasound-induced cavitation in a flowing state. Figure 2.56a–c corresponds to images of the first pulse, final pulse, and the accumulated total of all 25 pulses, while figures (d–h) show the joint time–frequency-domain analyses of five points of interest. The main differences between the first and last pulse are embodied in the drastic decrease of the source energy at the focus and the attenuation of downstream microbubbles, which also can be observed in the total accumulated image. The strong broadband noise and strong harmonic components are clearly viewed in (e, f), both of which are contained within focal ROI and represent upstream microbubbles. And yet, downstream microbubbles, which are characterized by (g, h), seemed to produce few broadband recordings.

This technique allows us to adjust the position of HIFU relative to the linear array in order to investigate cavitation in the more complicated situation where the transmitting and receiving systems are separated; thus, this system provides much more convenience and flexibility for experiments or applications. Even though it is hard to confirm the focal position very precisely, this system affords a possible approach to continuously monitor cavitation during the whole procedure of HIFU exposure. And thus, it can allow cavitation to be monitored in opaque tissues in real time.

Furthermore, Gyöngy and Coussions (2010b) implemented another design whereby the linear array was inserted into the middle of the HIFU, such that the imaging plane is parallel to the axial direction of the HIFU transducer. The position of the HIFU and linear array receiver were relatively fixed (Fig. 2.57) in this situation. This novel system employs a circular hole whose diameter equals the size of the linear array, whereby the linear array can rotate along an arbitrary direction. Thus, it can detect cavitation from every angle, providing more information about the distribution of microbubbles. Such a system has many possible uses in clinical applications.

Figure 2.58 shows localization of several cavitation sources by the PCI algorithm (Jensen et al. 2013). Because passive beamformed map of original data shown in Fig. 2.58a includes non-cavitational components, data should be filtered harmonic frequency components to reserve broadband noise, then spatial distribution map of cavitation can be obtained, as shown in Fig. 2.58b. Compared with Fig. 2.58b, c allows explicitly observing number and size of cavitation source through extracting outline of the former, and this method can effectively avoid interference and imaging artefacts in the maps.

Compared with conventional B-mode imaging, passive methods have excellent SNR because bright regions in PCM images are only caused acoustically by active bubbles and there is no speckle. Also, detection of cavitation by using this technique is more sensitive (Gyöngy and Coussios 2010b).

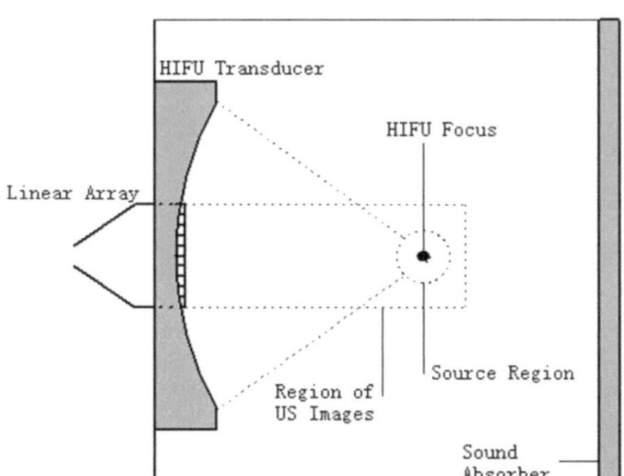

Fig. 2.57 Coaxial placement of an HIFU transducer and a linear array for PCM. Linear array is fixed in the middle of HIFU transducer and can passively receive emissions emanating from bubbles. The imaging plane is parallel to the axis of the HIFU transducer

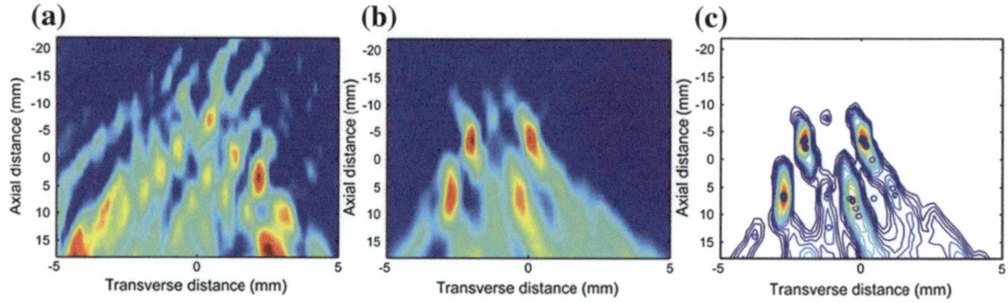

Fig. 2.58 Passive image of original data, passive image of data that has been filtered harmonic components, and contour map of filtered data. Reproduced with permission from C.R. Jensen et al. © 2013 IOP Publishing

Although broadband electrical noise may influence the detection of cavitation, distinction in the harmonics can be used to map cavitation. This allows us to localize the focus of low-intensity HIFU, which can injure biological tissue, and create 3D maps of cavitation to provide real-time feedback (Collin et al. 2013).

2.7.2 *Fourier-Domain PCM*

Salgaonkar (2009a) investigated passive cavitation maps from another distinct perspective, whereby the source intensity was characterized in the Fourier domain. The theory, which is called analytic expression, is similar to the theory of cavitation mapping based on DAS. The difference is that in the Fourier-domain theory, the signal should be processed in the frequency domain to portray its features. This is in direct contrast to the typical methods where signals are processed in the time domain (discussed above).

Specifically, as mentioned before, the pressure field at a position r for the point source radiating in a medium with density ρ and sound speed c is given by Blackstock (2000):

$$p(r,t) = \frac{\rho Q(t - |r|/c)}{4\pi|r|} \tag{2.20}$$

where $Q(t)$ is the volume flow of fluid from the source. Then, the received emission signal $s_n(t)$ for an element n centered at r_0 (region S_0) from a bubble at r_s is obtained by integrating the radiated pressure over the element surface. This can be finally expressed by the Rayleigh–Sommerfeld integral (Salgaonkar et al. 2009a):

$$s_n(t) = \int \frac{\rho Q(t - |r_0 - r_s|/c)}{4\pi|r_0 - r_s|} \mathrm{d}S_0 \tag{2.21}$$

which can be transformed into the frequency domain as:

$$S_m(\omega) = \frac{-i\omega\rho Q(\omega)}{4\pi} \int \frac{\exp(i|r_0 - r_s|\omega/c)}{|r_0 - r_s|} \mathrm{d}S_0 \quad (2.22)$$

where ω denotes radial frequency and $\mathrm{d}S_0$ is an area element on the receiver surface.

In order to produce passive cavitation maps, the brightness value for a passive image point (Y, Z) is defined as the total beamformed energy from a group of acoustic sources sensed by a transducer which is focused at that point:

$$I(Y, Z) = \sum_{\forall\omega} |S(\omega, Y, Z)|^2 \quad (2.23)$$

where ω is the radial frequency of a signal radiation frequency component and the symbol $\forall$ denotes summation over all radiating frequencies. When focusing an array with N elements on its axis at depth Z, the time delay applied to the nth element is expressed as:

$$\tau_n = \frac{-Z - \sqrt{(y_n - Y)^2 + Z^2}}{c} \quad (2.24)$$

The beamformed emission signal, corresponding to an image brightness value, is expressed as:

$$I(\omega, Y, Z) = \sum_{\forall\omega} \left| \sum_{\forall n} e^{j\omega T_n} S_n(\omega) \right|^2 \quad (2.25)$$

Salgaonkar et al. (2009a) designed a plan for evaluating signals in the Fourier domain (analytic expression) of PCM. In this setup (Fig. 2.59), the green rectangular box represents the linear array, while the gray cylinder represents the continuous-wave (CW) source. The orange test tube shape represents samples that are fixed in the middle. In their study, a 1-mm steel wire, a phosphate-buffered saline (PBS) solution, and a bovine liver were sonicated by CW ultrasound.

After sonicating PBS solution at 520 kHz with a PNP of 0.137 MPa, a constant 64-element subaperture (total 192 elements) was used to generate passive images from the ultraharmonic and broadband frequency components (Fig. 2.60). The cavitation cloud is clearly mapped (left image) in B-scan mode, and the result shows that the azimuth angle of the bubble clusters corresponds with the region where bubbles acted most dynamically. The center image was formed from the ultraharmonic emissions (6.5 MHz, or 12.5 × 520 kHz), while the right image was formed from the broadband emissions (6.3–6.7 MHz).

Processing signals in the frequency domain is based on the analytic expression, which was derived by numerically solving the Rayleigh–Sommerfeld integral under

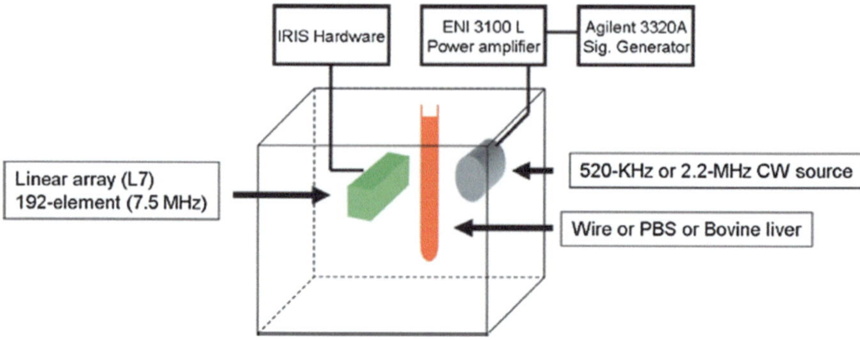

Fig. 2.59 Experimental setup for analytical expression in the frequency domain. Reproduced with permission from Salgaonkar et al. © Acoustical Society of America

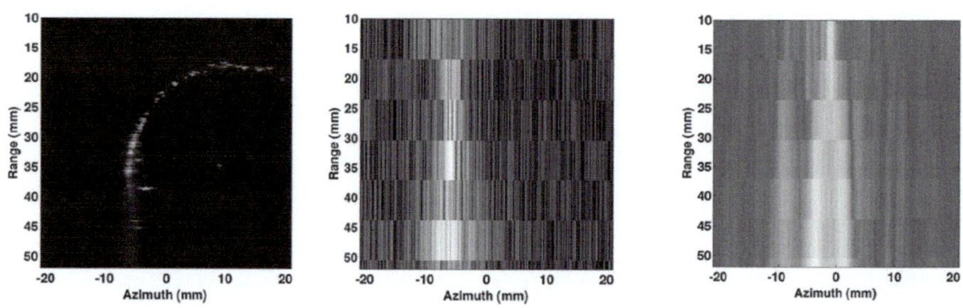

Fig. 2.60 Representative passive cavitation mapping in a saline solution for sonication at 520 kHz with 0.137 MPa peak negative pressure. Reproduced with permission from Salgaonkar et al. © Acoustical Society of America

the Fresnel approximation (Salgaonkar et al. 2009b). The results substantiate that passively receiving acoustic emissions can separate images for different frequency ranges. Consequently, this method provides a better way for focusing on information related to the frequency of interest. However, the B-scan mode does not have the capacity to distinguish the change in the size and position of cavitation events. In addition, this method allows cavitation sources to be localized more accurately in the array direction than in the range direction.

PCM may be the most effective way to monitor ultrasound ablation, because it has high sensitivity and a very high SNR compared with conventional B-mode images. Furthermore, this technique can provide higher frame frequency than MRI-guided imaging techniques. However, PCM cannot yet satisfactorily distinguish the distribution of a cavitation bubble cloud, mostly because of tissue heterogeneity and nonlinearity, especially in biological tissues. There are two possible ways to address this problem: increasing the array aperture and increasing the source frequency. In addition, a more complicated and effective imaging algorithm must be introduced for mapping and locating cavitation, thereby providing more information for clinical applications. But, the complexity of the imaging algorithm must not impede the imaging speed of PCM. Further investigations should be focused on maintaining imaging speed while improving the resolution of passive maps.

2.8 Summary

In this chapter, cavitation detection and mapping techniques in water, including traditional methods and the latest developments, have been introduced. These methods provide technical support in cavitation research and can be improved for cavitation imaging in tissue, as cavitation mapping in water underpins how cavitation is studied in tissues. High-speed photography can be used to capture images of bubble dynamics and of the spatial–temporal distribution of the bubble cloud when ultrasound exposure is on. Also, SL and SCL provide more information about the spatial distribution of chemically active cavitation bubbles. However, these optical methods can only be employed in transparent media. PCD and PCM are performed by passively recording acoustic emissions from cavitation bubbles and thus can be employed to monitor cavitation activity during ultrasound exposure. Conversely ACD and ACM, which are based on pulse-echo mode, can only be used when ultrasound exposure is turned off, ensuring that they avoid interfering with the interrogating pulse.

Because acoustic cavitation is an evolutionary process, generally starting with the nucleation of a single bubble and often culminating with the formation of a bubble cloud, detecting and mapping methods fall into two classes: single-bubble detection and bubble cloud detection. Single-bubble dynamics may be studied by high-speed photography, SL and ACD. Lauterborn et al. investigated single-bubble dynamics with laser-produced bubbles and high-speed photography using framing rates up to 20.8 million fps. Single-bubble SL occurs when an acoustically trapped and periodically driven gas bubble collapses so strongly that the energy focusing at collapse leads to light emission. Additionally, ACD with high-frequency and narrowband transmitting pulses has been introduced for transient microcavitation detection, and it affords sufficient sensitivity and temporal resolution to noninvasively detect a single bubble of 1 μm. Also, the frequency shift signal obtained by the Doppler method is much higher than the signal caused by translation motion of a solid particle, so the rapid bubble growth and collapse characteristic of cavitation events can be distinguished via Doppler methods.

When a bubble cloud, rather than an individual bubble, has formed, a detectable amount of generated light or scattered acoustic emission may be produced, and thus, the limitations of the detector sensitivity can be solved. SL and SCL can be employed to map chemically active cavitation bubbles. However, the most common detection method to use with ultrasound exposure is PCD of the acoustic emissions generated by bubbles. Subharmonic and ultraharmonic emissions of the main excitation frequency can be used to reflect stable cavitation, while the broadband noise emissions appearing above the background noise level indicate the presence of inertial cavitation. Also, the ICD and SCD can be calculated to reflect the relative intensity of inertial and stable cavitation. In order to increase the coverage of ACD and PCD, ACM and PCM have been developed by using a linear array connected to the ultrasound system in pulse-echo and "receive-only" mode, respectively. Here, we can achieve precise ACM by a modified ultrasound line-by-line scanning, and ultrafast ACM with plane wave beam. Although precise ACM is time consuming and can limited to a medium that offers a repeatable cavitation distribution, it can give good details with high sensitivity and spatial–temporal resolution, and it may develop as a standard cavitation mapping technique. Another technique, MVCF-based UACM, can map transient bubble clouds with relatively high SNR and good spatial–temporal resolution by combining plane wave transmission, MV beamforming, and CF weighting. Finally, using a passive beamforming algorithm on PCM data simultaneously acquired from all receiver elements of the linear array, it is feasible to map cavitation activity with time, supplying real-time cavitation maps during ultrasound treatment.

References

Abeyratne UR, Petropulu AP, Reid JM. Higher-order spectra based deconvolution of ultrasound images. IEEE Trans Ultrason Ferroelectr Freq Control. 1995;42:1064–75.

Akhatov I, Lindau O, Topolnikov A, Mettin R, Vakhitova N, Lauterborn W. Collapse and rebound of a laser-induced cavitation bubble. Phys Fluids. 2001;13:2805–19.

Ammi AY, Cleveland RO, Mamou J, Wang GI, Bridal SL, O'Brien WD. Ultrasonic contrast agent shell rupture detected by inertial cavitation and rebound signals. IEEE Trans Ultrason Ferroelectr Freq Control. 2006;53:126–36.

Apfel R. Acoustic cavitation inception. Ultrasonics. 1984;22:167–73.

Asl BM, Mahloojifar A. Minimum variance beamforming combined with adaptive coherence weighting applied to medical ultrasound imaging. IEEE Trans Ultrason Ferroelectr Freq Control. 2009;56:1923–31.

Arvanitis CD, Clement G, McDannold N. Transcranial passive cavitation mapping with a linear array: a simulation study with clinical datasets. J Acoust Soc Am. 2014;135:2210.

Atchley A, Frizzell L, Apfel R, Holland C, Madanshetty S, Roy R. Thresholds for cavitation produced in water by pulsed ultrasound. Ultrasonics. 1988;26:280–5.

Barnett S. Nonthermal issues: cavitation– its nature, detection and measurement. Ultrasound Med Biol. 1998;24:S11–21.

Blackstock DT. Fundamentals of physical acoustics. New York: Wiley; 2000.

Brenner MP, Hilgenfeldt S, Lohse D. Single-bubble sonoluminescence. Rev Mod Phys. 2002;74:425–84.

Cao H, Wan MX, Qiao YZ, Zhang SY, Li RX. Spatial distribution of sonoluminescence and sonochemiluminescence generated by cavitation bubbles in 1.2 MHz focused ultrasound field. Ultrason Sonochem. 2012;19:257–63.

Cao H, Yin H, Qiao YZ, Zhang SY, Wan MX. Sonochemiluminescence observation and acoustic detection of cavitation induced by pulsed HIFU at a tissue–fluid interface. Ultrason Sonochem. 2013;20:1370–5.

Capon J. High-resolution frequency-wavenumber spectrum analysis. Proc IEEE. 1969;57:1408–18.

Chambers L. The emission of visible light from pure liquids during acoustic excitation. Phys Rev. 1936;49:881.

Chen H, Kreider W, Brayman AA, Bailey MR, Matula TJ. Blood vessel deformations on microsecond time scales by ultrasonic cavitation. Phys Rev Lett. 2011;106:034301.

Chen H, Li XJ, Wan MX. The inception of cavitation bubble clouds induced by high-intensity focused ultrasound. Ultrasonics. 2006a;44:e427–9.

Chen H, Li XJ, Wan MX. Spatial–temporal dynamics of cavitation bubble clouds in 1.2 MHz focused ultrasound field. Ultrason Sonochem. 2006b;13:480–6.

Chen H, Li XJ, Wan MX, Wang SP. High-speed observation of cavitation bubble clouds near a tissue boundary in high-intensity focused ultrasound fields. Ultrasonics. 2009;49:289–92.

Chen W-S, Brayman AA, Matula TJ, Crum LA. Inertial cavitation dose and hemolysis produced in vitro with or without Optison®. Ultrasound Med Biol. 2003a;29:725–37.

Chen W-S, Brayman AA, Matula TJ, Crum LA, Miller MW. The pulse length-dependence of inertial cavitation dose and hemolysis. Ultrasound Med Biol. 2003b;29:739–48.

Chen W-S, Matula TJ, Crum LA. The disappearance of ultrasound contrast bubbles: observations of bubble dissolution and cavitation nucleation. Ultrasound Med Biol. 2002;28:793–803.

Choi JJ, Coussios C-C. Spatiotemporal evolution of cavitation dynamics exhibited by flowing microbubbles during ultrasound exposure. J Acoust Soc Am. 2012;132:3538–49.

Chomas JE, Dayton P, May D, Ferrara K. Threshold of fragmentation for ultrasonic contrast agents. J Biomed Opt. 2001;6:141–50.

Cole RH. Underwater explosions. New York: Dover Publications; 1965.

Coleman AJ, Saunders JE, Crum LA, Dyson M. Acoustic cavitation generated by an extracorporeal shockwave lithotripter. Ultrasound Med Biol. 1987;13:69–76.

Collin J, Coviello C, Lyka E. Real-time three-dimensional passive cavitation detection for clinical high intensity focused ultrasound systems. J Acoust Soc Am. 2013;133:3263.

Cox H, Zeskind RM, Owen MM. Robust adaptive beamforming. IEEE Trans Acoust Speech Sign Process. 1987;35:1365–76.

Crum LA. Acoustic cavitation. In: Proceeding of 1982 IEEE ultrasonics symposium. New York: IEEE; 1982. pp 1–11.

Crum LA, Roy RA, Dinno MA, Church CC, Apfel RE, Holland CK, Madanshetty SI. Acoustic cavitation produced by microsecond pulses of ultrasound: a discussion of some selected results. J Acoust Soc Am. 1992;91:1113–9.

Deng CX, Xu Q, Apfel RE, Holland CK. In vitro measurements of inertial cavitation thresholds in human blood. Ultrasound Med Biol. 1996;22:939–48.

Ding T, Zhang SY, Fu QY, Xu ZA, Wan MX. Ultrasound line-by-line scanning method of spatial–temporal active cavitation mapping for high-intensity focused ultrasound. Ultrasonics. 2013;54:147–55.

Everbach EC, Francis CW. Cavitational mechanisms in ultrasound-accelerated thrombolysis at 1 MHz. Ultrasound Med Biol. 2000;26:1153–60.

Farny CH, Holt RG, Roy RA. Temporal and spatial detection of HIFU-induced inertial and hot-vapor cavitation with a diagnostic ultrasound system. Ultrasound Med Biol. 2009;35:603–15.

Fernandez Rivas D, Ashokkumar M, Leong T, Yasui K, Tuziuti T, Kentish S, Lohse D, Gardeniers HJ. Sonoluminescence and sonochemiluminescence from a microreactor. Ultrason Sonochem. 2012;19:1252–9.

Frenzel J, Schultes H. Zeit. für Physical Chemistry 1934;421.

Frommhold L. Electron-atom bremsstrahlung and the sonoluminescence of rare gas bubbles. Phys Rev E. 1998;58:1899–905.

Gaitan DF, Crum LA, Church CC, Roy RA. Sonoluminescence and bubble dynamics for a single, stable, cavitation bubble. J Acoust Soc Am. 1992;91:3166–83.

Gateau J, Aubry JF, Pernot M, Fink M, Tanter M. Combined passive detection and ultrafast active imaging of cavitation events induced by short pulses of high-intensity ultrasound. IEEE Trans Ultrason Ferroelectr Freq Control. 2011a;58:517–32.

Gateau J, Aubry JF, Chauvet D, Boch A, Fink M, Tanter M. In vivo bubble nucleation probability in sheep brain tissue. Phys Med Biol. 2011b;56:7001.

Greenspan HP, Nadim A. On sonoluminescence of an oscillating gas bubble. Phys Fluids A. 1993;5:1065–7.
Gyöngy M. Passive cavitation mapping for monitoring ultrasound therapy. Oxford: Oxford University; 2010.
Gyöngy M, Arora M, Noble JA, Coussios CC. Use of passive arrays for characterization and mapping of cavitation activity during HIFU exposure. In: Proceeding of 2008 IEEE ultrasonics symposium. Beijing: IEEE; 2008. pp. 871–74.
Gyöngy M, Coussios C-C. Passive cavitation mapping for localization and tracking of bubble dynamics. J Acoust Soc Am. 2010a;128:EL175–80.
Gyöngy M, Coussios C-C. Passive spatial mapping of inertial cavitation during HIFU exposure. IEEE Trans Biomed Eng. 2010;57:48–56.
Gyöngy M, Coviello CM. Passive cavitation mapping with temporal sparsity constraint. J Acoust Soc Am. 2011;130:3489–97.
Hallez L, Touyeras F, Hihn JY, Klima J, Guey JL, Spajer M, Bailly Y. Characterization of HIFU transducers designed for sonochemistry application: cavitation distribution and quantification. Ultrasonics. 2010;50:310–7.
Hammer D, Frommhold L. Electron-ion bremsstrahlung spectra calculations for sonoluminescence. Phys Rev E. 2002;66:056303.
Hatanaka S-I, Yasui K, Tuziuti T, Mitome H. Difference in threshold between sono-and sonochemical luminescence. Jpn J Appl Phys. 2000;39:2962–6.
Hoerig CL, Serrone JC, Burgess MT, Zuccarello M, Mast TD. Prediction and suppression of HIFU-induced vessel rupture using passive cavitation detection in an ex vivo model. J Ther Ultrasound. 2014;2:14.
Holland CK, Roy R, Apfel R, Crum L. In vitro detection of cavitation induced by a diagnostic ultrasound system. IEEE Trans Ultrason Ferroelectr Freq Control. 1992;39:95–101.
Hollman K, Rigby K, O'Donnell M. Coherence factor of speckle from a multi-row probe. In: Proceeding of 1999 IEEE ultrasonics symposium. Caesars Tahoe: IEEE; 1999. pp. 1257–1260.
Holt M. Underwater explosions. Annu Rev Fluid Mech. 1977;9:187–214.
Hu H, Xu SS, Yuan Y, Liu RN, Wang SP, Wan MX. Spatial–temporal ultrasound imaging of residual cavitation bubbles around a fluid-tissue interface in histotripsy. J Acoust Soc Am. 2015;137:2563–2572.
Jarman P. Sonoluminescence: a discussion. J Acoust Soc Am. 2005;32:1459–62.
Jensen CR, Cleveland RO, Coussios CC. Real-time temperature estimation and monitoring of HIFU ablation through a combined modeling and passive acoustic mapping approach. Phys Med Biol. 2013;58:5833–50.
Kennedy JE. High-intensity focused ultrasound in the treatment of solid tumours. Nat Rev Cancer. 2005;5:321–7.
Khokhlova T, Li T, Sapozhnikov O, Hwang JH. The use of twinkling artifact of Doppler imaging to monitor cavitation in tissue during high intensity focused ultrasound therapy. In: Proceedings of meetings on acoustics. Montreal: Acoustical Society of America; 2013. p. 075034.
Lauterborn W, Ohl C-D. Cavitation bubble dynamics. Ultrason Sonochem. 1997;4:65–75.
Leighton TG. The acoustic bubble. London: Academic Press; 1994.
Leighton TG, Walton A, Field J. High-speed photography of transient excitation. Ultrasonics. 1989;27:370–3.
Li T, Khokhlova T, Sapozhnikov O, O'Donnell M, Hwang JH. A new active cavitation mapping technique for pulsed HIFU applications-bubble doppler. IEEE Trans Ultrason Ferroelectr Freq Control. 2014;61:1698–708.
Luther S, Mettin R, Koch P, Lauterborn W. Observation of acoustic cavitation bubbles at 2250 frames per second. Ultrason Sonochem. 2001;8:159–62.
Mast TD, Salgaonkar VA, Karunakaran C, Besse JA, Datta S, Holland CK. Measurements of cavitation dose, echogenicity, and temperature during ultrasound ablation. In: 6th international symposium on therapeutic ultrasound. Oxford: AIP Conference Proceedings; 2007. pp. 335–41.

McLaughlan J, Rivens I, Haar GT, Shaw A, Leighton T, Humphrey V, Birkin P, Vian C. The design and implementation of a passive cavitation detection system for use with ex vivo tissue. In: 5th international symposium on therapeutic ultrasound. New York: AIP Conference Proceedings; 2006. pp. 338–42.

McLaughlan J, Rivens I, ter Haar G. Cavitation detection in ex vivo bovine liver tissue exposed to high intensity focused ultrasound (HIFU). In: 4th IEEE international symposium on biomedical imaging: from nano to macro. Arlington: IEEE; 2007. pp. 1124–27.

McLaughlan J, Rivens I, Leighton T, ter Haar G. A study of bubble activity generated in ex vivo tissue by high intensity focused ultrasound. Ultrasound Med Biol. 2010;36:1327–44.

McMurray H, Wilson B. Mechanistic and spatial study of ultrasonically induced luminol chemiluminescence. J Phys Chem A. 1999;103:3955–62.

Moss WC, Clarke DB, White JW, Young DA. Hydrodynamic simulations of bubble collapse and picosecond sonoluminescence. Phys Fluids. 1994;6:2979–85.

Moussatov A, Granger C, Dubus B. Cone-like bubble formation in ultrasonic cavitation field. Ultrason Sonochem. 2003;10:191–5.

Neppiras E, Noltingk B. Cavitation produced by ultrasonics: theoretical conditions for the onset of cavitation. Proc Phys Soc Sect B. 1951;64:1032–38.

Nguyen MM, Kopechek JA, Hasjim B, Villanueva FS, Kim K. Passive cavitation imaging with nucleic acid-loaded microbubbles in mouse tumors. J Acoust Soc Am. 2014;136:2302.

Noltingk BE, Neppiras EA Cavitation produced by ultrasonics. Proc Phys Soc Sect B. 1950;63:674–85.

Ohl S-W, Ow DS-W, Klaseboer E, Wong VV, Camattari A, Ohl C-D. Creation of cavitation activity in a microfluidic device through acoustically driven capillary waves. Lab Chip. 2010;10:1848–55.

Pecha R, Gompf B. Microimplosions: cavitation collapse and shock wave emission on a nanosecond time scale. Phys Rev Lett. 2000;84:1328–30.

Putterman S, Weninger K. Sonoluminescence: how bubbles turn sound into light. Annu Rev Fluid Mech. 2000;32:445–76.

Rabkin BA, Zderic V, Vaezy S. Hyperecho in ultrasound images of HIFU therapy: involvement of cavitation. Ultrasound Med Biol. 2005;31:947–56.

Robinson P, Blake J, Kodama T, Shima A, Tomita Y. Interaction of cavitation bubbles with a free surface. J Appl Phys. 2001;89:8225–37.

Roy RA, Madanshetty SI, Apfel RE. An acoustic backscattering technique for the detection of transient cavitation produced by microsecond pulses of ultrasound. J Acoust Soc Am. 1990;87:2451.

Salgaonkar VA, Datta S, Holland CK, Mast TD. Passive cavitation imaging with ultrasound arrays. J Acoust Soc Am. 2009a;126:3071–83.

Salgaonkar VA, Datta S, Holland CK, Mast TD. Passive imaging of cavitational acoustic emissions with ultrasound arrays. In: AIP conference proceedings, 2009b. pp. 73–7.

Shi WT, Forsberg F, Tornes A, Ostensen J, Goldberg B. Experimental investigation of contrast microbubble destruction. In: Proceeding of 2001 IEEE ultrasonics symposium, Atlanta: IEEE; 2001. pp. 1691–94.

Shi WT, Forsberg F, Tornes A, Ostensen J, Goldberg BB. Destruction of contrast microbubbles and the association with inertial cavitation. Ultrasound Med Biol. 2000;26:1009–19.

Suslick KS. Sonochemistry. Science. 1990;247:1439–45.

Suslick KS, Didenko Y, Fang MM, Hyeon T, Kolbeck KJ, McNamara WB, Mdleleni MM, Wong M. Acoustic cavitation and its chemical consequences. Philos Trans R Soc Lond Ser A Math Phys Eng Sci. 1999;357:335–53.

Suslick KS, Hammerton DA, Cline RE. Sonochemical hot spot. J Am Chem Soc. 1986;108:5641–2.

Synnevag JF, Austeng A, Holm S. Adaptive beamforming applied to medical ultrasound imaging. IEEE Trans Ultrason Ferroelectr Freq Control. 2007;54:1606–13.

Taylor K, Jarman P. The spectra of sonoluminescence. Aust J Phys. 1970;23:319–34.

Tu J, Matula TJ, Brayman AA, Crum LA. Inertial cavitation dose produced in *ex vivo* rabbit ear arteries with optison@ by 1-MHz pulsed ultrasound. Ultrasound Med Biol. 2006;32:281–8.

Vaezy S, Andrew M, Kaczkowski P, Crum L. Image-guided acoustic therapy. Annu Rev Biomed Eng. 2001a;3:375–90.

Vaezy S, Martin R, Mourad P, Crum L. Hemostasis using high intensity focused ultrasound. Eur J Ultrasound. 1999;9:79–87.

Vaezy S, Shi X, Martin RW, Chi E, Nelson PI, Bailey MR, Crum LA. Real-time visualization of high-intensity focused ultrasound treatment using ultrasound imaging. Ultrasound Med Biol. 2001b;27:33–42.

Vaezy S, Vaezy S, Starr F, Chi E, Cornejo C, Crum L, Martin RW. Intra-operative acoustic hemostasis of liver: production of a homogenate for effective treatment. Ultrasonics. 2005;43:265–9.

Vazquez G, Camara C, Putterman S, Weninger K. Sonoluminescence: nature's smallest blackbody. Opt Lett. 2001;26:575–7.

Wang ZS, Li J, Wu RB. Time-delay-and time-reversal-based robust Capon beamformers for ultrasound imaging. IEEE Trans Med Imaging. 2005;24:1308–22.

Xu N, Wang L, Hu XW. Numerical study of electronic impact and radiation in sonoluminescence. Phys Rev E. 1998;57:1615–20.

Xu N, Wang L, Hu XW. Extreme electrostatic phenomena in a single sonoluminescing bubble. Phys Rev Lett. 1999;83:2441.

Xu N, Wang L, Hu XW. Bremsstrahlung of nitrogen and noble gases in single-bubble sonoluminescence. Phys Rev E. 2000;61:2611–6.

Xu Z, Raghavan M, Hall TL, Mycek M-A, Fowlkes JB, Cain CA. Evolution of bubble clouds induced by pulsed cavitational ultrasound therapy-histotripsy. IEEE Trans Ultrason Ferroelectr Freq Control. 2008;55:1122–32.

Yasui K. Mechanism of single-bubble sonoluminescence. Phys Rev E. 1999;60:1754–8.

Young FR. Sonoluminescence. Boca Raton: CRC Press; 2004.

Zhang SY, Li C, Yin H, Wang SP, Wan MX. Surface vibration and nearby cavitation of an ex vivo bovine femur exposed to high intensity focused ultrasound. J Acoust Soc Am. 2013;134:1656–62.

Chapter 3
Size Distribution of Cavitation Bubbles

Shanshan Xu, Xiaodong Liu and Mingxi Wan

Keywords Dissolution kinetics and condensation · Single-/two-/multi-component system · Sonoluminescence method · Acoustic method · Electromagnetic method · Acoustic droplet vaporization

3.1 Introduction

Acoustic cavitation is defined as the formation of a vapor cavity and bubbles in response to an acoustic field. One can generalize this definition to encompass any observable activity involving a bubble or a population of bubbles stimulated into motion by an acoustic field (Apfel 1984). There are two basic stages of acoustic cavitation: the nucleation (or inception) of the cavity; and the subsequent bubble dynamics involving growth, oscillation, and collapse. For acoustic cavitation to occur, an acoustic field—which is made up of an alternating pressure field—must be present. After the acoustic field vanishes, the remaining bubbles undergo dissolution or even settling in the crack of dirt. Cavitation bubbles can be characterized using various parameters. For example, they are characterized by their dynamic behaviors in response to an acoustic field, their spatial distribution, and the maximum temperatures and pressures they reach upon collapse. Most of these parameters have been extensively investigated theoretically and experimentally. However,

S. Xu · X. Liu · M. Wan (✉)
The Key Laboratory of Biomedical Information Engineering of Ministry of Education, Department of Biomedical Engineering, School of Life Science and Technology, Xi'an Jiaotong University, 710049 Xi'an, People's Republic of China
e-mail: mxwan@mail.xjtu.edu.cn

S. Xu
e-mail: xushanshan@mail.xjtu.edu.cn

X. Liu
e-mail: lxdmwz123@163.com

M. Wan et al. (eds.), *Cavitation in Biomedicine*,
DOI 10.1007/978-94-017-7255-6_3

another fundamental parameter of cavitation for a bubble population is bubble size distribution, but this has not been widely investigated.

In bubble population, the bubble size determines the dynamics of how bubbles will behave in a sonicated medium (Apfel 1981). Different bubble dynamics will lead to specific physical effects, such as microstreaming, microjetting, and shock wave formation, which are all strongly correlated with a wide range of applications in hydrodynamics, ultrasonic cleaning, sonochemical reaction, and biomedical engineering. Therefore, knowledge about the size distribution of cavitation bubbles is critical for promoting or suppressing the aforementioned physical effects in order to optimize or minimize, respectively, the corresponding cavitation activity. Earlier studies have demonstrated that the equilibrium radius of a cavitation bubble encompasses a range of radii rather than a single linear resonance radius calculated using the linear equation for a given frequency (Yasui et al. 2008; Yasui 2002). Furthermore, because of nonlinearity, the actual equilibrium radii of cavitation bubbles are always smaller than the calculated single value (Yasui 2002; Luther et al. 2001; Lee et al. 2005). Therefore, the size distribution of cavitation bubbles must be evaluated experimentally.

While measuring this parameter is difficult due to the complex nature of acoustic cavitation, several studies have estimated cavitation bubble size and size distributions using different techniques. In this chapter, we will present in detail some recently developed methods, based on bubble dissolution, to measure this parameter. We will examine the principles, experimental setups, measurement processes, data analysis methods, and typical results associated with each method. Overall, these methods can be classified into two categories: In the first category, the methods determine the bubble size distribution during when the source ultrasound is applied, while the methods in the second category determine the bubble size distribution once the source ultrasound has been removed. Since the methods described in this chapter are all based on bubble dissolution in solution, we first introduce the kinetics of free bubble dissolution in liquid.

3.2 The Kinetics of Free Bubble Dissolution

3.2.1 Single-Component System

If a bubble initially contains only a single gas which is the same as that dissolved in solution, the system is called a single-component system. For a spherical bubble in such system, its dissolution kinetics are described in Eq. (3.1) by Epstein and Plesset (1950).

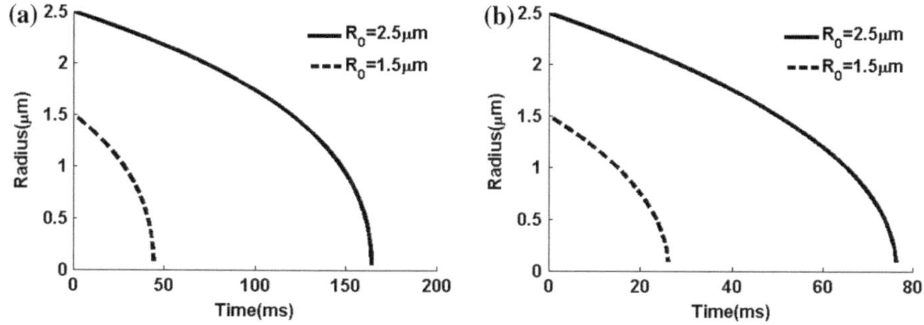

Fig. 3.1 Predicted dissolution curves for 1.5- and 2.5-μm-radius air bubbles calculated using Eq. (3.1) in **a** saturated water and **b** degassed water

$$\frac{dR}{dt} = -DL\frac{1 - \frac{C_i}{C_s} + \frac{\tau}{R\rho_g}}{1 + \frac{2\tau}{3R\rho_g}}\left(\frac{1}{R} + \frac{1}{\sqrt{\pi D t}}\right)$$
$$\tau = \frac{2M\sigma}{BT} \tag{3.1}$$

where R is the instantaneous bubble radius at the time instant t, D is the diffusion coefficient of the gas, L is the Ostwald coefficient of the gas (ratio of the solubility of the gas in the liquid to the gas density), C_i is the dissolved gas concentration in liquid far from the bubble, C_s is the saturated dissolved gas concentration, M is the molecular weight of the gas, ρ_g is the density of gas, σ is the surface tension, T is the temperature, and B is the universal gas constant.

If the solution is undersaturated or saturated, the bubble will dissolve due to gas diffusion from the bubble into the solution. Further, the larger the Ostwald coefficient of the gas within the bubble, the sooner the bubble will dissolve. Given the initial radius of a bubble and the solution conditions, we can obtain the dissolution process of a bubble filled with air in air-saturated water, as shown in Fig. 3.1.

In this process, the radius of bubble obviously decreases with time as dissolution occurs. Also, bubbles with a smaller initial radius or those present in a solution with less dissolved gas will have shorter (faster) complete dissolution times.

3.2.2 *Two-Component System*

If a bubble in solution is filled with one type of gas, but a different type of gas is dissolved within the solution, a two-component system should be considered. Kabalnov et al. (1998) discussed the dissolution kinetics of a bubble filled with two gases—air and a sparingly water-soluble gas (from now on referred to as "less-soluble gas" for convenience). The Ostwald coefficient of the less-soluble gas is far less than that of air.

For a spherical bubble containing two gases in an aqueous solution, there will be a pressure balance existing between the partial pressures of gases and the ambient pressure, including the atmospheric pressure and the Laplace pressure ($2\sigma/R$). This can be expressed as follows (Kabalnov et al. 1998):

$$p_{\mathrm{A}} + p_{\mathrm{F}} = \frac{2\sigma}{R} + P_0 = (C_{\mathrm{A}} + C_{\mathrm{F}})BT \tag{3.2}$$

where p_{A} and p_{F} are the partial pressures of air and the less-soluble gas in the bubble, respectively; P_0 is the ambient pressure; and C_{A} and C_{F} are the concentrations of air and the less-soluble gas in the bubble (mol/m^3), respectively, both of which are the function of time. The other parameters are the same as in Eq. (3.1). The diffusion equations for air and the less-soluble gas are described below.

The dissolution rate of the bubble in terms of the air concentration (Kabalnov et al. 1998) is given as follows:

$$\frac{\mathrm{d}}{\mathrm{d}t}\left(R^3 C_{\mathrm{A}}\right) = -3RD_{\mathrm{A}}L_{\mathrm{A}}\left(C_{\mathrm{A}} - \frac{P_0}{BT}\right) \tag{3.3}$$

The dissolution rate of the bubble in terms of the less-soluble gas concentration (Kabalnov et al. 1998) is given as follows:

$$\frac{\mathrm{d}}{\mathrm{d}t}\left(R^3 C_{\mathrm{F}}\right) = -3RD_{\mathrm{F}}L_{\mathrm{F}}C_{\mathrm{F}} \tag{3.4}$$

where D_{A} and D_{F} are the diffusion coefficients of air and the less-soluble gas (m^2/s), respectively; and L_{A} and L_{F} are the Ostwald coefficients of air and the less-soluble gas, respectively.

We can incorporate Eqs. (3.2)–(3.4) to describe the kinetics of bubble dissolution in a two-component system. These three equations are satisfied with the initial condition below (Kabalnov et al. 1998):

$$R(t = 0) = R_0 \tag{3.5}$$

$$X_{\mathrm{F}} = \frac{C_{\mathrm{F}}(t = 0)}{C_{\mathrm{A}}(t = 0) + C_{\mathrm{F}}(t = 0)} \tag{3.6}$$

$$(C_{\mathrm{A}}(t = 0) + C_{\mathrm{F}}(t = 0))BT = \frac{2\sigma}{R_0} + P_0 \tag{3.7}$$

The first condition gives the initial radius of a mixed-gas bubble. The second defines the volume fraction of the less-soluble gas in the bubble at the initial time (X_{F}). If the bubble is initially filled only with the less-soluble gas, $X_{\mathrm{F}} = 1$; for a bubble filled only with air, $X_{\mathrm{F}} = 0$. The third condition describes the mechanical equilibrium of the bubble in the solution at the initial time.

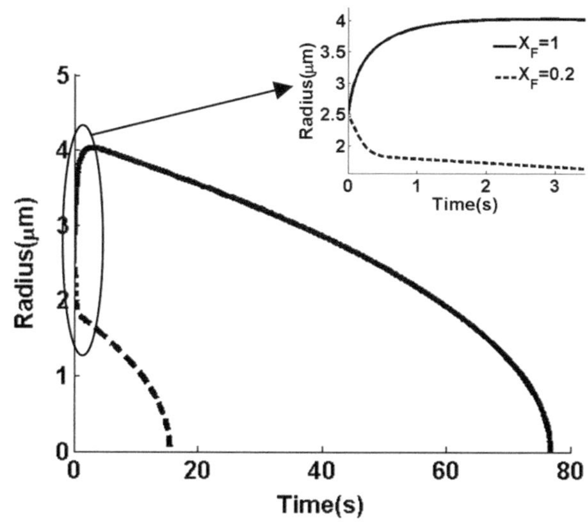

Fig. 3.2 Dissolution curves for a 2.5-μm-radius bubble with $X_F = 1$ and $X_F = 0.2$

For example, if we take a bubble filled with an air–perfluoropentane mixture in saturated water, we can determine the numerical simulations of its dissolution process based on the kinetics of a bubble in a two-component system, as shown in Fig. 3.2.

From Fig. 3.2, we can see that the complete dissolution time of a bubble filled with gaseous perfluoropentane is much longer than that of an air bubble within a single-component system. The greater the proportion of perfluoropentane within the bubble, the longer the bubble's dissolution time will be. To be more precise though, we can separate the dissolution of the bubble into three stages.

1. In the first stage of dissolution, we can see that the bubble's radius undergoes rapid swelling (see the insert in Fig. 3.2) when the initial partial concentration of perfluoropentane is greater than 0.5 (e.g., $X_F = 1$), which occurs as air from the surrounding solution diffuses into the bubble. However, when the initial partial concentration of perfluoropentane is less than 0.5 (e.g., $X_F = 0.2$), the initial stage of dissolution is characterized by rapid bubble shrinking, caused by the diffusion of air out of the bubble.
2. In the second stage of dissolution, the radius decreases quickly. The partial pressure of air in the bubble is established at ambient pressure in water and stays essentially constant during further dissolution. Meanwhile, the partial pressure of perfluoropentane increases with time, because it must keep up with the increasing Laplace pressure inside the shrinking bubble. This stage contributes the most to the overall lifetime of the bubble.
3. In the last stage of dissolution, the partial pressure of perfluoropentane becomes so high that the perfluoropentane condenses. The size and compressibility of the bubble exhibit a sharp drop, and the bubble becomes invisible to ultrasound.

3.2.3 Multi-component System

A microbubble can also be suspended in a bulk medium containing multiple dissolved gases. In such cases, the initial microbubble may comprise a single gas, but the surrounding fluid may have several dissolved gases. For example, this occurs when lipid-coated microbubbles used in ultrasonic diagnosis and therapy are exposed to a multi-gas environment when they are injected into the blood, which has oxygen, nitrogen, carbon dioxide, and possibly anesthetic gases as well. In other cases, naturally occurring microbubbles formed in the ocean and synthetic microbubbles used in bioreactors, microfluidic devices, and microfloatation columns may also experience multi-gas environments. To determine the fate of microbubbles in these situations, several models and experiments have been developed. Burkard and Van Liew (1994) developed a numerical model to predict multi-gas growth and dissolution of microbubbles within tissue, which incorporates the effects of blood perfusion on the influx and efflux of the gas and uses an iterative technique to solve the model. Furthermore, expanding the EP model, Kwan (Kwan and Borden 2010) proposed a new modified model that does not limit the number of gases in the system.

The numerical model describing the multi-component system can be separated into two components: the mass balance coupled with diffusion and the Laplace pressure equation. Assuming that there is no convection occurring due to the efflux of gas or bulk flow around the microbubble, the component balance is as follows:

$$\frac{\mathrm{d}n_\mathrm{i}}{\mathrm{d}t} = -4\pi R D_\mathrm{i} K_{\mathrm{H,i}} \left(\frac{2\sigma}{R} - P_{\infty,\mathrm{i}} f_\mathrm{i} + P_0 - \frac{3BT}{4\pi R^3} \sum_{j=1}^{N} n_j \right) \quad \text{where } j \neq i \tag{3.8}$$

where n_i is the moles of component i; D_i is diffusivity of component i; $K_{\mathrm{H,i}}$ is the molar Henry's constant; $P_{\infty,\mathrm{i}}$ is the saturation pressure far from the microbubble in bubble medium; and f_i is the saturation fraction of component i.

Considering a bubble of radius R in the bulk medium, the Laplace pressure equation is as follows:

$$0 = 2\sigma R^2 + P_0 R^3 - \frac{3BT}{4\pi} \sum_{i=1}^{N} n_\mathrm{i} \tag{3.9}$$

Using the initial microbubble radius and gas composition as an initial condition, the future molar gas compositions of the microbubble interior can be estimated as follows:

$$n_\mathrm{i}^{\tau+1} = -4\pi R D_\mathrm{i} K_{\mathrm{H,i}} \left(\frac{2\sigma}{R} - P_{\infty,\mathrm{i}} f_\mathrm{i} + P_0 - \frac{3BT}{4\pi R^3} \sum_{j=1}^{N} n_j \right) + n_\mathrm{i} \quad \text{where } \mathrm{j} \neq i \tag{3.10}$$

where τ is the numerical time step and h is the difference in time between n_i and $n_i^{\tau+1}$. After determining the new composition of the microbubble, a new radius can be calculated, which gives

$$0 = \frac{8\pi\sigma}{3BT}\left(R^{\tau+1}\right)^2 + \frac{4\pi P_0}{3BT}\left(R^{\tau+1}\right)^3 - \frac{3BT}{4\pi}\sum_{i=1}^{N} n_i^{\tau+1} \tag{3.11}$$

Using Eqs. (3.10) and (3.11), Kwan and Borden (2010) simulated the change in diameter of sulfur hexafluoride (SF_6) microbubbles coated with sodium dodecyl sulfate (SDS) over time in a blood environment, and they observed the same results through experimentation. Their experimental and theoretical results are shown in Fig. 3.3. Just like in the two-component system, the bubbles here also grew at first and then underwent dissolution (Fig. 3.3a, b). However, in the multi-component system, bubbles took much longer to totally disappear.

Now that the dissolution kinetics of free gas bubbles in liquid have been briefly introduced, we will next present the various methods based on bubble dissolution that have been developed to estimate the size distribution of cavitation bubbles. First, we will discuss two methods that use different techniques to characterize the equilibrium radii of the cavitation bubbles when an ultrasound pulse is applied (when it is "on"). Both techniques use an ultrasonic pulsing strategy, but one

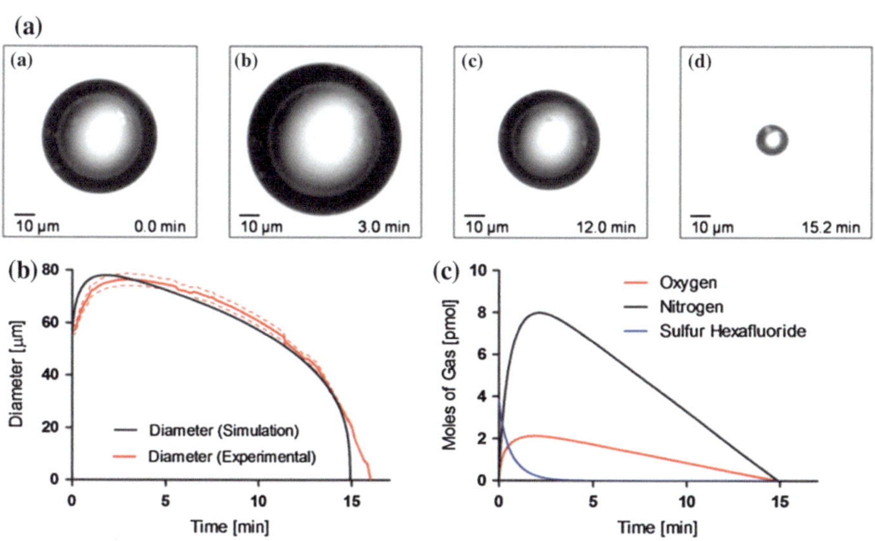

Fig. 3.3 An SDS-coated SF_6-filled microbubble exposed to an air-saturated environment. **a** Growth and dissolution of a microbubble observed by a CCD camera. **b** Theoretical and experimental diameter–time curve. **c** Predicted gas content of microbubble interiors. Reproduced with permission from Kwan (2010). © American Chemical Society

method uses the averaged sonoluminescence (SL) (Lee et al. 2005) and the other uses the cavitation noise (CN) (Hauptmann et al. 2013) as a function of the pulse interval. In the following sections, each method is described in detail.

3.3 Sonoluminescence Method

Under certain conditions, bubble collapse can result in light emission, called SL, originating from the core of the bubble during the final stages of collapse (Suslick 1998).

If acoustic pulse trains are transmitted into an aqueous solution with a certain pulse-on time (T) and pulse-off time (T_0), a "steady-state active bubble population" can be reached within the periods of the pulse-on time (T), and the SL intensity within this population depends largely upon the sizes of the bubbles that form in the pulse-on time. If the pulse-off time (T_0) is prolonged, the SL intensity arising during the pulse-on time (T) tends to decrease due to bubbles dissolution between pulses (Lee et al. 2005). This scenario is displayed in Fig. 3.4.

Thus, as T_0 increases and the SL intensity also decreases due to the dissolution of a group of bubbles, each T_0 therefore corresponds to the total dissolution time of a group of bubbles of a particular size. This size can be calculated using the transformation of the Epstein and Plesset equation described in Sect. 3.2.1.

$$\left(\frac{DC_s}{\rho_g R_0^2}\right) t = \frac{1}{3}\left(\frac{BT\rho_g R_0}{2M\sigma} + 1\right) \tag{3.12}$$

The parameters in Eq. (3.12) are identical to those in Eq. (3.1). In this equation, we can replace t with T_0 to calculate the initial radius of the bubble.

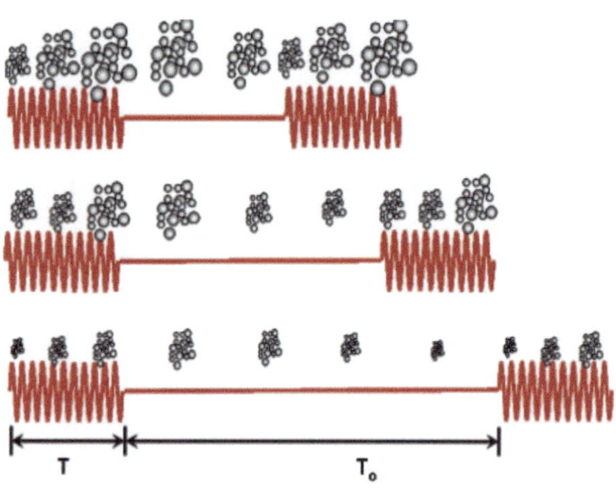

Fig. 3.4 The effect of increasing the pulse-off time (T_0) on the size of bubble cloud while keeping the pulse-on time (T) fixed. Reproduced with permission from Lee (2005). © American Chemical Society

Assuming that the SL intensity decreases linearly with a decrease in bubble population, we can estimate the size distribution of the SL bubbles. Therefore, we can use the SL intensity measured in an experiment as a function of T_0 to calculate the bubble size distribution of SL bubbles.

The ultrasound source used in these experiments was an Undatim Ultrasonics D-Reactor modified in-house and operating at 515 kHz (Henglein et al. 1989). The modifications to the instrument allowed the delivery of pulsed ultrasound with a variable pulse-on time between 0.1 and 25 ms and pulse-off time between 0.8 ms and 6 s. The reactor was placed in a light-shielded housing, which also contained an end-on photomultiplier (Hamamatsu) for recording the SL from the insonated solutions (Ashokkumar et al. 1997).

In a series of experiments, the SL intensities at different pulse-off times are measured. As an example, Fig. 3.1a shows individual intensity versus time curves for different T_0 experiments, averaged over approximately 250 pulses and normalized to the SL intensity at T_0 = 12 ms.

Furthermore, Fig. 3.5b shows the relative SL intensities at pulse-off times ranging from 12 to 408 ms for water and SDS.

From Fig. 3.5a, we see that for T_0 = 12 ms or T_0 = 200 ms, the SL intensity increases until reaching its maximum value within 4 ms and then decreases rapidly, indicating that a steady-state active bubble population has been reached within the 4-ms pulses. However, we do not see this for T_0 = 250 ms and T_0 = 290 ms. In these cases, the steady-state active bubble population was not reached because these T_0 times are so long that some bubbles have dissolved during the pulse-off times and cannot recover to the active state by the time the next pulse-on occurs. Thus, whether or not the steady-state active bubble population can be achieved strongly depends on T_0. In Fig. 3.5b, we see that the SL intensity initially remains constant when the pulse-off time was below 200 ms. As the pulse-off times continue to increase, the SL signal decreased until it was no longer detectable at a time of about 408 ms. This phenomenon reveals that when the pulse-off time exceeds a critical

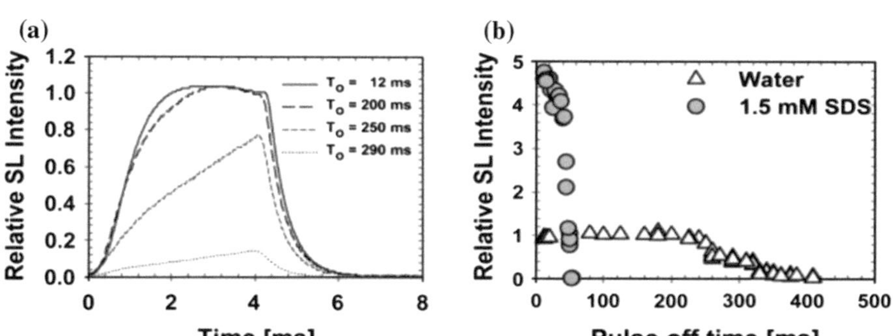

Fig. 3.5 **a** Individual SL pulse shapes in water averaged over ~250 pulses for different pulse-off times (T_0). **b** Relative SL intensity as a function of pulse-off time for water and 1.5 mm SDS. Reproduced with permission from Lee (2005). © American Chemical Society

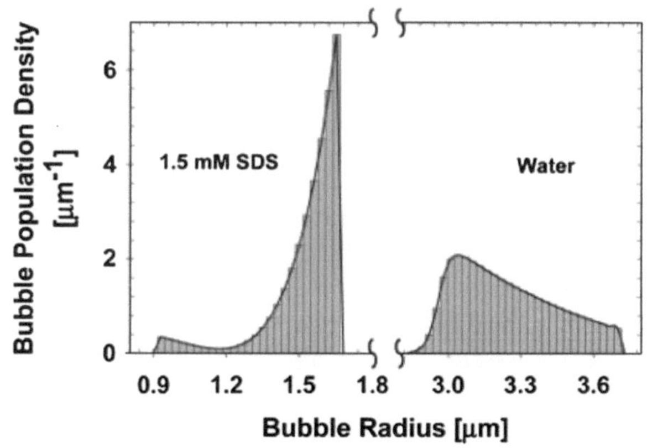

Fig. 3.6 The size distribution of SL bubbles for water and 1.5 mM SDS; the bubble radius interval is 0.03 μm. Reproduced with permission from Lee (2005). © American Chemical Society

value, some bubbles dissolve too much to recover their active population size, which causes the SL intensity to decrease.

According to the data (Fig. 3.5) measured in experiments and the analysis above, the size distribution of SL bubbles in water can be achieved using a sigmoid equation with five parameters to fit the decay of the SL intensity. For SDS, a cubic fit was used. The results are shown in Fig. 3.6.

It is important to notice that only a monotonic decay of the SL signal that is observed for large pulse-off time (T_0) can be explained by this model. That is, the mathematical model proposed in the SL method cannot solve the situation when the SL intensity increases with an initial increase in the pulse-off time, which the developers of this method encountered in their follow-up work (Brotchie et al. 2009). Therefore, recently, a modified methodology has been proposed to settle this problem.

3.4 PCD-Based Cavitation Noise Method

In this section, we introduce a modified methodology (Hauptmann et al. 2013) to evaluate the active bubble size distribution by means of measuring the ultraharmonic cavitation noise (UHCN). UHCN signals can be utilized when SL measurements are difficult to carry out. This is generally the case at lower powers, at which SL does not occur anymore. This modified method is centered around defining an upper size limit, which corresponds to when insufficient bubble dissolution causes bubbles to grow beyond the active size range during subsequent pulses. Hence, if the pulse-off time is too short for larger bubbles to dissolve back to their original size, they will subsequently grow out of the active size distribution. As a result, they will eventually become inactive and subsequently get removed from the liquid. On the other hand, for smaller bubbles, the same pulse-off time

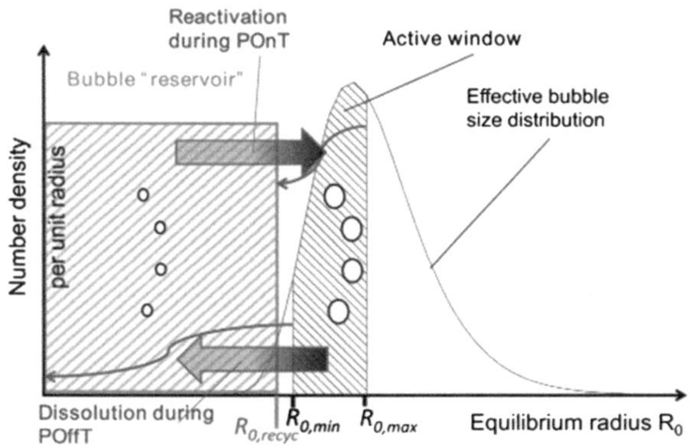

Fig. 3.7 Scheme of the bubble recycling mechanism. The upper size limit of the bubble reservoir corresponds to an upper threshold parameter $R_{0,\mathrm{recyc}}$ for the fitting of the model distributions to the experimental data. The lower size limit is set to zero. Reproduced with permission from Hauptmann (2013). © AIP Publishing

might allow those smaller bubbles to undergo total dissolution. The resulting situation is depicted in Fig. 3.7.

In Fig. 3.7, bubbles that are part of the "active" window at the end of the pulse-on time (right black hatched area) dissolve back to sizes corresponding to the depicted bubble "reservoir" (left gray hatched area) during the pulse-off time. During the next pulse-on time, these bubbles can grow again (being "reactivated") and thus contribute to the CN signal. Over a series of pulses, this leads to an accumulation of cavitation activity up to a certain steady state, as more and more bubbles become involved in the dissolution–reactivation cycle between the reservoir and the active window. For different pulse-off times, this active window corresponds to different equilibrium radii. Measuring CN as a function of the pulse-off time (for a fixed pulse-on time, acoustic intensity, gas content, etc.) can be used to map the effective bubble size distribution.

This experiment was performed in an ultrasonic cleaning system with a volume of 13 L. The tank was supplied with ultrapure water (UPW) at a flow rate of 6.5 L/min. The acoustic field was emitted from a flat plate transducer (ProSys, Inc.) that was attached to one side of the tank. The transducer has a surface area of 59.6 cm^2 and was driven at its resonant frequency of 928 kHz through a combination of a programmable function generator (HP 33120A) and a commercial megasonic amplifier (based on a modified KMA2040 amplifier, AR Worldwide) (Fig. 3.8).

Cavitation activity was measured using a needle hydrophone (Onda Corp., HNR-0500). The hydrophone signal was amplified using a homemade amplifier based on the AD8066 FastFET operational amplifier (Analog Devices) (Hauptmann et al. 2012).

The hydrophone was scanned for 45 s through the main acoustic field along a vertical line perpendicular to the propagation direction of the acoustic waves.

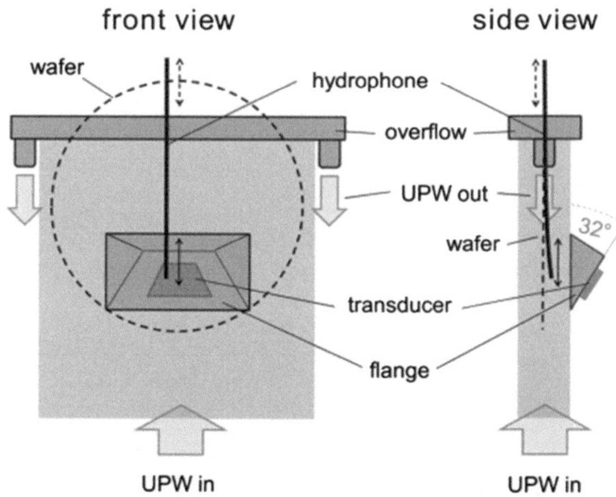

Fig. 3.8 Schematic *front-* and *side-view* depiction of the megasonic setup, in which the hydrophone was scanned. Reproduced with permission from Hauptmann (2013). © AIP Publishing

The scanned distance amounted to 9 cm, which was slightly larger than the width of the source transducer. The liquid was sonicated in a pulsed fashion, at a nominal acoustic intensity of 0.33 W/cm^2 and a constant pulse-on time of 10, 50, and 250 ms, respectively, while the pulse-off time was varied between 0 and 1 s.

The effective bubble size distribution can be deduced from pulse-off time versus the UHCN data. However, to obtain this distribution, a model distribution must be fitted to the UHCN versus pulse-off time data. Here, the Weibull distribution was used (Hauptmann et al. 2013):

$$f(R_0) = \frac{\partial}{\partial R_0} F(R_0) = \frac{\partial}{\partial R_0} \left(1 - \exp\left(-\left(\frac{R_0}{\lambda}\right)^k\right)\right)^\alpha \tag{3.13}$$

The CN signal intensity is assumed to be proportional to the (steady state) number of active bubbles within the active window. Thus, the CN intensity for a certain pulse-off time is proportional to the integration of the bubbles present over the corresponding active window. So, Eq. (3.13) links the CN to the bubble size distribution (Hauptmann et al. 2013):

$$\begin{aligned} \mathrm{UHCN(POffT)} &= C \cdot \int\limits_{R_{0,\min(\mathrm{POffT})}}^{R_{0,\max(\mathrm{POffT})}} f(R_0)\mathrm{d}R_0 + \Delta\mathrm{UHCN(POffT)} \\ &= C \cdot \left(1 - \exp\left(-\left(\frac{R_0}{\lambda}\right)^k\right)\right)^\alpha \Bigg|_{R_{0,\min(\mathrm{POffT})}}^{R_{0,\max(\mathrm{POffT})}} \\ &\quad + \Delta\mathrm{UHCN(POffT)} \end{aligned} \tag{3.14}$$

The three distribution parameters λ, k, and α represent the width, skewness, and center position of the distribution function, respectively. This yields—taking into account an additional factor C and the upper reservoir limit size $R_{0,recyc}$—five fitting parameters. $R_{0,min}$ is the smallest radius of a bubble that reduces completely to the lower boundary of the reservoir (set as 0 μm here); $R_{0,max}$ is determined by $R_{0,recyc}$, which defines the upper limit of the active window. $R_{0,min}$ and $R_{0,max}$ can be deduced from the bubble dissolution equation. By fitting Eq. (3.14) with the experimental UHCN versus pulse-off data using nonlinear least-squares fitting to minimize the square sum of the residues ΔUHCN (POffT), we can obtain the values for λ, k, α, $R_{0,recyc}$, and C.

Figure 3.9 shows the experimental measurement results along with the model fits. For a pulse-on time of 10 ms, the pulse-off time range over which a signal can still be observed is narrow, while the overall signal strength is low compared to when a pulse-on time of 50 ms is used. For a pulse-on time of 250 ms, the pulse-off time range is very broad and contains UHCN signals that are considerably higher than in the 10 ms case, but they are still lower than for when the pulse-on time is 50 ms. These trends are reflected by the resulting bubble size distributions displayed in Fig. 3.9b, which show the calculated effective bubble size distribution for each pulse-on time. For longer pulse-on times, the obtained bubble size distributions correspondingly extend to contain larger bubble radii. For instance, the bubble size distribution for a pulse-on time of 10 ms cuts off rather sharply at a linear resonance radius of 3.4 μm, whereas the bubble size distribution for a pulse-on time of 250 ms spreads out to contain bubble radii larger than twice this resonance radius.

Clearly, a longer duration of the sonication per pulse leads to larger bubble radii, as reflected by the data in Fig. 3.9. These data suggest that there exists an optimal pulse-on time for which the measured UHCN signals are the strongest, and this optimal pulse-on time occurs when the bubble size distribution is centered on the linear resonance size $R_{0,recyc}$. If the pulse-on time is too short, then not all partially dissolved bubbles will be able to be reactivated. Conversely, if the pulse-on time is too long, bubbles might grow too big to be recycled, and cavitation activity will be

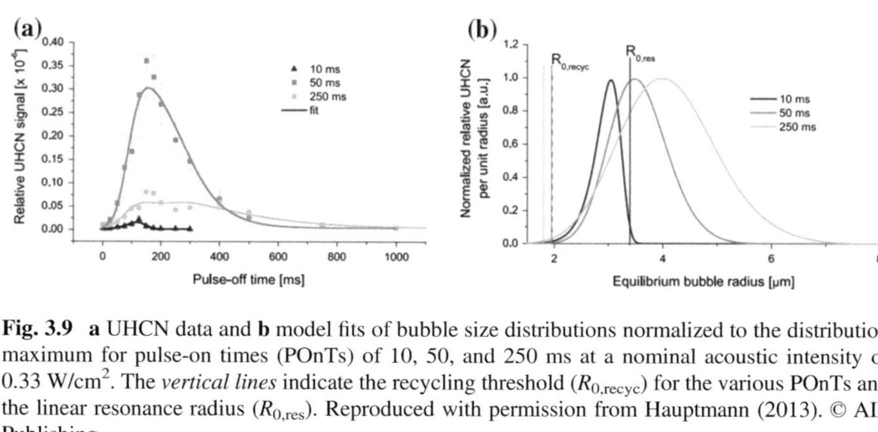

Fig. 3.9 **a** UHCN data and **b** model fits of bubble size distributions normalized to the distribution maximum for pulse-on times (POnTs) of 10, 50, and 250 ms at a nominal acoustic intensity of 0.33 W/cm^2. The *vertical lines* indicate the recycling threshold ($R_{0,recyc}$) for the various POnTs and the linear resonance radius ($R_{0,res}$). Reproduced with permission from Hauptmann (2013). © AIP Publishing

diminished. This demonstrates again that introducing an upper size limit in the model to account for the loss of bubbles due to growth is very important. Notably though, the SL method discussed in the previous section represents a special case in this model where $R_{0,\mathrm{recyc}}$ is located at infinity, effectively ignoring bubble loss due to growth.

Up to now, we have discussed methods to determine the equilibrium bubble size distribution when an ultrasound pulse is on. However, another way to determine size distribution is to evaluate what happens after the pulse has been removed. Thus, the next three methods we discuss determine the equilibrium size distribution of bubbles by measuring the bubble dissolution process after the pulse has been removed.

3.5 Void Rate Method by Shifting the Electromagnetic Resonant Frequency

The presence of cavitation bubbles in a resonant electromagnetic cavity will induce a shift of the cavity electromagnetic resonant frequency. The relative shift in electromagnetic resonant frequency caused by the volume of one bubble is proportional to both the bubble volume, expressed in terms of void rate, and the square of the electric field amplitude at the bubble position (Labouret and Frohly 2002):

$$\left(\frac{\delta f_\gamma}{f_\gamma}\right)_{1\mathrm{bulle}} = A \iiint_{V_{\mathrm{bulle}}} |E(x,y,z)|^2 \mathrm{d}V \tag{3.15}$$

where A is a constant determined by liquid and gas permittivity and by the electromagnetic resonant mode, and δf_γ is the resonant frequency shift in the electromagnetic cavity.

The total electromagnetic resonant frequency shift caused by the whole bubble cloud is the sum of the frequency shifts caused by each bubble.

Incorporating the radius evolution law into the initial size distribution, we can calculate the total bubble volume measured in the cavity at the instant t (Labouret and Frohly 2002):

$$V_{\mathrm{t}}(t) = \int_0^{+\infty} n(R_0)\frac{4}{3}\pi R^3(R_0,t)\,\mathrm{d}R_0 \tag{3.16}$$

where R_0 is the initial radius of bubble; R (R_0, t) is the radius at instant t of the bubble with an initial radius of R_0; and n (R_0) is the number of bubbles whose radii are included within the [R_0, R_0 + dR_0] range.

If the liquid is gas saturated, taking the derivative of Eq. (3.16) with respect to time and neglecting the term $(1/\pi \mathrm{D}t)^{1/2}$ produces the following:

$$\frac{\mathrm{d}V_{\mathrm{t}}(t_1)}{\mathrm{d}t} = -\psi \int_{R_{0\mathrm{d}}(t_1)}^{+\infty} n(R_0)\mathrm{d}R_0 \tag{3.17}$$

where ψ is defined as follows (Labouret and Frohly 2002):

$$\psi = 8\pi \frac{C_{\mathrm{s}} D\sigma}{\rho_{\mathrm{g}} P_0} \tag{3.18}$$

Here, $R_{0\mathrm{d}}$ (t) is the initial radius of a bubble whose complete dissolution duration is t, and it replaces 0 as the lower boundary.

We can then deduce the number of bubbles that have completely dissolved between two instants by using the variation of the void rate dissipation speed between the two instants. For example, if $t_1 < t_2$ and we use Eq. (3.17), we find the following (Hauptmann et al. 2013):

$$\begin{aligned} \frac{\mathrm{d}V_{\mathrm{t}}(t_1)}{\mathrm{d}t} - \frac{\mathrm{d}V_{\mathrm{t}}(t_2)}{\mathrm{d}t} &= -\psi \left(\int_{R_{0\mathrm{d}}(t_1)}^{+\infty} n(R_0)\mathrm{d}R_0 - \int_{R_{0\mathrm{d}}(t_2)}^{+\infty} n(R_0)\mathrm{d}R_0 \right) \\ &= -\psi \int_{R_{0\mathrm{d}}(t_1)}^{R_{0\mathrm{d}}(t_2)} n(R_0)\mathrm{d}R_0 \end{aligned} \tag{3.19}$$

where the boundaries of the integral are:

$$R_{0\mathrm{d}}(t_1) = \sqrt[3]{\frac{3}{4\pi}\psi t_1} \quad \text{and} \quad R_{0\mathrm{d}}(t_2) = \sqrt[3]{\frac{3}{4\pi}\psi t_2} \tag{3.20}$$

Thus, using Eqs. (3.19) and (3.20), we can calculate the number of bubbles whose initial radii were within the [$R_{0\mathrm{d}}$ (t_1): $R_{0\mathrm{d}}$ (t_2)] interval. By reproducing this calculation for other instants of time, we can theoretically achieve a distribution of the bubble radii in the cavity after the ultrasonic irradiation has been removed.

The experimental setup is shown in Fig. 3.10. The electromagnetic cavity is a $6 \times 6 \times 3$ cm^3 rectangular parallelepiped. A transducer is glued onto the upper cavity, driven by a power amplifier (ENI, A300-1381) and a function generator. The transducer emits a 308 kHz frequency and is as large as the cavity base. An antenna is fixed on the bottom base of the cavity, which enables it to work as a resonant electromagnetic resonance cavity. The electric field is theoretically constant with respect to the height of the cavity and only depends on the variables x and z. In addition, the metallic bellow attached to the bottom of the foundation maintains the cavity's tightness.

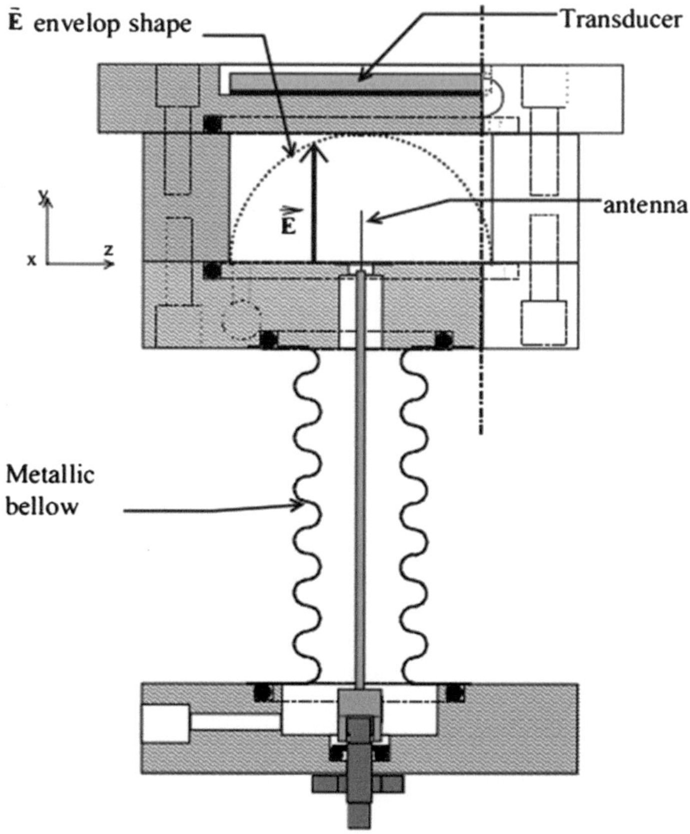

Fig. 3.10 The experimental setup showing the electromagnetic cavity, antenna, and metallic bellow. Reproduced with permission from Labouret (2002). © EDP Sciences

The void rate evolution was achieved under a 30-s acoustic irradiation with a 0.21-bar acoustic pressure in air-saturated water. The observed void rate evolution is displayed in Fig. 3.11.

From the figure, we can see that the void rate increases rapidly once the acoustic irradiation starts, and it reaches a maximum when the ultrasound is switched off. Once the acoustic irradiation "interruption" stops, the void rate decreases exponentially until it becomes zero.

After the 30-s "interruption" of acoustic irradiation, the void rate dissipates due to bubble dissolution. More precisely, the size distribution of the bubbles relates to the concavity of the void rate dissipation curve. For dissolving bubble clouds, the concavity is positive. We can observe this concavity in more detail during the first 60 s after the ultrasound is switched off, as depicted in Fig. 3.12. As seen here, during the first 0–15 s, the void rate dissipation curve has a positive concavity (concave), and then, it becomes temporary negative (convex) until 40 s, after which

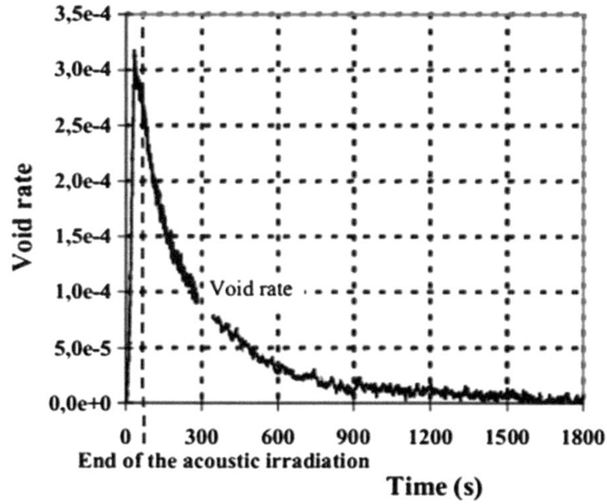

Fig. 3.11 Void rate evolution: acoustic pressure, 0.21 bar; acoustic irradiation duration, 30 s; air-saturated water at 20 °C. Reproduced with permission from Labouret (2002). © EDP Sciences

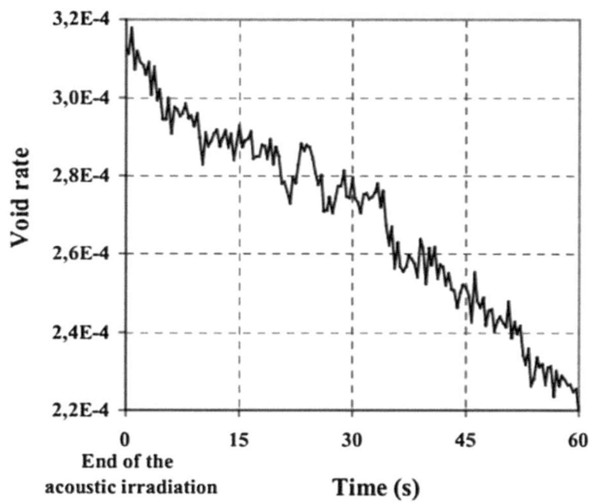

Fig. 3.12 The void rate dissipation curve during the first 60 s after the ultrasound was switched off. Reproduced with permission from Labouret (2002). © EDP Sciences

it becomes positive again. Thus, this method can only start analyzing void rate dissipation 40 s after the acoustic irradiation "interruption" stops.

In Fig. 3.13, we can see the size distribution of bubbles measured from the void rate dissipation curve. The bubble sizes (radii) range from 19 to 72 μm in 2-μm increments. This size distribution does not encompass bubbles with radii <19 μm because they dissolve completely within the first 40 s. Similarly, the distribution does not include bubbles with radii >59 μm because after 1000 s, the variations in the void rate are no longer large enough compared to the measurement noise.

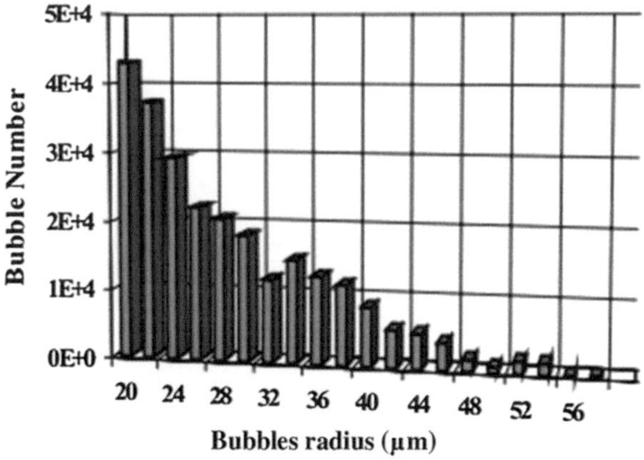

Fig. 3.13 Size distribution of bubbles calculated from the void rate dissipation curve with a step of 2 μm. Reproduced with permission from Labouret (2002). © EDP Sciences

3.6 ACD-Based Scattering Method

According to the kinetics of free bubbles, the radius of a bubble at any time can be calculated. After the instantaneous radius is known, we can also find the scattering cross section of a bubble (σ_s) as a function of time (Chen et al. 2002) via Eq. (3.21), based on the assumptions that the wavelength $\lambda \gg R$.

$$\sigma_s = 4\pi R^2 \frac{\Omega^4}{\left(\Omega^2 - 1\right)^2 + \Omega^2 \delta^2} \tag{3.21}$$

where $\Omega = \omega/\omega_0$, δ is the total damping constant, ω is the frequency, and ω_0 is the resonant frequency. For more details about the damping factors, see (Church 1995; Hoff et al. 2000).

After obtaining the scattering cross section versus time, we can then determine the expected scattering signal amplitude, which is proportional to the square root of the scattering cross section (σ_s).

If the region of interest contains a population of bubbles, the scattering cross section will be given by:

$$\sigma_f = W(R_0)\sigma_s H(f) \tag{3.22}$$

where $W(R_0)$ is the size distribution of the bubbles, which is generally assumed as a Gaussian function with a mean radius of R_{mean} and a variance represented by the standard deviation (SD) for simplicity, and $H(f)$ is the frequency response of the system, which is determined by transducer characteristics.

Next, by comparing the theoretical square root of the scattering cross section versus time with the scattered signal intensity obtained from the experiment, we can estimate the size distribution of the bubbles (Chen et al. 2002).

All of these experiments have been performed in an acrylic tank containing deionized and filtered water. The nuclei examined in the experiment were the free bubbles released from contrast agents—including biSphere (Point Biomedical, San Carlos, CA) or Optison (Amersham Health, Princeton, NJ)—diluted to 1–5 μL/L. The biSphere contrast agent is an experimental agent that contains air enclosed by an inner polymer shell and an outer albumin shell. Conversely, Optison is commercially available and contains octofluoropentane (OFP) gas enclosed within an albumin shell. These two shelled agents rupture and release their gas bubbles when insonified with high-intensity acoustic pulses beyond a certain pressure threshold.

The experimental setup is shown in Fig. 3.14. Mounted onto one side of the tank is a 1.1-MHz (7.0 cm diameter, 5.5 cm focal distance) single-element, high-intensity focused ultrasound (HIFU) transducer, which is used to destruct the contrast agents within a focal region. Two identical 5-MHz bistatic transducers (1.2 cm diameter, 4.6 cm focal length) constitute the active cavitation detection system: One transducer interrogates the HIFU focal region with low-amplitude acoustic pulses, and the other collects the scattering signals from the focal zone. The measured −3-dB profile of these transducers is approximately 0.4 cm wide by 5 cm long.

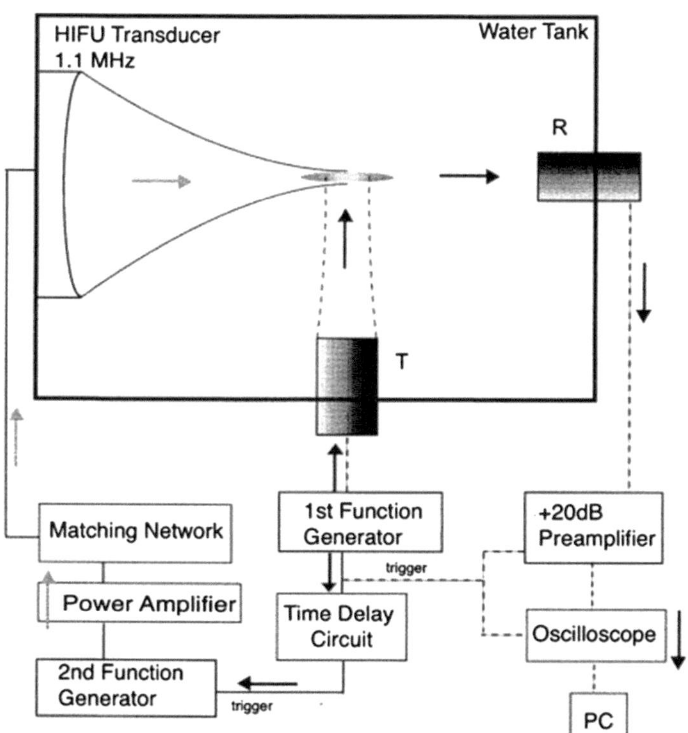

Fig. 3.14 The experimental setup. All three transducers are focused on the same region; the circuit below the reactor is used to control transducers and data processing. Reproduced with permission from Chen (2002). © Elsevier B.V.

The experiment consists of two independent experiments: one using the biSphere contrast agent and the other using Optison. For each experiment, the shelled contrast agent was first injected into the water and stirred into solution using a low-speed magnetic stirrer to distribute the agent.

The pulse-repetition frequency (PRF) for the HIFU transducer was set at 2–10 Hz. Each HIFU pulse was followed by 500–2500 interrogating pulses (PRF = 5 kHz). A time delay circuit was used to synchronize each device and provide the necessary delay.

In the experiment, the HIFU transducer emitted a 1.1 MHz frequency, high-amplitude burst to destruct the contrast agent, releasing its interior gas bubbles. Interrogating pulses were then turned on to detect the released bubbles. To ensure that the bubbles remained relatively stationary over short timescales, the pulse length of the HIFU transducer is set to five cycles, which minimized the possibility of streaming. For Optison agents, a slightly different protocol was used because it contains a less-soluble gas, so detecting it is more difficult. In the Optison protocol, the HIFU PRF was reduced to 1 Hz and the PRF of the interrogating pulse was 500 Hz.

For data analysis, the scattering signals were first received by one of the bistatic transducers, and the recorded waveforms were fitted to simulations based on the kinetics of free gas bubbles. The simulations used the initial radius R_0, size distribution $W(R_0)$, SD of the distribution, and total relative number of bubbles as fitting parameters.

The scattering signals of the bubbles released from biSphere are shown in Fig. 3.15. A steady, high baseline signal indicates that the bubbles were in the focus of the HIFU bistatic detector system. After the first HIFU pulse, the signal immediately jumped to a high level (labeled A), followed by a quickly decreasing amplitude that stabilized near the baseline value. Then, the signal slowly decreased for about 20 ms before undergoing a secondary increase, followed by decay to the noise background level.

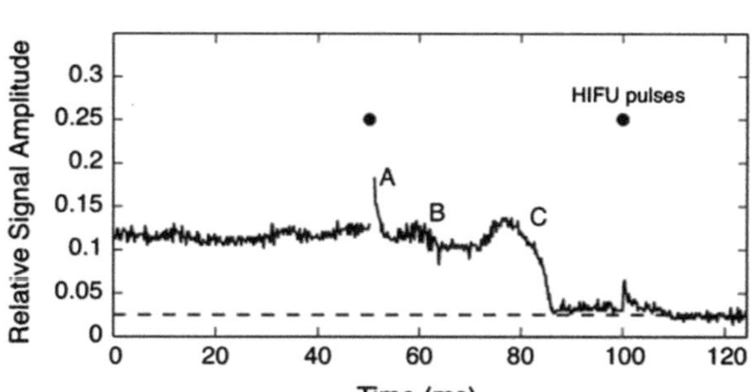

Fig. 3.15 The scattered signal from biSphere bubbles before and after the contrast agent spheres were destroyed. The main signal features are labeled as *A*, *B*, and *C*. Reproduced with permission from Chen (2002). © Elsevier B.V.

Assuming that each major feature (labeled A, B and C) in Fig. 3.15 is associated with independent scattering from a particular bubble size distribution, we can estimate these individual size distributions of bubbles, as shown in Fig. 3.16, where the individual size distributions are labeled (1a), (2a), and (3a). By summing the individual size distributions, we obtain the total size distribution of the bubbles, as shown in Fig. 3.16(4a), (4b).

The scattering signals of the bubbles released from Optison were also recorded, but the OFP bubble dissolution was not observable because the dissolution time of OFP was so long that the bubbles moved past the focal region of the interrogating transducer during these long periods. Thus, the bubble size distribution could not be determined.

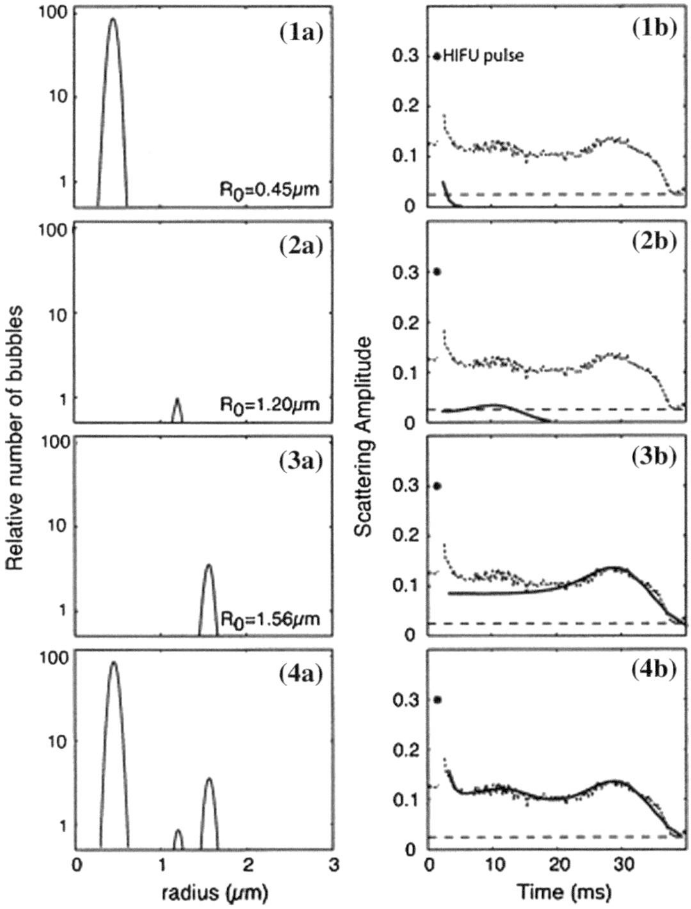

Fig. 3.16 **a** The bubble size distribution of bubbles for each particular feature. **b** The associated scattering amplitude from each bubble size distribution and the data curve obtained from the experiment. Reproduced with permission from Chen (2002). © Elsevier B.V.

3.7 Wide-Beam ACD-Based Backscattering Method

The method proposed in the previous section can estimate the size distribution of bubbles released from encapsulated ultrasound contrast agents using a tightly focused ultrasound (FUS) transducer. However, it cannot determine the size distribution of bubbles that have long dissolution times, such as bubbles containing less-soluble gasses. As mentioned above, during the long dissolution period, these bubbles can move out of the focal region of the tightly focused transducer. Here though, we introduce an improved method (Xu et al. 2013) using a wide-beam ultrasound with low pressure to acquire the time-intensity curve (TIC) of the dissolution process for a cavitation bubble population, after which the size distribution of the bubbles can be determined.

The improved method for measuring the size distribution of cavitation bubbles is based on experimentally measuring the backscattering signals from a bubble cloud. If we ignore the interaction between bubbles and take the concentration of bubbles to be sufficiently low to prevent the acoustic shielding effect, we can express the intensity of backscattering signals at any time (which are attributed to the bubbles that have not yet completely dissolved) by Eq. (3.23) (Xu et al. 2013).

$$I(t) = \sum_{i=1}^{m} n(R_{0\mathrm{d}}(t_\mathrm{i}))\sigma_\mathrm{s}(R_{0\mathrm{d}}(t_\mathrm{i}), t) \tag{3.23}$$

where $R_{0\mathrm{d}}(t)$ is the initial radius of a bubble whose complete dissolution duration is t. If we divide the size of a bubble cloud into m, t_1 is the complete dissolution time of a bubble with the minimum initial radius in the bubble cloud, t_m is the complete dissolution time of a bubble with the maximum initial radius in the bubble cloud, $n(R_{0\mathrm{d}}(t_\mathrm{i}))$ is the number of bubbles with an initial radius of $R_{0\mathrm{d}}(t_\mathrm{i})$, and $\sigma_\mathrm{s}(R_{0\mathrm{d}}(t_\mathrm{i}), t)$ is the acoustic scattering cross section of a bubble with the initial radius $R_{0\mathrm{d}}(t_\mathrm{i})$ at time instant t. The bubble radius at the end of the FUS exposure was considered to be the initial radius.

According to Eq. (3.23), the population of bubbles with different initial radii (namely, a size distribution of bubbles) can be inversely calculated from the measured TIC combined with the dissolution kinetics of the cavitation bubbles in the liquid.

A concave and spherically focused single-element transducer (1.2 MHz, focal width 1.6 mm, focal length 8 mm and aperture diameter 15 cm, Imasonic, Besancon, France) driven by a power amplifier (AG1017, T&G Power Conversion Inc., Rochester, NY, USA) and a dual-channel arbitrary waveform generator (AWG 420, Tektronix, Beaverton, OR, USA) was mounted onto one side of the tank, as illustrated in Fig. 3.17a, which was used to initiate cavitation. A 5-MHz linear array (L14-5/38, Ultrasonix, Richmond, Canada) aligned vertical to the axis of the FUS transducer was used to detect the dissolution of the cavitation bubble population in plane wave transmitting and receiving mode (Sonix DAQ, Ultrasonix, Richmond, Canada) at around 90 kPa (Fig. 3.17b), which is sufficiently weak to prevent

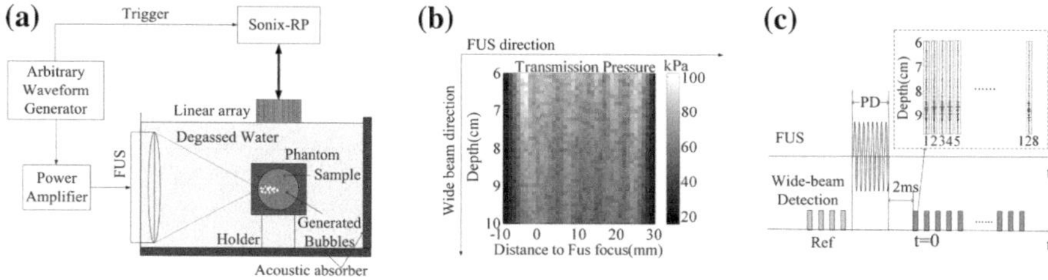

Fig. 3.17 **a** Experimental setup for the experiments performed in undegassed saline. **b** The acoustic pressure distribution of wide-beam ultrasound used to detect cavitation bubbles. **c** Synchronization sequence inserted with an example of prebeamformed radio frequency (*RF*) data of the first frame received in each channel. Reproduced with permission from Xu (2013). © Elsevier B.V.

acoustic-driven diffusion, yet strong enough to maintain a reasonable signal-to-noise ratio.

The acoustic pressure distribution of the interrogating wide beam was measured in a degassed water tank with a needle hydrophone (0.5 mm diameter, Precision Acoustics Ltd., Dorchester, UK) when the excitation voltage of the ultrasound scanner (Sonix RP, Ultrasonix, Richmond, Canada) was 18 V (this voltage was applied through all experiments in this study). The linear array was synchronized with a pulser–receiver (5800 PR, Panametrics Inc., Waltham, MA, USA), and the hydrophone output was connected to the receiver input of the emitter/receiver, amplified, and digitized at 100 ms/s with the resolution of 8 bits (Qiao et al. 2012).

The experiment was performed in a tank filled with degassed water, in which an acoustically transparent phantom containing undegassed saline was placed at the focus region of the concave and spherically focused single-element transducer.

In the experiment, the FUS transducer was first used to transmit an acoustic pulse to initiate cavitation at the focus region. The wide beam transmitted from the imaging transducer was started 2 ms after the FUS was turned off, and it was applied at a PRF of 100 Hz, as shown in Fig. 3.17c, to detect the intensity of backscattering signals from the bubble cloud. The intensity, defined as the integral of the squared amplitude of the noise-subtracted prebeamformed RF data from each channel, was calculated for each frame, giving a TIC. The experimental TIC, which was under an acoustic duration of 20 μs and an acoustic power of 60 W, is shown in Fig. 3.18a.

Here, the 2-ms delay between the FUS and the start of wide-beam detection ensured that reverberations from the FUS would not interfere with wide-beam detection until after 2 ms, and the PRF of 100 Hz was deemed optimal for tracking rapidly dissolving cavitation bubbles of low density while also preventing acoustic-driven diffusion.

After the experimental TIC was acquired, the solution of Eq. (3.23), namely the distribution of the bubble cloud, was calculated. First, the complete dissolution time $t_{\mathrm{m}}, t_{\mathrm{m-1}} \ldots t_1$ of bubbles with initial radii of $R_{0\mathrm{d}}\ (t_{\mathrm{m}}), R_{0\mathrm{d}}\ (t_{\mathrm{m-1}}) \ldots R_{0\mathrm{d}}\ (t_1)$ was

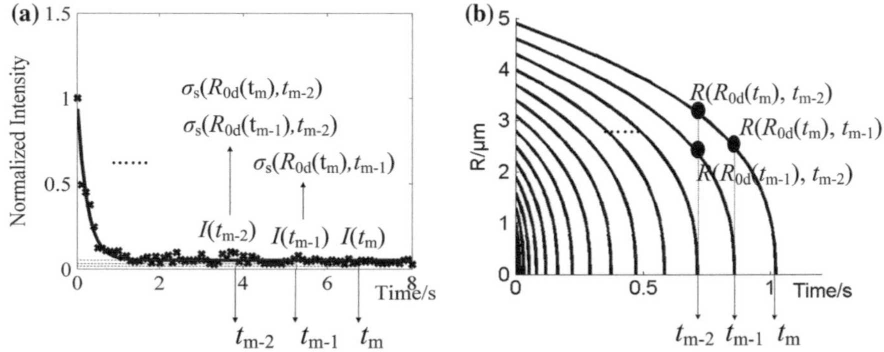

Fig. 3.18 **a** Diagram depicting the contribution of bubble populations with different initial radii to intensities of TIC at specific times; the data shown as *crossed dots* are the normalized intensities of the backscattering signals from the bubbles, and the *solid line* is the corresponding fitted curve. **b** Predicted time evolutions of air bubble radii during their dissolution in air-saturated water, with the *black solid dots* indicating the reached radii of larger bubbles when the smaller bubbles had dissolved completely. Reproduced with permission from Xu (2013). © Elsevier B.V.

derived from the dissolution kinetics of cavitation bubbles. The intensities on the TIC at these times were labeled as $I\ (t_{\mathrm{m}})$, $I\ (t_{\mathrm{m}-1})$ … $I\ (t_1)$, as shown in Fig. 3.18a.

It is reasonable to presume that $I\ (t_{\mathrm{m}-1})$ can be attributed to the bubbles with an initial radius of $R_{0\mathrm{d}}\ (t_{\mathrm{m}})$, as illustrated in Fig. 3.18a. Assuming that m is sufficiently large and that the backscatter of bubbles within the initial radius range of $[R_{0\mathrm{d}}(t_{\mathrm{m}-1}),\ R_{0\mathrm{d}}(t_{\mathrm{m}})]$ is not significantly different from that of $R_{0\mathrm{d}}\ (t_{\mathrm{m}})$, then Eq. (3.24) follows (Xu et al. 2013):

$$I(t_{\mathrm{m}-1}) = n(R_{0\mathrm{d}}(t_{\mathrm{m}}))\sigma_{\mathrm{s}}(R_{0\mathrm{d}}(t_{\mathrm{m}}), t_{\mathrm{m}-1}) \tag{3.24}$$

Based on the dissolution kinetics of a bubble in liquid, $\sigma_{\mathrm{s}}\ (R_{0\mathrm{d}}(t_{\mathrm{m}}),\ t_{\mathrm{m}-1})$ can be calculated using the radius $R_{0\mathrm{d}}(t_{\mathrm{m}})$ as the initial bubble radius at time $t_{\mathrm{m}-1}$, as illustrated in Fig. 3.18b. Then, $n\ (R_{0\mathrm{d}}(t_{\mathrm{m}}))$ can be obtained from Eq. (3.24). As shown in Fig. 3.18a, this leads to (Xu et al. 2013):

$$\begin{aligned} I(t_{\mathrm{m}-2}) = {} & n(R_{0\mathrm{d}}(t_{\mathrm{m}}))\sigma_{\mathrm{s}}(R_{0\mathrm{d}}(t_{\mathrm{m}}), t_{\mathrm{m}-2}) \\ & + n(R_{0\mathrm{d}}(t_{\mathrm{m}-1}))\sigma_{\mathrm{s}}(R_{0\mathrm{d}}(t_{\mathrm{m}-1}), t_{\mathrm{m}-2}) \end{aligned} \tag{3.25}$$

Here, $n(R_{0\mathrm{d}}(t_{\mathrm{m}-1}))$ can be easily determined using the previously obtained $n\ (R_{0\mathrm{d}}(t_{\mathrm{m}}))$ and calculations of the relevant scattering cross sections. Finally, we can get $n(R_{0\mathrm{d}}(t_{\mathrm{m}}))$, $n(R_{0\mathrm{d}}(t_{\mathrm{m}}))$ … $n(R_{0\mathrm{d}}(t_2))$, from which we can calculate the size distribution of the cavitation bubble cloud.

According to the TIC measured from the experiment (Fig. 3.18a), we can see the bubble size distribution in Fig. 3.19.

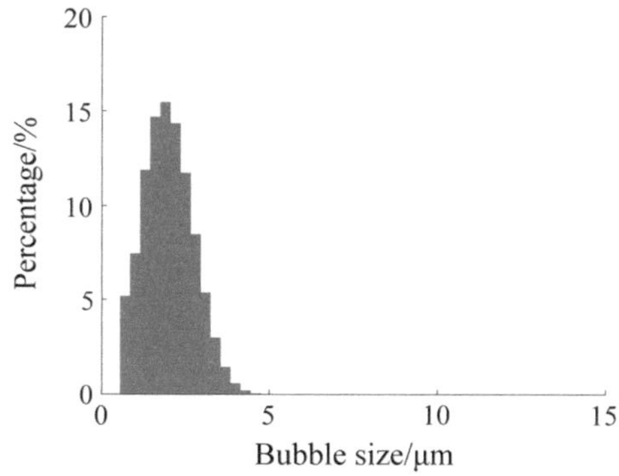

Fig. 3.19 The size distribution of bubbles estimated from experimental data. Acoustic power, 60 W; acoustic irradiation duration, 20 μs; air-saturated saline at room temperature under atmospheric pressure. Reproduced with permission from Xu (2013). © Elsevier B.V.

It is clear that the size distribution follows a Gaussian distribution. From the figure, we see that the mean bubble size of the cavitation bubble cloud, under sonication with an acoustic pulse duration of 20 μs at an acoustic power of 60 W, was 1.9 μm.

3.8 Effects of Acoustic Pressure, Frequency, and Duration

In biological and industrial applications, such as histotripsy, sonothrombolysis, and ultrasonic cleaning, periodically switching a sound field on and off (pulsed ultrasound) can increase the efficiency of cavitation-enhanced applications (Ashokkumar 2011). This approach works by stabilizing the bubble size distribution within the active radius range, since active bubbles oscillate strongly as a response to the applied acoustic pressure. According to Yasui et al. (2008), the active radius range spans from the Blake threshold to radii on the order of linear resonance radii. However, Hauptmann et al. (2013) reported the strongest cavitation activity when the bubble size distribution was centered on the linear resonance size. Overall, the goal of all these methods is to experimentally predict the size distribution of the bubbles as feedback in order to potentially optimize the experimental parameters so the bubble size distribution could be controlled to remain within the active range. To achieve this control, the experimental parameters to optimize are the acoustic pulse amplitude, the pulse duration, and also the frequency.

In the previous sections, we introduced several acoustic methods for estimating the size distribution of cavitation bubbles. However, we have not yet discussed how source ultrasound parameters, such as ultrasonic frequency, pressure, and irradiate duration, affect the size distribution of cavitation bubbles. We examine these effects in the next section.

3.8.1 The Effect of Acoustic Pressure and Power

Brotchie et al. (2009) studied the effect of acoustic power on the size distribution of cavitation bubbles arising in acoustic cavitation. Figure 3.20 shows the mean bubble radius as a function of acoustic power at a constant acoustic frequency of 1056 kHz.

From the figure, we see that the mean bubble radius increases with increasing acoustic power and reaches the maximum, approximately 4.5 μm, at an acoustic power of ~6 W. Other studies performed at other frequencies also observed this trend of increasing bubble size with higher power. In addition, Matula (1999) verified this trend in a single-bubble system. However, Yasui et al. (2008) suggested that the bubble radius should decrease with an increase in acoustic pressure because multi-bubble studies showed that higher acoustic powers produce stronger nonlinear pulsations. Also, another study showed that no change in bubble radius occurred with a 50 % change in acoustic power (Lee et al. 2005). Furthermore, in a study by Xu et al. (2013), acoustic power was found to hardly affect the bubble size distribution at all. However, this particular study employed an acoustic power at a frequency of 1.2 MHz and a pulse duration of 20 μs. Thus, the acoustic powers in this study were much larger than those Brotchie et al. (2009) used.

When acoustic power increases, the negative acoustic pressure will also increase because acoustic power is proportional to the square of the acoustic pressure. As such, this increasing negative pressure will induce a larger tensile strength on the cavitation nuclei, making bubbles expand. Thus, the mean bubble size increases with power. However, this increase has an upper limit because the volume of a bubble will not increase further after acoustic power reaches a critical value. Therefore, Xu et al. (2013) and Lee et al. (2005) found results that conflict with those found by Brotchie et al. (2009) because the range of acoustic power in their studies differed, the first two studies applied larger powers, while Brotchie et al. (2009) used relative small powers.

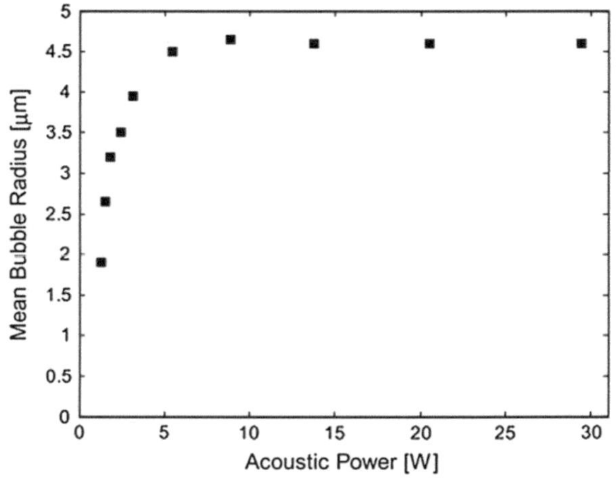

Fig. 3.20 Mean bubble radius as a function of acoustic power at an acoustic insonification of 1056 kHz. Reproduced with permission from Brotchie (2009). © The American Physical Society

Therefore, when relatively small powers are applied, the mean bubble size increases rapidly with the acoustic power; conversely, when the acoustic power is so large that it exceeds the critical value, bubble sizes remain nearly unchanged with increasing power.

3.8.2 The Effect of Acoustic Frequency

Brotchie et al. (2009) also studied the effect of acoustic frequency on the size distribution of bubbles. Figure 3.21 shows some results from this study, namely the size distributions of bubbles when various ultrasonic frequencies were applied at an acoustic power of about 1.5 ± 0.4 W.

From this figure, we see that the mean bubble size becomes smaller with increasing acoustic frequency, and the distribution itself becomes narrower with increasing frequency. Thus, bubble size distribution can be significantly affected by ultrasonic frequency, and this relationship has been explored (Yasui 2002; Brotchie et al. 2009; Merouani et al. 2013). From both theory and experimental investigations, we know that not only do bubble radii decrease with increasing frequency, but the range of the ambient bubble radius for cavitation bubbles also narrows as the frequency increases. This occurs because at higher frequencies, bubbles have a shorter amount of time to expand, causing the bubble collapses to become milder.

3.8.3 The Effect of Acoustic Pulse Duration

In the wide-beam method, if the acoustic frequency and power are kept constant, we can estimate the effect of pulse duration on the size distribution of bubbles. The

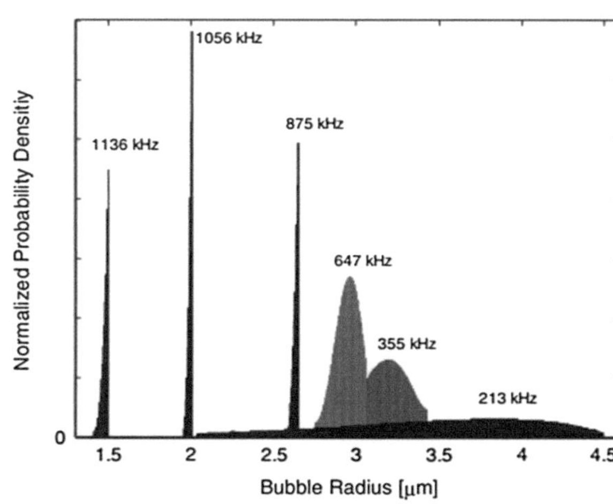

Fig. 3.21 The bubble size distribution for different frequencies. Reproduced with permission from Brotchie et al. (2009). © The American Physical Society

results for different pulse durations are shown in Fig. 3.22, where the frequency was 1.2 MHz and the power was 36 W.

Here, we see that larger bubbles are induced by prolonging the emitted pulse duration and small bubbles are induced by shortening it, which is in agreement with the conclusions made by Iida et al. (2010), who used the laser diffraction method. Moreover, Labouret and Frohly (2002) also found that increasing the duration of irradiation favors the proportion of the large bubbles, as shown in Fig. 3.23.

When cavitation bubbles are exposed to ultrasound with sufficient intensity, the bubbles will oscillate and grow. However, for short pulse durations, the exposure time is not long enough for a bubble to grow to its maximum size; thus, the bubbles remain small and have a narrow distribution. Conversely, for long pulse durations, a bubble has enough time to grow and reach its maximum size due to the rectified

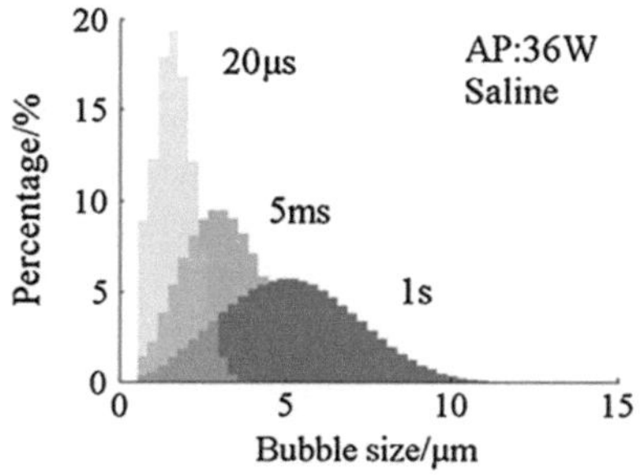

Fig. 3.22 The size distribution of bubbles for different pulse durations at an acoustic power of 36 W in saline. Reproduced with permission from Xu (2013). © Elsevier B.V.

Fig. 3.23 Bubble distributions for three durations of irradiation. Acoustic pressure: 0.21 bar, air-saturated water at 20 °C under atmospheric pressure. Reproduced with permission from Labouret (2002). © EDP Sciences

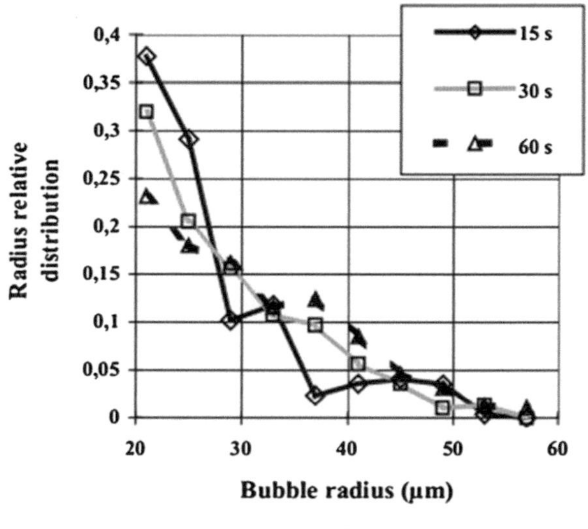

diffusion, which causes the size distribution of bubbles to be wider. Moreover, with longer pulse durations, some bubbles will coalesce to become secondary bubbles while the source transducer is on. Therefore, for longer pulses, the estimated bubble size distribution shifts toward larger bubbles because of rectified dissolution and bubble coalescence.

3.9 Bubble Size Distribution for Acoustic Droplet Vaporization

Another biomedical application where cavitation bubble size is a concern is in acoustic droplet vaporization (ADV) (Singh et al. 2012). This technique, which can be used as a diagnostic tool for cancers, works because acoustic droplets can leak out of poorly constructed blood vessels, such as those found in tumors (Rapoport et al. 2007). Once they leak out into the interstitial tissue and vaporize, the newly induced microbubbles can scatter ultrasound nonlinearly. This scattering can be detected with contrast-specific imaging techniques, such as pulse inversion (Reznik et al. 2011). For therapeutic purposes, ADV has been investigated for a wide range of applications, such as embolotherapy (Qamar et al. 2010), therapeutic drug delivery (Fabiilli et al. 2010), and HIFU thermal ablation (Zhang and Porter 2010). Clearly, the performance and efficiency of the imaging techniques used in diagnostic applications and the removal of vascular occlusions in therapeutic applications are related to size distribution of the induced gas bubbles, which are often much less soluble than air bubbles. Therefore, knowing about the size distribution of microbubbles produced for ADV is particularly important.

3.9.1 The Principle

The active cavitation detection method proposed in Sect. 3.6, using the tightly FUS transducer, cannot resolve the size distribution of bubbles that contain sparingly soluble gases. However, the improved acoustic method described in Sect. 3.7, which uses the wide-beam method, can be used to obtain the dissolution information of ADV bubbles.

During the ADV process, nano- or microdroplets are activated and then vaporized to gaseous bubbles upon ultrasound irradiation. Since the gases contained within the vaporized bubbles (these gasses are not very soluble) are different from the gasses that are often dissolved in liquid (usually air), the bubbles can be regarded as a two-component system, as described in Sect. 3.2.2. Therefore, using the improved wide-beam method and the dissolution kinetics of bubbles in a two-component system, the distribution of the vaporized gas bubbles can be estimated.

3.9.2 Experimental Methods

The experimental system is the same as that described in Sect. 3.7, and only the focus ultrasound (FUS) transducer mounted on the side of the tank was replaced by different one (5 MHz, focal width 1 mm, focal length 2 mm and aperture diameter 28.575 mm, Olympus NDT, Waltham, MA, USA). This transducer was used to generate cavitation bubbles in a phase-shift nanodroplet emulsion diluted either with degassed (PNE-DS) or with undegassed saline (PNE-S).

In the experiment, the emulsion of perfluoropentane nanodroplets was prepared as follows: 0.3 mL of perfluoropentane (F2 Chemicals Ltd., UK) and 9.7 mL of degassed, deionized water were mixed and emulsified with an ultrasonic liquid processor (VC 705, Sonic & Materials, Newton, CT, USA) for 1 min. The emulsion was then extracted with a pipette and transferred separately into undegassed or degassed saline at a volume concentration of 1:2000 as PNE-S and PNE-DS.

The measurement process was similar to that proposed in Sect. 3.7. The FUS transducer was first used to make the emulsion of perfluoropentane nanodroplets vaporized to bubbles in either PNE-S or PNE-DS. Then, the imaging transducer emitted the wide beam at a low acoustic pressure to detect the backscattering signals as a function of time from the dissolving bubbles; the resulting curve was the TIC. Figure 3.24 shows the experimental TICs in PNE-S and PNE-DS for two different ultrasonic pulse durations at the peak negative pressure of 4 MPa.

Because the dissolving rate of perfluorocarbon (PFC) gas is much slower than that of air, the observation timescale had to be extended, but this can cause bubbles to float and disrupt the TIC measurement. In this study, however, the induced low-density bubbles and the viscosity of saline allowed us to trap the initial key part

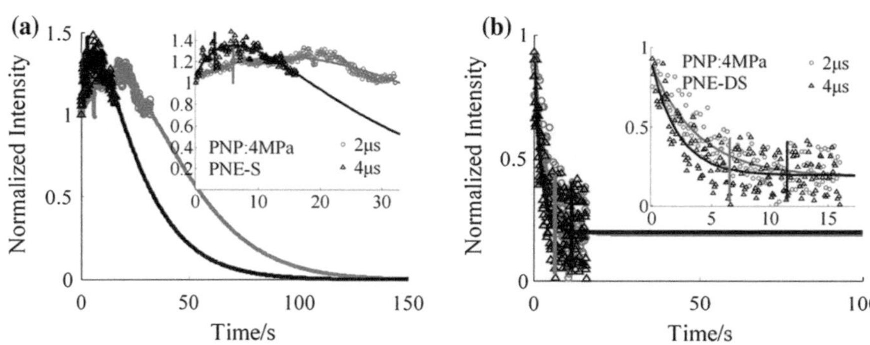

Fig. 3.24 The experimental TICs and corresponding *fitted curves* in **a** PNE-S and **b** PNE-DS for two different ultrasonic pulse durations under the same acoustic pressure of 4 MPa. A gamma fitting with four parameters was used to fit the experimentally derived TIC in (**a**) with $R^2 > 0.9$, and an exponential fitting with four parameters was used to fit the experimentally derived TIC in (**b**) with $R^2 > 0.8$. The *inserts* in (**a**) and (**b**) provide an expanded-scale view of the plot, and the bars represent the maximum SD for each curve. Reproduced with permission from Xu (2013). © Elsevier B.V.

of the dissolution process, as shown in Fig. 3.24. As such, floating bubbles were hardly observed on the beamformed images during this time. Therefore, this initial part of the dissolution process was used to resolve a curve fit depicting the whole dissolution process. This method offered a good solution for investigating the dissolution of less-soluble gas bubbles within an infinite space in liquid.

From Fig. 3.24a, we can see that the intensities of the two backscattering signals increase within the first several seconds due to the gradual bubble growth, which is the result of air diffusion from the surrounding medium into the bubbles. This phenomenon is in reasonable agreement with theoretical simulations for the dissolution of a bubble in a two-component system. After the concentration balance of air is achieved, the intensity decreases due to bubble dissolution.

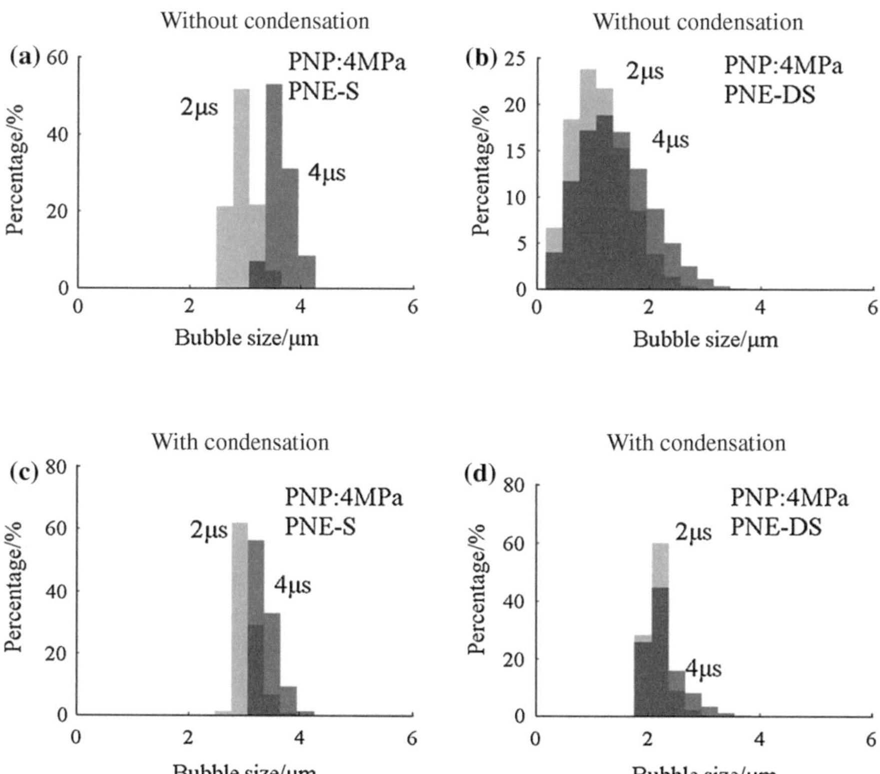

Fig. 3.25 The corresponding size distribution of cavitation bubbles induced in **a**, **c** PNE-S and **b**, **d** PNE-DS for two ultrasound pulse durations under the same peak negative pressure from the experimental TICs shown in Fig. 3.24. Distributions in **a**, **b** were calculated using the method that ignores vapor bubble condensation, and those in **c**, **d** were calculated using the method that considers vapor bubble condensation. Reproduced with permission from Xu (2013). © Elsevier B.V.

3.9.3 Typical Results

According to the experimental TICs and the wide-beam method proposed in Sect. 3.7, the size distribution of bubbles generated in PNE-S and PNE-DS can be estimated, and they are shown in Fig. 3.25.

The results in Fig. 3.25a, b show that with the same transmission parameters and optical observation techniques, the bubbles induced in PNE-DS are smaller than those induced in PNE-S (Sheeran et al. 2011). This may be because diluting the phase-shift emulsion with degassed saline could have substantially reduced the air diffusion into the vaporizing bubbles during the ultrasound sonication, perhaps because PNE-DS contained a considerably lower concentration of dissolved air than did PNE-S.

Importantly, the principles proposed here can be further extended to a multi-component system (Kwan 2010), such that experimentally derived TIC and bubble dissolution kinetics can be used to estimate bubble size distribution in a liquid that is saturated with more than two gases, as is the case for blood.

3.10 Considering Vapor Condensation for PNE-S/PNE-DS

3.10.1 A Corrected Approach on the Wide-Beam ACD Method

During the dissolution process of perfluoropentane (PFP) gas in liquid, the partial pressure of PFP gas increases with time in order to keep up with the increasing Laplace pressure inside the shrinking bubble, because it is only counterbalanced by this Laplace pressure ($2\sigma/R$; σ is the surface tension, R is the radius) (Kabalnov et al. 1998), as indicated in Fig. 3.26. Eventually, the partial pressure of the PFP gas becomes high enough such that it reaches the saturated vapor pressure of PFP gas at 25 °C (8.5×10^4 Pa) (Reed 1964), whereby it may condense into a liquid. The critical radius achieved at the condensation moment can be calculated to be 1.7 μm ($R = 2\sigma/8.5 \times 10^4$). In other words, when the radius of the dissolving bubble reaches 1.7 μm, the bubble may condense into a liquid and appear invisible to ultrasound. Furthermore, a bubble whose radius is >1.7 μm may condense immediately upon vaporization and not contribute to the TIC.

Considering vapor bubble condensation, the complete "dissolution" time points (t_m, t_{m-1} … t_1) of bubbles with initial radii of R_{0d} (t_m), R_{0d} (t_{m-1}) … R_{0d} (t_1) as analyzed in Sect. 3.7 can be replaced with the complete "condensation" time points namely, the intersection points of all the dissolution curves with the dashed-dotted horizontal line, as indicated in Fig. 3.27a, b, where the amplitude of the dashed-dotted vertical line is 1.7 μm. The physical meaning of this modification is that when the radius of a dissolving bubble reaches 1.7 μm, no matter what the

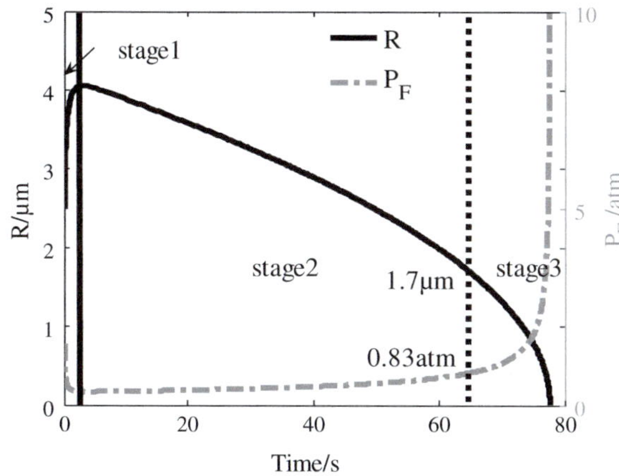

Fig. 3.26 Dissolution process of a C_5F_{12} bubble with an initial radius of 2.5 μm and $X_f = 1$ in air-saturated water. The radius of the bubble and the partial pressure of C_5F_{12} versus time are illustrated as the *black solid curve* and the *gray dashed-dotted curve*, respectively

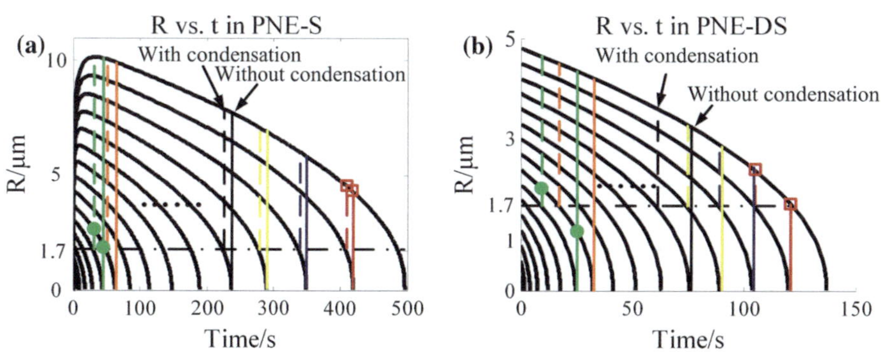

Fig. 3.27 **a** Dissolution curves for C_5F_{12} bubbles of initial radii of 0.3:0.3:5 μm with $X_f = 1$ in air-saturated water. **b** Dissolution curves for C_5F_{12} bubbles of initial radii of 0.3:0.3:5 μm in degassed water. The intersection points of all *dissolution curves* with the *vertical solid lines* were used in the method without considering condensation, while those of all *dissolution curves* with the *vertical dashed lines* were used in the method considering condensation. Each pair of *vertical lines* (*solid* and *dashed*) with the *same color* is for a bubble of a given initial diameter. Reproduced with permission from Xu (2013). © Elsevier B.V.

initial radius is, the bubble will condense into liquid and will no longer contribute to the TIC. In addition, the R_{min} has to be larger than 1.7 μm. Then, the subsequent calculations involved the intensities on TIC at these complete "condensation" time points, the reached radii of bubbles with different initial radii at these time points, and the intersection points of all the dissolution curves with the vertical dashed lines in Fig. 3.27a, b.

3.10.2 The Effect of Vapor Condensation in the Bubble Size Estimation

Considering condensation of PFP bubbles, the bubble size distributions in PNE-S and PNE-DS were recalculated, as shown in Fig. 3.25c, d. When we consider condensation, the proportion of large bubbles increases compared to when condensation is ignored. This increase is especially obvious for the small bubbles (size distribution displayed with light gray histogram in Fig. 3.25c, d. Thus, the calculations that consider condensation will cause greater differences for the reached radii of bubbles with small initial radii, which can be simply observed from the amplitude difference between the two green solid circles and that between the two red open rectangles in Fig. 3.27a, b. Nevertheless, the effect of PD and PNP on the bubble size distribution remained undisturbed.

3.11 Summary

Size distribution of cavitation bubbles is a fundamental parameter that describes a multi-bubble cavitation field. However, the size of cavitation bubbles is difficult to measure due to the complex nature of acoustic cavitation. In this chapter, we presented several recently developed methods to estimate bubble size that are based on bubble dissolution.

For the estimation methods that involved experiments for when the source ultrasound was on, cavitation activity was able to be detected through SL or UHCN intensity during the pulse-on time of the source transducer. Then, through feedback, the acoustic parameters for controlling the cavitation bubble size were optimized. However, acquiring the SL or UHCN intensity versus pulse interval requires many additional experiments that have various pulse-off times between pulse trains. Thus, it is necessary to maintain good reproducibility between each experiment. With regard to the SL method, it is limited to transparent medium and a SL signal decay that is observed for large pulse-off times must be explained. In addition, the SL may not be detected at low acoustic powers. The PCD-based CN method can be performed in the non-transparent mediums and is applicable to the non-monotonic model for acoustic emission by collapsing bubbles.

For estimation methods that involved the pulse-off times of the source ultrasound, bubbles are prone to floating to the top of the liquid surface after the source ultrasound is removed. So, because these methods must detect the natural dissolution of bubbles, a corrected approach to deal with the floating bubbles might be useful to expand the application scope of this class of methods. As for the void rate method, the measured smallest bubble radius and the step size are relatively large, since this method can only handle the void rate dissipation curve with a positive concavity. Regarding the ACD-based scattering method that employs a tightly focused transducer, this method had difficulty in tracing the dissolution process of the PFC gas bubble because it has a limited sensing region and the PFC moved past the region during its long dissolution time. For the wide-beam ACD-based back-scattering method, it is clear that the concentration of the induced bubbles should be relatively low in order to prevent acoustic shielding effects and multiple scattering effects.

In the future, these methods will be applied in various fields, such as hydro-dynamics, ultrasonic cleaning, sonochemical reactions, and biomedical engineering, to investigate the size distribution of cavitation bubbles in order to further control cavitation activity. However, when utilized in a specific circumstance, such as when bubbles are induced into a fluid or at an interface or even within a tube or vessel, these methods might have to be modified further as needed.

References

Apfel RE. Acoustic cavitation. Methods Exp Phys. 1981;19:355–411.

Apfel RE. Acoustic cavitation inception. Ultrasonics. 1984;22(4):167–73.

Ashokkumar M. The characterization of acoustic cavitation bubbles—an overview. Ultrason Sonochem. 2011;18(4):864–72.

Ashokkumar M, Hall R, Mulvaney P, Grieser F. Sonoluminescence from aqueous alcohol and surfactant solutions. J Phys Chem B. 1997;101(50):10845–50.

Brotchie A, Grieser F, Ashokkumar M. Effect of power and frequency on bubble-size distributions in acoustic cavitation. Phys Rev Lett. 2009;102(8):084302.

Burkard ME, Van Liew HD. Simulation of exchanges of multiple gases in bubbles in the body. Respir Physiol. 1994;95(2):131–45.

Chen W-S, Matula TJ, Crum LA. The disappearance of ultrasound contrast bubbles: observations of bubble dissolution and cavitation nucleation. Ultrasound Med Biol. 2002;28(6):793–803.

Church CC. The effects of an elastic solid surface layer on the radial pulsations of gas bubbles. J Acoust Soc Am. 1995;97:1510–21.

Epstein P, Plesset M. On the stability of gas bubbles in liquid–gas solutions. J Chem Phys. 1950;18:1505–9.

Fabiilli ML, Haworth KJ, Sebastian IE, Kripfgans OD, Carson PL, Fowlkes JB. Delivery of chlorambucil using an acoustically-triggered perfluoropentane emulsion. Ultrasound Med Biol. 2010;36(8):1364–75.

Hauptmann M, Brems S, Struyf H, Mertens P, Heyns M, De Gendt S, Glorieux C. Time-resolved monitoring of cavitation activity in megasonic cleaning systems. Rev Sci Instrum. 2012;83 (3):034904.

Hauptmann M, Struyf H, De Gendt S, Glorieux C, Brems S. Evaluation and interpretation of bubble size distributions in pulsed megasonic fields. J Appl Phys. 2013;113(18):184902.

Henglein A, Ulrich R, Lilie J. Luminescence and chemical action by pulsed ultrasound. J Am Chem Soc. 1989;111(6):1974–9.

Hoff L, Sontum PC, Hovem JM. Oscillations of polymeric microbubbles: effect of the encapsulating shell. J Acoust Soc Am. 2000;107:2272–80.

Iida Y, Ashokkumar M, Tuziuti T, Kozuka T, Yasui K, Towata A, Lee J. Bubble population phenomena in sonochemical reactor I: estimation of bubble size distribution and its number density with pulsed sonication—laser diffraction method. Ultrason Sonochem. 2010;17 (2):473–9.

Kabalnov A, Klein D, Pelura T, Schutt E, Weers J. Dissolution of multicomponent microbubbles in the bloodstream 1: theory. Ultrasound Med Biol. 1998;24(5):739–49.

Kwan JJ, Borden MA. Microbubble dissolution in a multigas environment. Langmuir. 2010;26 (9):6542–8.

Labouret S, Frohly J. Bubble size distribution estimation via void rate dissipation in gas saturated liquid: application to ultrasonic cavitation bubble fields. Eur Phys J Appl Phys. 2002;19 (01):39–54.

Lee J, Ashokkumar M, Kentish S, Grieser F. Determination of the size distribution of sonoluminescence bubbles in a pulsed acoustic field. J Am Chem Soc. 2005;127(48):16810–1.

Luther S, Mettin R, Koch P, Lauterborn W. Observation of acoustic cavitation bubbles at 2250 frames per second. Ultrason Sonochem. 2001;8(3):159–62.

Matula TJ. Inertial cavitation and single—bubble sonoluminescence. Philos Trans R Soc Lond Ser A Math Phys Eng Sci. 1999;357(1751):225–49.

Merouani S, Hamdaoui O, Rezgui Y, Guemini M. Effects of ultrasound frequency and acoustic amplitude on the size of sonochemically active bubbles—theoretical study. Ultrason Sonochem. 2013;20(3):815–9.

Qamar A, Wong ZZ, Fowlkes JB, Bull JL. Dynamics of acoustic droplet vaporization in gas embolotherapy. Appl Phys Lett. 2010;96(14):143702.

Qiao Y, Cao H, Zhang S, Yin H, Wan M. Sonochemiluminescence observation of lipid-and polymer-shelled ultrasound contrast agents in 1.2 MHz focused ultrasound field. Ultrason Sonochem. 2012;20(1):162–70.

Rapoport N, Gao Z, Kennedy A. Multifunctional nanoparticles for combining ultrasonic tumor imaging and targeted chemotherapy. J Natl Cancer Inst. 2007;99(14):1095–106.

Reed T. Physical chemistry of fluorocarbons. Fluorine Chem. 1964;5:133–221.

Reznik N, Williams R, Burns PN. Investigation of vaporized submicron perfluorocarbon droplets as an ultrasound contrast agent. Ultrasound Med Biol. 2011;37(8):1271–9.

Sheeran PS, Wong VP, Luois S, McFarland RJ, Ross WD, Feingold S, Matsunaga TO, Dayton PA. Decafluorobutane as a phase-change contrast agent for low-energy extravascular ultrasonic imaging. Ultrasound Med Biol. 2011;37(9):1518–30.

Singh R, Husseini GA, Pitt WG. Phase transitions of nanoemulsions using ultrasound: experimental observations. Ultrason Sonochem. 2012;19(5):1120–5.

Suslick KS. Sonochemistry: Kirk-Othmer encyclopedia of chemical technology. New York: Wiley; 1998.

Xu S, Zong Y, Li W, Zhang S, Wan M. Bubble size distribution in acoustic droplet vaporization via dissolution using an ultrasound wide-beam method. Ultrason Sonochem. 2013;21(3): 975–83.

Yasui K. Influence of ultrasonic frequency on multibubble sonoluminescence. J Acoust Soc Am. 2002;112:1405–13.

Yasui K, Tuziuti T, Lee J, Kozuka T, Towata A, Iida Y. The range of ambient radius for an active bubble in sonoluminescence and sonochemical reactions. J Chem Phys. 2008;128:184705.

Zhang P, Porter T. An in vitro study of a phase-shift nanoemulsion: a potential nucleation agent for bubble-enhanced HIFU tumor ablation. Ultrasound Med Biol. 2010;36(11):1856–66.

Chapter 4
Cavitation-Enhanced Thermal Effects and Applications

Siyuan Zhang, Mingxi Wan and Gail ter Haar

Keywords Cavitation-enhanced heating mechanisms · Bioheat transfer · High-intensity focused ultrasound · Multi-frequency ultrasound · Boiling histotripsy · Nanoscale thermometry

4.1 Introduction

Cavitation and heating are the primary physical mechanisms by which various therapeutic applications of ultrasound are brought about (ter Haar 1999, 2001a, 2007; Kennedy 2005; Bailey et al. 2003; Bohannon 2008; Shaw et al. 2014; Ahmadi et al. 2012; Kollmann et al. 2013). High-intensity focused ultrasound (HIFU) is currently being developed as a noninvasive technique capable of selective and localized destruction of tissue volumes within the body while sparing intervening and surrounding tissue from harmful exposure. For example, the thermal effects of HIFU are used to ablate tumors in a number of tissues (ter Haar 2007, 2001b; Kennedy 2005; Bailey et al. 2003). As an ultrasound wave propagates through the tissue, a portion of the ultrasonic energy is absorbed and converted to heat (Pierce 1994; Vilensky et al. 2012). However, how tissues respond to this heat generation must be predicted and quantified in order to ensure the safety of

S. Zhang · M. Wan
The Key Laboratory of Biomedical Information Engineering of Ministry of Education, Department of Biomedical Engineering, School of Life Science and Technology, Xi'an Jiaotong University, Xi'an 710049, People's Republic of China
e-mail: xjtusyzhang@mail.xjtu.edu.cn

M. Wan
e-mail: mxwan@mail.xjtu.edu.cn

G. ter Haar (✉)
Division of Radiotherapy and Imaging, The Institute of Cancer Research and the Royal Marsden NHS Foundation Trust, 123 Old Brompton Road, London SW7 3RP, UK
e-mail: Gail.terHaar@icr.ac.uk

M. Wan et al. (eds.), *Cavitation in Biomedicine*,
DOI 10.1007/978-94-017-7255-6_4

therapeutic ultrasound in the clinic and in the potential future applications. Models which quantify tissue responses to ultrasound have therefore been developed. For example, acoustic propagation in biological tissues can be described by linear or nonlinear wave propagation equations, while the transfer of energy from acoustic to thermal forms in biological tissues can be described by the Pennes bioheat transfer equation (Pennes 1948). Linear equations, such as proposed by O'Neil, and the linear Westervelt wave equation, are able to describe the propagation of ultrasound in tissue at low acoustic pressure and are used to investigate the temperature response in tissue exposed to ultrasound (Meaney et al. 1998). However, at high acoustic pressure, the value of the temperature rise calculated by these theoretical predictions is always a little lower than that measured in experiments because of nonlinear effects (Solovchuk et al. 2014). Thus, nonlinear equations, such as the Khokhlov–Zabolotskaya–Kuznetsov (KZK) equation and the nonlinear Westervelt wave equation, are better methods for simulating ultrasound propagation in tissue at higher acoustic pressure (Bakhvalov et al. 1987; Hoffelner et al. 2001; Hamilton and Blackstock 1998). Simulations which include nonlinearity are, as such, more consistent with experimental observations than those calculated using linear equations. Nevertheless, if the ultrasound power is higher than a certain value, there is still a small difference between the temperature rise calculated by theoretical predictions and the actual values measured in thermal response experiments on tissues (Holt and Roy 2001). The size of this difference increases with increasing acoustic pressures (Holt and Roy 2001). Some researchers have proposed that the difference between the simulation and experimental results might be caused by cavitation resulting from ultrasound exposure (Hilgenfeldt et al. 2000; Holt and Roy 2001). Cavitation may enhance local heating when driven acoustically, and thus, cavitation-enhanced thermal effects should be taken into consideration when predicting and quantifying the thermal response of a tissue to focused ultrasound (FUS). This is especially important for therapeutic ultrasound applications, because their primary mechanism of action relies on raising tissue temperatures rapidly within the target treatment region (ter Haar 1999, 2001a, 2007; Kennedy 2005; Bailey et al. 2003; Bohannon 2008; Shaw et al. 2014; Ahmadi et al. 2012; Kollmann et al. 2013). In certain tissues, such as those that are highly perfused like the liver, temperature elevation can be limited by the presence of vascular structures because of convective transport phenomena (Chen et al. 1993; Curra et al. 2000; Davis et al. 2006; Hariharan et al. 2007; Huang et al. 2010a). Thus, to achieve therapeutic effects in these tissues, ultrasound may need to be applied at very high acoustic intensities and for longer durations, which can cause unintended, and undesired, damage to the intervening and surrounding tissue. Efficient methods for increasing heating and reducing treatment times would therefore improve therapeutic applications of HIFU. One way to enhance the thermal effects of therapeutic HIFU might be found through investigation of the interaction between ultrasound and cavitation bubbles. Recently, different encapsulated gas-filled microbubbles (MBs) and/or phase-shift nanodroplets (NDs) have been used to enhance local cavitation and heating. This has spurred the development of ultrasound therapy methods that use cavitation-enhanced thermal effects. This chapter will introduce the principles of cavitation-enhanced thermal effects, present various investigations that have been carried out in this area, and describe potential applications for these effects in biomedical ultrasound therapies.

4.2 Principles of Cavitation-Enhanced Thermal Effects

4.2.1 Heating Due to Primary Absorption of an Ultrasound Field

As an ultrasound wave propagates through tissue, a portion of its energy is absorbed and converted into heat (Pierce 1994; Vilensky et al. 2012). The energy deposition per unit volume and unit time, q_{pri}, due to ultrasound is (Edson 2001; Clarke and ter Haar 1997):

$$q_{\text{pri}} = 2\alpha I_0 \tag{4.1}$$

where α is the amplitude attenuation coefficient and I_0 is the spatial peak, time-averaged acoustic intensity at the focus of the ultrasound field. The temperature T may be written as a function of time, t, as follows (Clarke and ter Haar 1997):

$$T = \frac{\alpha I_0 G^2}{2k_0 \rho_0 C_0} \ln\left[1 + \frac{4k_0 t}{G^2}\right] \tag{4.2}$$

where k_0, ρ_0, and C_0 are, respectively, the thermal diffusivity, density, and heat capacity of the medium. T and G are the temperature rise at the focus and the Gaussian radius of the acoustic beam.

4.2.2 Viscous Damping of Cavitation Bubble

4.2.2.1 Viscous Dissipation of Free Gas Bubble

Viscosity is a characteristic of a fluid that depends on the friction between neighboring particles that are moving at different velocities within the fluid and is a measure of a fluid's resistance to gradual deformation by shear or tensile stresses. Viscous damping reflects the stress at the boundary of a cavitation bubble and its neighboring medium. In a relatively thin layer between the host medium and the cavitation bubble, viscous dissipation causes some of the bubble's mechanical energy to be converted into thermal energy, thus heating the surrounding medium (Poritsky 1951; Hilgenfeldt et al. 2000; Holt and Roy 2001; Okita et al. 2013). To understand how this heating, which depends on the bubble size and the acoustic characteristics of the host medium, occurs better, we must also investigate how bubbles behave when driven acoustically and how their motion translates into energy dissipation. Many equations describe the radial response of a gas bubble, in terms of its compressibility, in an incompressible and inviscid liquid under ultrasound excitation. These equations are usually derivatives of the Rayleigh–Plesset–Noltingk–Neppiras–Poritsky (RPNNP) equation (Lauterborn 1976; Edson 2001):

$$R\ddot{R} + \frac{3}{2}\dot{R}^2 = \frac{1}{\rho}\left(\left(P_0 + \frac{2\sigma}{R} - P_v\right)\left(\frac{R_0}{R}\right)^{3\gamma} + P_v - \frac{2\sigma}{R} - \frac{4\mu\dot{R}}{R} - P_0 + P_{ac}\right) \quad (4.3)$$

where R and $\dot{R}$ are the bubble radius (with an initial, stationary value R_0) and the velocity of the bubble wall, respectively. The terms $\ddot{R}$, P_v, and ρ are the acceleration of the bubble wall, the vapor pressure, and the density of the liquid. The terms γ, σ, μ, P_{ac}, and P_0 are the adiabatic exponent of the gas, the surface tension, the shear viscosity of the liquid, the pressure due to the acoustic wave, and the hydrostatic pressure of the liquid, respectively. This equation was later improved by Keller and Miksis who took into account sound radiated by the bubble, including not only the effects of acoustic radiation from the bubble, but also the polytropic exponent of the gas, k (Keller and Miksis 1980; Edson 2001):

$$\left(1 - \frac{\dot{R}}{c}\right)R\ddot{R} + \frac{3}{2}\dot{R}^2\left(1 - \frac{\dot{R}}{3c}\right) = \left(1 + \frac{\dot{R}}{c}\right)\frac{P(\dot{R}, R, t)}{\rho} + \frac{R}{\rho c}\frac{\partial P(\dot{R}, R, t)}{\partial t} \quad (4.4)$$

$$P(\dot{R}, R, t) = \left[\left(P_0 - P_v - \frac{2\sigma}{R}\right)\left(\frac{R_0}{R}\right)^{3k} + P_v\right] - P_0 - \frac{2\sigma}{R} - \frac{4\mu\dot{R}}{R} - P_{ac} \quad (4.5)$$

where c is the speed of sound in the medium. Earlier, Poritsky (1951) proposed a term representing this boundary stress; this term has thereafter been used to indicate heating due to the energy deposited. For an oscillation cavitation bubble, the stress (S_{wall}) at the bubble wall caused by the viscosity of the neighboring liquid is (Edson 2001; Hilgenfeldt et al. 2000):

$$S_{wall} = \frac{4\mu\dot{R}}{R} \quad (4.6)$$

Because of the stress over the entire surface of the bubble, the term $S_{wall} \times 4\pi R^2$ indicates a force at the bubble interface. Mechanical energy is dissipated at a rate of $S_{wall} \times 4\pi R^2 \times \dot{R}$ as the interface moves at a velocity $\dot{R}$. Here, the time-averaged energy deposited per unit time due to viscous dissipation, termed the viscous power deposition, P_{vis}, is as follows (Edson 2001; Hilgenfeldt et al. 2000):

$$P_{vis} = \left\langle S_{wall} \times 4\pi R^2 \times \dot{R}\right\rangle_t = 16\pi\mu\left\langle \dot{R}^2 R\right\rangle_t \quad (4.7)$$

where $\langle\cdot\rangle_t$ indicates time-averaged values. The power deposition occurs in a relatively small volume around the bubble when the viscous dissipation occurs at the bubble interface. Edson (2001) and Yang et al. (2004) determined numerically that heating from stable cavitation occurred for larger bubbles and in fluids with high viscosities. However, for micron-sized cavitation bubbles in a host with biologically relevant rheological properties whose shear viscosities are much greater than water, viscous dissipation is expected to be significant at MPa pressures and MHz frequencies; in some cases, it may even be the dominant mechanism of heat deposition (Prosperetti 1977).

4.2.2.2 Viscous Dissipation of Encapsulated Bubble

The viscous dissipation coefficient of various encapsulated gas-filled MBs or phase-shift NDs is much higher than that of free cavitation bubbles, due to the high viscosities of the encapsulating shells, which are made of materials such as albumin, lipids, polymers, etc. Encapsulated bubbles will dissipate energy as heat as a result of viscous friction between the surrounding liquid and the coating material, and this heat dissipation leads to enhanced heating due to the viscous damping of bubble oscillations by the tissue or body fluids (Sboros 2008; Stride and Saffari 2004). Figure 4.1 demonstrates that there is a significant difference in the temperature rise predicted by the encapsulated and free bubble models for identical ultrasound exposure.

4.2.3 *Acoustic Energy Radiation from a Cavitation Bubble*

4.2.3.1 Secondary Acoustic Emission from a Cavitation Bubble

A cavitation bubble is considered to radiate acoustic energy as a sound source, and the radiated pressure wave from the cavitation bubble, i.e., the "secondary acoustic emission," is absorbed and converted into heat in the same manner as the primary absorption of energy from the incident acoustic field (Edson 2001; Farny et al. 2010). To develop an expression for evaluating the secondary acoustic power deposition by a cavitation bubble where the wavelength of the radiated sound is much larger than the bubble, the radiated pressure due to the secondary acoustic emission as a function of distance r_b from the bubble undergoing a perfectly spherical collapse is as follows (Edson 2001; Farny et al. 2010):

$$P(r, r_b, t) = \frac{R\rho}{r_b}(2\dot{R}^2 + R\ddot{R}) \tag{4.8}$$

where the bubble is located at r and ρ is the density of the host medium surrounding the bubble. The radiated acoustic energy is estimated by evaluating the time-averaged radiated intensity $I_b(r, r_b)$ as follows (Farny et al. 2010):

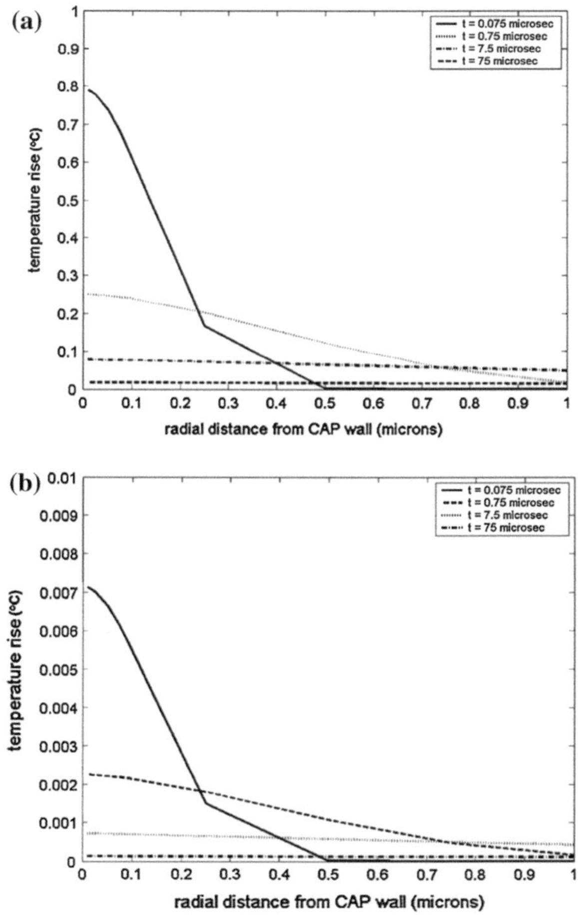

Fig. 4.1 The temperature rise produced by a single pulse of center frequency 3 MHz and peak amplitude 0.2 MPa in the fluid surrounding, **a** an Albunex bubble and **b** a free air bubble of equal dimensions, for various times after the end of the pulse. Reproduced with permission from Stride and Saffari (2004). © Elsevier

$$I_b(r, r_b) = \frac{\langle P^2(r, r_b, t) \rangle_\tau}{\rho c} \tag{4.9}$$

where τ is the time interval over which the root-mean-square (rms) pressure is estimated. The power, D, deposited by the bubble within a volume defined by a sphere of radius r' will depend on the path length over which the acoustic emission is absorbed and the loss due to attenuation (Hilgenfeldt et al. 2000):

$$D = 4\pi r_b^2 \times I_b(r, r_b)(1 - e^{-2\alpha(f)r'}) \tag{4.10}$$

The spatial average thermal power density over this volume, q, is estimated as (Edson 2001; Farny et al. 2010):

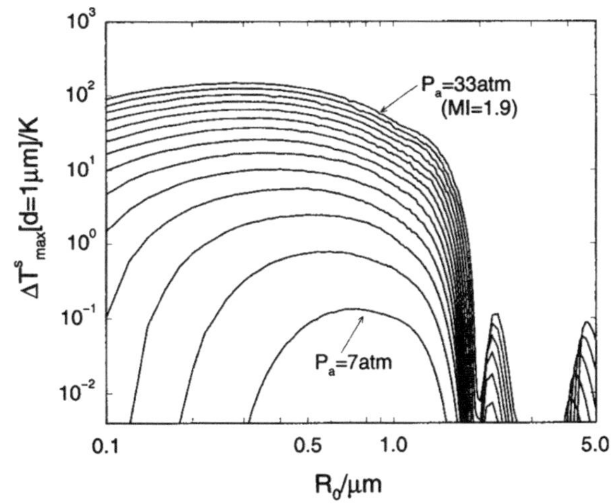

Fig. 4.2 Maximum temperature rise due to acoustic heating as a function of the ambient radius and the pressure at a distance of 1 μm from the bubble wall. Pa is the amplitude of the driving acoustic wave. Reproduced with permission from Hilgenfeldt et al. (2000). © Acoustical Society of America

$$q = \frac{4}{3\pi r'^3} D \tag{4.11}$$

Compared with the calculation of the viscous dissipation when the time-averaged power deposition is considered, the maximum "secondary acoustic power deposition" is evaluated by considering the total sound radiated by the bubble (Edson 2001). Hilgenfeldt et al. (2000) showed that for small, nonlinearly driven bubbles, the energy absorbed from the emitted sound could be greater than the energy dissipated by viscosity, and they described how this can affect the liquid temperature very near the bubble. Figure 4.2 displays the resulting maximum temperature rises near a bubble when it was assumed that all of the absorbed heat from the emitted acoustic pulse, up to a distance of 1 μm, was delta-distributed in a shell at the initial, static radius. They demonstrated that the values of maximum temperature rises dropped quickly when the distance from the bubble was even slightly larger. They also concluded that any damage which could be done to biomolecules by the passage of temperature pulses through the liquid would be limited to an extremely small fluid volume around the bubble.

4.2.3.2 Higher Harmonics from a Cavitation Bubble

When exposed to ultrasound at a certain acoustic pressure, a cavitation bubble will expand to several times its initial size and then undergo extremely rapid collapse. The nonlinear acoustic oscillation of the cavitation bubble will produce signals at a broad band of frequencies, which include the harmonics of the driving frequency, which is usually the resonance frequency of the ultrasound transducer. Thus, the acoustic signals from cavitation bubbles can be detected as broadband noise arising from inertial bubble collapse (Bull et al. 2013). Any acoustic energy that is radiated from a bubble will be absorbed in the same way as the incident acoustic wave. In tissues, the acoustic energy in the higher harmonics that are scattered or emitted by a cavitation bubble can be more effectively absorbed and converted to heat than can energy at the excitation frequency (Holt and Roy 2001; Coussios et al. 2007; Khokhlova et al. 2006). The appearance of higher harmonics has an effect on the amount of energy absorbed in tissue to some degree, especially in the focal region. In most materials, including tissue, the rate of absorption increases with frequency and thus the presence of bubbles can significantly enhance heating effects due to this harmonic generation.

Unlike the heating that arises from viscous dissipation in a relatively small region near the cavitation bubble, the absorption of radiated acoustic energy from the cavitation bubble will occur at a distance from the bubble, where this distance depends on the magnitude of the acoustic absorption coefficient.

4.2.4 *Increased Local Acoustic Absorption by Cavitation Bubble*

Cavitation bubbles tend to increase the local attenuation coefficient of ultrasound, and this stronger acoustic attenuation occurring in the presence of cavitation bubbles can enhance heating related to ultrasound absorption (Asai et al. 2013;

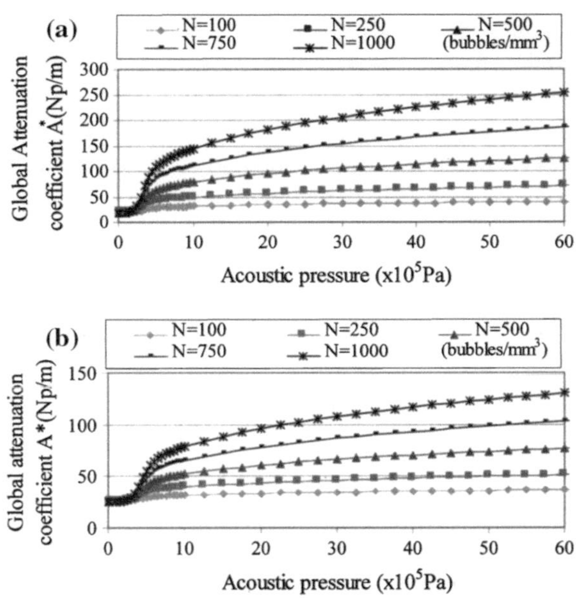

Fig. 4.3 Global attenuation coefficient versus acoustic pressure for different bubble densities: **a** frequency = 2.25 MHz; **b** frequency = 3.0 MHz. Reproduced with permission from Chavrier et al. (2000). © Acoustical Society of America

Moriyama et al. 2012; Chavrier et al. 2000). Cavitation bubbles capture part of the ultrasound energy, which is then redistributed in the form of spherical waves. To take into account the global attenuation coefficient for cavitation bubbles (Fig. 4.3), the volumetric heat rate can be calculated. Through a thermal simulation, Moriyama et al. (2012) investigated the enhancement of heating due to cavitation by increasing the acoustic absorption coefficient in the region of the generated cavitation bubbles. The absorption coefficient was calculated by fitting curves obtained from the temperature rise at the focal point experimentally with those from simulation. The results showed that an increased absorption coefficient could simulate the effect of enhanced heating by cavitation.

4.2.5 Bioheat Transfer Equation

The induction of ultrasound-induced thermal effects is based on the ability of biological tissues to absorb part of the ultrasound energy, and on the resulting localized temperature increase created by this absorbed energy. Theoretically, the thermal effects of ultrasound in biological tissues can be calculated from solution of the bioheat transfer equation (BHTE), defined as follows (Pennes 1948):

$$\rho_t c_t \frac{\partial \theta_p}{\partial t} = \nabla \cdot (k_t \nabla \theta_p) + \omega_b c_b (\theta_a - \theta_p) + Q_p \tag{4.12}$$

where θ_p is the temperature at point P, ρ_t is the tissue density, c_t is the tissue-specific heat, k_t is the tissue thermal conductivity, ω_b is the volumetric blood perfusion rate, c_b is the specific heat of blood, and θ_a is the arterial blood temperature. Here, Q_p represents the volumetric rate of heating generated at point P due to ultrasound exposure. In the presence of cavitation though, the presence of cavitation bubbles tends to increase the local acoustic attenuation coefficient, and the higher harmonics scattered or emitted by these cavitation bubbles are more readily absorbed and converted into heat in the tissue. Therefore, the parameter Q_p is calculated using a different approach, which takes cavitation bubbles into account when cavitation-enhanced heating is being considered.

4.3 Simultaneously Measuring Cavitation Activity and Temperature Rise

In order to investigate the heating effects of cavitation, cavitation activity and temperature rise can be measured simultaneously during HIFU exposures applied to different substances, such as transparent tissue-mimicking phantoms or even ex vivo and in vivo tissues. The temperature rise measured is then related to the cavitation activity by accounting for, and then subtraction of, the heating that is predicted to occur from absorption of the primary ultrasound, leaving the temperature rise associated with cavitation-enhanced heating.

4.3.1 Cavitation and Temperature Rise in Phantoms

Watmough et al. (1993) observed an increased temperature rise in a transparent phantom made of bovine hide gelatin. Gaseous inclusions of about resonant size artificially introduced into the sample led to a mean temperature rise of about 41 °C when the gel was exposed to continuous wave (CW) ultrasound at 0.75 MHz with an intensity of 1 W/cm^2 (spatial average). They observed that the temperature increase was as much as six times greater for samples that were not degassed compared with those that were, and therefore they concluded that the enhanced heating in the nondegassed samples was the result of MBs trapped at the inclusions in these phantoms. Later, Holt and Roy (2001) measured cavitation-enhanced heating in a tissue-mimicking phantom at 1 MHz HIFU exposure in CW mode,

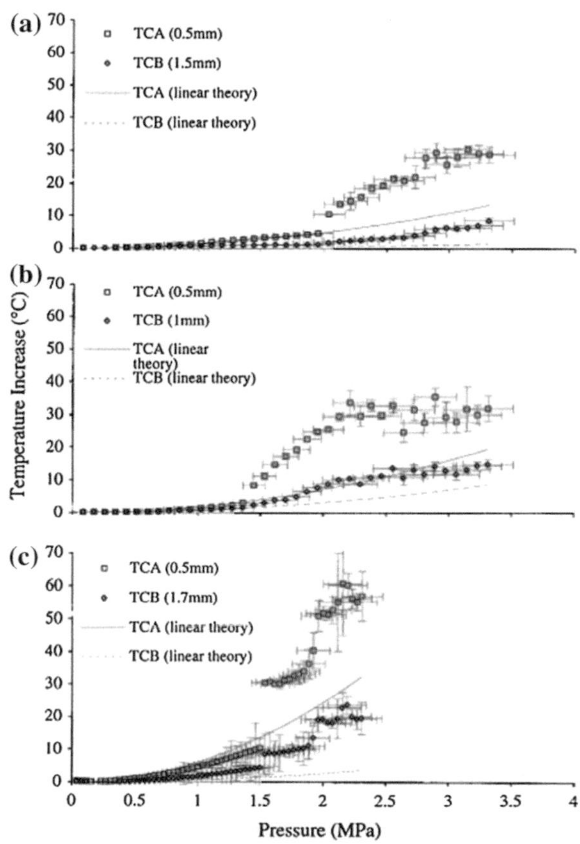

Fig. 4.4 Peak temperature rise increase from ambient versus insonation pressure for the agar phantom at three increasing exposure durations: **a** 0.7 s, **b** 1.2 s, and **c** 10.2 s. TCA and TCB are two different thermocouples situated at the distances shown from the focal peak. Reproduced with permission from Holt and Roy (2001). © Elsevier

where exposures lasting between 0.5 and 10 s were performed at a range of pressures. They measured the temperature rise in several positions using thermocouples embedded close to the focus. Above a critical acoustic pressure threshold, they observed a larger increase in the temperature rise than that predicted by the classical linear model, which does not account for cavitation-enhanced heating (Fig. 4.4). The onset of the enhanced heating coincided closely with the onset of raised subharmonic and broadband signals arising from cavitation. They suggested not only that the viscous dissipation and the acoustic emission from cavitation bubbles were the possible heating mechanisms, but also argued that cavitation-enhanced heating could improve the efficiency of HIFU treatment if controlled properly.

Farny et al. (2010) evaluated the relationship between inertial cavitation emissions and heat generation in a controlled setting. Cavitation and temperature rise near the focus were measured at a 1.1-MHz CW HIFU 5.5-s exposure applied to a tissue-mimicking phantom (Fig. 4.5). Their results demonstrated that cavitation power correlates well with the rate of temperature rise associated with energy deposition from inertial cavitation when HIFU heating is accounted for and when the nonlinear effects are small. They also suggested that careful processing of the cavitation signals may serve as a proxy for measuring the heating contribution from inertial cavitation.

4.3.2 Cavitation and Temperature Rise in Tissues

4.3.2.1 Ex Vivo Investigations

As early as 1987, Lele (1987) measured temperature rise and acoustic emission associated with cavitation simultaneously in a fresh calf liver exposed to a 2.7 MHz HIFU source over a range of intensities with 0.2- and 0.3-s tone bursts. At low intensities, the temperature rise was modest and predictable. However, due to the subharmonic and broadband acoustic emissions detected at higher intensities, they found that the temperature rise was higher than expected. They hypothesized that the enhanced heating they observed was due to the occurrence of cavitation activity. Later, Clarke and ter Haar (1997) observed enhanced heating in liver samples from cows, pigs, and lambs, as well as in one dog prostate ex vivo during 1 MHz HIFU at varying pressures with 10-s CW tone bursts. They argued that the heat deposition was correlated with cavitation and/or boiling bubbles. Melodelima et al. (2004) presented a method that combined cavitation with a thermal effect to obtain deeper lesions using a plane ultrasound transducer rather than a focused one. All tests were performed ex vivo on parallelepiped-shaped samples of pig liver. Cavitation was produced by delivering a 0.5-s exposure with 60 W/cm^2 at the face of the transducer. Immediately afterward, the ultrasound applicator was rotated on its axis through angular openings of 90°. During this rotation, a thermal effect was produced in the ex vivo tissue exposed to an acoustic intensity of 14 W/cm^2, which worked to combine the cavitation effect with a thermal effect. The depth of lesion

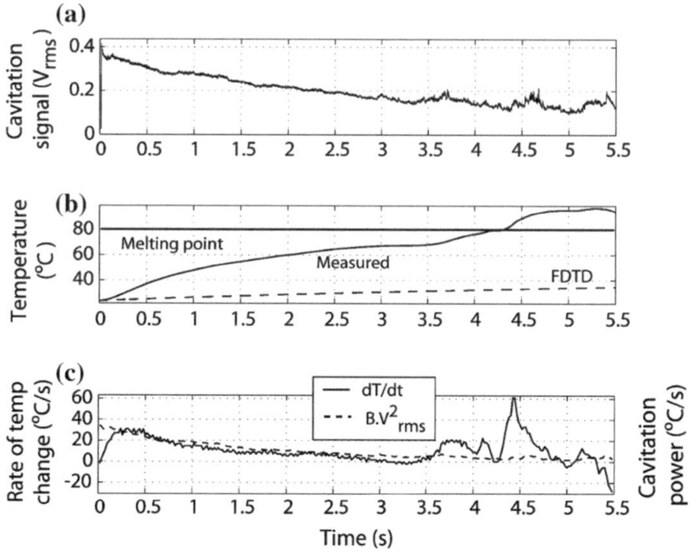

Fig. 4.5 Comparison of cavitation power and heating rates in an agar–graphite phantom for 5.5-s HIFU exposures at 1.1 MHz. **a** Inertial cavitation emissions measured by the passive cavitation detector. **b** Temperature rise measured and predicted from a finite-difference time-domain (FDTD) solution 0.5 mm lateral to the focus. **c** Bubble-generated heating rate and cavitation power. Cavitation power is measured by the mean square voltage output of the passive cavitation detector. Reproduced with permission from Farny et al. (2010). © IEEE

through the combination of cavitation and thermal effects was twice as deep as that produced from thermal effects alone (Fig. 4.6).

In another study, Kyriakou et al. (2011) quantified cavitation activity and heating induced by ultrasound in adipose tissue. Cavitation activity was investigated using the passive cavitation detection (PCD) method, and the focal temperature rise was measured using minimally invasive needle thermocouples. HIFU exposures were performed in excised subcutaneous porcine fat for exposure durations of 2–4 s at four different frequencies over a range of pressure amplitudes across several tissue samples from different donor animals. Experimental results showed that the inertial cavitation activity initiated at 0.5 MHz by pressures greater than a rarefactional pressure of 1.6 MPa could enhance focal heat deposition. They found a good correlation between the energy of broadband emissions received by the PCD and the resulting focal temperature rise (Fig. 4.7) and proposed that this could be exploited as a noninvasive indicator for monitoring successful treatment delivery.

4.3.2.2 In Vivo Investigations

Hynynen (1991) investigated transient cavitation and temperature rise in vivo in a dog's thigh muscle exposed to FUS at frequencies from 0.246 to 1.68 MHz using

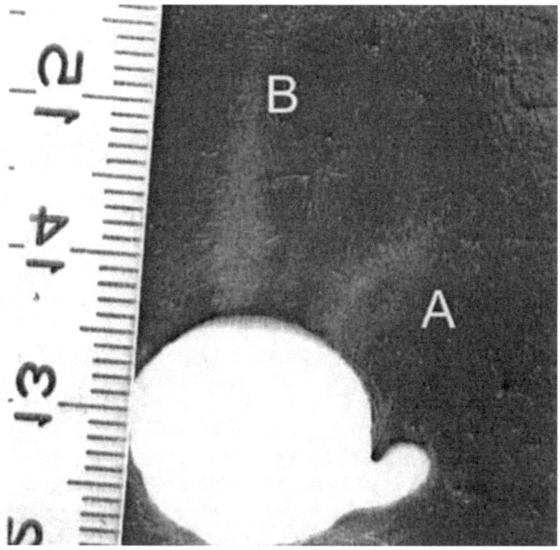

Fig. 4.6 Two elementary lesions: *lesion A* was induced by an US exposure of 14 W/cm^2 over 20 s; *lesion B* was induced by an US exposure of 60 W/cm^2 over 0.5 s, followed by an exposure at 14 W/cm^2 for 20 s. Reproduced with permission from Melodelima et al. (2004). © Elsevier

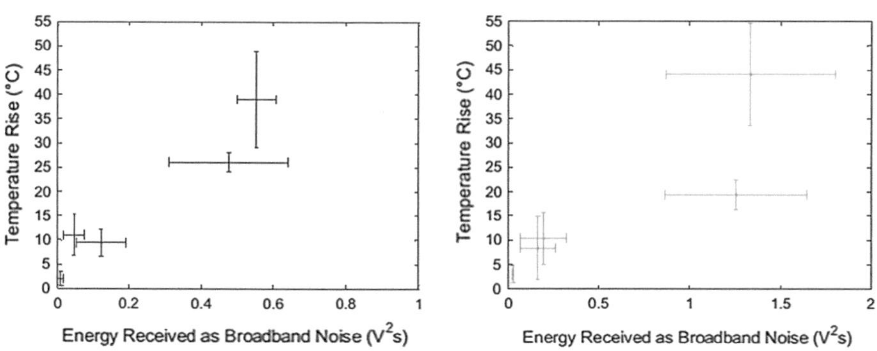

Fig. 4.7 Temperature rise achieved during HIFU exposures at 0.5 MHz as a function of the total energy received as broadband noise by PCD over 2 s (*left*) and 4 s (*right*). Reproduced with permission from Kyriakou et al. (2011). © Elsevier

1-s tone bursts and a range of acoustic pressures. As expected, the temperature rise had a linear dependence on acoustic intensity at low acoustic pressures, but the onset of enhanced temperature rise was correlated with the onset of the broadband emissions (monitored by a hydrophone) at high acoustic pressures. They concluded that the acoustically induced cavitation in the focal region led to the enhanced temperature response. In a different study, Umemura et al. (2005) measured the

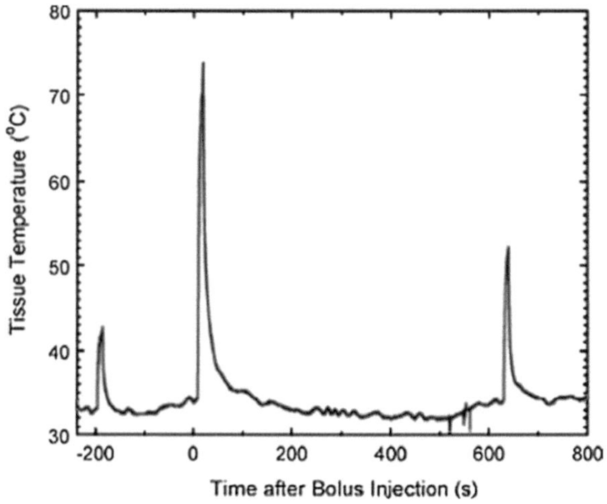

Fig. 4.8 Temperature rise with a series of three exposures plotted against time after injection of MBs. Reproduced with permission from Umemura et al. (2005). © IEEE

temperature rise in vivo in murine kidney samples exposed to FUS at 3.2 MHz when microbubble agents (Optison) were introduced into the body by means of an intravenous bolus injection. Compared with the temperature rise before the MBs injection, the temperature elevation immediately after the bolus injection was 3–5 times higher and gradually decreased as the MBs were eliminated from the body. Figure 4.8 shows the higher temperature rise for the exposure immediately after the bolus injection, compared with that recorded 200 s before or 650 s after this introduction of MBs.

4.4 High-Intensity Focused Ultrasound Ablation

HIFU refers to a technique where high-energy FUS beams are accurately aimed, through intact skin and tissues, at a deep-seated target volume (ter Haar 1995; Kennedy et al. 2003). Within seconds, the energy level at the focus of the beam is great enough to cause a temperature rise sufficient to kill cells instantaneously. Tissue ablation using HIFU can be achieved by raising the in situ temperature above a coagulation threshold in just a few seconds (ter Haar 2001b, 2007; Kennedy 2005; Bailey et al. 2003; ter Haar and Coussios 2007). Thus, a defined tissue volume can be thermally destroyed, while the ultrasound-penetrated skin and tissue layers outside the ablated area remain comparatively unaffected. HIFU works by causing tissue coagulation and necrosis from both thermal and cavitation effects. Clinical HIFU applications deliver energy at frequencies of 0.8–3.5 MHz, and the energy levels carried in HIFU beams are several orders of magnitude greater than those of a standard diagnostic ultrasound beam (Lu et al. 2014; Civale et al. 2006). Also, degassed water contained within a balloon is sometimes used to provide efficient coupling of the transducer to the patient. Although HIFU was first proposed in the middle of the last century, advances in medical imaging techniques for targeting and monitoring the procedure, such as magnetic resonance imaging (MRI) (Hynynen 2010; Wharton et al. 2007; Ong et al. 2003) and ultrasound (US) imaging (Zhang et al. 2009; Civale et al. 2013; Miller et al. 2004; Zhang et al. 2012a), have now enabled HIFU to become a practical possibility as a clinical intervention. Clinical investigations are now underway to use HIFU for hemostasis (Zderic et al. 2006), neurosurgery (Hertzberg et al. 2010; Tyler et al. 2010), obstetrics and gynecology (Griffiths et al. 2012; Denbow et al. 2000; Rivens et al. 1999), and treating tumors in various tissues (Gelat et al. 2012; Visioli et al. 1999) including the uterus (Chapman and ter Haar 2007; McDannold et al. 2006a), liver (Illing et al. 2005; ter Haar et al. 1989; Leslie et al. 2008; Kennedy et al. 2004b), kidney (Ritchie et al. 2010; Hacker et al. 2005), prostate (Murat and Gelet 2009), breast (Wu et al. 2007), bone (Napoli et al. 2013; Zhang et al. 2013), and brain (Damianou et al. 2009; Alkins et al. 2013; Chauvet et al. 2013).

4.4.1 In Vitro Methods of Evaluating Cavitation-Enhanced Thermal Effects

4.4.1.1 Tissue-Mimicking Phantom

The potential enhancement of cavitation to heighten the therapeutic efficiency of HIFU ablation has been evaluated in various tissue-mimicking phantoms. The effects of thermal enhancement on lesion formation around a vessel were determined when applying flowing encapsulated MBs (SonoVue) in a wall-less flow phantom (Zhang et al. 2011). The HIFU focus was positioned in the middle of the vessel, and no thermal lesion was observed around the vessel when saline that did not contain encapsulated MBs flowed through the vessel at 10 cm/s (72 W). However, when the MB solution (0.001 %) flowed through the vessel at the same acoustic power level and velocity, lesions appeared on both sides of the vessel. At a flow velocity of 20 cm/s, a smaller thermal lesion was observed at the front of the vessel, but no lesion was observed at the back of the vessel. Thermal lesions on both sides of the vessel became larger at an acoustic power of 96 W and a flow velocity of 20 cm/s. When the MB concentration was increased to 0.1 %, the thermal lesion appeared in front of the vessel and migrated toward the focused ultrasound (FU) transducer, due to the enhanced ultrasonic scattering and reflection caused by the MBs. These findings are shown in Fig. 4.9.

Tung et al. (2006) used ultrasound MBs (Definity) to enhance HIFU (1.85 MHz) ablation in a transparent tissue phantom. Their results confirmed that the lesion size depends strongly on electrical power and the concentration of MBs. When they used MBs, they could substantially either increase the size of the thermal lesion or produce a lesion of the same size by reducing the power by about 30 %, which could shorten the overall treatment time. However, using MBs of higher concentrations might shift the greatest heating position away from the transducer focus,

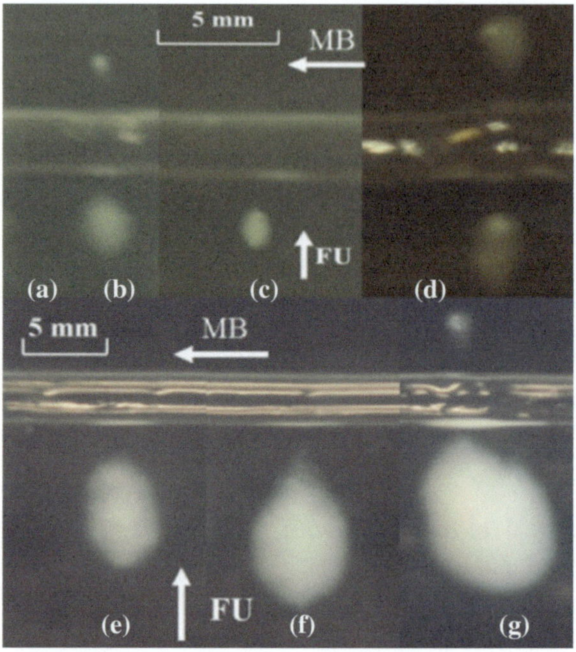

Fig. 4.9 Thermal lesions induced by HIFU exposure around a wall-less vessel in transparent phantoms. In (**a**) the flowing solution in the vessel was 0.9 % saline without MBs at a flow velocity of 10 cm/s. In (**b**) and (**d**) the MB concentration was 0.001 % and the flow velocities were 10, 20, and 20 cm/s, respectively. The MB concentration in (**e**) (**g**) was 0.1 %, and the flow velocities were 10, 20, and 30 cm/s, respectively. The *arrows* indicate the direction of flow of the MBs and of the HIFU beam. The acoustic power in (**a**) and (**c**) was 72 W, in (**d**) was 96 W, and in (**e**)–(**g**) was 120 W, with 80 % duty cycle for 15 s. Reproduced with permission from Zhang et al. (2011). © Acoustical Society of America

which might cause unwanted side effects during clinical applications. By compromising between the potential enlargement of lesion size and the shift in heating position, the authors obtained an optimal condition when using 0.001 % MBs at 50 W (Fig. 4.10). The thermal lesions were 12 times larger than those in control phantoms without MBs, and had an acceptable shift (less than half the lesion length). These experimental results suggest that MBs can effectively increase the size of the thermal lesion, but the potential shift of the lesion must be carefully considered when performing HIFU ablations.

Chen et al. (2003b) investigated the changes in lesion size, the pressure threshold and the mechanism of lesion formation after the addition of ultrasound MBs in transparent tissue phantoms exposed to 1.1 and 3.5 MHz acoustic fields. They suggested that MBs would lower the pressure threshold of HIFU ablation and enlarge the thermal lesions, which might be beneficial for HIFU therapy (Fig. 4.11).

While MBs have considerable potential for enhancing HIFU ablation, ultrasound propagation can become disturbed when too many MBs lie between the transducer

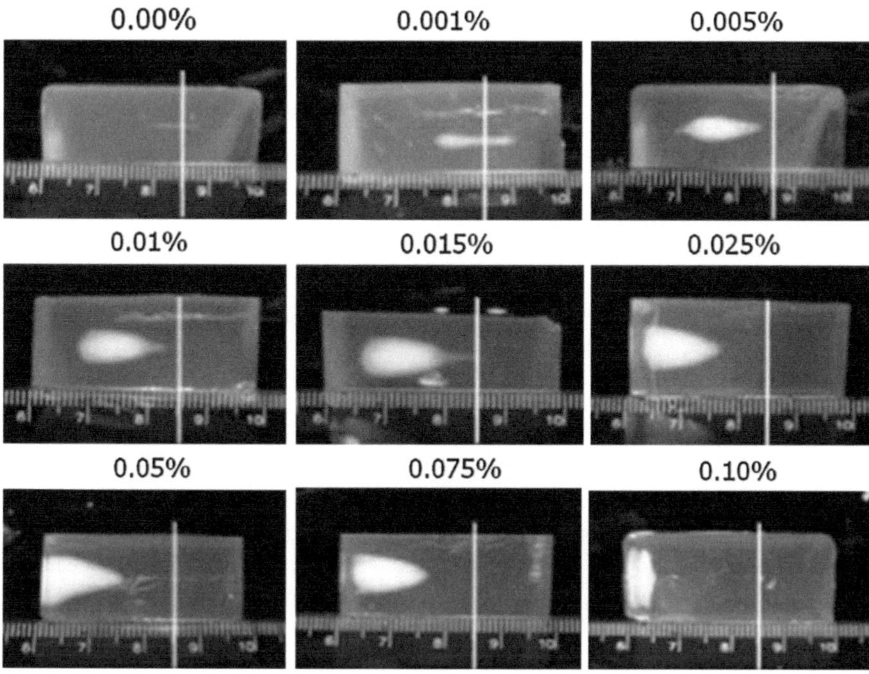

Fig. 4.10 Lesions formed after introducing MBs of different concentrations at 50 W. The total heating time was 30 s and the transducer was on the *left-hand side*. Reproduced with permission from Tung et al. (2006). © Elsevier B.V.

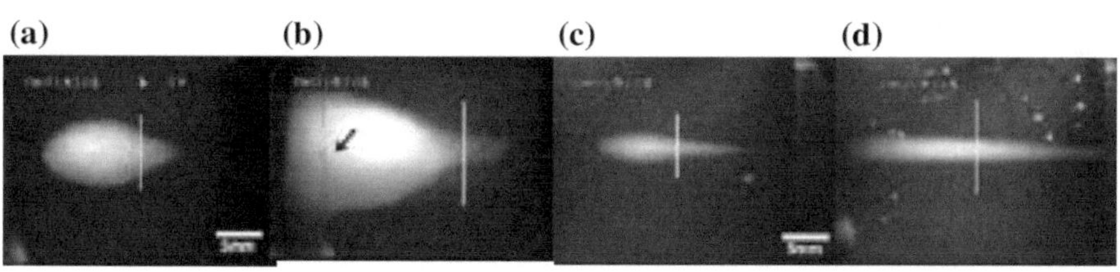

Fig. 4.11 Lesions formed in phantoms without MBs (**a**, **c**) and with MBs (**b**, **d**) for exposures at 1.1 (**a**, **b**) and 3.5 MHz (**c**, **d**) at $P-$ = 4.8 MPa. Reproduced with permission from Chen et al. (2003b). © Acoustical Society of America

and the focus. A high concentration of MBs between the transducer and the focus would introduce a high attenuation reducing the penetration of the ultrasound energy and scattering it, shifting the position of maximal heating from the transducer focus. This would be detrimental to HIFU therapy because control of the coagulated region is critical for a successful HIFU treatment, and the thermal effects are expected to form around the geometric focus of HIFU. However, when too many MBs are introduced, the disturbance of the HIFU acoustic field between the transducer and the focus may cause the lesion to migrate or to change its shape. For example, a lesion might form in front of the transducer focus or even at the surface of the phantom, which may produce unwanted side effects during clinical applications.

To avoid such unwanted consequences, Kajiyama et al. (2010) proposed a method of achieving accurate position control of enhanced heating at the HIFU focus. They destroyed MBs in front of the HIFU focus on the transducer side by using intense pulses for microseconds, to clear the path, before using the high-intensity ultrasound beam to heat the target region. They thus prevented unintentional heating in the path from the transducer to the focus. The experiment was performed in a polyacrylamide phantom containing MBs, and a temperature-sensing liquid crystal sheet was set in the focus to measure the temperature distribution. As the number of intense burst waves for destroying MBs increased, the heated region moved from the transducer toward the focus, indicating that MBs were effectively destroyed in front of the focus (Fig. 4.12). This study suggests that when using MBs to enhance HIFU, the heating position can be controlled by introducing a sufficient number of short pulses prior to HIFU ablation.

In another study aimed at controlling the ablation location of HIFU, Phillips et al. (2013) proposed that their new nanodroplet perfluorocarbon (PFC) formulation would thermally and spatially enhance HIFU ablation at the intended target site (Figs. 4.13 and 4.14). They investigated and compared the effects on enhanced heating of encapsulated gas-filled MBs with NDs in a

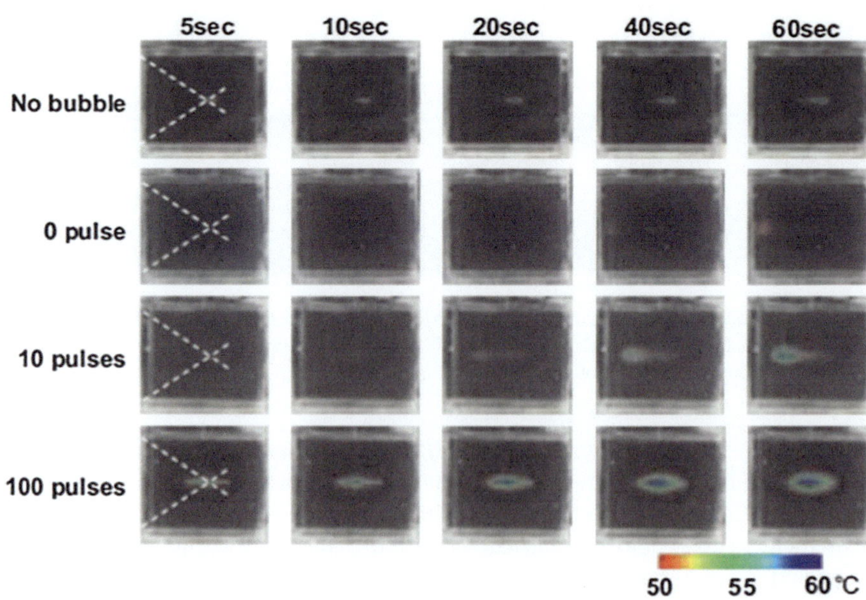

Fig. 4.12 Snapshots of the temperature distribution profile for different acoustic pulse numbers. Reproduced with permission from Kajiyama et al. (2010). © Elsevier B.V.

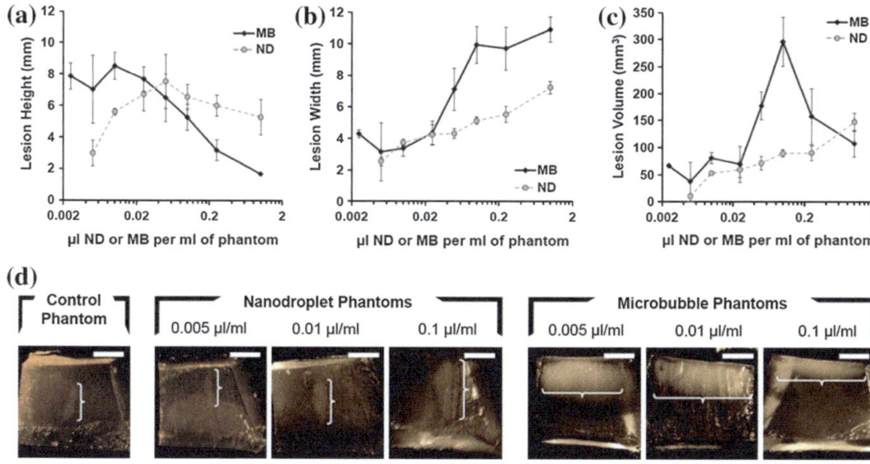

Fig. 4.13 Ablation lesion height (**a**), width (**b**), and corresponding volume (**c**) as a function of ND or MB concentration in tissue-mimicking phantoms following HIFU exposure for 20 s at 4 MPa (**d**). Representative images of lesions formed inside the albumin-acrylamide phantoms containing no agents, ND, or MB at various concentrations. The location and size of each lesion is indicated by the *superimposed white bracket*. The *white scale bar* in the *upper right corner* of each image represents 5 mm. Reproduced with permission from Phillips et al. (2013). © Acoustical Society of America

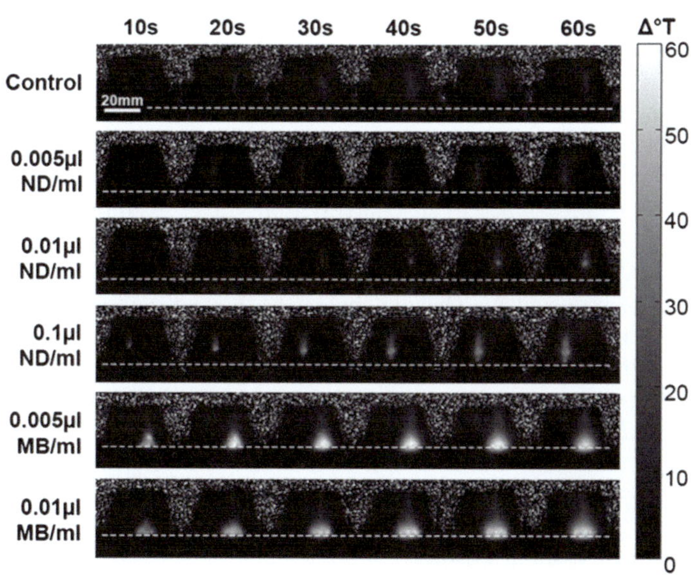

Fig. 4.14 MR thermometry maps of phantoms containing no agents (*top row*), NDs (*second to fourth rows*), or MBs (*fifth and sixth rows*). Reproduced with permission from Phillips et al. (2013). © Acoustical Society of America

tissue-mimicking acrylamide–albumin phantom. The dual perfluorocarbon droplet had PFC cores containing a mixture of decafluorobutane and dodecafluoropentane encapsulated in a phospholipid shell. After the NDs and MBs were incorporated into the tissue-mimicking phantom, magnetic resonance (MR) thermometry mapping during HIFU exposure showed that for the phantom containing MBs, the maximal temperature rise occurred near the surface of the phantom. However, for the phantom containing NDs, the maximal temperature rise occurred near the acoustic focus in the center of the phantom. The maximum change in temperature rise at the acoustic target site was enhanced by 16.9 and 37.0 % using MBs and NDs, respectively. Thus, these perfluorocarbon NDs may allow shorter procedure times to ablate tumors during FUS surgery and may also safely enhance thermal deposition at the target site, potentially offering several advantages over the use of MBs with HIFU ablation.

In another tissue-mimicking phantom study, Arvanitis and McDannold (2013) integrated an ultrasound imaging array into an MRI-guided FUS (MRgFUS) system to visualize temperature increases using MR temperature imaging (MRT1) and detect cavitation activity using passive acoustic mapping (PAM) simultaneously. They tested the combined system with ultrasound exposure that induced cavitation-enhanced heating in a tissue-mimicking phantom and in a phantom-filled ex vivo macaque skull. They also compared PAM with standard B-mode ultrasound images and tested the feasibility of the system for mapping cavitation activity during low-intensity burst ultrasound exposure of a mock vessel (filled with an ultrasound contrast agent) in the skull phantom. In the absence of broadband emissions, only a moderate temperature rise was observed on MRI. When inertial cavitation was seen, localized cavitation activity was present in PAM and the temperature rise produced was several times higher than would be expected from the increase in acoustic power. After registration of the two imaging modalities, the location of the maximum cavitation activity in the passive acoustic images agreed with the center of the focal heating in MRI on average, but some variation was observed. This was presumed to arise largely from image distortion and other MRI issues. Although there was substantial variation between individual measurements, a nonlinear relationship between the strength of cavitation activity and the peak temperature rise was evident. Thus, as this study showed successful integration of an ultrasound imaging array into a clinical MRgFUS system to map localized cavitation activity and temperature rise simultaneously, the authors suggest that this integrated approach may offer a promising way for developing controllers for cavitation-enhanced ablation and optimizing the development of ultrasound therapies.

4.4.1.2 In Vitro Human Plasma

While many in vitro studies have been performed using tissue-mimicking phantoms, only a few have explored other media. Takegami et al. (2005) performed an in vitro investigation of the temperature rise in human plasma with various

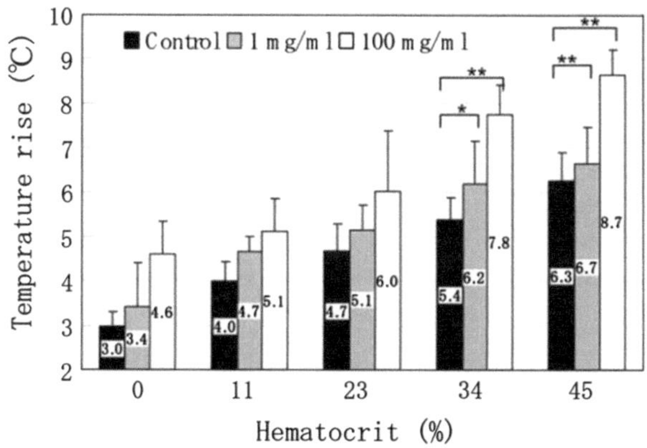

Fig. 4.15 Temperature rise after 2.17 MHz, 60-s HIFU irradiation of human blood with different concentrations of MBs. The average spatial peak intensity was 30 W/cm^2 in all experiments. Reproduced with permission from Takegami et al. (2005). © Elsevier B.V.

concentrations of human erythrocytes in combination with or without MBs (Levovist). They observed a significant correlation between the HIFU-induced temperature rise and the hematocrit level, with a correlation coefficient of 0.998 ($p = 0.0001$); furthermore, the temperature rise was significantly increased by adding MBs to the suspension ($p < 0.01$) (Fig. 4.15).

4.4.2 *In Vivo Applications of Cavitation-Enhanced Thermal Effects*

The introduction of cavitation MBs before HIFU ablation of tumor can effectively increase lesion size and reduce either the treatment time or the required ultrasound intensity, or both. Thus, a number of in vivo experiments on a variety of organs have investigated the use of MBs to enhance HIFU volume ablation rates.

4.4.2.1 Effects in the Liver

The potential enhancement of cavitation using MBs combined with HIFU ablation has been evaluated in livers of various animals. Fourteen days after implantation of VX2 tumors in the livers of 50 rabbits, Luo et al. (2007, 2008, 2009) randomly separated the animals into two groups. Boluses of encapsulated MBs (SonoVue) were injected into ear veins of rabbits in group II, and a normal saline solution was administrated in group I. The volumes of thermal lesions in the livers of rabbits with MBs were generally larger than those found in the control group. Also, in the control group, residual vessels remained in the periphery of ablated areas, while no residual vessels were found in the livers of animals that received the MBs (Fig. 4.16). The results also showed that by using MBs with HIFU, a larger tissue volume can be ablated in a shorter time and the residual vessels within the tumors can be reduced.

In another study with liver tissue, He et al. (2011) found that MB-assisted HIFU ablation could lower the required intensity of HIFU and destroy targeted tumors more effectively and with fewer residual viable tumor nodules than could HIFU ablation alone. Li et al. (2009) also showed that HIFU ablation combined with MBs destroyed cell ultrastructure more effectively and produced a larger ablation region than with HIFU ablation alone. These findings suggest that the therapeutic efficiency of HIFU ablation on rabbit liver tumors can be enhanced using encapsulated MBs. Yu et al. (2006a) investigated the effect of cavitation MBs on HIFU ablation in goat livers (Fig. 4.17). When MBs were delivered intravenously prior to HIFU, the tissue necrosis rate was higher, and pathological examination revealed that there was no residual unaffected tissue within the exposed volume. When they examined the therapeutic efficiency of using MBs with HIFU ablation in rabbit livers (New

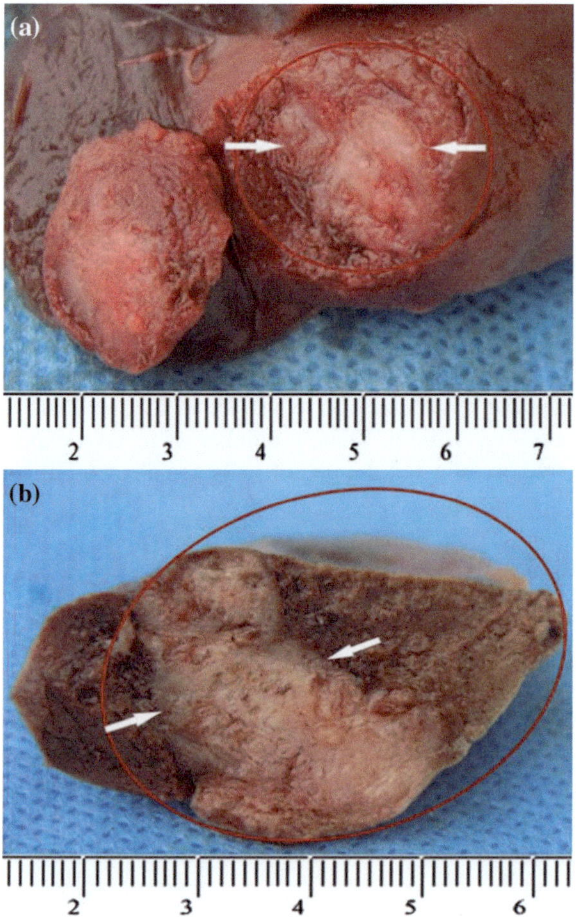

Fig. 4.16 Macroscopic images of the ablated areas (*red ring*) in VX2 liver tumors (*arrows*) in (**a**) group I and (**b**) group II. Reproduced with permission from Luo et al. (2009). © Society of Surgical Oncology

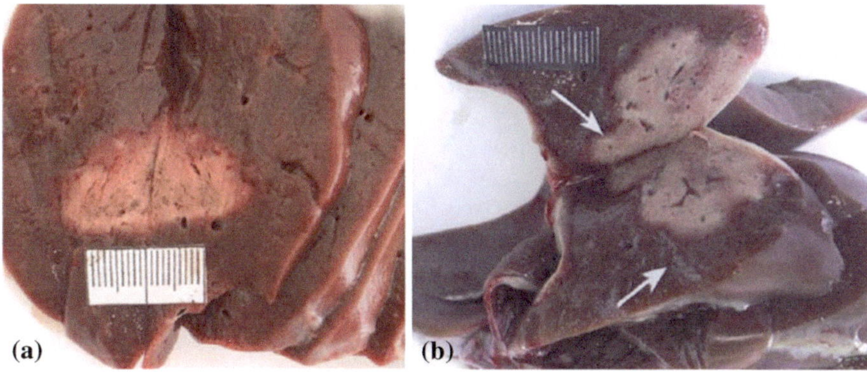

Fig. 4.17 Macroscopic inspection of liver ablation (**a**) without and (**b**) with MBs. Administration of MBs led to a larger volume destroyed. Reproduced with permission from Yu et al. (2006a). © Springer-Verlag

Zealand rabbits), they again found a higher necrosis rate with MBs (Yu et al. 2006b). Similar findings have also been reported in rat livers (Hanajiri et al. 2006).

In addition to studies on the thermal lesions induced by HIFU, the temperature rise has also been measured when using encapsulated MBs to improve HIFU ablation efficiency. Kaneko et al. (2005) studied the effect of MBs (Levovist) on the temperature rise and tissue necrosis induced by HIFU in rabbit livers (Fig. 4.18). Their results indicate that the volume of HIFU-induced thermal lesions increased in animals that were given encapsulated MBs and that the temperature rise in these animals was also higher.

Many studies have shown that using MBs with HIFU for liver tissue ablation produces a larger necrosis volume and requires less time to perform ablation, but not many have examined the shape of thermal lesion after ablation. In one study by Chung et al. (2012), the authors found that a pyramidal shape was more prevalent in

Fig. 4.18 Temperature rise in the liver during HIFU exposure. Reproduced with permission from Kaneko et al. (2005). © Springer-Verlag

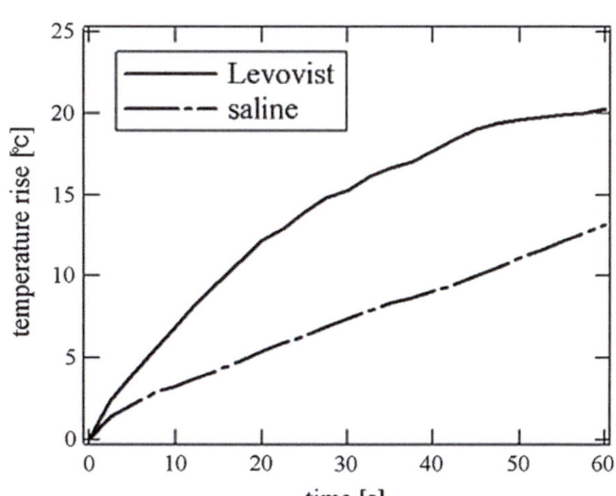

tissues that underwent HIFU without MBs than for those that underwent the combination therapy ($P < 0.05$). They also found evidence of more cavitary tissue necrosis for the combination therapy group than when MBs were not used.

4.4.2.2 Effects in the Kidney

Another organ in which the effects of HIFU have been tested is the kidney. Yu et al. (2004) investigated the effect of lipid-coated perflupropane MBs on the therapeutic efficiency of HIFU ablation in rabbit kidneys. Compared with the control group, necrosis rates in the kidneys of New Zealand rabbits were increased in both the left and right kidneys when ablated using HIFU in combination with lipid-coated perflupropane MBs administrated intravenously prior to HIFU. They also investigated the effects of MBs on the therapeutic efficiency of HIFU ablation in goat kidneys and found that necrosis rates were increased when MBs were used (Yu et al. 2008). By enlarging the ablated tissue volume in a shorter time, these findings suggest that MBs can improve the therapeutic efficiency of HIFU. In another study, Tran et al. (2003) explored the use of stabilized MBs (Optison) to aid in tissue ablation of canine kidneys in vivo. They showed that these MBs significantly lowered the threshold for tissue damage: Both the required intensity and exposure duration were reduced when MBs were present. They also evaluated the threshold for macroscopic tissue damage with MB infusion rates (IR) to determine is the level that was effective for HIFU ablation in the kidneys of adult female dogs (Tran et al. 2005). In this study, canine kidneys were exposed to HIFU with 1.44-MHz tone bursts. As the IR increased, the threshold intensity decreased by up to a factor of 3 and the threshold duration decreased by up to a factor of 200 (Fig. 4.19).

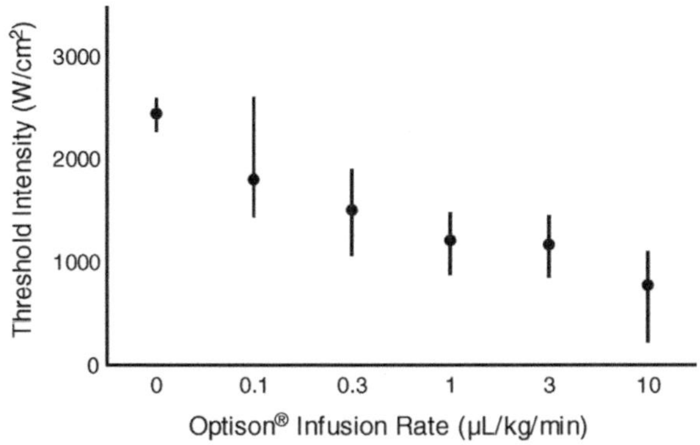

Fig. 4.19 Threshold intensities with 95 % fiducial intervals. The threshold intensity of a 250-ms exposure with Optison was significantly lower than that without Optison. Reproduced with permission from Tran et al. (2005). © IEEE

4.4.2.3 Effects in the Brain

HIFU is potentially also important for treating tumors in brain and so a few studies have examined this organ specifically. The efficiency of MBs (Optison) in creating brain lesions using 1.5-MHz HIFU ablation with a range of exposure parameters was evaluated in rabbit brain (McDannold et al. 2006b; Vykhodtseva et al. 2006). MR imaging was used to map the temperature distribution, and the threshold temperature rise for thermal lesion formation was compared for HIFU ablation performed with and without MBs (Fig. 4.20). Their results showed that the temperature rise was statistically larger ($P < 0.01$) when MBs were used than when they were not. The authors found that MBs enhanced the heating and reduced the acoustic power threshold required to induce lesions by more than a factor of 10.

In a different study, Huang et al. (2013) used low-intensity FUS with MBs to produce brain lesions in rats at 558 kHz. Compared with the control group, acoustic power levels could be significantly reduced when MBs were used with FUS, and problems with heating the skull could be avoided. The authors proposed that FUS combined with MBs could produce brain lesions with low power at a frequency suitable for transcranial application and could thus potentially extend the treatable volume of transcranial FUS to tissues lying close to the skull.

4.4.2.4 Effects in the Thigh

HIFU studies that employ MBs have also been performed on thigh tissue. In a study by Sokka et al. (2003), a high-intensity burst well above the cavitation threshold

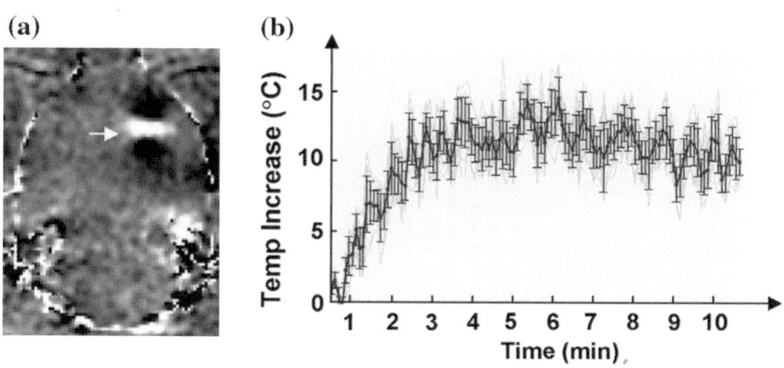

Fig. 4.20 Magnetic resonance thermometry measurement of one 5-min sonication at 444 kPa, 10 ms/Hz. **a** Arrow indicates temperature change at the focus (bright), surrounded by dipole pattern artifacts (dark). **b** Temporal profile of the temperature elevation and subsequent cooling at the focal volume. Reproduced with permission from Huang et al. (2013). © Elsevier

was used to generate cavitation MBs at the beginning of the sonication. Then, conventional CW heating was used immediately thereafter to achieve typical HIFU ablation. The cavitation-enhanced exposures showed a faster rise in temperature within the first 4 s and a higher overall temperature rise than those without MBs. For tissues exhibiting cavitation enhancement, their lesions were 2–3 times larger by volume, were more spherical, and lay closer to the transducer than lesions in the control exposures (Fig. 4.21). This suggests that in thigh tissue, cavitation can again lower the required power of HIFU and be used to create significantly larger lesions in vivo.

In another study that explored FUS in thigh tissue, Kopechek et al. (2013) examined the accumulation of phase-shift nanoemulsions (PSNE) in intramuscular rabbit tumors in vivo using MRgFUS as a noninvasive technology to treat solid tumors (Fig. 4.22). Before experiment, VX2 cells were injected directly into the left thigh of New Zealand white rabbits, and PSNE were injected intravenously through a catheter in the ear vein and flushed with 0.5 ml saline. MR signal enhancement was observed in rabbit tumors, indicating that PSNE accumulated in the tumors. The brightness of the B-mode ultrasound image increased at the HIFU focus after the acoustic droplet vaporization (ADV) pulse, which indicated the presence of bubbles that were formed by PSNE vaporization. Furthermore, MR thermometry measurements indicated that PSNE-enhanced the rate of HIFU-mediated heating. In

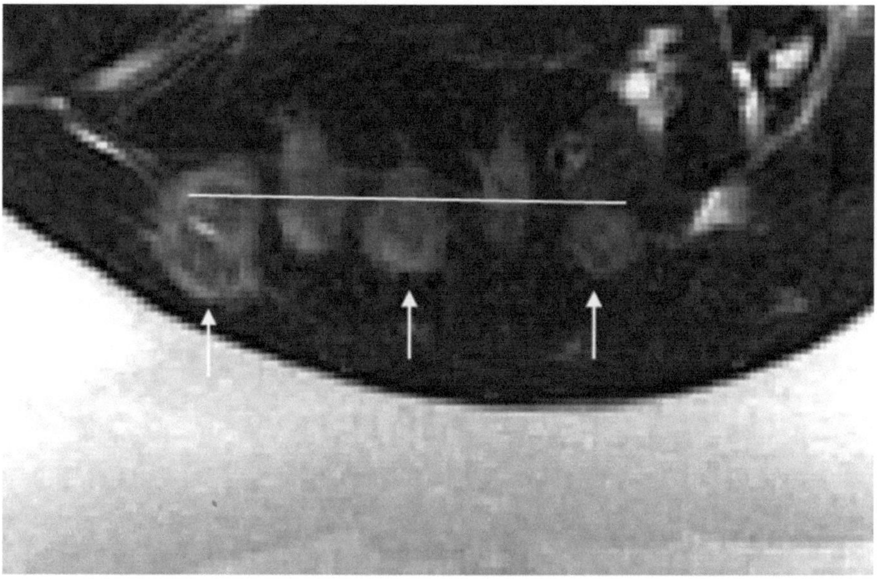

Fig. 4.21 T2-weighted images of the thermal lesions after five consecutive sonications at 1.7 MHz. From left to right, the *arrows* indicate the 300–21 W, 300–14 W, and 300–7 W sonications, respectively. The control lesions for the first two bubble-enhanced sonications are also visible immediately to the right of the respective experimental sonications. Reproduced with permission from Sokka et al. (2003). © IOP Publishing Ltd.

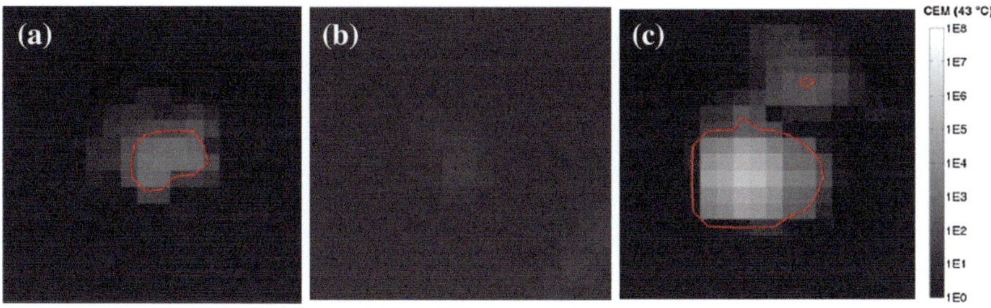

Fig. 4.22 Thermal dose maps from HIFU treatments in tumors with and without PSNE, plotted in units of cumulative equivalent minutes (CEM) at 43 °C. **a** Tumor with PSNE, sonicated for 10 s at an acoustic power of 15.7 W. **b** Tumor without PSNE, sonicated for 10 s at an acoustic power of 16.5 W. **c** Tumor without PSNE, sonicated for 10 s at an acoustic power of 38.5 W. Reproduced with permission from Kopechek et al. (2013). © Multi Science Publishing

their follow-up work (Kopechek et al. 2014), the authors showed that PSNE could be vaporized into MBs in tumors in order to nucleate cavitation activity and that cavitation emissions were well correlated with enhanced HIFU-mediated heating in tumors. This suggests that PSNE, as many MBs, may be able to improve the clinical feasibility of MRgHIFU by reducing both treatment times and the required acoustic intensities.

4.4.3 Preclinical Evaluation of Cavitation-Enhanced Thermal Effects

Cavitation MBs enhance local heating, and many experiments have shown increased therapeutic effects when stabilized, gas-filled MBs are introduced prior to, or during, HIFU procedures. These findings suggest that MBs may be able to aid clinical treatment during surgical application of ultrasound therapy and may therefore spark interest in developing new ultrasound therapy methods that employ cavitation MBs. This section explores the preclinical investigations of cavitation MBs used with HIFU to treat uterine fibroids.

In order to evaluate the feasibility of HIFU-associated microbubble-enhancing therapeutic ablation of uterine fibroids in patients, a total of 291 patients with a solitary uterine fibroid were treated with HIFU in China (Peng et al. 2012). Of these, 129 patients in the Suining Central Hospital of Sichuan were treated without encapsulated MBs, and 162 patients in the First Hospital of Chongqing Medical University and Chongqing Haifu Hospital were treated with encapsulated MBs (SonoVue). The term "median fractional ablation" was defined as the nonperfused volume divided by the fibroid volume immediately after HIFU treatment. The median fractional ablation values in the groups with and without the MBs were 86 and 83 %, respectively. In this study, the sonication time required to ablate 1 cm^3 of fibroid was 16 s in the group with MBs and increased to 26 s in the group without

MBs. Thus, the authors suggested that the encapsulated MBs enhanced the outcome of HIFU ablation and can be used to assess the extent of treatment. In a similar study, Jiang et al. (2014) evaluated the clinical value of encapsulated MBs in the treatment of uterine fibroids with HIFU ablation in patients with a history of pelvic operation. A total of 80 patients with solitary uterine fibroids at a single center were randomly assigned to a control or SonoVue group (half-and-half) and were treated with HIFU. This study suggested that encapsulated MBs may yield a high success rate and reduce the sonication time and acoustic energy required for HIFU treatment of uterine fibroids (Fig. 4.23).

Other preclinical investigations have been performed using MBs with HIFU for treating liver tumors. In one such study, Elbes et al. (2013, 2014) demonstrated the feasibility of increasing the heating efficiency for treating liver tumors by exploiting cavitation-enhanced heating in pig livers using a clinical HIFU platform (Fig. 4.24). This study compared the heating patterns and thermal lesion characteristics resulting from continuous sonication or sonication that included a burst pulse. The results showed that the cavitation group exhibited a near-spherical heating with a 40 % mean enhancement of the maximal temperature rise as compared to the

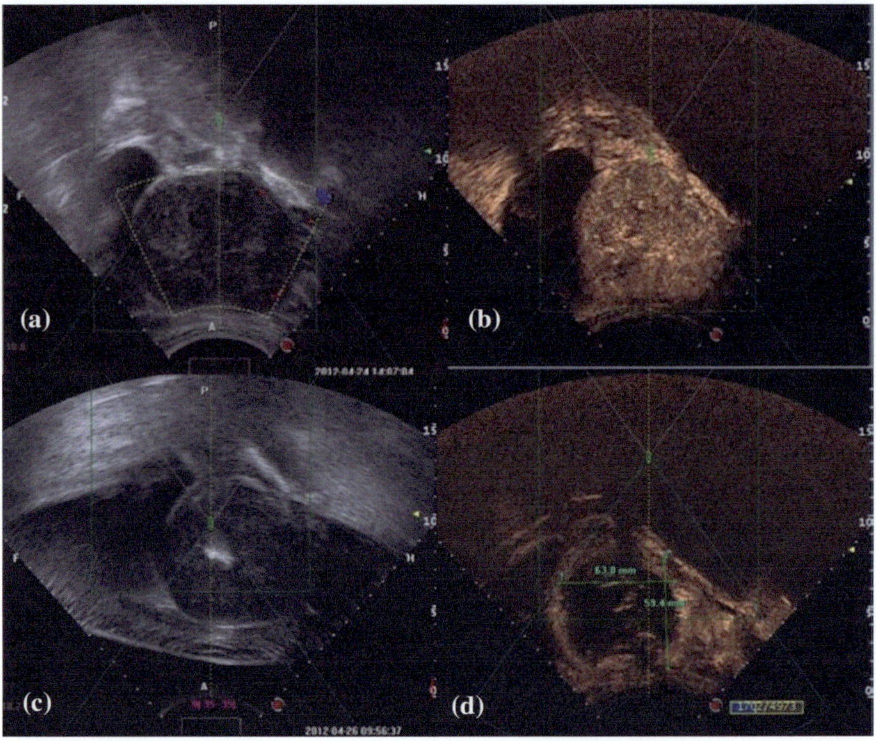

Fig. 4.23 **a**, **c** Ultrasound images and **b**, **d** contrast-enhanced ultrasound images **a**, **b** before and **c**, **d** after a successful treatment of a uterine fibroid using MBs to enhance the HIFU treatment. Reproduced with permission from Jiang et al. (2014). © Springer Science

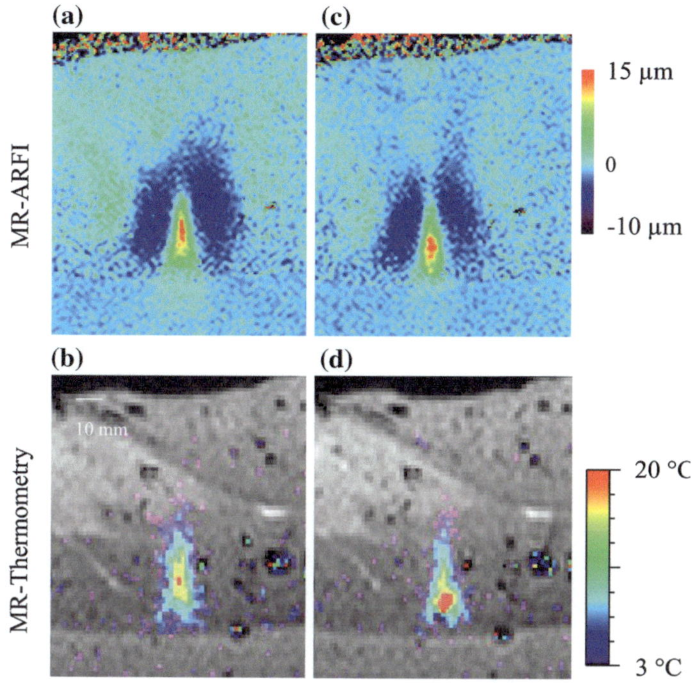

Fig. 4.24 Local displacement maps (**a**, **b**) and temperature images (**c**, **d**) for the conventional group (continuous, 20 W, 30 s, *left column*) and composite group (*right column*). In the composite group, a HIFU burst pulse (0.5 s in duration, 250–800 W) was delivered before continuous sonication identical to that in the conventional group. Displacement images were obtained at an acoustic power of 150 W, and temperature images were obtained at the end of identical sonications at 20 W during 30 s. Displacement and temperature scales are indicated on the *right*. MR-ARFI = magnetic resonance acoustic radiation force imaging. Reproduced with permission from Elbes et al. (2014). © Elsevier B.V.

conventional sonication group. This suggests that more efficient heating can be achieved by using acoustic cavitation-enhanced heating while still protecting tissues located beyond the focal point.

4.5 Thermal Effects of Flowing Microbubbles

4.5.1 Effects of Blood Flow on Heating

In tissues heated by FUS, the accuracy of the heating prediction and quantification of the temperature distribution is affected by the structure of the organ. Vessels adjacent to the focal region of FUS are resistant to the temperature rise due to the heat-sink effect of blood flow. Compared with the results obtained when both the hepatic artery and the portal vein were ligated, Chen et al. (1993) found that the diameter of the thermal lesions in rat livers exposed to FUS was reduced by 14 % with normal blood flow and that the reduction was more than 20 % for longer exposure durations. In a different study, Curra et al. (2000) found that both the heating rate and temperature rise are affected by the blood flow in vessel. This convective transport phenomenon is related to many factors, such as the vessel diameter, blood flow velocity, vessel location, and operating parameters of therapeutic ultrasound.

4.5.1.1 Effects of Vessel Diameter and Blood Flow Velocity

Huang et al. (2010a) used a countercurrent blood vessel network model to calculate the temperature rise in a tissue. They found that as the ratio of diameters in successive vessel generations increased, there was about a 0.8 % decrease in total optimized absorbed power in the focal region. They also found that hyperthermia treatments improved significantly as the velocity of blood flow in the treated region decreased. Finally, another study by Heydari and Jahed (2009) also demonstrated that there are higher temperature rises in treatment regions adjacent to smaller vessels.

4.5.1.2 Effects of Vessel Location on Heating

The location of blood vessels can also affect the temperature distribution, and this has been shown in simulation. Hariharan et al. (2007) built a three-dimensional computational model to study the efficacy of FUS procedures performed near a large blood vessel. They demonstrated that the lesion volume was reduced by about 40 % when the vessel was parallel to the acoustic beam, while it was reduced about 20 % when the blood vessel was perpendicular to the beam. These simulations also showed that when the heating duration was shorter than 2 s, or the gap between the large blood vessel and the tumor was >2 mm, the location of the large blood vessel did not significantly affect the temperature distribution or the thermal dose profile.

4.5.1.3 Effects of Focused Ultrasound Parameters

Although the temperature rise in tissue can be reduced by blood flow in vessel, this reduction can be minimized by varying the therapeutic ultrasound exposure parameters. Davis et al. (2006) used the New Zealand rabbit ear model and varied the ultrasound parameters using a dose-escalation approach. They found that when the vessel had minimal adjacent connective tissue, such as in the rabbit ear, the temperature rise during FUS treatments at lower frequencies was less affected by blood flow than at higher frequencies. Moreover, they suggested that it was possible to achieve favorable results using lower duty cycle exposures. The simulations by Heydari and Jahed (2009) indicate that longer exposure durations can produce higher temperature rises when the heat-sink effect of blood flow influences the treatment. However, when the vessel diameter is larger than 2 mm, neither longer exposure duration nor higher exposure power results in complete ablation of tumors. In addition, Hariharan et al. (2007) suggested that the temperature rise during multiple-exposure procedures may be affected by blood flow less than that during single exposure procedures.

4.5.2 *Cavitation-Enhanced Heating by Flowing Microbubbles*

Increased therapeutic effects have been shown both theoretically and experimentally when encapsulated MBs are introduced into the body by means of an intravenous bolus injection. These increased therapeutic effects occur because of the enhanced local cavitation and heating caused by the ultrasound exposure. The encapsulated MBs, used as ultrasonic contrast agents, have been successfully employed in diagnostic ultrasonography because they intensify the reflection and scattering of ultrasonic waves, resulting in improved images (Qin et al. 2009; Bhatia and Senior 2008; Kennedy et al. 2004a). The first-generation commercial agents contained an air core and were stabilized by a coating of albumin, but were not stable enough to ensure consistent pulmonary passage (Sboros 2008; Kaul 2008). A second generation of agents was developed. These contained a fluorinated gas core and a thin lipid monolayer shell (Kaul 2008; Sboros 2008). Recently, many more studies have investigated the use of polymer-shelled MBs loaded with targeting ligands and drugs (Cavalieri et al. 2005) and have explored how they concentrate selectively and disrupt the ultrasound focal regions (Von Zur Muhlen et al. 2008). These findings suggest that MBs may be able to aid treatment during surgical applications of ultrasound therapy.

4.5.2.1 Minimizing Thermal Losses Using Flowing Encapsulated MBs

The presence of vascular structures can limit HIFU's ability to elevate the tissue temperature because of blood flow. Conversely, when a solution containing MBs is static, the cavitation induced by these MBs is reduced as the encapsulated MBs are destroyed by the ultrasound exposure. Thus, a flowing MB solution might replenish the MBs destroyed by ultrasound and maintain the cavitation activity and tissue heating. However, as mentioned previously, as blood flow velocity increases through a vessel, convective heat losses increase. The magnitude of these losses also depends on the temperature around the vessel, but the heat losses can be prevented by the additional heating effects that occur when MBs are replenished. Thus, studies have systematically assessed how to use flowing MBs with acoustic cavitation to minimize thermal losses from perfusion in a HIFU field (Zhang et al. 2011).

As shown in Fig. 4.25, a wall-less flow phantom with embedded thermocouples was constructed and exposed to FUS at various acoustic powers. The cavitation activity and temperature rise were measured simultaneously when MBs flowed through the vessel with varying flow velocities and concentrations. Experiments were performed in a Lucite tank filled with degassed and deionized water. Initially, each experiment was maintained at room temperature and the sample was allowed to cool completely between each FUS exposure. The appropriate amount of SonoVue was slowly injected into a separate vial containing 100 ml of degassed saline, giving concentrations of MB solution: from 0.001 % (v/v, equivalent to 2×10^3 bubbles/ml) to 0.1 %. The solution was gently shaken to produce a uniformly distributed MBs solution and injected into the phantom reservoir. The flow of MBs was sustained in a cylindrical, wall-less flow vessel of the phantom. Transparent polyacrylamide phantoms containing bovine serum albumin (BSA) were used to simulate tissue having acoustic properties close to those of soft tissue (Zhang et al. 2014). Cavitation activity was monitored using a PCD, and broadband noise was used to quantify the inertial cavitation dose (ICD) (Chen et al. 2003a; Bull et al. 2013; McLaughlan et al. 2010). In order to investigate the enhanced heating and cavitation created by the flowing MBs, temperature and cavitation were measured as a function of flow velocity of the MB solution at different concentrations and exposed to FUS at a range of acoustic power levels. Pure controls at the same exposure parameters and flow velocity were also studied.

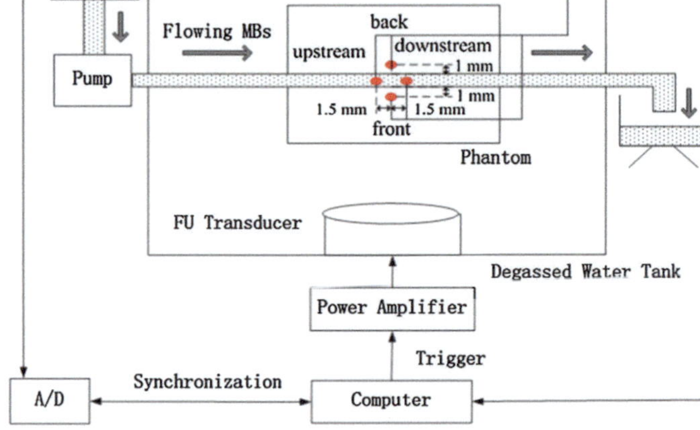

Fig. 4.25 Schematic of the experimental apparatus. Reproduced with permission from Zhang et al. (2011). © Acoustical Society of America

The temperature rise at the front of the vessel and the normalized ICD were measured simultaneously with varying flow velocities, from 10 to 55 cm/s, of the MB solution at a concentration of 0.02 %. The temperature increased at the start of the FUS exposure and the values of peak temperature at the end of the FUS exposure were recorded (Fig. 4.26). At the lowest flow velocity of 10 cm/s, the temperature rose from an initial temperature of 23 °C to a peak temperature 50 °C at the end of the FUS exposure. In the range of flow velocities from 10 to 30 cm/s, the values of the peak temperature increased from 50 to 56 °C. When the flow velocity exceeded 30 cm/s and increased to 55 cm/s, the peak temperature decreased to 47 °C. Meanwhile, cavitation increased as the velocity increased from 10 to 55 cm/s. Results from the pure controls showed that the peak temperature constantly decreased with increasing velocities of the MB solution, and there were no significant changes in cavitation levels with increasing flow velocities.

In the same study, the values of the peak temperatures at the end of the FUS exposures were recorded at the outer (front and back) regions of the vessel with varying flow velocities from 2 to 55 cm/s (Fig. 4.27). The peak temperature initially increased with flow velocity. At a flow velocity of 10 cm/s, the peak temperatures at the front and at back of the vessel were 52 and 50 °C, respectively. As the flow velocity increased to 35 cm/s, the peak temperatures in front and at back of the vessel increased to 55 and 53 °C, respectively. When the flow velocity exceeded 35 cm/s, the peak temperature decreased with increasing flow velocity. The peak temperatures at the front and at back of the vessel decreased to 51 and 49 °C when the flow velocity increased to 55 cm/s. In the case of the pure controls, the values of peak temperature were lower than those measured in the presence of MBs at the

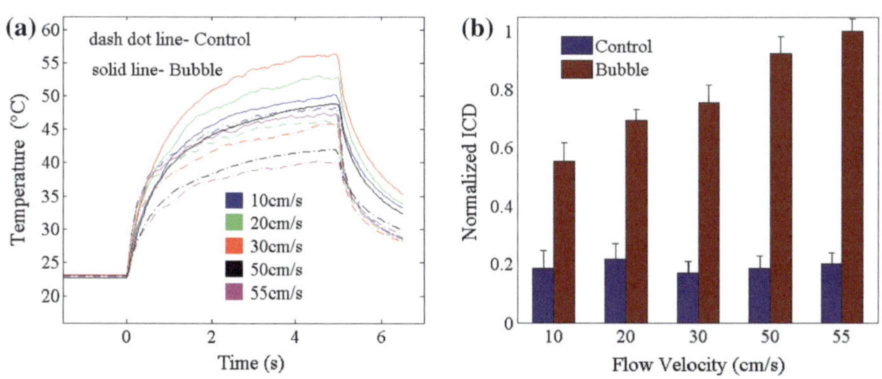

Fig. 4.26 **a** Measured temperature rise versus time at the front of a 3-mm diameter vessel in the phantom for different flow velocities of MB solution at a concentration of 0.02 %. **b** Normalized ICD versus flow velocities of solution with and without MBs (pure controls). The FUS focus was positioned in middle of the vessel, with an acoustic power of 27 W and an exposure time of 5 s. In **b**, the values shown represent the average of five measurements for each velocity; error bars show the standard deviation. Reproduced with permission from Zhang et al. (2011). © Acoustical Society of America

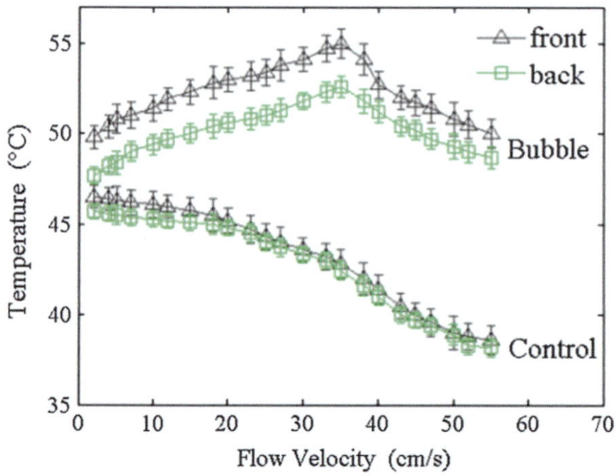

Fig. 4.27 Peak temperatures at the end of the FUS exposures at the front and at back of a 3-mm vessel with varying flow velocities from 2 to 55 cm/s at an acoustic power 25 W for 5 s and an MB concentration of 0.02 %. Results from pure controls (without MBs) are also shown. The FUS focus was positioned in the middle of the vessel. The values shown are the average and standard deviation of five measurements for each velocity. Reproduced with permission from Zhang et al. (2011). © Acoustical Society of America

same exposure parameters. The temperature of the pure controls decreased steadily as the flow velocity of the MB solution increased from 2 to 55 cm/s.

The term "inflection velocity" was defined as a threshold velocity taken as the velocity in the curve at which the temperature at the end of the FUS exposures is highest (see Fig. 4.27). For a given acoustic power and MB concentration, when the flow velocity of the MB solution was lower than the inflection velocity, the peak temperature at the end of the FUS exposures increased with increasing flow velocity of the MB solution. When the flow velocity was larger than the inflection velocity, the peak temperature decreased with increasing velocity of the MB solution. The values of the inflection velocity and the normalized ICD increased with increasing acoustic power and concentration of the MB solution (Fig. 4.28). Furthermore, the increases of inflection velocity and normalized ICD were greater at higher acoustic powers and higher concentrations of the MB solution. When the acoustic power increased to 30 W, the inflection velocity increased nearly 9 fold over the MB concentration range of 0.001–0.1 %. Thus, flowing MBs in a FUS field minimized convective thermal losses because the MBs enhanced cavitation and heating. The significance of the inflection velocity was also that when the flow velocity of the MBs solution was lower or higher than this inflection velocity, the additional heat induced by the flowing MBs was either more than or less than the thermal losses due to convective heat transport. Thus, this approach might be useful when generating a thermal lesion, such as with HIFU and ultrasonic hyperthermia treatments, in tissues that are associated with convective vascular structures.

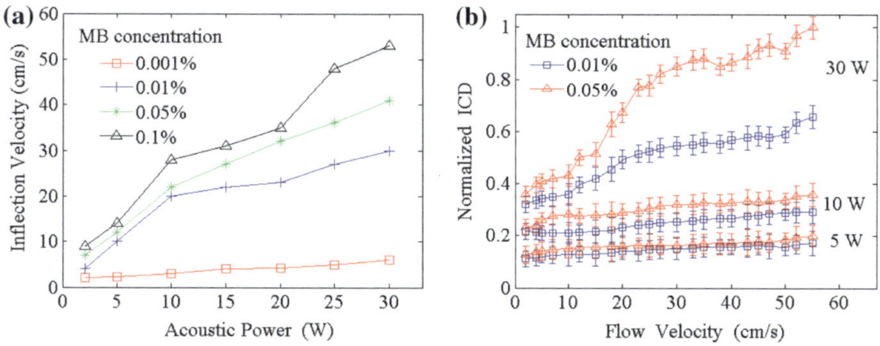

Fig. 4.28 **a** Inflection velocity and **b** normalized ICD at various concentrations of MB solution and different acoustic power levels for 5 s in a 3-mm vessel. In **b**, the *upper two curves* indicate the ICD obtained at an acoustic power of 30 W, the *middle* and *lower two curves* illustrate the ICD at 10 W and 5 W, respectively. The values shown are the average of five measurements for each velocity, and the error bars represent the standard deviation. Reproduced with permission from Zhang et al. (2011). © Acoustical Society of America

4.5.2.2 Comparison of Cavitation with Flowing Polymer MBs and Lipid-Shelled MBs

The variation in components of a MB shell and the different types of gases inside a bubble may mean that different MBs exhibit different efficiencies in heating and cavitation. Thus, comparing the efficiencies in heating and cavitation between different shelled MBs is important when using them for treatments that involve thermal effects. Zhang et al. (2012b) compared ultrasound-induced heating and cavitation from flowing polymer-shelled and lipid-shelled MBs during FUS exposures. Temperature and cavitation activity were measured when the polymer- and lipid-shelled MB solutions flowed through a vessel in the phantom shown in Fig. 4.25 with varying flow velocities and HIFU exposures. Two different types of encapsulated MBs were studied. The first was SonoVue (2×10^8 bubbles/ml and mean size was 2.5 μm, Bracco SpA, Milan, Italy), which contains sulfur hexafluoride (SF6) gas stabilized by a lipid. The second was an SF_6-gas-filled poly (lactic-co-glycolic acid) (PLGA) MB prepared by a modified double emulsion-solvent evaporation method (mean diameter 1.6 μm and a concentration of $\sim 1.5 \times 10^8$ bubbles/ml). The appropriate volume of each MB was slowly injected into a separate vial containing 100 ml of natural saline, giving a 0.01 % concentration of the MB (v/v, $\sim 10^4$ bubbles/ml). The solution was gently shaken to uniformly distribute the MBs, after which it was injected into the phantom reservoir. The flow velocity was controlled by the rotation speed of a pump with three impellers and was monitored using the flow meter.

The temperature rise at the front of the vessel and the normalized ICD were recorded for the saline, polymer-shelled, and lipid-shelled MB in a static solution, as shown in Fig. 4.29. The temperature rise and cavitation activity increased with

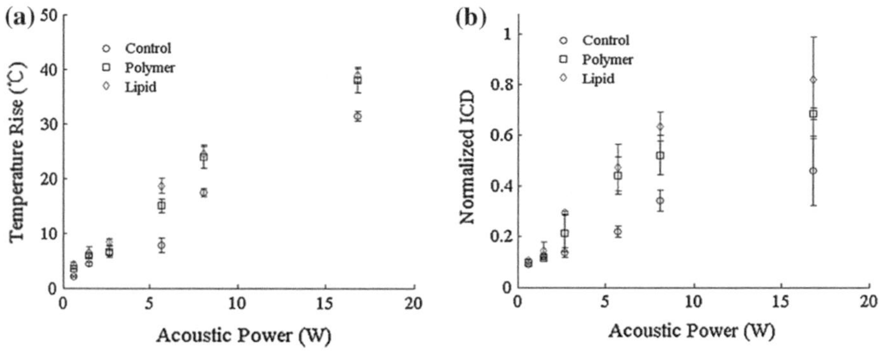

Fig. 4.29 Statistical results of **a** temperature rise and **b** normalized ICD for the saline, polymer-shelled, and lipid-shelled MBs as the static solution in the vessel of the phantom was exposed to FUS at various acoustic power levels from 0.6 to 16.8 W for a 5-s exposure. The values shown are the average of five measurements, and the error bars show the standard deviation. Reproduced with permission from Zhang et al. (2012b). © Acoustical Society of America

increasing acoustic power for the saline and for both types of shelled MBs. At an acoustic power of 5.7 W, the mean value of the temperature rise reached about 18 °C for the lipid-shelled MBs and was greater than 15 °C for the polymer-shelled MBs. For higher acoustic powers, the temperature rise at the end of the FUS exposure was very similar for polymer-shelled and lipid-shelled MBs.

The values of temperature rise at the front of the vessel and the normalized ICD in the phantom were recorded at various acoustic power levels when the saline, polymer-shelled, and lipid-shelled MB solutions flowed through the phantom vessel at velocities ranging from 3 to 20 cm/s (Figs. 4.30, 4.31 and 4.32). At the lowest acoustic power tested (1.5 W), the temperature rise decreased with increasing flow velocities for the saline solution and both types of shelled MBs. At the higher acoustic power of 5.7 W, the temperature rise was ~15 °C at a flow velocity of 3 cm/s for the polymer-shelled MBs and increased to ~18 °C at a flow velocity of 10 cm/s. As the flow velocity increased to 20 cm/s, the temperature rise decreased to ~12 °C. For the lipid-shelled MBs, the same trend of temperature response was observed. The temperature rise was ~16° C at a flow velocity of 3 cm/s, which increased to ~21 °C and then decreased to ~14 °C with flow velocities of 10 and 20 cm/s, respectively. At an acoustic power of 11.1 W, the temperature rise for the lipid-shelled MBs increased from 31 to 41 °C with increasing flow velocities from 3 to 20 cm/s. For the polymer-shelled MBs at this acoustic power, however, the temperature rise increased from 24 to 33 °C with increasing flow velocities from 3 to 15 cm/s, but then decreased ~30 °C when the flow velocity increased to 20 cm/s. For the saline, the values of temperature rise decreased as the flow velocity of the MB solution increased from 3 to 20 cm/s at all three acoustic powers. Cavitation tended to increase with increasing flow velocities from 3 to 20 cm/s for the solution of polymer-shelled MBs at acoustic powers of 1.5, 5.7 and 11.1 W

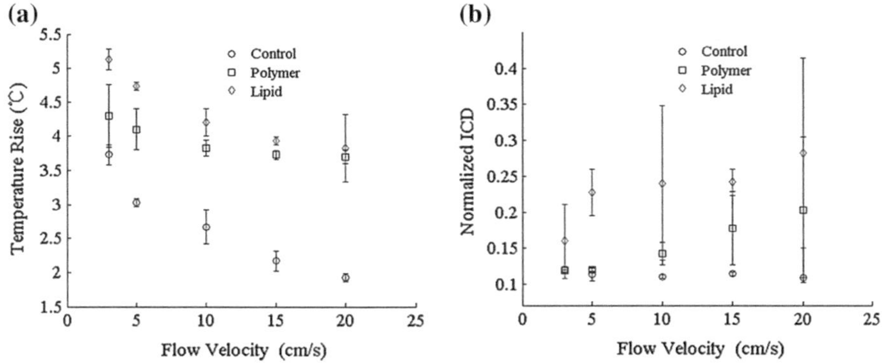

Fig. 4.30 Statistical results of **a** temperature rise and **b** normalized ICD for saline, polymer-shelled, and lipid-shelled MBs in solution with varying flow velocities, from 3 to 20 cm/s, at an acoustic power 1.5 W for 5 s. The values shown are the averages and standard deviations of five measurements for each velocity. Reproduced with permission from Zhang et al. (2012b). © Acoustical Society of America

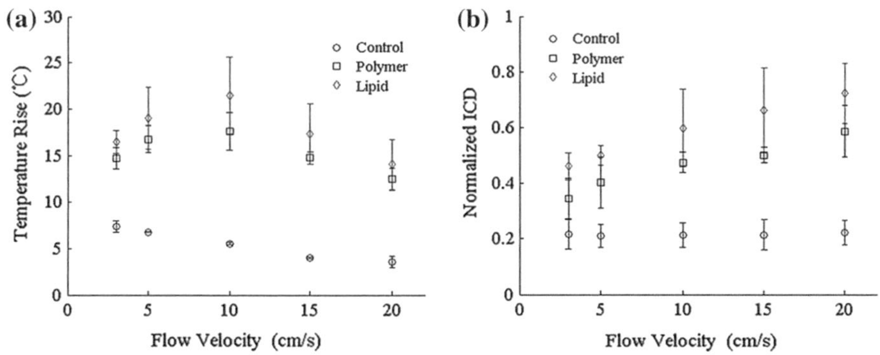

Fig. 4.31 Statistical results of **a** temperature rise and **b** normalized ICD for the saline, polymer-shelled, and lipid-shelled MBs in solution with varying flow velocities, from 3 to 20 cm/s, at an acoustic power 5.7 W for 5 s. The values shown are the averages and standard deviations of five measurements for each velocity. Reproduced with permission from Zhang et al. (2012b). © Acoustical Society of America

($R > 0.95$). For lipid-shelled MBs in solution, cavitation tended to increase with increasing flow velocities at acoustic powers of 5.7 and 11.1 W ($R < 0.95$), but this trend was not obvious at 1.5 W ($R = 0.8665$). Meanwhile, the saline solution showed no significant changes in cavitation levels with increasing flow velocities at any acoustic power (p-value > 0.05). The temperature rise and normalized ICD values for the lipid-shelled MBs in solution was slightly higher than those for the polymer-shelled MBs in solution. These results suggest that lipid-shelled MBs may be more efficient than polymer-shelled MBs in inducing heating and cavitation during FUS exposure.

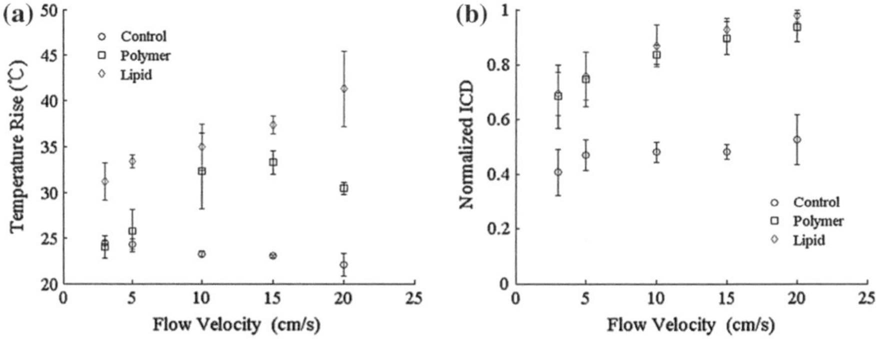

Fig. 4.32 Statistical results of **a** temperature rise and **b** normalized ICD for the saline, polymer-shelled, and lipid-shelled MBs in solution with varying flow velocities, from 3 to 20 cm/s, at an acoustic power 11.1 W for 5 s. The values shown are the averages and standard deviations of five measurements for each velocity. Reproduced with permission from Zhang et al. (2012b). © Acoustical Society of America

4.5.2.3 Comparing Cavitation Between Flowing Phase-Shift NDs and Lipid-Shelled MBs

Another type of encapsulated bubble that can be used in cavitation applications is phase-shift droplet emulsions. These contain micron- or nanometer-sized liquid droplets containing low-boiling-point perfluorocarbon (PFC) compounds encapsulated in shells made of lipids, albumin, or PLGA. When in the body, NDs become superheated at body temperature and can be vaporized into gas bubbles at a desired location once ultrasound pulses are applied; this process is known as ADV (Kripfgans et al. 2000; Qamar et al. 2010; Zhang and Porter 2010). When used as cavitation nuclei, phase-shift NDs a number of therapeutic applications, such as the targeted delivery of therapeutic agents (Rapoport et al. 2010), enhanced HIFU therapy for cancer treatment (Zhang and Porter 2010), and gas embolotherapy, in which gas MBs selectively formed from liquid droplets are used to prevent blood flow to tumors (Qamar et al. 2010). Additional applications for phase-shift NDs potentially lie in ultrasound molecular imaging, phase aberration correction, and intraoperative assessment of cancer ablation margins produced by thermal techniques (Haworth et al. 2008; Huang et al. 2010b; Sheeran et al. 2013). Here, we compare how flowing phase-shift NDs and lipid-shelled MBs affect cavitation during FUS exposures. Cavitation was monitored using the PCD method as solutions of either phase-shift NDs or lipid-shelled MBs flowed at varying velocities through a 5-mm diameter wall-less vessel in a transparent tissue-mimicking phantom (Fig. 4.25) and were exposed to FUS at a range of acoustic power levels. Pure controls (saline) were also investigated with the same exposure parameters and flow velocities. Phase-shift NDs were filled with dodecafluoropentane (DDFP, C_5F_{12}) and the shell material was albumin. For comparison, the lipid-shelled MBs used were commercially available SonoVue ultrasound contrast agents (UCAs).

To compare cavitation between the phase-shift NDs and lipid-shelled MBs after they were exposed to FU, two discrete identical FUS exposures were performed when the phase-shift ND, lipid-shelled MB and saline solutions were static in the vessel of the phantom (Fig. 4.33). The normalized ICD was calculated after each exposure. The intervals between the two discrete exposures were not strictly consistent: the next exposure was started (manually) once the temperature returned to room temperature, which was the initial temperature before each exposure. The results showed that ICD values for the phase-shift ND and lipid-shelled MB solutions were higher than those for the saline (p-value <0.05), as expected. Interestingly, while the normalized ICD value for the lipid-shelled MB solution was higher than that for the saline in the first exposure (p-value <0.05), it decreased to almost the same level in the second exposure. For the phase-shift ND solution, the normalized ICD was 0.35 in the first exposure and increased to 0.49 in the second exposure. No significant changes in cavitation levels were observed in the saline solution between the two exposures (p-value >0.05).

The normalized ICD values in the phantom were calculated when the saline, lipid-shelled MB or phase-shift ND solutions flowed through the vessel of the phantom at flow velocities ranging from 5 to 30 cm/s. At an acoustic power of 2 W (Fig. 4.34), the normalized ICD value for the phase-shift NDs was ~0.182 at a flow velocity of 5 cm/s, and this increased to ~0.188 at a flow velocity of 15 cm/s. As the flow velocity increased to 20 cm/s, the normalized ICD was ~0.185, and this decreased to 0.178 at a flow velocity of 30 cm/s. Meanwhile, for the lipid-shelled MBs, the normalized ICD values increased from 0.183 to 0.198 with increasing flow velocities from 5 to 30 cm/s at an acoustic power of 2 W. At an acoustic power of 15 W (Fig. 4.35), the normalized ICD for phase-shift NDs was ~0.62 at a flow velocity of 5 cm/s, and this increased to ~0.76 as the flow velocity increased to 15 cm/s. At a flow velocity of 20 cm/s, the normalized ICD was ~0.86 and increased to 0.93 for a velocity of 30 cm/s. For the lipid-shelled MBs, when the acoustic power was 15 W, the normalized ICD was ~0.58 at a flow velocity of 5 cm/s, and this increased to ~0.61 at a flow velocity of 15 cm/s. At a flow velocity of 20 cm/s, the normalized ICD was ~0.63, and this increased to 0.74 at 30 cm/s.

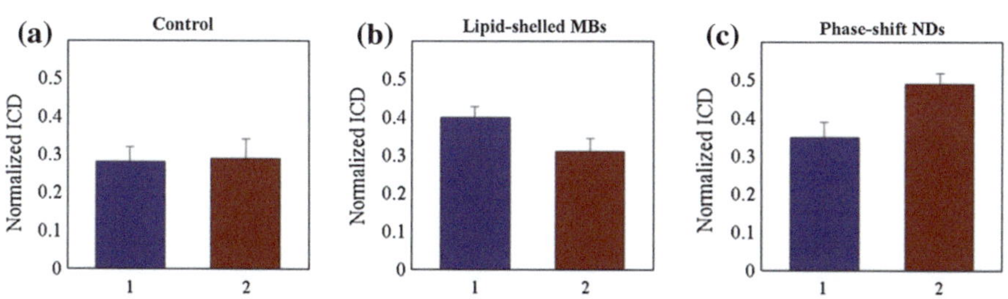

Fig. 4.33 Normalized ICD values for two same FUS exposures on the static **a** control (saline), **b** lipid-shelled MB, and **c** phase-shift ND solutions in the vessel of the phantom. The FUS focus was positioned in the middle of the vessel at an acoustic power of 12 W, and each exposure lasted 4 s. The values shown are the average and standard deviation of five measurements

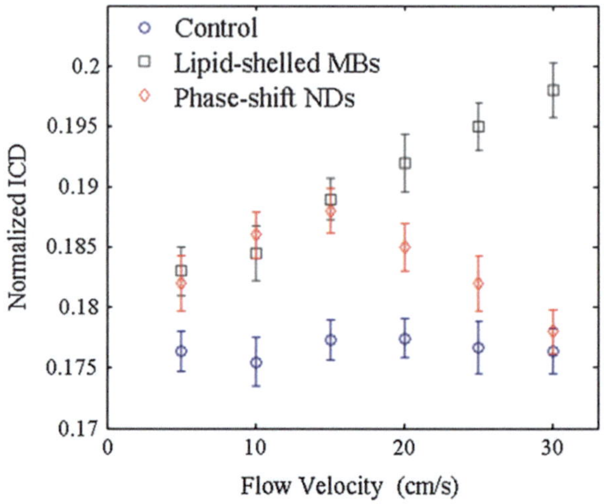

Fig. 4.34 Statistical results of normalized ICD values for the saline, lipid-shelled MB, and phase-shift ND solutions with varying flow velocities, from 5 to 30 cm/s, at an acoustic power of 2 W for 4 s. The values shown are the averages and standard deviations of five measurements for each velocity

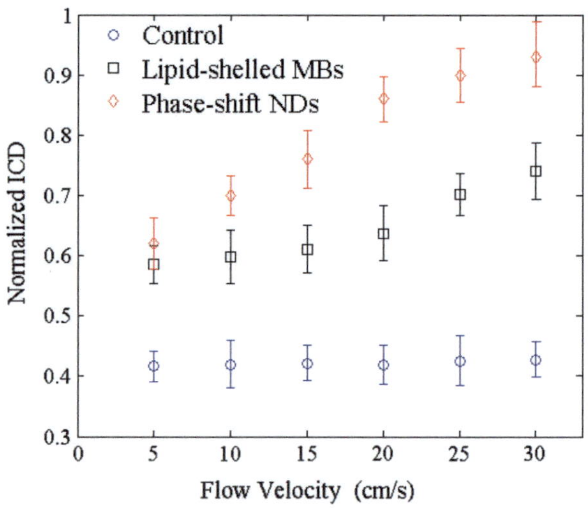

Fig. 4.35 Statistical results of normalized ICD values for the saline, lipid-shelled MB, and phase-shift ND solutions with varying flow velocities, from 5 to 30 cm/s, at an acoustic power of 15 W for 4 s. The values shown are the averages and standard deviations of five measurements for each velocity

The trends in the change in normalized ICD values with different flow velocities were different for lipid-shelled MBs and phase-shift NDs at different acoustic powers. At a low acoustic power, the normalized ICD values for the lipid-shelled MBs tended to increase with increasing velocities from 5 to 30 cm/s ($r > 0.95$). Meanwhile, at low acoustic power, the normalized ICD values for the phase-shift NDs also tended to increase with increasing flow velocities from 5 to 15 cm/s ($r > 0.95$), but then decreased as flow velocities increased from 20 to 30 cm/s ($r > 0.95$). At high acoustic power, the normalized ICD values for both the lipid-shelled MBs and the phase-shift NDs increased with increasing flow velocities from 5 to 30 cm/s ($r > 0.95$). For saline, no significant changes in cavitation levels were observed with increasing flow velocities at all acoustic powers (p-value >0.05), and the associated normalized ICD values were lower than those for the lipid-shelled MBs and the phase-shift NDs, as expected.

4.6 Enhancing Acoustic Cavitation with Multi-frequency Ultrasound

A potential way of substantially enhancing heating in a HIFU treatment is to use a broadband split-focus transducer, which introduces a dual frequency and allows split-focus control. In one implementation of this novel technique, a broadband split-focus transducer (Imasonic, Besançon, France) that has an enlarged spatial focal volume formed from split foci and can produce enhanced heating effects using a confocal twofold dual-frequency excitation of 1.2 and 2.4 MHz is used (Lu et al. 2014) (Fig. 4.36).

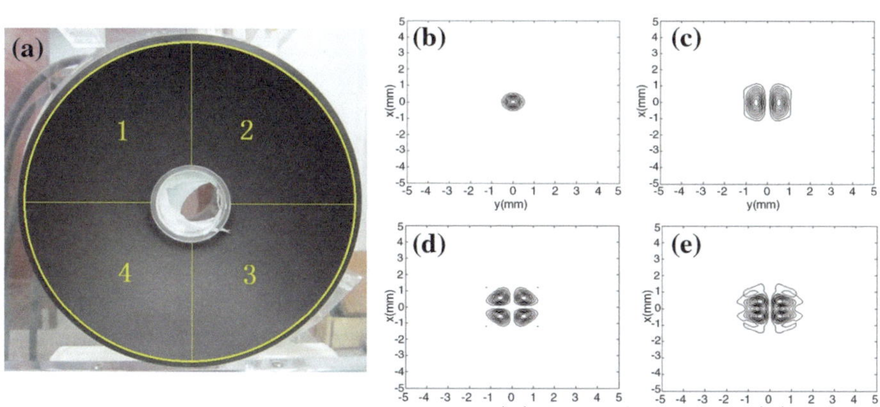

Fig. 4.36 **a** Prototype broadband split-focus approach with four elements. The four split-focus intensity profiles in the focus plane are shown in: **b** a single focus with a 0-degree shift, single 1.6-MHz frequency; **c** two split foci with two elements, a 180-degree shift, 1.6 MHz; **d** four split foci with a 180-degree phase shift, 1.6 MHz, and **e** four split foci with a 180-degree phase shift at twofold dual frequency of 1.2 and 2.4 MHz. Reproduced with permission from Lu et al. (2014). © IEEE

To achieve enhanced cavitation heating using this split-focus technique, one study created split-focus lesions in BSA gel phantoms using a single 1.6-MHz frequency and a dual frequency of 1.2 and 2.4 MHz. The acoustic power was set as high as possible to ensure that the focal zone pressure was much higher than the cavitation threshold. The onset time of lesion formation using the twofold dual-frequency technique was shorter than that when using a single frequency, and the lesion size was also larger.

Next, the authors recorded ultrasound images during HIFU treatments using either a single 1.6-MHz frequency or the dual frequency of 1.2 and 2.4 MHz. The bright hyperechoic region was observed when using the dual-frequency mode was larger than that using the single-frequency mode at the same treatment time. Here, the ratio of hyperechoic area increase was similar to that of the geometrical lesion size increase. When cavitation activity was measured, the authors found that its level using the dual-frequency technique was higher than when using a single frequency, where this activity was similar to the lesion size ratio Fig. 4.37. The results show that the dual-frequency and split-focus methods can enhance heating by cavitation activity at the focal zone. Ex vivo liver and BSA-gel-phantom

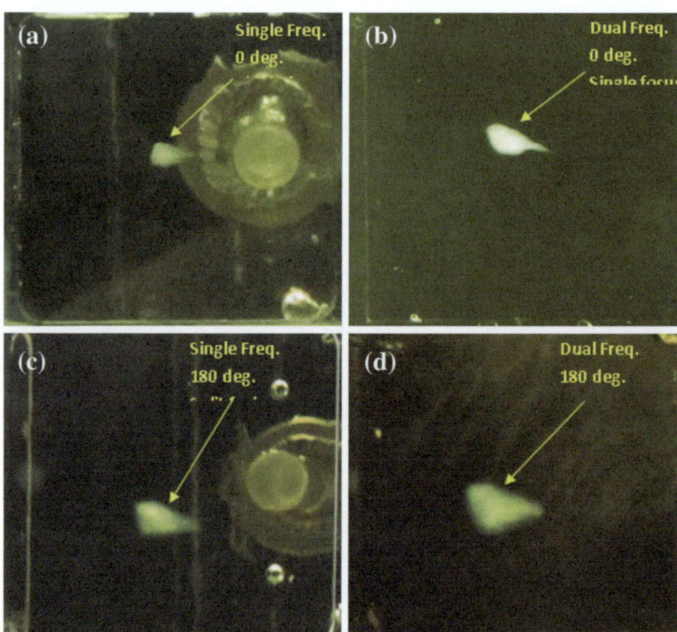

Fig. 4.37 Lesions formed in BSA gel phantoms showed cavitation-enhanced heating effects when the twofold dual-frequency, split-focus protocols were used. The lesions formed by a single focus were compared when using **a** a single 1.6-MHz frequency or **b** a twofold dual frequency of 1.2 and 2.4 MHz. The lesions formed under split-focus conditions were compared when using **c** a single 1.6 MHz or **d** twofold dual frequency of 1.2 and 2.4 MHz. Reproduced with permission from Lu et al. (2014). © IEEE

experiments performed using the dual frequencies of 1.2 and 2.4 MHz gave a larger thermal lesion than when using a single 1.6-MHz frequency (Figs. 4.38 and 4.39). Thus, the twofold dual-frequency method combined with the split-focus protocol may offer an advantage in temporally and spatially enhancing cavitation heating.

In another study, Liu et al. (2006) demonstrated another novel approach using multi-frequency ultrasound. In this study, they induced cavitation-enhanced heating in ex vivo fresh porcine muscles by creating acoustic cavitation using simultaneous sonication with low- and high-frequency ultrasound. A high-frequency spherical focused transducer (566 or 1155 kHz) was used to generate the thermal lesions, and a low-frequency planar transducer (40 or 28 kHz) was used to enhance the cavitation activity. Compared with the thermal lesion formed by the single focused transducer, the thermal lesion generated using this approach was larger in the axial and radial directions for the same power intensity of FUS (Fig. 4.40). The lesions generated were more symmetrical and less shifted, implying that deeper tissue regions can be treated with this technique. The authors suggest that this approach may be useful for achieving cavitation-enhanced heating and to generate large and

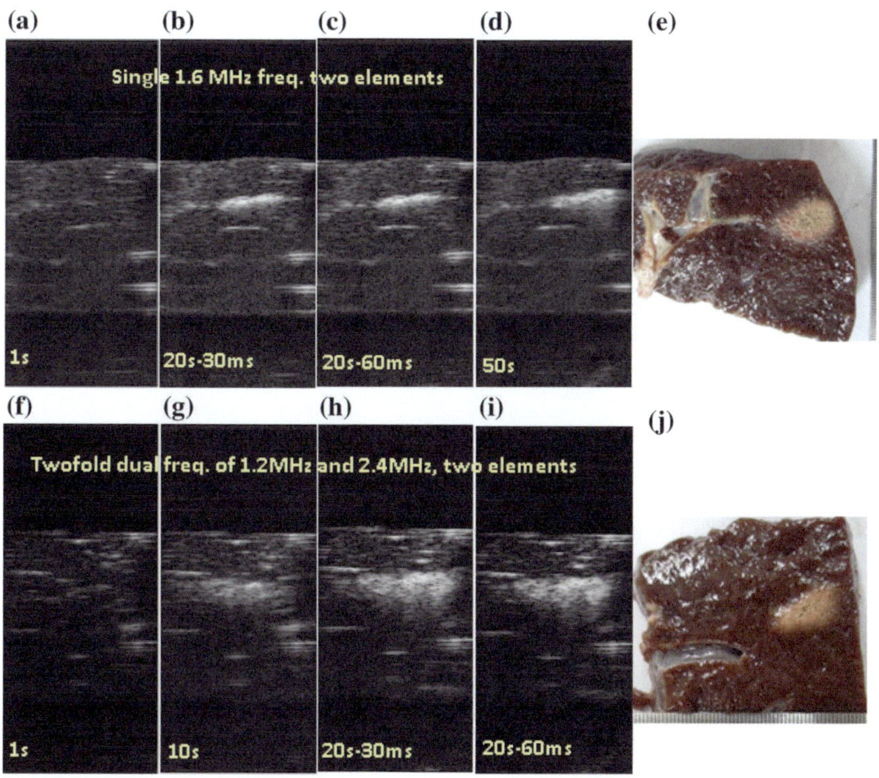

Fig. 4.38 Ultrasound image monitoring of ex vivo bovine livers treated using a single 1.6-MHz frequency (*top row*) or a dual frequency of 1.2 and 2.4 MHz (*bottom row*). Reproduced with permission from Lu et al. (2014). © IEEE

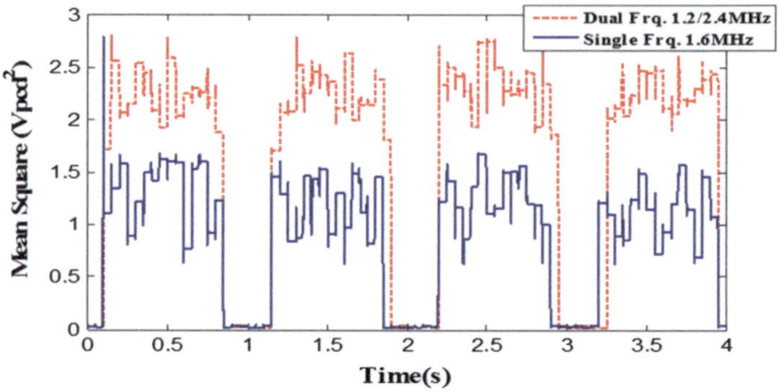

Fig. 4.39 Inertial cavitation detected in ex vivo livers treated by a single 1.6-MHz frequency or a dual frequency of 1.2 and 2.4 MHz. Reproduced with permission from Lu et al. (2014). © IEEE

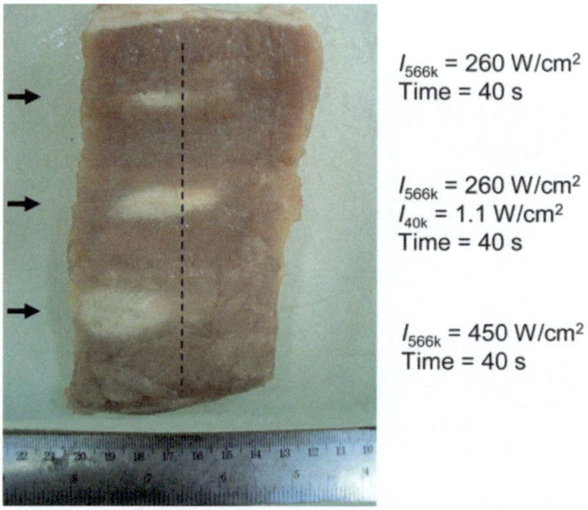

Fig. 4.40 Comparison of the thermal lesions generated under three conditions: 260 W/cm^2, 566-kHz ultrasound alone (*top*); combined 260 W/cm^2, 566 kHz, and 1.1 W/cm^2, 40 kHz ultrasound (*middle*); and 450 W/cm^2, 566-kHz ultrasound alone (*bottom*). Reproduced with permission from Liu et al. (2006). © Elsevier B.V.

deep-seated thermal lesions. Another study has also proposed that enhanced cavitation could be achieved using two consecutive ultrasound waves at different frequencies, suggesting that the enhanced cavitation activity may be due to the temporary gas supersaturation and the reduction of surface tension at an elevated temperature (Yang and Jo 2014).

In yet another study, Guo et al. (2013) compared cavitation activity and temperature rise in chicken breast for single- and multi-frequency HIFU exposures at controlled acoustic powers and exposure times (Figs. 4.41 and 4.42). They used

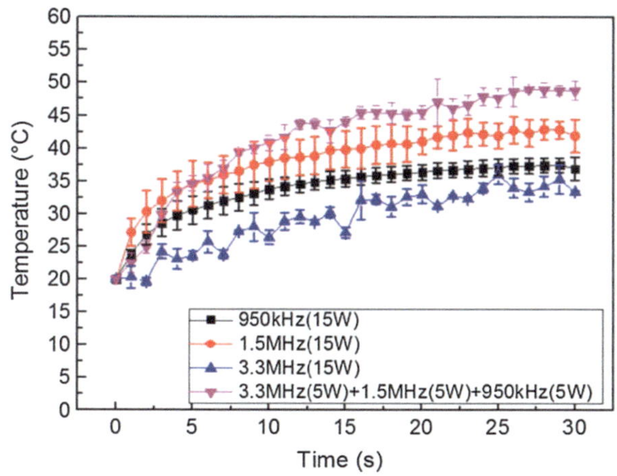

Fig. 4.41 Measured temperature in chicken breast as a function of time for single-frequency and tri-frequency exposures at 15 W total input power, and 5 mm depth of focus. Reproduced with permission from Guo et al. (2013). © IEEE

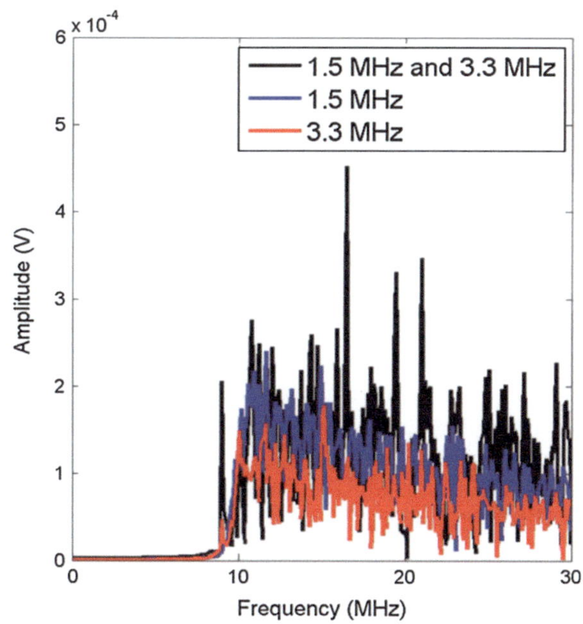

Fig. 4.42 Filtered cavitation detection signals from single- and dual-frequency exposures (1.5, 3.3 MHz, and the two frequencies combined) under 15-W total input power. Reproduced with permission from Guo et al. (2013). © IEEE

single-frequency, dual-frequency, and tri-frequency HIFU configurations (center frequencies were 950 kHz, 1.5 MHz, and 3.3 MHz) to heat the chicken breast, during which cavitation and temperature rise were recorded. They found increases of 5–10 % in temperature rise between single- and multi-frequency modes and suggested that multi-frequency HIFU probably caused enhanced cavitation and was more effective at producing tissue ablation.

In another study, Sasaki et al. (2014) combined dual-frequency and multiple-focus HIFU exposure to generate a cavitation bubble cloud and to enhance ultrasonic heating in polyacrylamide gels and excised chicken breast. Their results suggest that this combined method is effective in coagulating a large volume and is more efficient than the conventional triggered HIFU method. In the sequence "triggered HIFU," high-intensity and short pulses were used to generate and grow cavitation bubbles first. Then, low-intensity and long-duration bursts were used to vibrate the cavitation bubbles to enhance the heating effect.

4.7 Boiling Histotripsy

In addition to producing thermal effects, HIFU can be used to mechanically fractionate soft tissues into a liquid-like homogenate. This ultrasound-induced tissue fractionation opens up a new potential in clinical procedures in which mechanical subdivision and removal of soft tissue are required. This technique has been dubbed "histotripsy" as an analog to lithotripsy (Xu et al. 2004, 2005, 2009, 2010). Lin et al. (2014) used very short (<2 cycles) of histotripsy pulses to generate dense bubble clouds and a precise, subwavelength lesion, and called this procedure "microtripsy." Shock wave heating and millisecond boiling in HIFU fields, known as "boiling histotripsy," use millisecond pulses with lower pulse repetition frequency and acoustic pressures than used in cavitation cloud histotripsy to expand a millimeter-sized boiling bubble at the transducer focus explosively, resulting in mechanical fractionation or emulsification of tissue (Khokhlova et al. 2011; Wang et al. 2013; Simon et al. 2012; Maxwell et al. 2012). Effective tissue heating, caused by the absorption of shock fronts due to nonlinear propagation effects, can lead to temperatures greater than 100 °C and can produce repeat localized boiling in a tissue in just milliseconds within each pulse (Canney et al. 2010). Here, a miniature acoustic fountain forms because the millimeter-sized boiling bubble acts as a pressure release interface between the tissue and the air or vapor; thus, atomization occurs within the boiling bubble. The explosion of this boiling bubble and its interaction with subsequent shock waves causes mechanical fractionation of the tissue and ultimately its complete emulsification (Khokhlova et al. 2011; Wang et al. 2013; Simon et al. 2012; Maxwell et al. 2012) (Fig. 4.43).

Khokhlova et al. (2011) demonstrated that shock wave heating and millisecond boiling could effectively induce mechanical emulsification in ex vivo bovine heart and liver tissue using 1-, 2-, and 3-MHz focused transducers and different pulsing schemes, in which the pressure, duty factor, and pulse duration were varied (Fig. 4.44). They found that higher duty factors or longer pulses produced varying degrees of thermal denaturation combined with mechanical emulsification. They also found that the general shape of the lesions did not change with frequency, and larger lesions were obtained using lower ultrasound frequencies, suggesting that the length was determined by the axial length of the beam focus.

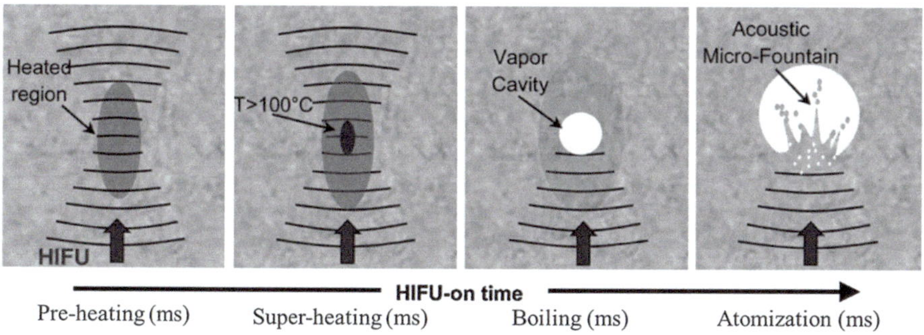

Fig. 4.43 Proposed mechanism of tissue fractionation by boiling histotripsy. Reproduced with permission from Simon et al. (2012). © IOP Publishing Ltd.

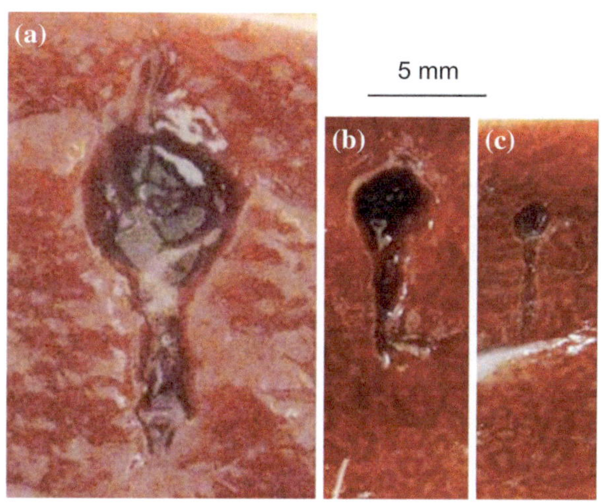

Fig. 4.44 Axial cross sections of liquefied lesions obtained in ex vivo bovine liver exposed to HIFU at different frequencies: **a** 1.1 MHz, **b** 2.158 MHz, and **c** 3.4 MHz. Reproduced with permission from Khokhlova et al. (2011). © Acoustical Society of America

Recently, another study explored boiling histotripsy in tissues. Wang et al. (2013) examined thermal and mechanical effects of boiling histotripsy in ex vivo bovine heart and liver using a 2-MHz single-element transducer operating at duty factors of 0.005–0.01, pulse durations of 5–500 ms and in situ shock amplitude of 73 MPa. The resulting lesions that appeared as liquids contained no cellular structures larger than a cell nucleus and had a sharp border one to two cells wide. Both protein analysis and nicotinamide adenine dinucleotide-diaphorase staining showed less thermal denaturation than seen on gross observation as whitening of its content. This work offered the first histologic and biochemical confirmation that boiling histotripsy can efficiently and noninvasively fractionate tissue into subcellular-sized debris with a predictable degree of thermal damage.

4.8 Microscale Cavitation Heating and Nanoscale Thermometry

Investigations of micro/nanoscale thermal effects enhanced by microscale cavitation are promising in various biomedical applications. However, while measuring temperature variations on the nanometer scale is important for many disciplines, it is challenging. So far, many promising techniques have been introduced to measure local temperatures, such as scanning probe microscopy, Raman spectroscopy, and fluorescence-based measurements using nanoparticles and organic dyes. However, many of these existing methods are limited by low sensitivity and systematic errors induced by the local chemical environment, as well as poor optical properties of the surrounding medium. Moreover, although they are promising, methods based on green fluorescent proteins rely on cellular transfection, which is difficult to achieve in certain primary cell types.

Recently, a new promising nanoscale thermometry approach has been introduced (Fig. 4.45). It uses coherent manipulation of the electronic spin associated with nitrogen-vacancy color centers in diamonds, such that temperature variations as small as 1.8 mK can be detected in an ultrapure bulk diamond sample (Kucsko et al. 2013). Using nitrogen-vacancy centers in diamond nanocrystals, the local thermal environment on length scales as short as 200 nm can be measured directly. When both nanodiamonds and gold nanoparticles are introduced into a single human embryonic fibroblast, temperature-gradient control and mapping at the subcellular level can be conducted simultaneously. This nitrogen-vacancy-based thermometry may provide a unique tool for many areas of biological, physical, and chemical research. For example, precisely control and measurement of temperature may be used to control gene expression and tumor metabolism, selectively treat diseased cells, or even study heat dissipation in integrated circuits.

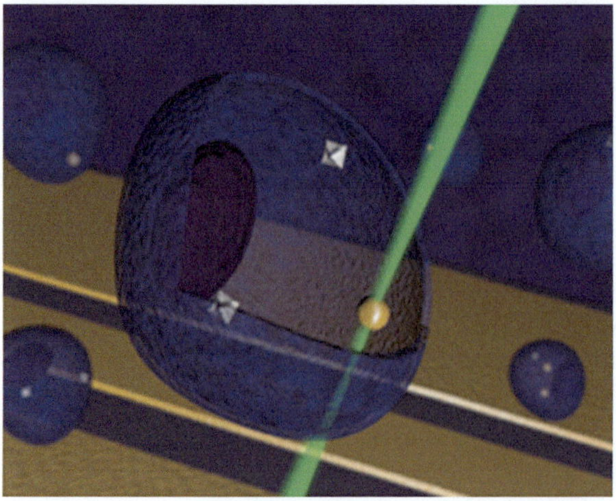

Fig. 4.45 Schematic image depicting nanodiamonds (*gray diamonds*) and a gold nanoparticle (*yellow sphere*) within a living cell (*central blue object*). Reproduced with permission from Kucsko et al. (2013). © Nature Publishing Group

4.9 Summary

The primary physical mechanism of many therapeutic applications of FUS is to elevate tissue temperature within the treated region rapidly. One way to increase the efficacy of therapeutic ultrasound may be to capitalize on the interaction between ultrasound and cavitation, which has been recently shown to enhance local heating when driven acoustically. The principles behind cavitation-enhanced thermal effects include viscous dissipation of cavitation bubbles, acoustic energy radiation from cavitation bubbles and increased local acoustic absorption caused by the cavitation bubbles. Viscous dissipation of the cavitation bubble energy produces cavitation-enhanced heating as the bubbles' mechanical energy is converted into thermal energy. This viscous dissipation occurs in the relatively thin layer of the host medium surrounding the oscillating cavitation bubble. Importantly, the viscous dissipation coefficient is much higher for encapsulated MBs than for free bubbles, given the high viscosity of the encapsulating bubbles' shell. Thus, higher harmonics scattered or emitted by cavitation are more readily absorbed and converted into heat in the tissue. The selective creation of cavitation bubbles tends to increase the attenuation coefficient locally and enhance local heating related to absorption of sound.

To evaluate the relationship between cavitation and heat generation, many studies have measured cavitation activity and temperature rise during HIFU exposure simultaneously for different parameters and in different mediums: transparent tissue-mimicking phantoms, ex vivo tissues, and in vivo tissues. Overall, the enhanced temperature rise correlates well with cavitation activity, and this can potentially be used to correct for the heating predicted to occur from primary ultrasound absorption, thus isolating the cavitation-enhanced heating. Various in vitro and in vivo studies have examined the therapeutic effects of cavitation-enhanced heating. One way to further enhance cavitation heating is by injecting encapsulated gas-filled MBs and phase-shift NDs before therapeutic ultrasound treatment, to act as cavitation nuclei. This technique has the potential to effectively increase lesion size and reduce the treatment time or the required ultrasound intensity. Also, cavitation-enhanced heating is exploited in HIFU thermal ablation to improve its efficiency. This has been investigated in vitro in tissue-mimicking phantoms and human plasma, and in vivo in tissues, including the liver, kidney, brain, and thigh muscle. Furthermore, preclinical evaluation has begun to evaluate cavitation-enhanced heating effects on thermal ablation, in which hundreds of patients with solitary uterine fibroids were treated by HIFU combined with encapsulated MBs in China. This study found that encapsulated MBs may offer high success rates and reduce the sonication time and acoustic energy required to treat uterine fibroids with HIFU. Overall, these and other recent findings suggest that cavitation-enhanced heating may be able to improve the efficiency of therapeutic ultrasound if properly controlled.

References

Ahmadi F, McLoughlin IV, Chauhan S, ter Haar GR. Bio-effects and safety of low-intensity, low-frequency ultrasonic exposure. Prog Biophys Mol Biol. 2012;108:119–38.

Alkins RD, Brodersen PM, Sodhi RNS, Hynynen K. Enhancing drug delivery for boron neutron capture therapy of brain tumors with focused ultrasound. Neuro-Oncology. 2013;15:1225–35.

Arvanitis CD, McDannold N. Integrated ultrasound and magnetic resonance imaging for simultaneous temperature and cavitation monitoring during focused ultrasound therapies. Med Phys. 2013;40:112901.

Asai A, Okano H, Yoshizawa S, Umemura S. Analysis of temperature rise induced by high-intensity focused ultrasound in tissue-mimicking gel considering cavitation bubbles. Jpn J Appl Phy. 1999;52:07HF02.

Bailey MR, Khokhlova VA, Sapozhnikov OA, Kargl SG, Crum LA. Physical mechanisms of the therapeutic effect of ultrasound (a review). Acoust Phys. 2003;49:369–88.

Bakhvalov NS, Zhileikin YM, Zabolotskaya EA, Beyer RT. Nonlinear theory of sound beams. New York: American Institute of Physics; 1987.

Bhatia VK, Senior R. Contrast echocardiography: evidence for clinical use. J Am Soc Echocardiogr. 2008;21:409–16.

Bohannon J. Acoustics' 08. Ultrasound uses in medicine heat up. Science. 2008;321:338–9.

Bull V, Civale J, Rivens I, ter Haar GR. A comparison of acoustic cavitation detection thresholds measured with piezo-electric and fiber-optic hydrophone sensors. Ultrasound Med Biol. 2013;39:2406–21.

Canney MS, Khokhlova VA, Bessonova OV, Bailey MR, Crum LA. Shock-induced heating and millisecond boiling in gels and tissue due to high intensity focused ultrasound. Ultrasound Med Biol. 2010;36:250–67.

Cavalieri F, El Hamassi A, Chiessi E, Paradossi G. Stable polymeric microballoons as multifunctional device for biomedical uses: synthesis and characterization. Langmuir. 2005;21:8758–64.

Chapman A, ter Haar GR. Thermal ablation of uterine fibroids using MR-guided focused ultrasound-a truly non-invasive treatment modality. Eur Radiol. 2007;17:2505–11.

Chauvet D, Marsac L, Pernot M, Boch AL, Guillevin R, Salameh N, Souris L, Darrasse L, Fink M, Tanter M, Aubry JF. Targeting accuracy of transcranial magnetic resonance-guided high-intensity focused ultrasound brain therapy: a fresh cadaver model laboratory investigation. J Neurosurg. 2013;118:1046–52.

Chavrier F, Chapelon JY, Gelet A, Cathignol D. Modeling of high-intensity focused ultrasound-induced lesions in the presence of cavitation bubbles. J Acoust Soc Am. 2000;108:432–40.

Chen LL, ter Haar GR, Hill CR, Dworkin M, Carnochan P, Young H, Bensted JP. Effect of blood perfusion on the ablation of liver parenchyma with high-intensity focused ultrasound. Phys Med Biol. 1993;38:1661.

Chen WS, Brayman AA, Matula TJ, Crum LA, Miller MW. The pulse length-dependence of inertial cavitation dose and hemolysis. Ultrasound Med Biol. 2003a;29:739–48.

Chen WS, Lafon C, Matula TJ, Vaezy S, Crum LA. Mechanisms of lesion formation in high intensity focused ultrasound therapy. Acoust Res Lett Online. 2003b;4:41–6.

Chung DJ, Cho SH, Lee JM, Hahn ST. Effect of microbubble contrast agent during high intensity focused ultrasound ablation on rabbit liver in vivo. Eur J Radiol. 2012;81:519–23.

Civale J, Clarke R, Rivens I, ter Haar GR. The use of a segmented transducer for rib sparing in HIFU treatments. Ultrasound Med Biol. 2006;32:1753–61.

Civale J, Rivens I, ter Haar GR, Morris H, Coussios C, Friend P, Bamber J. Calibration of ultrasound backscatter temperature imaging for high-intensity focused ultrasound treatment planning. Ultrasound Med Biol. 2013;39:1596–612.

Clarke RL, ter Haar GR. Temperature rise recorded during lesion formation by high-intensity focused ultrasound. Ultrasound Med Biol. 1997;23:299–306.

Coussios CC, Farny CH, ter Haar GR, Roy RA. Role of acoustic cavitation in the delivery and monitoring of cancer treatment by high-intensity focused ultrasound (HIFU). Int J Hyperthermia. 2007;23:327–327.
Curra FP, Mourad PD, Khokhlova VA, Cleveland RO, Crum LA. Numerical simulations of heating patterns and tissue temperature response due to high-intensity focused ultrasound. IEEE Trans Ultrason Ferroelectr Freq Control. 2000;47:1077–89.
Damianou C, Ioannides K, Hadjisavvas V, Mylonas N, Couppis A, Iosif D. In vitro and in vivo brain ablation created by high-intensity focused ultrasound and monitored by MRI. IEEE Trans Ultrason Ferroelectr Freq Control. 2009;56:1189–98.
Davis S, Raymond J, Church C. Acute effects of high intensity focused ultrasound on blood vessels in vivo. In: Therapeutic Ultrasound: 5th International Symposium on Therapeutic Ultrasound. Vol. 1. Boston: AIP Publishing, 2005. pp. 54–58.
Denbow ML, Rivens IH, Rowland IJ, Leach MO, Fisk NM, ter Haar GR. Preclinical development of noninvasive vascular occlusion with focused ultrasonic surgery for fetal therapy. Am J Obstet Gynecol. 2000;182:387–92.
Edson PL. The role of acoustic cavitation in enhanced ultrasound-induced heating in a tissue-mimicking phantom. Boston: Boston University; 2001.
Elbes D, Denost Q, Laurent C, Trillaud H, Rullier A, Quesson B. Pre-clinical study of in vivo magnetic resonance-guided bubble-enhanced heating in pig liver. Ultrasound Med Biol. 2013;39:1388–97.
Elbes D, Denost Q, Robert B, Kohler MO, Tanter M, Bruno Q. Magnetic resonance imaging for the exploitation of bubble-enhanced heating by high-intensity focused ultrasound: a feasibility study in ex vivo liver. Ultrasound Med Biol. 2014;40:956–64.
Farny CH, Holt RG, Roy RA. The correlation between bubble-enhanced HIFU heating and cavitation power. IEEE Trans Biomed Eng. 2010;57:175–84.
Gelat P, ter Haar GR, Saffari N. The optimization of acoustic fields for ablative therapies of tumours in the upper abdomen. Phys Med Biol. 2012;57:8471.
Griffiths A, ter Haar GR, Rivens I, Giussani D, Lees C. High-intensity focused ultrasound in obstetrics and gynecology: the birth of a new era of noninvasive surgery? Ultraschall Med. 2012;33:E8–15.
Guo SJ, Jing Y, Jiang XN. Temperature rise in tissue ablation using multi-frequency ultrasound. IEEE Trans Ultrason Ferroelectr Freq Control. 2013;60:1699–707.
Hacker A, Chauhan S, Peters K, Hildenbrand R, Marlinghaus E, Alken P, Michel MS. Multiple high-intensity focused ultrasound probes for kidney-tissue ablation. J Endourol. 2005;19:1036–40.
Hamilton MF, Blackstock DT. Nonlinear acoustics. San Diego: Academic Press; 1998.
Hanajiri K, Maruyama T, Kaneko Y, Mitsui H, Watanabe S, Sata M, Nagai R, Kashima T, Shibahara J, Omata M. Microbubble-induced increase in ablation of liver tumors by high-intensity focused ultrasound. Hepatol Res. 2006;36:308–14.
Hariharan P, Myers MR, Banerjee RK. HIFU procedures at moderate intensities—effect of large blood vessels. Phys Med Biol. 2007;52:3493.
Haworth KJ, Fowlkes JB, Carson PL, Kripfgans OD. Towards aberration correction of transcranial ultrasound using acoustic droplet vaporization. Ultrasound Med Biol. 2008;34:435–45.
He W, Wang W, Zhou P, Wang YJ, Zhou P, Li RZ, Wang JS, Ahuja AT. Enhanced ablation of high intensity focused ultrasound with microbubbles: an experimental study on rabbit hepatic VX2 tumors. Cardiovasc Intervent Radiol. 2011;34:1050–7.
Hertzberg Y, Volovick A, Zur Y, Medan Y, Vitek S, Navon G. Ultrasound focusing using magnetic resonance acoustic radiation force imaging: application to ultrasound transcranial therapy. Med Phys. 2010;37:2934–42.
Heydari M, Jahed M. Prediction of temperature distribution and volume of lesion during HIFU therapy. In: Sixth International Conference on Information Technology: New Generations (ITNG'09). Las Vegas: IEEE; 2009. pp. 1468–1473.
Hilgenfeldt S, Lohse D, Zomack M. Sound scattering and localized heat deposition of pulse-driven microbubbles. J Acoust Soc Am. 2000;107:3530–9.

Hoffelner J, Landes H, Kaltenbacher M, Lerch R. Finite element simulation of nonlinear wave propagation in thermoviscous fluids including dissipation. IEEE Trans Ultrason Ferroelectr Freq Control. 2001;48:779–86.

Holt RG, Roy RA. Measurements of bubble-enhanced heating from focused, MHz-frequency ultrasound in a tissue-mimicking material. Ultrasound Med Biol. 2001;27:1399–412.

Huang HW, Shih TC, Liauh CT. Predicting effects of blood flow rate and size of vessels in a vasculature on hyperthermia treatments using computer simulation. Biomed Eng Online. 2010a;9:18.

Huang JW, Xu JS, Xu RX. Heat-sensitive microbubbles for intraoperative assessment of cancer ablation margins. Biomaterials. 2010b;31:1278–86.

Huang Y, Vykhodtseva NI, Hynynen K. Creating brain lesions with low-intensity focused ultrasound with microbubbles: a rat study at half a megahertz. Ultrasound Med Biol. 2013;39:1420–8.

Hynynen K. The threshold for thermally significant cavitation in dog's thigh muscle in vivo. Ultrasound Med Biol. 1991;17:157–69.

Hynynen K. MRI-guided focused ultrasound treatments. Ultrasonics. 2010;50:221–9.

Illing RO, Kennedy JE, Wu F, ter Haar GR, Protheroe AS, Friend PJ, Gleeson FV, Cranston DW, Phillips RR, Middleton MR. The safety and feasibility of extracorporeal high-intensity focused ultrasound (HIFU) for the treatment of liver and kidney tumours in a Western population. Br J Cancer. 2005;93:890–5.

Jiang N, Xie B, Zhang X, He M, Li K, Bai J, Wang Z, He J, Zhang L. Enhancing ablation effects of a microbubble-enhancing contrast agent ("sonovue") in the treatment of uterine fibroids with high-intensity focused ultrasound: a randomized controlled trial. Cardiovasc Intervent Radiol. 2014;37:1321–8.

Kajiyama K, Yoshinaka K, Takagi S, Matsumoto Y. Microbubble enhanced HIFU. Phys Procedia. 2010;3:305–14.

Kaneko Y, Maruyama T, Takegami K, Watanabe T, Mitsui H, Hanajiri K, Nagawa H, Matsumoto Y. Use of a microbubble agent to increase the effects of high intensity focused ultrasound on liver tissue. Eur Radiol. 2005;15:1415–20.

Kaul S. Myocardial contrast echocardiography a 25-year retrospective. Circulation. 2008;118:291–308.

Keller JB, Miksis M. Bubble oscillations of large amplitude. J Acoust Soc Am. 1980;68:628–33.

Kennedy JE. High-intensity focused ultrasound in the treatment of solid tumours. Nat Rev Cancer. 2005;5:321–7.

Kennedy JE, ter Haar GR, Cranston D. High intensity focused ultrasound: surgery of the future? Br J Radiol. 2003;76:590–9.

Kennedy JE, ter Haar GR, Wu F, Gleeson FV, Roberts ISD, Middleton MR, Cranston D. Contrast-enhanced ultrasound assessment of tissue response to high-intensity focused ultrasound. Ultrasound Med Biol. 2004a;30:851–4.

Kennedy JE, Wu F, ter Haar GR, Gleeson FV, Phillips RR, Middleton MR, Cranston D. High-intensity focused ultrasound for the treatment of liver tumours. Ultrasonics. 2004b;42:931–5.

Khokhlova TD, Canney MS, Khokhlova VA, Sapozhnikov OA, Crum LA, Bailey MR. Controlled tissue emulsification produced by high intensity focused ultrasound shock waves and millisecond boiling. J Acoust Soc Am. 2011;130:3498–510.

Khokhlova VA, Bailey MR, Reed JA, Cunitz BW, Kaczkowski PJ, Crum LA. Effects of nonlinear propagation, cavitation, and boiling in lesion formation by high intensity focused ultrasound in a gel phantom. J Acoust Soc Am. 2006;119:1834–48.

Kollmann C, ter Haar GR, Dolezal L, Hennerici M, Salvesen KA, Valentin L. Ultrasound output: thermal (TI) and mechanical (MI) indices. Ultraschall Med. 2013;34:422–34.

Kopechek JA, Park E, Mei CS, McDannold NJ, Porter TM. Accumulation of phase-shift nanoemulsions to enhance MR-guided ultrasound-mediated tumor ablation in vivo. J Healthc Eng. 2013;4:109–26.

Kopechek JA, Park E, Zhang YZ, Vykhodtseva N, McDannold NJ, Porter TM. Cavitation-enhanced MR-guided focused ultrasound ablation of rabbit tumors in vivo using phase shift nanoemulsions. Phys Med Biol. 2014;59:3465–81.

Kripfgans OD, Fowlkes JB, Miller DL, Eldevik OP, Carson PL. Acoustic droplet vaporization for therapeutic and diagnostic applications. Ultrasound Med Biol. 2000;26:1177–89.

Kucsko G, Maurer PC, Yao NY, Kubo M, Noh HJ, Lo PK, Park H, Lukin MD. Nanometre-scale thermometry in a living cell. Nature. 2013;500:54–8.

Kyriakou Z, Corral Baques MI, Amat A, Coussios CC. HIFU-induced cavitation and heating in ex vivo porcine subcutaneous fat. Ultrasound Med Biol. 2011;37:568–79.

Lauterborn W. Numerical investigation of nonlinear oscillations of gas-bubbles in liquids. J Acoust Soc Am. 1976;59:283–93.

Lele PP. Effects of ultrasound on “solid” mammalian tissues and tumors in vivo. In: Repacholi MH, Grandolfo M, Rindi A, editors. Ultrasound. Germany: Springer; 1987. p. 275–306.

Leslie TA, Kennedy JE, Illing RO, ter Haar GR, Wili F, Phillips RR, Friend PJ, Roberts ISD, Cranston DW, Middleton MR. High-intensity focused ultrasound ablation of liver tumours: can radiological assessment predict the histological response? Br J Radiol. 2008;81:564–71.

Li QY, Du JF, Yu M, He GB, Luo W, Li HL, Zhou XD. Transmission electron microscopy of VX2 liver tumors after high-intensity focused ultrasound ablation enhanced with SonoVue®. Adv Ther. 2009;26:117–25.

Lin K, Kim Y, Maxwell AD, Wang T, Hall TL, Xu Z, Fowlkes B, Cain CA. Histotripsy beyond the “intrinsic” cavitation threshold using very short ultrasound pulses: “Microtripsy”. IEEE Trans Ultrason Ferroelectr Freq Control. 2014;61:251–65.

Liu HL, Chen WS, Chen JS, Shih TC, Chen YY, Lin WL. Cavitation-enhanced ultrasound thermal therapy by combined low-and high-frequency ultrasound exposure. Ultrasound Med Biol. 2006;32:759–67.

Lu M, Shi Y, Fang L, Guan Y, Jiang H, Wan M. Enhanced-cavitation heating protocols in focused ultrasound surgery with broadband split-focus approach. IEEE Trans Ultrason Ferroelectr Freq Control. 2014;61:631–46.

Luo W, Zhou XD, Ren XL, Zheng MJ, Zhang J, He GB. Enhancing effects of SonoVue, a microbubble sonographic contrast agent, on high-intensity focused ultrasound ablation in rabbit livers in vivo. J Ultrasound Med Official J Am Inst Ultrasound Med. 2007;26:469–76.

Luo W, Zhou XD, Yu M, He GB, Zheng XY, Li QY, Liu Q, Han ZH, Zhang J, Qian YQ. Ablation of high intensity focused ultrasound combined with SonoVue on rabbit VX2 liver tumors: assessment with conventional gray-scale US, conventional color/power Doppler US, contrast-enhanced color Doppler US, and contrast-enhanced pulse-inversion harmonic US. Ann Surg Oncol. 2008;15:2943–53.

Luo W, Zhou XD, Yu M, He GB, Zheng XY, Li QY, Liu Q, Han ZH, Zhang J, Qian YQ. Ablation of high-intensity focused ultrasound assisted with SonoVue on Rabbit VX2 liver tumors: sequential findings with histopathology, immunohistochemistry, and enzyme histochemistry. Ann Surg Oncol. 2009;16:2359–68.

Maxwell A, Sapozhnikov O, Bailey M, Crum L, Xu Z, Fowlkes B, Cain C, Khokhlova V. Disintegration of tissue using high intensity focused ultrasound: two approaches that utilize shock waves. Acoust Today. 2012;8:24–37.

McDannold N, Tempany CM, Fennessy FM, So MJ, Rybicki FJ, Stewart EA, Jolesz FA, Hynynen K. Uterine leiomyomas: MR imaging-based thermometry and thermal dosimetry during focused ultrasound thermal ablation. Radiology. 2006a;240:263–72.

McDannold N, Vykhodtseva NI, Hynynen K. Microbubble contrast agent with focused ultrasound to create brain lesions at low power levels: MR imaging and histologic study in rabbits. Radiology. 2006b;241:95–106.

McLaughlan J, Rivens I, Leighton T, ter Haar GR. A study of bubble activity generated in ex vivo tissue by high intensity focused ultrasound. Ultrasound Med Biol. 2010;36:1327–44.

Meaney PM, Clarke RL, Ter Haar GR, Rivens IH. A 3-D finite-element model for computation of temperature profiles and regions of thermal damage during focused ultrasound surgery exposures. Ultrasound Med Biol. 1998;24:1489–99.

Melodelima D, Chapelon JY, Theillere Y, Cathignol D. Combination of thermal and cavitation effects to generate deep lesions with an endocavitary applicator using a plane transducer: ex vivo studies. Ultrasound Med Biol. 2004;30:103–11.

Miller NR, Bamber JC, ter Haar GR. Imaging of temperature-induced echo strain: preliminary in vitro study to assess feasibility for guiding focused ultrasound surgery. Ultrasound Med Biol. 2004;30:345–56.

Moriyama T, Yoshizawa S, Umemura SI. Thermal simulation of cavitation-enhanced ultrasonic heating verified with tissue-mimicking gel. Jpn J Appl Phys. 2012;51:07GF27.

Murat FJ, Gelet A. Visually directed transrectal high intensity focused ultrasound for the treatment of prostate cancer: a preliminary report on the italian experience comments. J Urology. 2009;181:111.

Napoli A, Mastantuono M, Marincola BC, Anzidei M, Zaccagna F, Moreschini O, Passariello R, Catalano C. Osteoid osteoma: MR-guided focused ultrasound for entirely noninvasive treatment. Radiology. 2013;267:514–21.

Okita K, Sugiyama K, Takagi S, Matsumto Y. Microbubble behavior in an ultrasound field for high intensity focused ultrasound therapy enhancement. J Acoust Soc Am. 2013;134:1576–85.

Ong JT, d'Arcy JA, Collins DJ, Rivens IH, ter Haar GR, Leach MO. Sliding window dual gradient echo (SW-dGRE): T1 and proton resonance frequency (PRF) calibration for temperature imaging in polyacrylamide gel. Phys Med Biol. 2003;48:1917–31.

Peng S, Xiong Y, Li KQ, He M, Deng YB, Chen L, Zou M, Chen WZ, Wang ZB, He J. Clinical utility of a microbubble-enhancing contrast ("SonoVue") in treatment of uterine fibroids with high intensity focused ultrasound: a retrospective study. Eur J Radiol. 2012;81:3832–8.

Pennes HH. Analysis of tissue and arterial blood temperatures in the resting human forearm. J Appl Physiol. 1948;1:93–122.

Phillips LC, Puett C, Sheeran PS, Dayton PA, Miller GW, Matsunaga TO. Phase-shift perfluorocarbon agents enhance high intensity focused ultrasound thermal delivery with reduced near-field heating. J Acoust Soc Am. 2013;134:1473–82.

Pierce AD. Acoustics. Woodbury: Acoustical Society of America; 1994.

Poritsky H. The collapse or growth of a spherical bubble or cavity in a viscous fluid. J Appl Mech Trans Asme. 1951;18:332–3.

Prosperetti A. Thermal effects and damping mechanisms in the forced radial oscillations of gas bubbles in liquids. J Acoust Soc Am. 1977;61:17–27.

Qamar A, Wong ZZ, Fowlkes JB, Bull JL. Dynamics of acoustic droplet vaporization in gas embolotherapy. Appl Phys Lett. 2010;96:143702.

Qin SP, Caskey CF, Ferrara KW. Ultrasound contrast microbubbles in imaging and therapy: physical principles and engineering. Phys Med Biol. 2009;54:R27.

Rapoport N, Christensen DA, Kennedy AM, Nam KH. Cavitation properties of block copolymer stabilized phase-shift nanoemulsions used as drug carriers. Ultrasound Med Biol. 2010;36:419–29.

Ritchie RW, Leslie T, Phillips R, Wu F, Illing R, ter Haar GR, Protheroe A, Cranston D. Extracorporeal high intensity focused ultrasound for renal tumours: a 3-year follow-up. BJU Int. 2010;106:1004–9.

Rivens IH, Rowland IJ, Denbow M, Fisk NM, ter Haar GR, Leach MO. Vascular occlusion using focused ultrasound surgery for use in fetal medicine. Eur J Ultrasound Official J Eur Fed Soc Ultrasound Med Biol. 1999;9:89–97.

Sasaki H, Yasuda J, Takagi R, Miyashita T, Goto K, Yoshizawa S, Umemura S. Highly efficient cavitation-enhanced heating with dual-frequency ultrasound exposure in high-intensity focused ultrasound treatment. Jpn J Appl Phys 2014;53:07KF11.

Sboros V. Response of contrast agents to ultrasound. Adv Drug Deliv Rev. 2008;60:1117–36.

Shaw CJ, ter Haar GR, Rivens IH, Giussani DA, Lees CC. Pathophysiological mechanisms of high-intensity focused ultrasound-mediated vascular occlusion and relevance to non-invasive fetal surgery. J Royal Soc Interface. 2014;11.

Sheeran PS, Streeter JE, Mullin LB, Matsunaga TO, Dayton PA. Toward ultrasound molecular imaging with phase-change contrast agents: an in vitro proof of principle. Ultrasound Med Biol. 2013;39:893–902.
Simon JC, Sapozhnikov OA, Khokhlova VA, Wang YN, Crum LA, Bailey MR. Ultrasonic atomization of tissue and its role in tissue fractionation by high intensity focused ultrasound. Phys Med Biol. 2012;57:8061–78.
Sokka SD, King R, Hynynen K. MRI-guided gas bubble enhanced ultrasound heating in in vivo rabbit thigh. Phys Med Biol. 2003;48:223.
Solovchuk MA, Hwang SC, Chang H, Thiriet M, Sheu TWH. Temperature elevation by HIFU in ex vivo porcine muscle: MRI measurement and simulation study. Med Phys. 2014;41:052903.
Stride E, Saffari N. The potential for thermal damage posed by microbubble ultrasound contrast agents. Ultrasonics. 2004;42:907–13.
Takegami K, Kaneko Y, Watanabe T, Maruyama T, Matsumoto Y, Nagawa H. Erythrocytes, as well as microbubble contrast agents, are important factors in improving thermal and therapeutic effects of high-intensity focused ultrasound. Ultrasound Med Biol. 2005;31:385–90.
ter Haar GR. Ultrasound focal beam surgery. Ultrasound Med Biol. 1995;21:1089–100.
ter Haar GR. Therapeutic ultrasound. Eur J Ultrasound Official J Eur Fed Soc Ultrasound Med Biol. 1999;9:3–9.
ter Haar GR. Acoustic surgery. Phys Today. 2001a;54:29–34.
ter Haar GR. High intensity focused ultrasound for the treatment of tumors. Echocardiography. 2001b;18:317–22.
ter Haar GR. Therapeutic applications of ultrasound. Prog Biophys Mol Biol. 2007;93:111–29.
ter Haar GR, Coussios C. High intensity focused ultrasound: physical principles and devices. Int J Hyperth. 2007;23:89–104.
ter Haar GR, Sinnett D, Rivens I. High-intensity focused ultrasound—a surgical technique for the treatment of discrete liver tumors. Phys Med Biol. 1989;34:1743–50.
Tran BC, Seo J, Hall TL, Fowlkes JB, Cain CA. Microbubble-enhanced cavitation for noninvasive ultrasound surgery. IEEE Trans Ultrason Ferroelectr Freq Control. 2003;50:1296–304.
Tran BC, Seo J, Hall TL, Fowlkes JB, Cain CA. Effects of contrast agent infusion rates on thresholds for tissue damage produced by single exposures of high-intensity ultrasound. IEEE Trans Ultrason Ferroelectr Freq Control. 2005;52:1121–30.
Tung YS, Liu HL, Wu CC, Ju KC, Chen WS, Lin WL. Contrast-agent-enhanced ultrasound thermal ablation. Ultrasound Med Biol. 2006;32:1103–10.
Tyler WJ, Tufail Y, Pati S. Pain: noninvasive functional neurosurgery using ultrasound. Nat Rev Neurol. 2010;6:13–4.
Umemura SI, Kawabata KI, Sasaki K. In vivo acceleration of ultrasonic tissue heating by microbubble agent. IEEE Trans Ultrason Ferroelectr Freq Control. 2005;52:1690–8.
Vilensky G, ter Haar GR, Saffari N. A model of acoustic absorption in fluids based on a continuous distribution of relaxation times. Wave Motion. 2012;49:93–108.
Visioli AG, Rivens IH, ter Haar GR, Horwich A, Huddart RA, Moskovic E, Padhani A, Glees J. Preliminary results of a phase I dose escalation clinical trial using focused ultrasound in the treatment of localised tumours. Eur J Ultrasound: Official J Eur Fed Soc Ultrasound Med Biol. 1999;9:11–8.
Von Zur Muhlen C, Von Elverfeldt D, Moeller JA, Choudhury RP, Paul D, Hagemeyer CE, Olschewski M, Becker A, Neudorfer I, Bassler N, Schwarz M, Bode C, Peter K. Magnetic resonance imaging contrast agent targeted toward activated platelets allows in vivo detection of thrombosis and monitoring of thrombolysis. Circulation. 2008;118:258–267.
Vykhodtseva N, McDannold N, Hynynen K. Induction of apoptosis in vivo in the rabbit brain with focused ultrasound and optison. Ultrasound Med Biol. 2006;32:1923–9.
Wang YN, Khokhlova T, Bailey M, Hwang JH, Khokhlova V. Histological and biochemical analysis of mechanical and thermal bioeffects in boiling histotripsy lesions induced by high intensity focused ultrasound. Ultrasound Med Biol. 2013;39:424–38.

Watmough DJ, Lakshmi R, Ghezzi F, Quan KM, Watmough JA, Khizhnyak E, Pashovkin TN, Sarvazyan AP. The effect of gas-bubbles on the production of ultrasound hyperthermia at 0.75 mhz—a phantom study. Ultrasound Med Biol. 1993;19:231–41.
Wharton IP, Rivens IH, ter Haar GR, Gilderdale DJ, Collins DJ, Hand JW, Abel PD, Desouza NM. Design and development of a prototype endocavitary probe for high-intensity focused ultrasound delivery with integrated magnetic resonance imaging. J Magn Reson Imaging. 2007;25:548–56.
Wu F, ter Haar GR, Chen WR. High-intensity focused ultrasound ablation of breast cancer. Expert Rev Anticancer Ther. 2007;7:823–31.
Xu Z, Fan ZZ, Hall TL, Winterroth F, Fowlkes JB, Cain CA. Size measurement of tissue debris particles generated from pulsed ultrasound cavitational therapy—histotripsy. Ultrasound Med Biol. 2009;35:245–55.
Xu Z, Fowlkes JB, Ludomirsky A, Cain CA. Investigation of intensity thresholds for ultrasound tissue erosion. Ultrasound Med Biol. 2005;31:1673–82.
Xu Z, Ludomirsky A, Eun LY, Hall TL, Tran BC, Fowlkes JB, Cain CA. Controlled ultrasound tissue erosion. IEEE Trans Ultrason Ferroelectr Freq Control. 2004;51:726–36.
Xu Z, Owens G, Gordon D, Cain C, Ludomirsky A. Noninvasive creation of an atrial septal defect by histotripsy in a canine model. Circulation. 2010;121:742–9.
Yang XM, Roy RA, Holt RG. Bubble dynamics and size distributions during focused ultrasound insonation. J Acoust Soc Am. 2004;116:3423–31.
Yang XM, Jo J. Enhanced cavitation by using two consecutive ultrasound waves at different frequencies. Appl Phys Lett. 2014;105:193701.
Yu TH, Fan XL, Xiong SH, Hu K, Wang ZB. Microbubbles assist goat liver ablation by high intensity focused ultrasound. Eur Radiol. 2006a;16:1557–63.
Yu TH, Hu DB, Xu CS. Microbubbles improve the ablation efficiency of extracorporeal high intensity focused ultrasound against kidney tissues. World J Urol. 2008;26:631–6.
Yu TH, Wang GY, Hu K, Ma P, Bai J, Wang ZB. A microbubble agent improves the therapeutic efficiency of high intensity focused ultrasound: a rabbit kidney study. Urol Res. 2004;32:14–9.
Yu TH, Xiong SH, Mason TJ, Wang ZB. The use of a microbubble agent to enhance rabbit liver destruction using high intensity focused ultrasound. Ultrason Sonochem. 2006b;13:143–9.
Zderic V, Keshavarzi A, Noble ML, Paun M, Sharar SR, Crum LA, Martin RW, Vaezy S. Hemorrhage control in arteries using high-intensity focused ultrasound: a survival study. Ultrasonics. 2006;44:46–53.
Zhang P, Porter T. An in vitro study of a phase-shift nanoemulsion: a potential nucleation agent for bubble-enhanced HIFU tumor ablation. Ultrasound Med Biol. 2010;36:1856–66.
Zhang SY, Wan MX, Zhong H, Xu C, Liao Z, Liu H, Wang SP. Dynamic changes of integrated backscatter, attenuation coefficient and bubble activities during high-intensity focused ultrasound (HIFU) treatment. Ultrasound Med Biol. 2009;35:1828–44.
Zhang SY, Ding T, Wan MX, Jiang HJ, Yang X, Zhong H, Wang SP. Minimizing the thermal losses from perfusion during focused ultrasound exposures with flowing microbubbles. J Acoust Soc Am. 2011;129:2336–44.
Zhang SY, Zhou FY, Wan MX, Wei M, Fu QY, Wang X, Wang SP. Feasibility of using Nakagami distribution in evaluating the formation of ultrasound-induced thermal lesions. J Acoust Soc Am. 2012a;131:4836–44.
Zhang SY, Zong YJ, Wan MX, Yu XJ, Fu QY, Ding T, Zhou FY, Wang SP. Compare ultrasound-mediated heating and cavitation between flowing polymer and lipid-shelled microbubbles during focused ultrasound exposures. J Acoust Soc Am. 2012b;131:4845–55.
Zhang SY, Li C, Yin H, Wang SP, Wan MX. Surface vibration and nearby cavitation of an ex vivo bovine femur exposed to high intensity focused ultrasound. J Acoust Soc Am. 2013;134:1656–62.
Zhang SY, Li C, Zhou FY, Wan MX, Wang SP. Enhanced lesion-to-bubble ratio on ultrasonic Nakagami imaging for monitoring of high-intensity focused ultrasound. J Ultrasound Med Official J Am Inst Ultrasound Med. 2014;33:959–70.

Chapter 5
Cavitation-Enhanced Mechanical Effects and Applications

Yujin Zong, Shanshan Xu, Tom Matula and Mingxi Wan

Keywords Microstreaming · Liquid jetting · Shock wave · Lithotripsy · Histotripsy · Sonothrombolysis · Drug delivery

5.1 Introduction

In a broad sense, acoustic cavitation refers to ultrasonically induced bubble activity occurring in a biological material. If gaseous microbubbles, including cavitation bubbles or encapsulated contrast agent bubbles, are placed in a sound field, they might be powerful actuators, and their response to ultrasonic irradiation may involve growing to a resonant size and undergoing oscillatory action. A series of secondary motions can result from the oscillation of microbubbles, such as acoustic streaming, shock waves, and liquid microjets, thereby producing mechanical effects on surrounding tissue or cells.

If these microbubbles are surrounded by a liquid medium, the stable and fast vibration of the oscillating bubble wall, which has a displacement amplitude much higher than that of the surrounding liquid medium, may produce an eddying motion

Y. Zong · S. Xu
The Key Laboratory of Biomedical Information Engineering of Ministry of Education, Department of Biomedical Engineering, School of Life Science and Technology, Xi'an Jiaotong University, Xi'an 710049, People's Republic of China
e-mail: yjzong@mail.xjtu.edu.cn

S. Xu
e-mail: xushanshan@mail.xjtu.edu.cn

T. Matula · M. Wan (✉)
Center for Industrial and Medical Ultrasound, Applied Physics Laboratory, University of Washington, Seattle, WA 98105, USA
e-mail: mxwan@mail.xjtu.edu.cn

T. Matula
e-mail: matula@apl.washington.edu

M. Wan et al. (eds.), *Cavitation in Biomedicine*,
DOI 10.1007/978-94-017-7255-6_5

in the surrounding medium and generate a time-independent flow; this phenomenon is called acoustic microstreaming. The steady flow of acoustic microstreaming leads to local increases in shear stresses around the gas bubbles and exerts shear forces on the surrounding tissue or cells, thus modifying their local environment. Additionally, if this bubble oscillation occurs near a boundary, the bubble tends to transform, shrink, and finally collapse asymmetrically due to differential pressures inside and outside of the bubble. The extremely large implosive pressure generated at the moment when the bubble collapses leads to the formation of high-amplitude shock waves. At the same time, the collapsing bubbles experience non-uniformities in their surroundings that cause high-velocity microjets to form. The potential energy of the expanded bubble is converted into the kinetic energy of the liquid microjet that extends through the bubble interior and penetrates the opposite bubble wall. The microjet can penetrate into the tissue or generate secondary stress waves in the tissue. Thus, as jets and shock waves impinge on the surrounding tissue, these forces can also create localized mechanical effects.

These cavitation-enhanced mechanical effects are the physical means by which some bubble-mediated ultrasound therapies work, so they are paramount to therapeutic outcomes (Mitragotri 2005). In this chapter, we offer some examples of these therapeutic applications after we introduce the main mechanical effects associated with cavitation.

5.2 Acoustic Microstreaming and Stress Field Created by Oscillating Bubble

Cavitation microstreaming is a small-scale acoustic streaming generated by bubble oscillation (Kolb and Nyborg 1956; Elder 1959), and it usually occurs when a cavitation bubble or contrast microbubble, surrounded by a liquid media, undergoes direct oscillatory action when exposed to ultrasound. During bubble oscillation, the displacement of liquid around the gas bubble causes a bulk fluid flow around it, causing the cavitation microstreaming or acoustic microstreaming.

The microstreaming flow pattern depends on the oscillation mode of the bubble in the sound field. Bubbles may undergo several different types of oscillatory motions, including translating oscillations, volume oscillations, and higher order modes of oscillation. Translating oscillations refer to bubbles moving periodically along one or more different axes. Volume oscillations mean that the bubbles expand and contract radially while maintaining their spherical shape; this is also known as the "breathing" mode. This behavior is a consequence of the compressibility of the gas and is forced by the fluctuating acoustic pressure field. This oscillatory mode is most pronounced when the bubble is excited near its natural frequency. Consider that a spherical gas bubble in a liquid medium, whose radius is R_0 in the absence of sound, undergoes spherically symmetric vibration within a sound field. Its linear resonant frequency f (Hz) is given by the equation (Minnaert 1933)

$$2\pi R_0 f = \sqrt{3\gamma P_0/\rho} \quad (5.1)$$

where R_0 is the bubble's equilibrium radius (cm), γ is the ratio of specific heats for the gas, P_0 is the hydrostatic pressure (dyn/cm^2), and ρ is the density of the liquid (g/cm^3). Finally, bubbles may also be excited into higher order modes of oscillation where the surface of the bubble deforms into polygonal shapes.

In this chapter, we pay more attention to the problem of acoustic microstreaming that develops in liquids very near a vibrating gas bubble in an ultrasound field at megahertz frequencies. Similar to macroscopic acoustic streaming, cavitation microstreaming is a bulk non-periodic motion of fluid; it is a second-order nonlinear velocity gradient in the first-order oscillatory sound field. Microstreaming relies on the large velocity gradient due to small length scales of the bubble. Overall, cavitation microstreaming belongs to a class of acoustical manifestations that includes movements, forces, strains, and stresses, in which the geometrical scale of these manifestations is small, i.e., microscopic (Nyborg 1982).

Kolb and Nyborg (1956) were the first to study cavitation microstreaming. They found that cavitation microstreaming is orderly at low driving amplitudes and is the most pronounced for bubbles undergoing volume oscillation when the insonation frequency drives the bubbles at their resonance frequency for pulsation, especially when the bubbles are situated on solid boundaries. Acoustic microstreaming arising around a single bubble excited close to resonance produces strong liquid circulation flow. Therefore, bubble-induced streaming is strongly dependent on frequency for a given bubble radius, and on bubble radius for a given frequency. Additionally, microstreaming decreases in size as the ultrasonic frequency is increased (NCRP 2002); that is, less microstreaming occurs at higher frequencies. Microstreaming, therefore, occurs in response to microbubble oscillations driven by low-frequency, low-pressure ultrasonic fields.

5.2.1 Theoretical Calculation of Cavitation Microstreaming

Acoustic streaming can be theoretically described by quadratic convective terms in the governing equations for fluid flow. Cavitation microstreaming results from vorticity generated within the oscillatory boundary layer surrounding the bubble, and it is a consequence of second-order nonlinear terms in the vorticity transport equation. Nyborg (1958) developed an approximate solution for sonically induced steady flow near a fluid–solid interface (boundary), and derived theoretical expressions for acoustic streaming from the basic equations of hydrodynamics, making use of the method of successive approximations. Based on his method, the theoretical approach to calculate acoustic microstreaming is based on successive approximations to solutions of nonlinear hydrodynamic equations. One must obtain two solutions (of first and second order, respectively), each of which must satisfy boundary conditions. The first-order linear scattered field of the bubble is calculated first, and then the velocity of the acoustic streaming is calculated by solving the equations of acoustic streaming used in this scattered field and taken with accuracy up to second-order terms.

Associated with microstreaming is a time-independent boundary layer that sets up near the surface, through the interaction of oscillatory fluid motion. This acoustic boundary layer is the region near the boundary over which the tangential component of the acoustic particle velocity drops to zero. The boundary layer is characterized by the thickness δ, defined as follows:

$$\delta = \sqrt{\frac{\mu}{\pi f \rho}} \tag{5.2}$$

where f is the frequency, μ is the shear viscosity of the medium, and ρ is the density of the medium (Coakley and Nyborg 1978).

The basic equation for streaming (Nyborg 1958) is as follows:

$$\mu \nabla^2 \mathbf{u}_2 - \nabla p_2 + \mathbf{F} = 0 \tag{5.3}$$

where the nonlinear quantity $\mathbf{F} = -\rho_0 \langle \mathbf{u}_1 \cdot \nabla \mathbf{u}_1 + \mathbf{u}_1 \nabla \cdot \mathbf{u}_1 \rangle$ is the exact analog of a driving force field. Here, the brackets <> indicate a time average over a large number of cycles. The terms $\mathbf{u}_2$ and p_2 are the velocity and pressure fields of acoustic streaming, respectively, and both $\mathbf{u}_2$ and p_{2-} are the time-independent second-order quantities. Finally, ρ_0 and μ are the ambient fluid density and viscosity, respectively. It is also assumed that $\nabla \cdot \mathbf{u}_2 \approx 0$. The first-order velocity $\mathbf{u}_1$ is the oscillatory—i.e., "acoustic"—particle velocity.

Consider the first-order approximation $\mathbf{u}_1$ to the velocity. The governing equation for oscillatory flow in a viscous medium is as follows (Nyborg 1958):

$$(\nabla^2 + h^2)\mathbf{u}_1 = (1 - h^2 k^{-2}) \nabla \nabla \cdot \mathbf{u}_1 \tag{5.4}$$

where the wave number $k = (\omega/c)$, while ω is the angular frequency and c is the speed of sound; $h = (1-i)\beta$, $i^2 = -1$; and $\beta = (\omega/2v)^{1/2}$, where $v = \mu/\rho_0$, and ρ_0 and μ are the ambient fluid density and viscosity, respectively. Equation (5.4) is satisfied by any vector function $\mathbf{u}_1$ for which both Eqs. (5.5) and (5.6) are true:

$$\nabla \times \mathbf{u}_1 = 0, \quad (\nabla^2 + k^2)\mathbf{u}_1 = 0 \tag{5.5}$$

and

$$\nabla \cdot \mathbf{u}_1 = 0, \quad (\nabla^2 + h^2)\mathbf{u}_1 = 0 \tag{5.6}$$

The first-order flow velocity is defined as $\mathbf{u}_1 = -\nabla \varphi_1$, where φ_1 is the velocity potential. From Eq. (5.5), we obtain the following:

$$(\nabla^2 + k^2)\varphi_1 = 0 \tag{5.7}$$

Therefore, within the boundary layer, there are two possible types of solutions for $\mathbf{u}_1$, i.e., $\mathbf{u}_1 = \mathbf{u}_a + \mathbf{u}_b$. Here, $\mathbf{u}_a$ are solutions of Eq. (5.5), and $\mathbf{u}_b$ are solutions of the vector Helmholtz equation, Eq. (5.6). In spherical coordinates, both the r component and the θ component of $\mathbf{u}_b$ should satisfy the Eq. (5.6).

In biomedical applications of acoustics, the bubbles are usually freely suspended in a liquid, where they are away from any boundary, or near a boundary, but are not resting on the boundary. Thus, several groups have thus developed detailed theories for the streaming velocities near pulsating and translating bubbles in these situations (Krasovitski and Kimmel 2004; Doinikov and Bouakaz 2010a, b, c; Wu and Du 1997; Wu 2002; Liu and Wu 2009).

5.2.1.1 Unshelled Free Gas Bubble

An oscillating bubble can be assumed to be a simple source, generating a spherical wave (Elder 1959; Nyborg 1958). For an isolated bubble of radius R_0 in the field of a traveling plane wave, Wu and Du (1997) developed an analytical approach to calculate streaming velocity inside and outside the bubble, where the streaming is produced by a bubble-scattered sound field under the condition that the radius of the bubble is much smaller than the ultrasound wavelength. They assumed that viscous effects are essential only within a thin boundary layer, while beyond the boundary layer the fluid can still be considered to be inviscid. They also assumed the thickness of the viscous boundary layer surrounding the bubble is much smaller than the radius of the bubble. The streaming in a liquid (outside the bubble) or in the gas (inside the bubble) generated by the bubble is therefore primarily a result of the presence of a boundary layer at the bubble's liquid–gas interface, as shown in Fig. 5.1.

Based on Nyborg's approach (Nyborg 1958), Wu and Du (1997) obtained approximate solutions for the first-order velocity field, which are valid only inside the boundary layer, and then the solutions were used to find the solutions to the equations of acoustic streaming, which is the second-order time-independent term. Under the influence of the ultrasound wave, the bubble can pulsate as a monopole (pulsation, $n = 0$ mode) and oscillate like a dipole (translational harmonic vibrations, $n = 1$ mode). The streamlines for the modes $n = 0$ and $n = 1$ are shown in Fig. 5.2. The streaming was more vigorous inside than outside the bubble, implying that convection inside the bubble may be enhanced by acoustic streaming. Wu ad Du also showed that the pulsation mode contributes more to the streaming velocities than the translational vibrations does.

The assumption that viscous effects are essential only within a thin boundary layer might underestimate the power of acoustic streaming, so Doinikov and Bouakaz (2010a) suggested an improved theory that corrects the errors of Wu et al.'s previous theory and extends its limits. The proposed theory treats the entire bulk of the liquid outside the bubble and the gas inside the bubble as viscous,

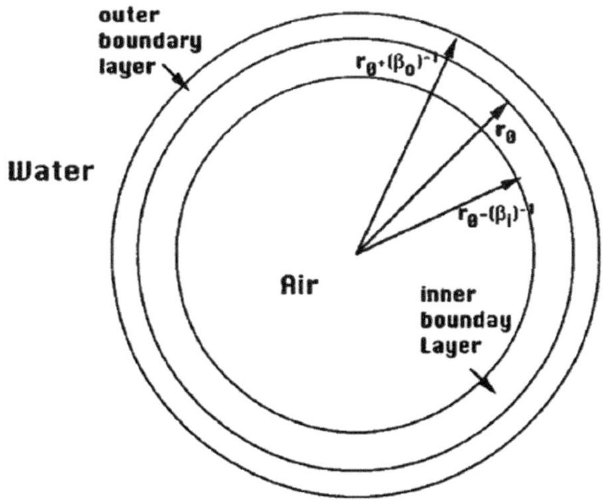

Fig. 5.1 Illustration of an air bubble and its boundary layer. Reproduced with permission from Wu et al. © Acoustical Society of America

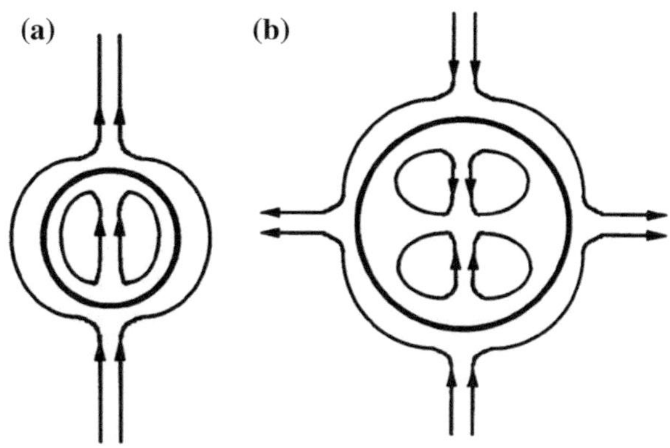

Fig. 5.2 Illustration of streamlines generated by the following modes: **a** $n = 0$ and **b** $n = 1$ modes. The incoming sound wave propagates from bottom to top. Reproduced with permission from Wu et al. © Acoustical Society of America

heat-conducting fluids. No restrictions are imposed on the size of the bubble relative to the viscous, thermal, and sound wavelengths in the ambient liquid or in the internal gas medium. All modes of the bubble's motion, pulsation, translation, and shape oscillation are taken into account.

The results of calculations for the radial and tangential streaming velocities are presented in Fig. 5.3. This figure shows that the intensity of acoustic streaming increases as the driving frequency approaches the resonant frequency of the bubble, which is about 607 kHz in the case under consideration. The behavior of bubbles in sound fields is determined largely by their resonance characteristics, such that

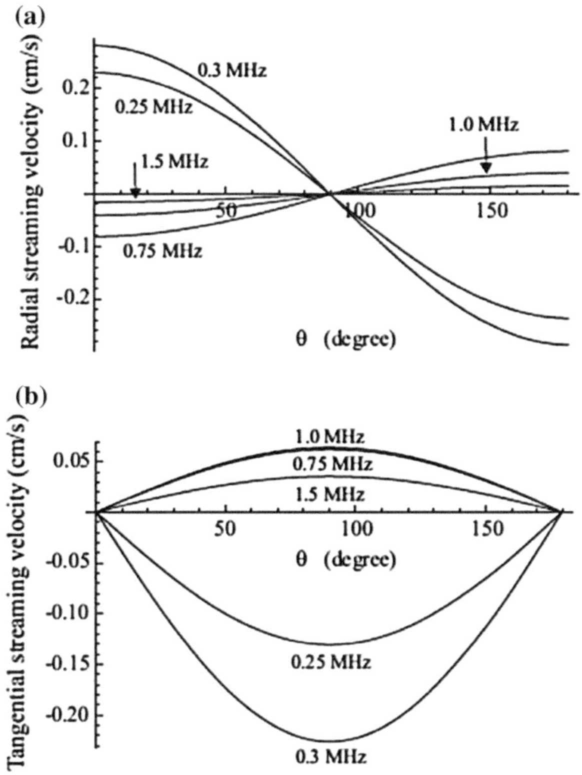

Fig. 5.3 Acoustic streaming velocity versus θ at different values of the driving frequency. The bubble radius is 5 μm and the acoustic pressure amplitude is 50 kPa. Reproduced with permission from Doinikov et al. © Acoustical Society of America

especially violent and chaotic motions appear near bubbles of resonant size. From Fig. 5.3, we can see that the direction of acoustic streaming around the bubble can change depending on whether the driving frequency is below or above the resonance frequency.

By means of numerical examples, it has been confirmed that the main contribution to the acoustic streaming generated by a gas bubble comes from the bubble's volume pulsation.

5.2.1.2 Shelled Microbubble

Compared to unshelled free gas bubbles, shelled microbubbles that are used as ultrasound contrast agents (UCAs) in diagnostic ultrasound imaging and various therapeutic applications have similar and more pronounced cavitation activity and mechanical effects. The main physical difference between a free gas bubble and a shelled UCA bubble is that the UCA bubble is encapsulated by lipids, proteins, or polymers that make the bubble more stable in liquids. At lower ultrasound levels, the UCA bubble in a liquid is forced to oscillate stably in volume, which can lead to local and relatively steady microstreaming.

Since the previous theory we discussed does not require input on the characteristics of the oscillating bubble, it can be applied to a shelled UCA as long as the initial radius and surface displacement of the UCA is known. Thus, the microstreaming field around an isolated UCA bubble in a viscous fluid excited by an ultrasonic field was theoretically studied using a similar mathematical procedure (Liu and Wu 2009; Doinikov and Bouakaz 2010b). First, solutions were obtained for the first-order flow (AC flow) velocity near an oscillating UCA. Then, the second-order non-oscillatory (DC flow)—i.e., acoustic microstreaming—was derived.

When an ultrasound wave propagates in a viscous liquid, the outer and inner boundary layers form at the interface between the liquid and the shell, as well as between the shell and core gas of the UCA. The calculated results show that the streaming velocity and stresses near a shelled bubble are functions of the mechanical properties of the shell and the gas. Overall, the streaming velocity and stresses for a shelled bubble are greater than those for a similar sized free bubble under the same ultrasound excitation.

An illustration of the streaming lines for modes $n = 0$ and $n = 1$ inside and outside a shelled bubble and a free bubble is shown in Fig. 5.4. Here, we see that the streaming line inside of the shelled bubble changes direction at $\theta = 90°$, while outside of the bubble the streaming line changes direction at $\theta = 180°$ for $n = 0$ mode. For $n = 1$, the streaming line changes direction at $\theta = 90°$ and $\theta = 180°$ for both the inside and outside of the bubble. Because of different boundary conditions, the streaming lines in a shelled bubble are quite different from those in free bubble.

When the bubble's equilibrium radius $R_0 \gg R_0 \gg \delta$, the streaming speeds near an isolated bubble are much smaller than those near a bubble resting on a solid boundary. However, when R_0 is comparable to δ, the streaming speeds for the two situations are of the same order of magnitude (Davidson and Riley 1971). For water as the surrounding medium, we find that δ equals 0.32 μm. The mean equilibrium radius for a commercial UCA is approximately 10× the boundary layer thickness, so the streaming near an isolated UCA is different with a UCA attached to a boundary.

5.2.2 *Experimental Observation of Cavitation Microstreaming*

Although the phenomenon of cavitation microstreaming has been studied since the 1950s, few microstreaming flow visualizations or experimental measurements of the velocity fields have been conducted. Experimental investigations of streaming motions around the oscillating bubble are commonly made by observing small tracer particles suspended in the streaming fluid, such as carmine red, aluminum dust, fluorescent microspheres, and Latex particles (Kolb and Nyborg 1956; Elder 1959; Gormley and Wu 1998; Tho et al. 2007; Collis et al. 2010).

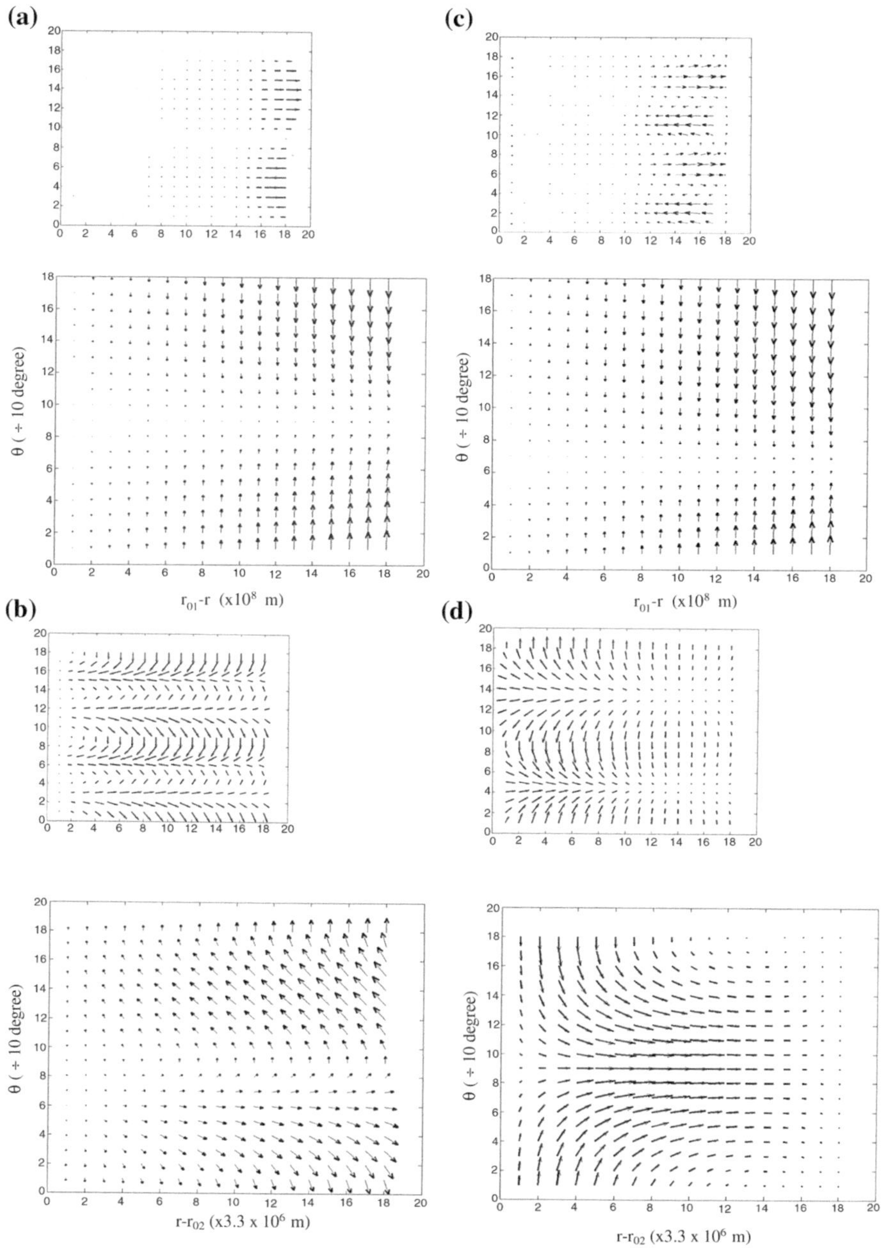

Fig. 5.4 Illustration of streaming lines **a** $n = 0$, inside the bubble; **b** $n = 0$, outside the bubble; **c** $n = 1$, inside the bubble; and **d** $n = 1$, outside the bubble. Inserts are plots for a free bubble. Reproduced with permission from Liu et al. © Acoustical Society of America

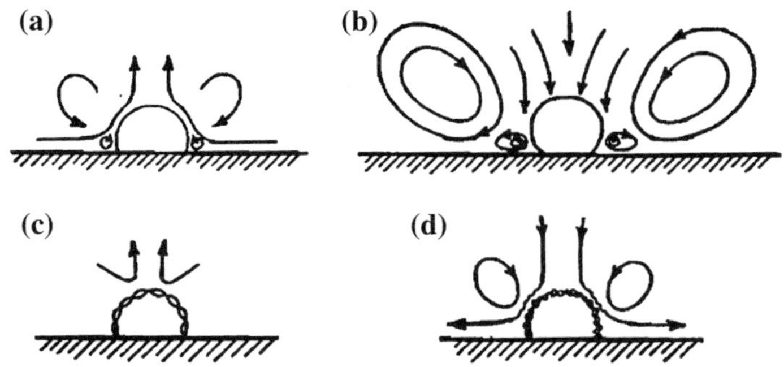

Fig. 5.5 Four regimes of cavitation microstreaming flow around an air bubble with a surface skin. Reproduced with permission from Elder et al. © Acoustical Society of America

In one such study, Elder (1959) performed a systematic experimental investigation of acoustic streaming near an ultrasound-excited bubble in a solution with varying kinematic viscosities. They used aluminum dust as a tracer particle and classified the patterns of bubble-associated streaming into four different types, as shown in Fig. 5.5. The first type is that observed near a surface-contaminated bubble in a liquid of low viscosity. At a low-amplitude, the streaming consisted of a large and a small vortex ring surrounding the bubble, as shown in Fig. 5.5a. The second type of streaming regime (shown in Fig. 5.5b) was predominant for medium and high viscosities, and it was observed to be stable over a wide range of amplitudes and viscosities. The third regime (Fig. 5.5c) was observed most readily in liquids of low viscosity, though it did occur at higher viscosities when sufficiently large driving amplitudes were applied. Its occurrence coincided with the onset of the first surface mode and the dissolution of the second regime at higher amplitudes. Finally, the fourth regime (Fig. 5.5d), which occurred only in the least viscous solutions, seems to represent a return to the pattern of the second regime, as the amplitude becomes too large to permit the existence of a single stable surface mode.

The microstreaming created by larger oscillating bubbles attached to a wall has also been investigated using a technique called microscopic particle–image velocimetry (micro-PIV) (Tho et al. 2007; Collis et al. 2010). The micro-PIV system consists of an inverted epifluorescent microscope, a light source, and a CCD camera imaging system, as shown in Fig. 5.6. Light at a certain wavelength is used to illuminate fluorescent particles in the microdevice through an epifluorescent inverted microscope. Then, a CCD camera records the particle images. Using this system, several different image series can be taken, including streak images depicting the streamlines of the flow field, micro-PIV image pairs in cross-correlation analysis to determine the instantaneous velocity fields, and also images depicting the oscillatory motion of the bubble.

Streak image results (Fig. 5.7) performed with micro-PIV show many different microstreaming patterns around a microbubble attached to a surface. Each pattern is

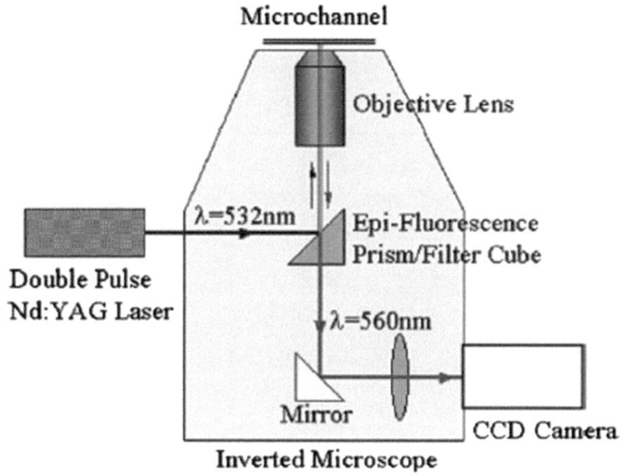

Fig. 5.6 Schematic of the micro-PIV system. Reproduced with permission from Collis et al. © Elsevier

associated with a particular oscillation mode of the bubble. Figure 5.7a is a "quadrupole" pattern that occurs via oscillatory linear translation, the frequency at which is well below the resonant frequency. The streaming flow pattern contains four vortices located roughly $2R_0$ from the center of the bubble. Fluid moves toward the bubble along the horizontal axis and away from the bubble along the vertical axis of oscillation. Figure 5.7b is a circular microstreaming flow that occurs via elliptical bubble motion; a single vortex centered on the bubble. Conversely, Fig. 5.7c is microstreaming in a "dipole" pattern. This streaming flow pattern is clearly directional with a plane of symmetry along the vertical axis with flow being recirculated from one end of the bubble to the other. This flow pattern is similar to the simulated streamline in Fig. 5.2. Near the bubble surface, fluid moves up and over the surface of the bubble to the other side with fluid velocities of approximately 0.3 mms^{-1}. In Fig. 5.7d, the bubble is undergoing a translating oscillation along a single axis and simultaneous volume oscillation. The streaming flow pattern is similar to the pure volume oscillating bubble (in panel c), except that the cores of the circulations appear to have been shifted away from the bubble. Changing between these patterns is achieved by changing the sound frequency.

Gormley and Wu (1998) have experimentally observed the acoustic microstreaming occurring near a UCA sphere, Albunex®, trapped at a pressure antinode of a standing wave field; the resulting microstreaming pattern is shown in Fig. 5.8. At low acoustic pressure amplitude of the standing wave, the sphere performed pulsating oscillation and the acoustic streaming around the sphere took a monopole pattern (Fig. 5.8a). As the acoustic pressure amplitude was increased gradually, the pulsation became unstable; translation oscillation then occurred. The sphere experienced both monopole and dipole oscillations (Fig. 5.8b, c, the pressure in c is higher than in b), and the amplitude of the translational (dipole) oscillation in c was greater than in b. The streaming velocity was estimated to be 50–100 m/s near the sphere. Additionally, it seemed that there were two jets leaving from the bubble in both b and c; the spacing between the two jets was greater in c than in b.

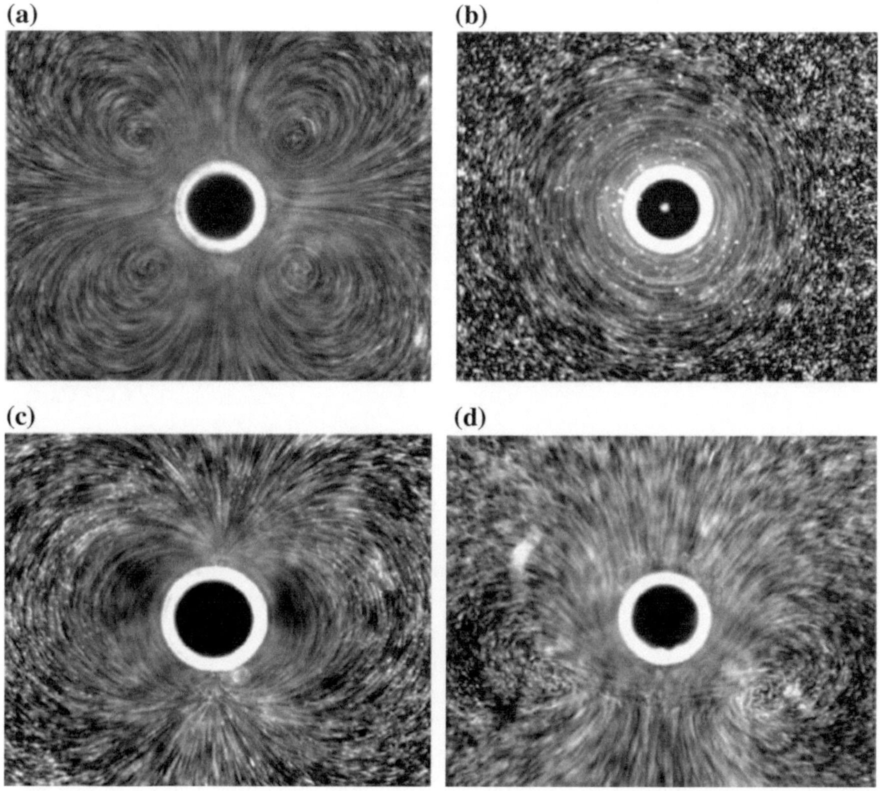

Fig. 5.7 Streaming patterns induced by **a** translating oscillation along a single axis ($R_0 = 232$ μm, $f = 2.422$ kHz); **b** circular orbiting oscillation ($R_0 = 224$ μm, $f = 1.188$ kHz); **c** volume oscillation ($R_0 = 267$ μm, $f = 8.658$ kHz); **d** translating and volume oscillation ($R_0 = 243$ μm, $f = 5.841$ kHz). Reproduced with permission from Collis et al. © Elsevier, Tho et al. © Cambridge University Press

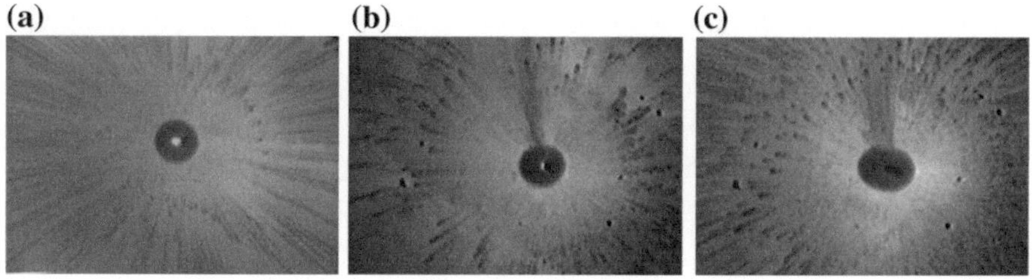

Fig. 5.8 Acoustic steaming pattern near an Albunex® sphere of 20 μm diameter in a 160 kHz standing wave field. Reproduced with permission from Gormley et al. © Acoustical Society of America

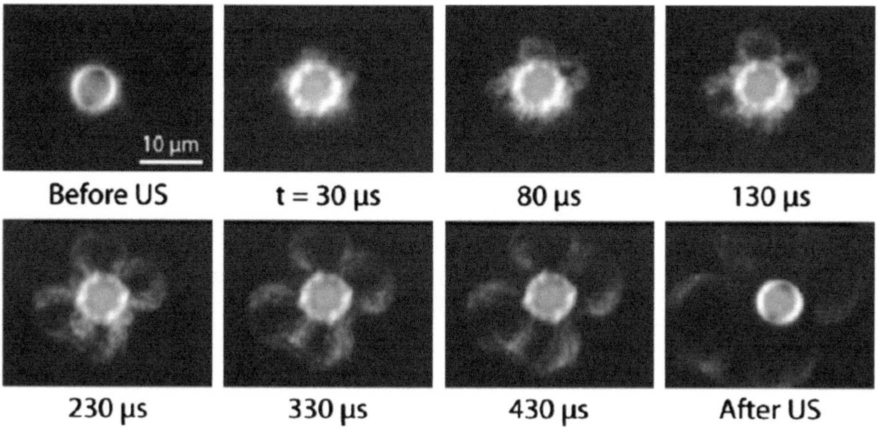

Fig. 5.9 Fluorescence high-frame rate recording of microstreaming around a liposome-loaded microbubble showing a mode oscillation. Excitation was a 1-Mhz tone burst of 80 kPa, 1000 cycles. Reproduced with permission from Kooiman et al. © Elsevier

The presence of such microstreaming patterns around clinically relevant microbubbles has been shown by using fluorescently labeled vesicles. Figure 5.9 shows the microstreaming pattern around an oscillating microbubble, obtained with high-frame rate fluorescence imaging (Kooiman et al. 2014). The flow around the microbubble enhances mixing of the fluid near the microbubble, which may increase drug-delivery efficacy; however, a direct relation between fluid motion and drug delivery has yet to be established.

5.2.3 Shear Stress Induced by Microstreaming

When objects are present in the microstreaming field around the oscillating bubbles, microstreaming can induce shear stress on the objects. In biomedical applications, the shear stress will potentially influence the biological medium. According to Nyborg's theory (Nyborg 1958), microstreaming established in the vicinity of a vibrating microbubble near a solid boundary produces a sharp velocity drop across the boundary layer. The velocity gradient G associated with the drop is approximated as follows (Rooney 1970):

$$G = \frac{2\pi f (R - R_0)^2}{R_0 \delta} \tag{5.8}$$

where f is the frequency, R is the instantaneous radius of a pulsating bubble, R_0 is the equilibrium radius of the bubble, and δ is the thickness of a boundary layer, which is defined in Eq. (5.2). Thus, the oscillatory viscous shear stress, S, can be calculated by multiplying the medium viscosity η by the acoustic streaming velocity gradient G:

$$S = \eta G = \frac{2\pi f \eta (R - R_0)^2}{R_0 \delta} \tag{5.9}$$

The bubble radius R changes with time during oscillation. Nonlinear differential equations of bubble oscillation can be used to describe the oscillation behavior of the microbubble, e.g., the well-known Rayleigh–Plesset equations:

$$R\ddot{R} + \frac{3}{2}\dot{R}^2 = \frac{1}{\rho_{\mathrm{L}}}\left[\left(P_0 + \frac{2\sigma}{R_0}\right)\left(\frac{R_0}{R}\right)^{3\gamma} - \frac{2\sigma}{R} - 4\eta_{\mathrm{L}}\frac{\dot{R}}{R} - P_0 - P_{\mathrm{ac(t)}}\right] \tag{5.10}$$

where R_0 is the bubble radius at rest, R is the actual bubble radius, γ is the polytropic exponent of the gas inside the bubble, ρ and η_{L} are the density and dynamic viscosity of the liquid, respectively, and σ is the surface tension between the liquid and gas/vapor phases at the bubble wall.

Depended upon the characteristics of the bubble's shell, the bubble displacement amplitudes ($R - R_0$) for various exposure conditions can be obtained by solving these nonlinear differential equations of microbubble oscillation, e.g., Equation (5.10), and then the associated oscillatory shear viscous stresses S can be obtained from the solution of Eq. (5.9). The shear stress resulting from microstreaming near a pulsating bubble resting on cell membrane (Lewin and Bjørnø 1982) and near a living cell (Wu 2002) can be numerically estimated by considering the cell to be a solid boundary.

Results from such numerical calculations indicate that shear stress associated with microstreaming surrounding encapsulated stable bubbles of contrast agents, near living cells driven by 0.12-MPa acoustic pressure amplitude ultrasound at 1 or 2 MHz may be large enough to generate reparable sonoporation of the cells (Wu 2002).

However, for shelled UCA bubbles used in diagnostic and therapeutic ultrasound applications, the bubbles are usually detached from boundaries (such as cell membranes or endothelial surfaces) rather than resting on them. Therefore, the effect of the boundary on bubble oscillation needs to be taken into account when calculating the potential field of a pulsating bubble detached from the boundary (Krasovitski and Kimmel 2004; Doinikov and Bouakaz 2010c, 2014).

Based on the oscillatory and non-oscillatory flow velocities associated with microstreaming, there are two types of shear stress in the boundary layer of an oscillatory bubble: the alternating (AC) shear stress related with AC fluid flow, and the direct (DC) shear stress caused by non-oscillatory microstreaming. Both of these shear stresses take place in the boundary layers. The stress generated by both AC and DC flows in boundary layers may play roles in applications of acoustic microstreaming. Suspended objects, such as cells or vehicles, may be brought

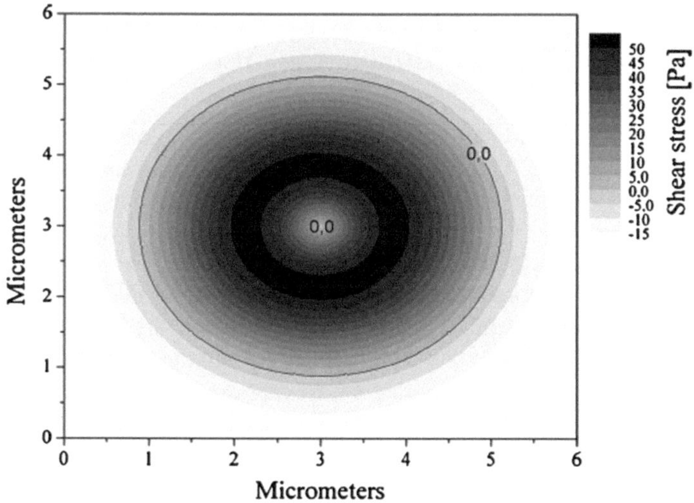

Fig. 5.10 Shear stress distribution created by a contrast microbubble on the cell membrane. Reproduced with permission from Dionikov et al. © Acoustical Society of America

momentarily into the boundary layer near a bubble and subjected to shear action, which might distort or fragment them.

Figure 5.10 exemplifies the shear stress distribution that is created by a contrast microbubble on the cell membrane. The resting radius of the bubble is $R_0 = 1.5\ \mu\text{m}$ and the separation distance between the bubble center and the cell membrane is $d = 2R_0$. The bubble is excited at 2.0 MHz and 200 kPa. Positive values of shear stress in Fig. 5.10 mean that the stress is directed away from the bubble, and vice versa. Figure 5.10 shows that at the point, which is the projection of the bubble center on the cell surface, the shear stress is zero, and then it increases, being directed away from the bubble, reaches a maximum, decreases down to zero, and then changes sign again, such that it is directed toward the bubble.

Additionally, when a vibrating bubble is in a suspension of particles, the particles tend to move toward the bubble, provided that the density of the particle exceeds the density of the suspending liquid (Miller 1988; Nyborg 1982). Thus, a radiation force exists on each particle and is directed toward the center of the bubble, and this causes cells and other particles to migrate toward the vibrating bubble into a region of relatively high stress.

Shear stress in a liquid can still be intense enough to lead to deformation of lipid vesicle membranes and lyse an artificial vesicle (Marmottant and Hilgenfeldt 2003), which resist shear stresses through membrane elasticity. Some encapsulated bubbles that have mechanically weak shells may break into free bubbles under the above-mentioned sound field. When that happens, the shear stress caused by microstreaming surrounding the free bubble increases dramatically and may play an important role in lethal sonoporation and fragmentation of cells during the early stage of ultrasound exposure.

5.3 Jet Formation and Shock Wave Emission During Ultrasound-Induced Bubble Collapse

Much research in the field of hydraulics has aimed to explore the reason of cavitation erosion at material surfaces in flowing fluids. From a hydrodynamic viewpoint, the cavitation bubbles that are produced in low-pressure regions may violently collapse when entering regions of pressure recovery, and this collapse might concentrate mechanical energy on very small areas of the solid walls, causing them to erode. Two characteristic effects are believed to be mainly responsible for the cavitation erosion: the emission of shock waves when the bubbles collapse and the generation of a high-speed liquid jet directed toward the solid surface. Similar to hydrodynamic cavitation, acoustic cavitation can also produce shock waves and liquid jets that concentrate energy onto nearby tissues or interfaces when the cavitation bubbles collapse. In fact, these phenomena are the main mechanical mechanisms by which the ultrasound achieves its therapeutic effects in biomedical applications. The jet often impacts the local boundary and can deposit energy at the site of impact, especially for large bubbles. The impingement of jets and shock waves on the surface creates the localized damage responsible for the desired (or sometimes undesired) biological effects.

5.3.1 High-Speed Observation of Liquid Jets and Shock Wave Emission

The most direct way to determine what happens during bubble collapse is to photograph the event. However, bubble dynamics are exceedingly fast when bubbles collapse, so visualization of the jet formation and shock wave emission can be recorded only with ultra-high-speed photography with very short exposure times and very high-speed framing. Single bubbles kept in a bubble trap and single laser-generated bubbles are now being investigated using ultrafast and high-speed photography in combination with hydrophones.

5.3.1.1 Shock Wave Emission During Symmetric Collapse

One mechanism of cavitation-induced mechanical effects involves shock waves created by bubble collapse in the liquid. Before collapsing adiabatically, a bubble expands to a maximum radius and then at the final stage of a symmetric collapse, the bubble content is compressed into a very small volume. Thus, the bubble rebounds due to the compressed interior gas and leads to the emission of a very high-pressure transient into the surrounding liquid that can evolve into a shock wave (Vogel et al. 1989; Brujan 2008; Ohl et al. 1999). A single spherical cavitation bubble can be easily produced by a focused laser, and these bubbles have

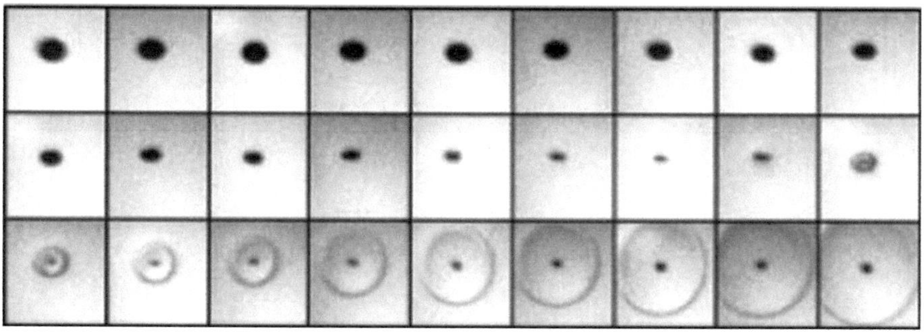

Fig. 5.11 A spherically symmetric pressure wave emitted by a laser-generated bubble upon first collapse. The frame size is 1.5 × ←1.8 mm^2 and the exposure time is ~5 ns. Reproduced with permission from Ohl et al. © The Royal Society

been used to study the dynamics of bubble collapse. Figure 5.11 shows that a free spherical laser-generated bubble in water undergoes a rapid collapse and emits a spherically symmetric shock wave; the image was taken at 20.8 million frames per second. The figure shows that after reaching its maximum expansion, the bubble starts to shrink, gaining speed at an ever increasing rate, driven by the static background fluid pressure and leading to a collapse with strong compression of the bubble contents (gas and vapor) in the final phase. The pressure within the bubble rises steeply, with the gas content acting as a hard spring, abruptly stopping the inward motion and driving the bubble back into expansion, i.e., rebound. As it is clearly visible from Fig. 5.11, in the stage of maximum compression, an outward-going pressure wave is emitted. The wave propagates at about the velocity of sound, ~1500 m/s in water.

Similar to laser-induced cavitation, a photographic series of the shock wave emitted by a sonoluminescing bubble into the surrounding liquid during acoustic collapse is shown in Fig. 5.12.

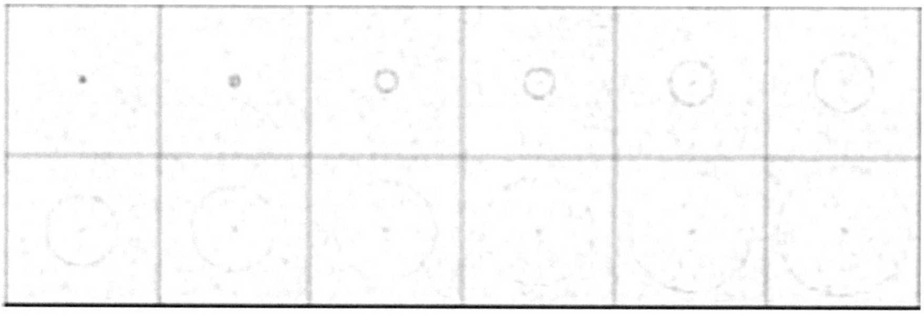

Fig. 5.12 Spherical shock wave emitted by a sonoluminescing bubble into the surrounding liquid. The temporal separation of adjacent frames is ~30 ns; the frame size corresponds to an area of 1.6 × 1.6 mm^2 in the object plane. Reproduced with permission from Ohl et al. © The Royal Society

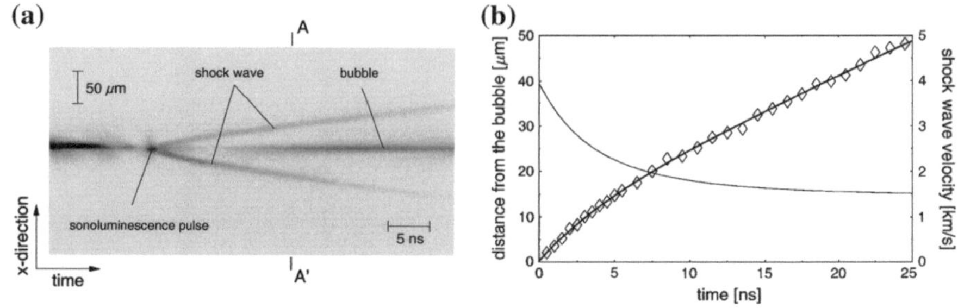

Fig. 5.13 **a** Streak image of the single-bubble sonoluminescence (SBSL) bubble collapse. Laser light is scattered at the bubble as well as at the emitted shock wave. **b** Distance of the shock wave from the bubble $r(t)$ determined from the streak image (open symbols) and corresponding fit (*solid line*). The shock wave velocity $v(t)$ is the derivative of the fit. Reproduced with permission from Pecha et al. © 2000 The American Physical Society

Pecha and Gompf (2000) used a streak camera with high spatial and temporal resolution to image the violent collapse dynamics of a single bubble under ultrasound exposure, as shown in Fig. 5.13a. From the figure, the high pressure in the last phase of the bubble collapse leads to the emission of a shock wave, which is launched with a shock velocity of almost 4000 ms. It has been shown that the propagation of cavitation-induced shock waves is nonlinear at the beginning of the emission (that is, near the bubble center), as shown in Fig. 5.13b. However, propagation of waves away from the bubble (>10 mm) can be described by linear acoustics. Under high-intensity ultrasound exposure, the shock amplitude decays much faster than $1/r$. From the strongly nonlinear propagation, the pressure in the vicinity of the bubble can be calculated to be in the range of 40–60 kbar.

5.3.1.2 Liquid Jet Formation and Shock Wave During Asymmetric Collapse

Whenever a bubble is near a boundary, the bubble oscillates under asymmetric boundary conditions; the asymmetry of the liquid motion during bubble collapse induces a deformation of the bubble. A high-speed liquid jet may form that projects through the bubble toward the boundary, as a result of the asymmetric behavior of the bubble in the presence of a pressure gradient near the boundary (Plesset and Chapman 1971; Barnett 1998). The potential energy of the expanded bubble is converted into the kinetic energy associated with the liquid jet that extends through the bubble interior and penetrates the opposite bubble wall.

Many researchers have used high-speed photography to observe laser-induced cavitation bubble collapse or forced ultrasound-induced inertial cavitation bubble collapse near solid surface, including the formation of a high-speed reentrant liquid jet and the emission of shock waves at the moment of collapse (Lauterborn and Bolle 1975; Lauterborn 1976; Vogel et al. 1989; Philipp and Lauterborn 1998; Brujan et al. 2005).

1. Near a Rigid Boundary

When a flat rigid surface is in the vicinity of a cavitation bubble, it causes the bubble to involute from the top and to develop a high-speed liquid jet toward the surface. When the jet hits the opposite bubble wall from the inside, it pushes the bubble wall ahead, causing a funnel-shaped protrusion with the jet inside, as shown in Fig. 5.14. In the figure, the funnel-shaped protrusion directed downward is the elongated bubble wall containing the jet that drives the elongation until its energy is used up. Then, the long tube of gas and vapor becomes unstable and decays into many tiny bubbles, while the main bubble surface snaps back to its former spherical shape.

The pressure gradient that causes the jet to form occurs because a region of low pressure develops between the bubble and the rigid wall during bubble collapse. During the initial collapse phase, the bubble acquires the form of a prolate spheroid. This shape also contributes to the formation of the liquid jet. The asymmetrical collapse causes the velocity of the upper bubble wall to exceed the velocity of the lower wall in order to conserve the impulse of the bubble–fluid system. Consequently, the fluid volume above the bubble is accelerated and focused during collapse, leading to the formation of a liquid jet directed toward the boundary. This jet hits the lower bubble wall, causing a funnel-shaped protrusion, and, finally, impacts the boundary, as shown in Fig. 5.15 (Postema et al. 2004).

The dynamic behavior of the jet characteristics depend not only on the properties of the bubbles and the acoustic field, but also on the distance between the bubble and boundary. The normalized parameter γ is an important parameter for classifying the bubble dynamics; it is defined as the distance s of the bubble center at its

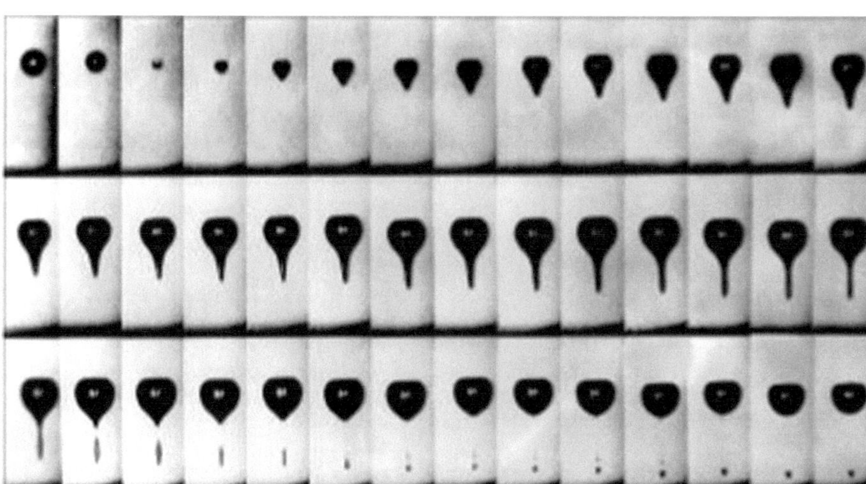

Fig. 5.14 Dynamics of a laser-generated spherical bubble in water of reduced surface tension in the vicinity of a plane solid boundary (at the *bottom*). Reproduced with permission from Ohl et al. © The Royal Society

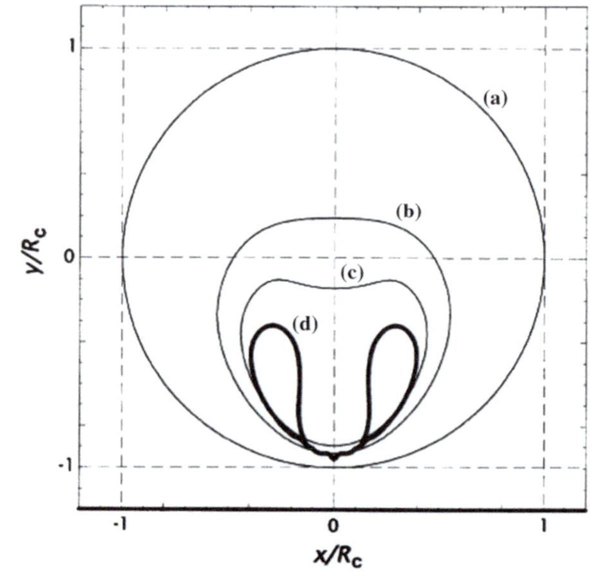

Fig. 5.15 Schematic evolution of jet development in a collapsing bubble near a boundary (*below*) **a** Bubble on the verge of collapse; **b** asymmetric collapse; **c** jet development; and **d** funnel-shaped protrusion. Reproduced with permission from Postema et al. © Elsevier

maximum expansion from the boundary scaled by the maximum bubble radius R_{max}, such that $\gamma = s/R_{max}$, as shown in Fig. 5.16. Bubbles with different R_{max} but the same γ value exhibit similar dynamics, thus allowing us to specify the degree of asymmetry of bubble collapse: cavitation bubbles with a small γ value are more influenced by the boundary, collapsing with a more pronounced shape variation, than those with a large γ value, which collapse in a more sphere-like fashion.

The wall shear stress from cavitation bubbles collapsing close to a rigid boundary was measured by Dijkink and Ohl (2008) with a constant temperature anemometer. The top row in Fig. 5.17 shows the dynamics of a focused laser-generated bubble observed with high-speed photography. The second row shows the shearing stress of the liquid close to the boundary, measured at the marked position in the first frame; the dashed vertical lines indicate the time of each

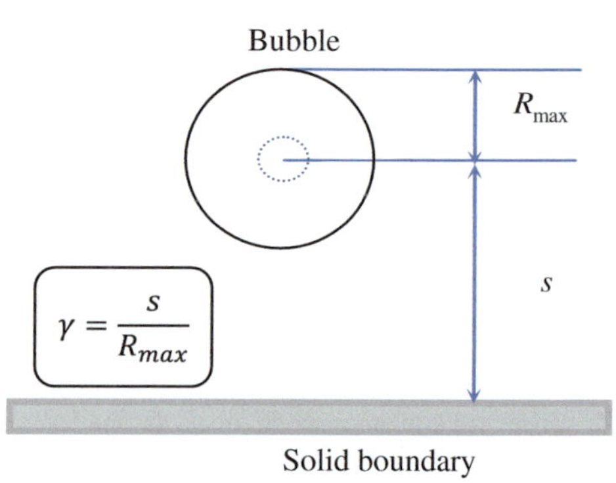

Fig. 5.16 Geometrical relations for the definition of the parameter γ

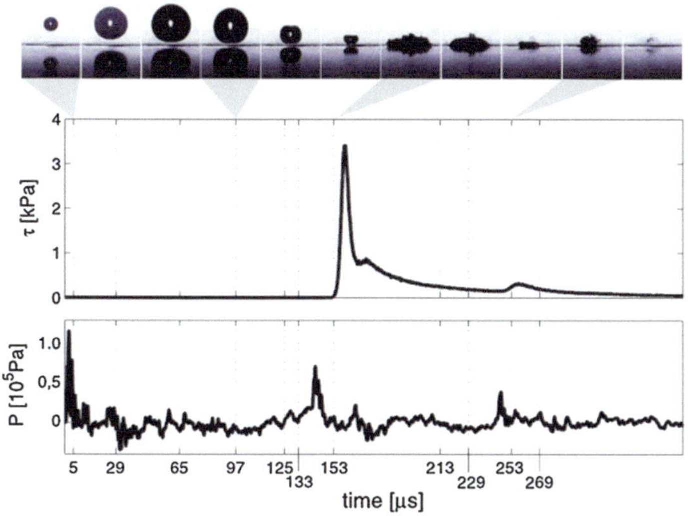

Fig. 5.17 Simultaneous measurement of the bubble shape, the acoustic emission, and the shear stress exerted on the boundary. Reproduced with permission from Dijkink et al. © American Institute of Physics

picture taken and is depicted in the top row. And the acoustic emission (third row) was recorded during the bubble creation and bubble collapses. The maximum bubble radius was 0.75 mm and the bubble was created 0.8 mm away from the boundary, i.e., $\gamma = 1.07$. From the figure, we see that the bubble collapses aspherically with the formation of a reentrant jet, and it deforms during re-expansion into a pancake shape. The intense shearing was only observed 17 μs after the first collapse, thus the maximum wall shear stress does not occur during bubble collapse but occurs when the radial spreading jet flows across the sensor. That is, the highest stresses are created after the impact of the jet, during its radial spreading on the surface. The highest amplitude of the signal of the wall shear stresses was found for bubbles oscillating close to the boundary, and it reached more than 3 kPa. Additionally, it contains as slowly decaying weaker component that may be generated by an expanding vortex ring.

To get a better understanding of the role of jet formation and shock wave emission during bubble collapse, which is important for the therapeutic applications of ultrasound, Brujan et al. (2005) investigated the dynamics of inertial cavitation bubbles produced by short high-intensity focused ultrasound (HIFU) pulses near a rigid boundary by high-speed photography and observed the bubble motion and the velocity of the liquid jet formed during bubble collapse. Additionally, they used hydrophone measurements to determine the pressure and the duration of the shock wave emitted during bubble rebound, thereby calculating the maximum velocity and the kinetic energy of the reentrant jet. Figure 5.18 shows the behavior of an ultrasound-induced cavitation bubble for $\gamma = 2$. Figure 5.18a shows the growth

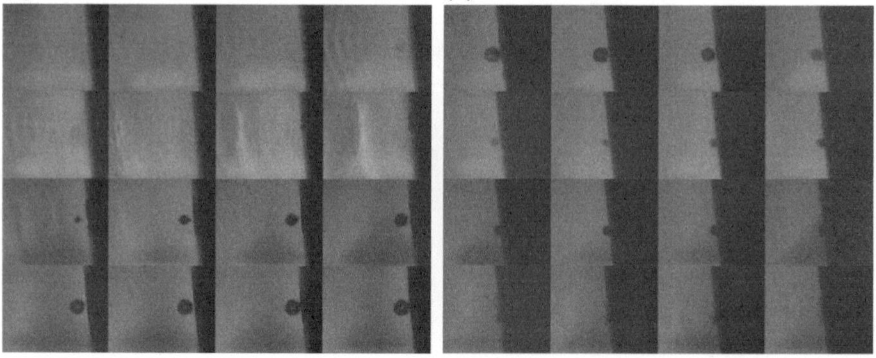

Fig. 5.18 Motion of an ultrasound-induced cavitation bubble near a solid boundary for $\gamma = 1.2$. **a** The growth phase, where the first frame was taken 30 μs after ultrasound exposure. **b** The collapse phase, where the first frame was taken 80 μs after ultrasound exposure. Frame interval, 3 μs. Frame width, 1.35 mm. Reproduced with permission from Brujan et al. © IOP Publishing Ltd.

phase of the bubble, and Fig. 5.18b shows the collapse phase and the second oscillation cycle of the bubble. After the bubble reached its maximum diameter during the first oscillation, the bubble started to collapse. The radial water flow was retarded by the solid boundary, which is visible on the right-hand side of each frame of Fig. 5.18. Therefore, the pressure at the bubble wall closer to the boundary was lower than the pressure at the bubble wall that was farther from the boundary during the whole collapse phase. This pressure gradient leads to different accelerations for different parts of the bubble and consequently to a movement of the center of the bubble toward the boundary during collapse. It can be seen that most of the second oscillation cycle of the bubble takes place in contact with the boundary surface.

These jets have the potential to impinge on the boundary at high speeds. Liquid jets can form in either the nonlinear oscillation regime of UCA microbubbles or during inertial cavitation. Postema et al. (2004) visualized the microjet development of the shelled UCA microbubble, one of which is shown in Fig. 5.19. The Quantison microbubble with a radius of 8.43 μm was exposed to an 8-cycle 1.7-MHz pulse at a peak rarefactional pressure of 2 MPa. The jet travelled a length of 26.2 μm in 0.33 μs, giving an average jet velocity of 79.4 m/s. However, they only observed 2 such occurrences of a jet, so these experiments gave no information regarding which exposure parameters induce jets, and they showed that microjet formation is possible with shelled UCAs.

Similar to for unshelled cavitation bubbles, a neighboring surface can induce non-spherical shape deformations, microliquid jets and translation of the encapsulated UCA microbubble toward and away from the wall during its oscillation. Vos et al. (2011) used an ultrafast framing camera to observe the non-spherical and translational behavior of individual-coated microbubbles of different sizes, which were in contact with a cellulose wall. They used 2.25-MHz ultrasound pulses with driving pressures from 80 to 325 kPa, and they recorded the images of

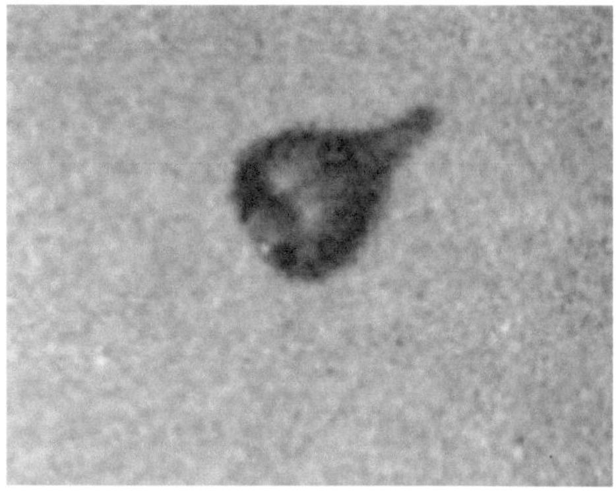

Fig. 5.19 Ultrasound-induced jet development in an experimental UCA microbubble (MI = 1.2). Each image frame corresponds to a 38 × 30 μm^2 area. Interframe time is 0.33 μs and exposure time is 10 ns. Reproduced with permission from Postema et al. © Elsevier

microbubbles from two orthogonally positioned microscopes. As shown in Fig. 5.20, they found larger non-spherical deformations of microbubbles. The frame zero in Fig. 5.20 shows the initial shape of the bubble and frames 1–15 depict the bubble oscillations. These frames started at the third bubble oscillation cycle and lasted for two cycles. The top view shows a spherical shape throughout, while the side view clearly shows non-spherical shapes of the vibrating bubble. An apparent toroidal shape of the bubble can be observed at frames 5–7, and frames 9–12 show an elongated shape with respect to the wall. Furthermore, the side-view images show that the vibrating bubble stayed in contact with the wall. Based on a

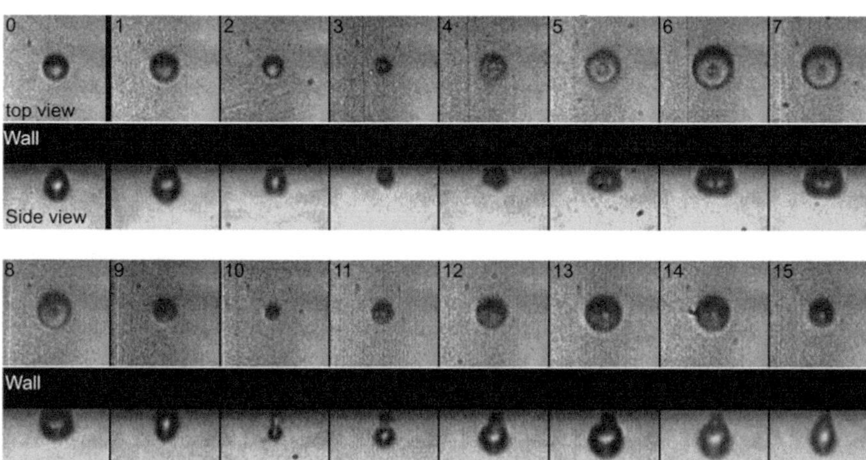

Fig. 5.20 Selected frames from the recording of a bubble with a radius of 2.3 μm. Ultrasound excitation pressure is 270 kPa, frequency $f = 2.25$ MHz. Sampling time is 65 ns. Reproduced with permission from Vos et al. © Elsevier

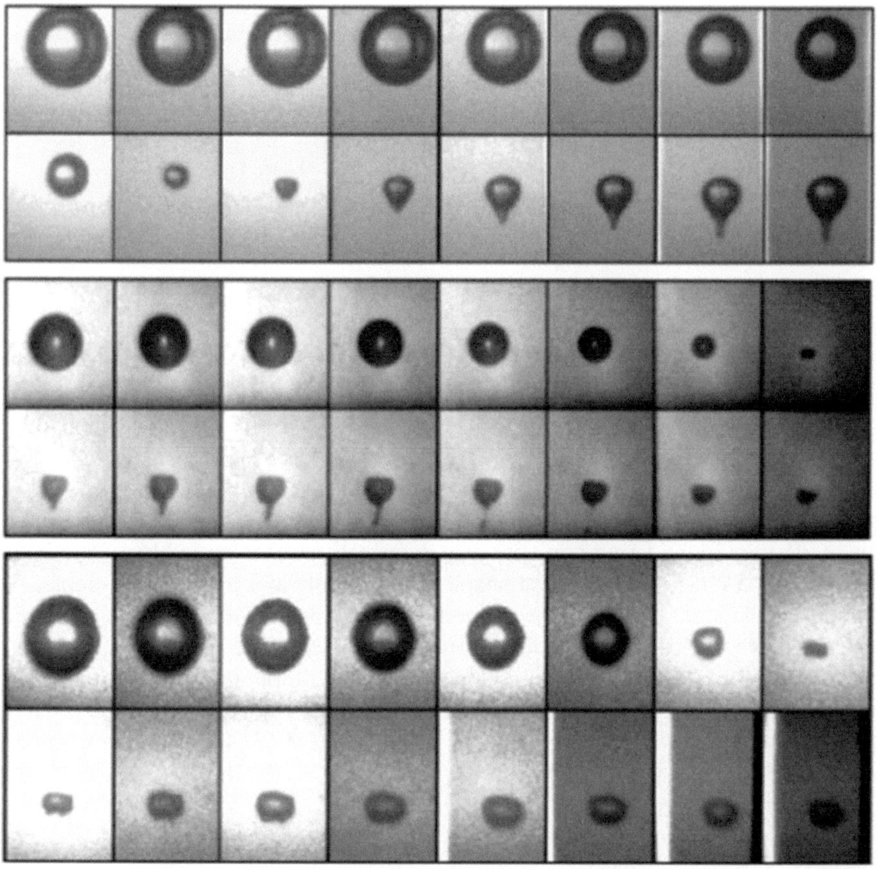

Fig. 5.21 Picture sequences of the behavior of a laser-induced bubble near a rigid wall in water and polymer solutions for $\gamma = 3.17$. *Top* water; *Middle* 0.5 % carboxymethyl cellulose solution; *Bottom* 0.5 % polyacrylamide solution. Frame interval, 4.8 μs; frame width, 1.7 mm. Reproduced with permission from Brujan © Springer-Verlag

parametric model combining potential flow theory with a viscous boundary layer at the wall, the peak shear stress of up to 300 kPa and normal stresses of up to 1 MPa are estimated when microbubbles are insonifed with a 2.25-MHz pulse at 325 kPa.

Thus far, it has been assumed that cavitation bubbles occur in Newtonian fluid. As non-Newtonian fluids are more prevalent in biomedicine and bioengineering application, in his book Brujan (2011) systemically discussed cavitation and bubble dynamics in non-Newtonian fluids from the standpoint of non-Newtonian fluid mechanics, physics, chemical engineering, and biomedical engineering. Figure 5.21 shows a series of high-speed photographic records of bubble motion in water, a 0.5 % carboxymethylcellulose (CMC) solution with a weak elastic component, and a 0.5 % polyacrylamide (PAM) solution with a strong elastic component for the case where $\gamma = 3.17$ (Brujan et al. 1996). The liquid jet, which developed on the upper side of the bubble leading to the protrusion of the lower bubble wall, can be seen in the case of bubbles situated in water (top sequence). A similar bubble shape was found in the CMC solution, but in this case, the jet was not as strong as the jet in water. The most interesting behavior for a bubble situated in the vicinity of a rigid boundary was found in the PAM solution. The liquid jet was not observed, and a flat form of the bubble shape was the dominant aspect of bubble motion after the first collapse.

2. **Near a Non-rigid Boundary**

In addition to the properties of the liquid surrounding the bubble and the distance to the boundary, the elastic properties of the nearby boundary material strongly influence bubble dynamics (Brujan 2011). In biomedical applications of cavitation, most boundaries near bubbles are non-rigid, e.g., vessel walls, cell membranes, and tissue surfaces. Figure 5.22 shows the interaction of a laser-produced cavitation bubble with an elastic boundary that has an elastic modulus E = 0.25 MPa and different γ values (Brujan et al. 2001). If a bubble is relatively far away from the boundary (γ = 1.14, Fig. 5.22a), the bubble motion is characterized by the generation of a high-speed liquid jet and migration of the bubble away from the boundary. A bubble produced closer to the boundary (γ = 0.62, Fig. 5.22b) generates an annular flow in the early stage of bubble collapse, leading to bubble splitting during the final stage of collapse and the formation of two axial jets flowing toward and away from the boundary. When the bubble is generated almost at the surface of the boundary (γ = 0.04, Fig. 5.22c), the boundary rebounds early during the growth phase of the bubble and a very strong jet develops, which, finally, penetrates the bubble wall opposite to the boundary. No bubble splitting occurs, and the bubble migrates away from the boundary during rebound.

Prentice et al. (2005) directly observed the energetic micrometer–scale interactions between violently cavitating shelled microbubbles and the membrane of individual cells. They exploited the spacing in their experiments to refine the interactions between a microbubble and the substrate, and they observed microbubble cavitation nearby cultured monolayers of live cells. They found active microjetting in 17 % of cell experiments. Figure 5.23a shows a quiescent 4-μm-diameter microbubble trapped 17 μm from a cell membrane; at t = 8 μs after cavitation inception, the microbubble underwent inertial cavitation and collapse, with the outermost bubble hemisphere undergoing involution to form a central reentrant jet directed at the cell, as shown in Fig. 5.23b.

High-speed photomicrography was also used to visualize the direct transient interactions between ultrasound-activated microbubbles and blood vessels within an ex vivo tissue (Chen et al. 2011a, b; Caskey et al. 2007). In these studies, a similar imaging system was designed to capture high-resolution images in tissue (Fig. 5.24), and it was used to directly observe the oscillation of microbubbles in small vessels. These studies found that microbubble expansion was significantly restricted when the microbubbles were constrained within an ex vivo capillary network, despite the larger compliance of the vessel wall. Also, the extent of vessel-wall invagination was greater than the corresponding distention, as shown in Fig. 5.25a, in which the sharp, notch-like shape of the vessel invagination implies

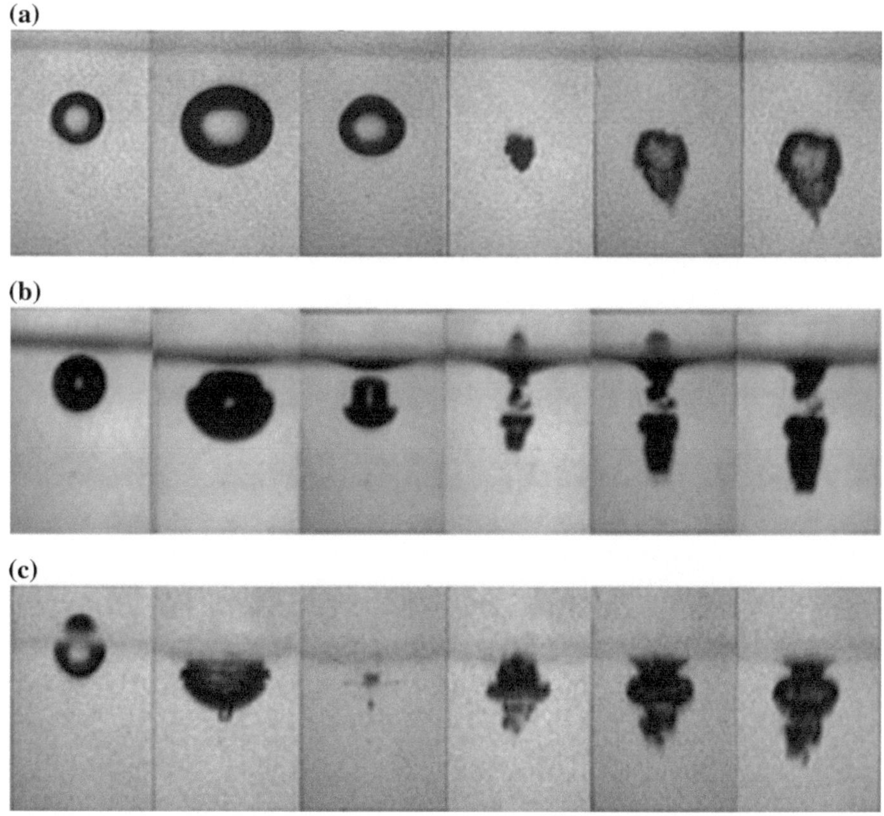

Fig. 5.22 Interaction of a laser-produced bubble with an elastic boundary (elastic modulus $E = 0.25$ MPa) for different γ values. The boundary is located at the top of each frame. Frame interval, 20 μs; frame width, 3.5 mm. Reproduced with permission from Brujan et al. © Cambridge University Press. **a** $\gamma = 1.14$, **b** $\gamma = 0.62$, **c** $\gamma = 0.04$

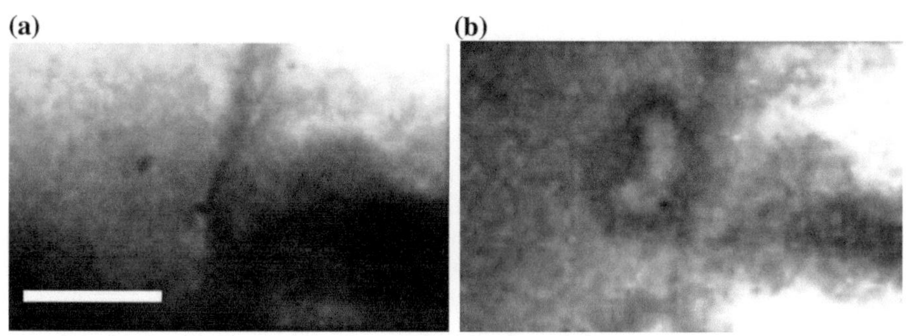

Fig. 5.23 Representative ultra-high-speed sequences acquired at a framing rate of 500 kHz showing microbubble cavitation in proximity to cell monolayers. Reproduced with permission from Prentice et al. © Nature

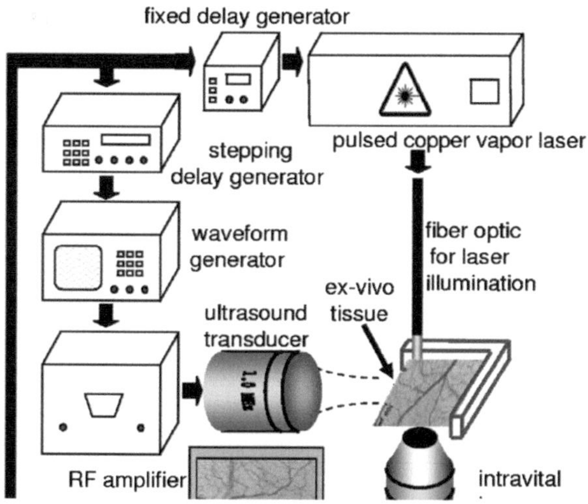

Fig. 5.24 Diagram of the experimental system. Reproduced with permission from Caskey et al. © Acoustical Society of America

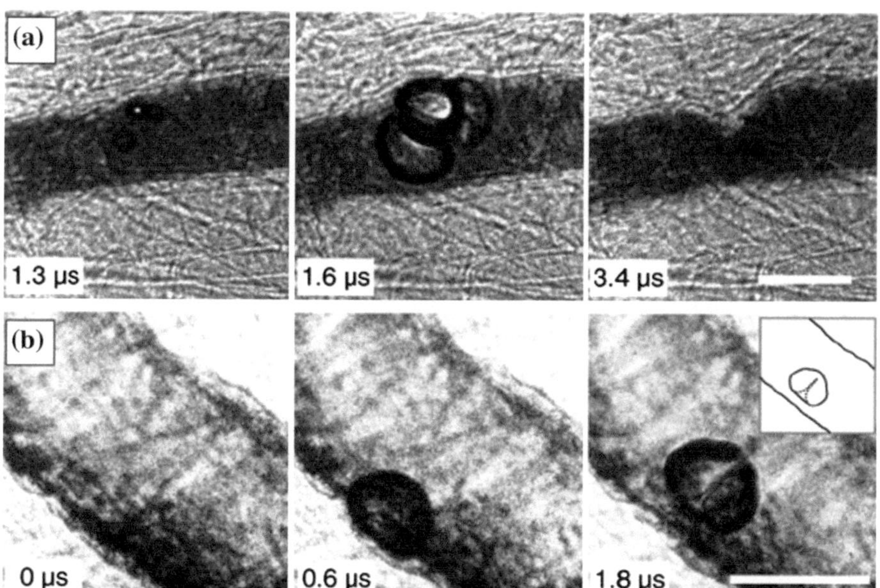

Fig. 5.25 Characteristics of observed bubble–vessel interactions. **a** Vessel-wall invagination (*right*) is greater than the distention (*middle*); **b** the vessel-wall distention (*middle*) and a liquid jet directed away from this wall (*right*, with inset showing a sketch for clarity). Reproduced with permission from Chen et al. © American Physical Society

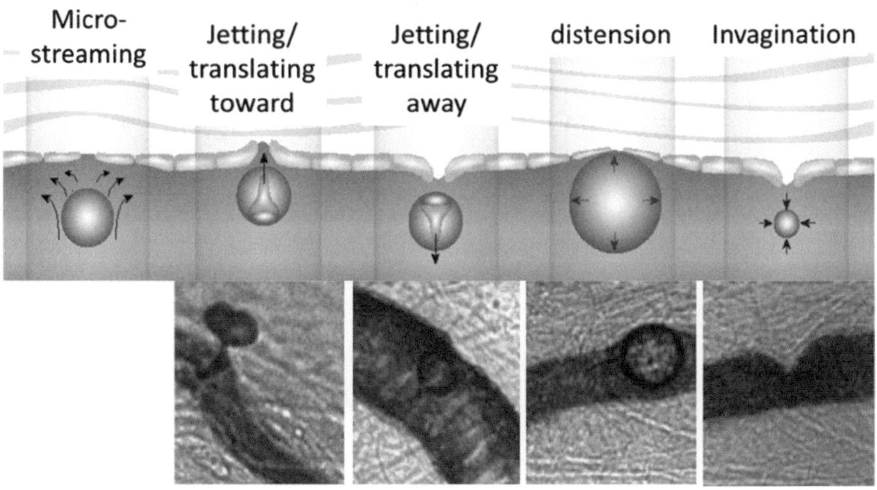

Fig. 5.26 Illustration of potential mechanisms for bubble-induced vascular injury, and observed interactions using high-speed photomicrography (the technique did not allow for imaging of microstreaming or shock waves). Reproduced with permission from Matula et al. © Nova Science Publishers

the presence of large mechanical strains. Additionally, Chen et al. observed asymmetric bubble collapses and the formation of liquid jets that penetrated through the bubble but were directed away from this wall, as shown in Fig. 5.25b. The jet was always directed away from the nearest vessel wall, which was different than seen with rigid boundaries, where the jet is usually directed toward the rigid boundary. Their observations suggest potential mechanisms of microvascular damage, as illustrated in Fig. 5.26 (Matula and Chen 2013).

Chen et al. (2012) were also able to show a direct correlation between UCA oscillations and endothelium damage. These particular observations suggest that invagination of the vessel wall can lead to endothelial damage. However, with short pulses, the damage did not penetrate the basement membrane.

5.3.2 Theoretical Modeling of Bubble Dynamics

The fundamental theory that describes the collapse of a spherical bubble is based on the well-known Rayleigh equation (Rayleigh 1917) under a number of assumptions, including that the bubble remains spherical and the fluid is incompressible, inviscid, and irrotational, thus neglecting compressibility and viscous effects. The Rayleigh collapse time T_c, which is the time needed for the bubble to shrink from the maximum expanded radius R_{max} to zero, is defined as follows:

$$T_c = 0.915 R_{max} \sqrt{\frac{\rho}{p_{stat}}} \quad (5.11)$$

where ρ is the liquid density, p_{stat} is the difference between the hydrostatic pressure p_∞, and the vapor pressure p_v. Rayleigh's theory shows how extremely large pressures may be generated during the collapse of a spherically symmetric bubble—this is now commonly known as the Rayleigh implosion mechanism. Because the theory assumes that a bubble contains no gaseous component, it leads to the conclusion that at the end of its collapse, the velocity of an empty bubble becomes infinite. In reality, the bubble will contain gas and vapor, and then this, as well as the surface tension, viscosity, and compressibility of the liquid, so this must be included in the spherically symmetric bubble growth and collapse model.

As previously mentioned, when a bubble near a boundary undergoes contraction due to an ultrasound wave, this contraction will be asymmetrical. Plesset and Chapman (1971) developed a numerical model to obtain an approximate description for a non-spherical bubble collapse near a solid boundary. They assumed that initially the bubble was spherical, at rest when collapse began, and nearly empty, containing only a small amount of vapor. They assumed that the jet developed at the very early stage in the bubble collapse history, so compressibility effects in the liquid and the vapor could be neglected. Calculations were stopped when the liquid jet reached the opposite wall of the bubble, since the assumption of incompressibility was no longer valid. They also neglected viscosity of the liquid and any other extraneous asymmetric effects, such as gravity.

Briefly, because the flow is irrotational, the velocity vector v can be written in terms of a velocity potential ϕ. The change in potential of a free surface point was approximated by

$$\Delta\phi = \left(\frac{\Delta p}{\rho} + \frac{1}{2}v^2\right)\Delta t \quad (5.12)$$

where ϕ is the velocity potential and $\Delta p = p_\infty - p_v$ where p_∞ is the ambient pressure and p_v is the vapor pressure inside the bubble. The solutions are characterized by the single parameter b/R_0, where b is the initial distance from the plane wall to the center of the bubble and R_0 is the initial radius of the bubble. The velocities are independent of the size of the bubble and are scaled like $(\Delta p/\rho)^{1/2}$.

Plesset and Chapman simulated two cases: a bubble initially in contact with the wall and a bubble initially half its radius from the wall at the closest point, as shown in Fig. 5.27. For a bubble in contact with the wall (Fig. 5.27a), the solid wall influences the bubble early in the collapse, chiefly by reducing the upward motion of the lower portion of the bubble. As a result, the bubble becomes elongated in the direction normal to the wall. For a bubble near a wall (distanced half the bubble radius from the wall, Fig. 5.27b), the bottom of the bubble still moves upward toward the bubble center, but since this upward motion is reduced, the centroid of the bubble moves toward the wall, displaying the Bjerknes effect. As the bubble

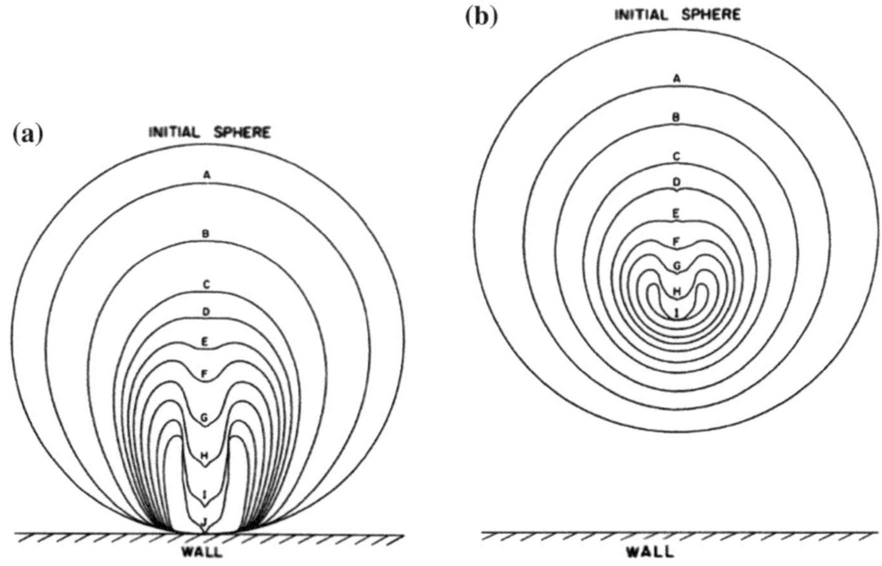

Fig. 5.27 The results of simulations tracking the surfaces of bubbles exposed to ultrasound. **a** A bubble initially in contact with the wall. **b** A bubble initially half its radius from the wall. Reproduced with permission from Plesset et al. © Cambridge University Press

acquires kinetic energy, this energy is concentrated in the upper portion of the bubble, which eventually flattens and forms a jet. Once the jet is formed, the speed of its tip remains very constant. The shapes of the two cases appear quite different because in the first case, the bottom of the bubble must remain in contact with the solid wall; in the second case, it is allowed mobility. Also, in the second case, the jet speed is somewhat larger than the speed in the bubble on the wall; however, the final jet must pass through the liquid for a distance greater than 5× its diameter before it reaches the solid wall. Therefore, it seems that the jet in the first case, which strikes the wall directly, is more capable of damage even though the jet speed is lower. The applicability of this numerical model has been verified by experimental studies (Lauterborn and Bolle 1975).

The radius of the jet R_j is approximately related to the radius of the bubble on the verge of collapsing R_c, as $\frac{R_j}{R_c} \approx 0.1$. The length of the jet l_j, defined as the full travel path of the propelled liquid, is approximately related to the R_c as $\frac{l_j}{R_c} \approx 3$. From these two ratios, the amount of liquid within the jet, V_j, is estimated as $V_j \approx 0.1R_c^3$ (Ohl and Ikink 2003). The impact of a jet on a surface generates a high-pressure region, and the pressure in that region is called the "water-hammer pressure." The water-hammer pressure of a cavitation jet was estimated by $p_{wh} \approx \frac{1}{2}\rho c v_j$, where p_{wh} is the water-hammer pressure, v_j is the jet velocity, ρ is the fluid density, and c is the speed of sound.

The damage caused by a jet can be estimated by comparing the water-hammer pressure with the maximum stress that the impact surface can withstand before rupture. The maximum stress $\tau_{\max}$ is given by:

$$\tau_{\max} = E\varepsilon_r \tag{5.13}$$

where E is the elastic (Young's) modulus of the material and r is the relative deformation of the surface before rupture.

5.4 Applications

The above-mentioned cavitation mechanical effects, i.e., microstreaming, shock waves, and jetting, play important roles in the therapeutic use of microbubble-mediated ultrasound in medicine.

5.4.1 Lithotripsy

In lithotripsy, high-energy shock waves or HIFU are used to fragment and disintegrate kid kidney stones or stones in the ureter through mechanical mechanisms. Lithotripsy using shock waves has a long history of nearly 30 years, and it is the preferred clinical modality for treating renal and upper ureteral calculi.

A shock wave lithotripter utilizes roughly a 1-μs duration and a 100-MPa pressure spike triggered at 1–2 Hz to fragment kidney stones through mechanical mechanisms (Sapozhnikov et al. 2002; Zhou et al. 2004). In addition to dynamically fracturing the stones as a result of the stress wave (Eisenmenger 2001), cavitation also helps destroy kidney stones and injure tissue during shock wave lithotripsy (SWL) (Zhu et al. 2002; Rassweiler et al. 2011; Bailey et al. 2003): stress waves and cavitation work synergistically to effectively disintegrate renal calculi. Cavitation is generated in the fluid near the stone surface by the negative-pressure phase of shock waves; the microjet resulting from the violent collapse of bubbles causes cavitation erosion of the anterior surface of the stone (Fig. 5.28). This cavitation activity works to break the stones down into finely passable fragments after the initial fragmentation by the shock wave. Therefore, some mechanical modifications to lithotripsy have focused on controlling cavitation in order to improve stone comminution while limiting injury to nearby tissues.

While the use of HIFU in lithotripsy is relatively recent (Ikeda et al. 2006; Yoshizawa et al. 2009; Duryea et al. 2011; Harper et al. 2014; Maxwell et al. 2015), it has some significant advantages over SWL. HIFU can be focused much more precisely than shock waves such that it can enhance the damage at the desired site while reducing collateral tissue damage. Cavitation is also an important factor in

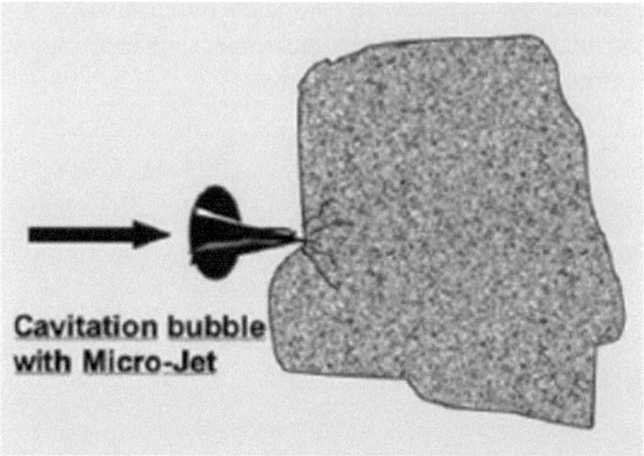

Fig. 5.28 Negative-pressure waves of high-speed shocks cause cavitation in liquids surrounding stones and within microcracks or cleavage interfaces by inducing microjets. Reproduced with permission from Rassweiler et al. © European Association of Urology

HIFU lithotripsy, and the cavitation generated at the focal point pulverizes the stones, ultimately reducing them to powder (Ikeda et al. 2006).

One method to further exploit cavitation in HIFU lithotripsy has been developed by a group from Japan. They proposed a two-frequency combined waveform method to concentrate very high pressures on stone surfaces within the localized cavitation area, and they utilized the collapse of a cavitation cloud to fragment the stones (Yoshizawa et al. 2009; Matsumoto et al. 2005; Ikeda et al. 2006). Figure 5.29 shows a schematic of their cavitation control lithotripsy (CCL) method. First, a higher frequency ultrasound (1–5 MHz, at a shorter wavelength than the length of the stone) is focused at the stone surface (Fig. 5.29-1) to generate a hemispherical cavitation bubble cloud only at the stone surface (Fig. 5.29-2). Immediately after the higher frequency is stopped, the bubble cloud is forced—by a short pulse of low-frequency focused ultrasound (100 kHz to 1 MHz) (Fig. 5.29-3)—to oscillate near resonance (Fig. 5.29-4). Accompanying the forced oscillation of the bubble cloud, a shock wave propagates inward from the hemispherical bubble cloud (Fig. 5.29-5). The bubbles near the center collapse violently and emit an extremely high-pressure wave at the stone surface to crush the stone (Fig. 5.29-6). The bottom panel of Fig. 5.29 shows the CCL ultrasound waveform; we can see that the high-frequency ultrasound (bubble cloud creator) is immediately followed by low-frequency ultrasound (cloud collapse inducer).

From the high-speed photographs of a cavitation bubble cloud that was forced to collapse (Fig. 5.30a), we can see that a stable bubble cloud was generated and forced into oscillation by a 545-kHz pulse of focused ultrasound immediately after a 100–200 μs irradiation of 2.74-MHz ultrasound. Because there is a strong relationship between the size of the bubble cloud and the ultrasound wavelength, the sizes of the area generated by the bubble cloud can be controlled with respect to the ultrasound frequency. For example, in the focused ultrasound field, acoustic

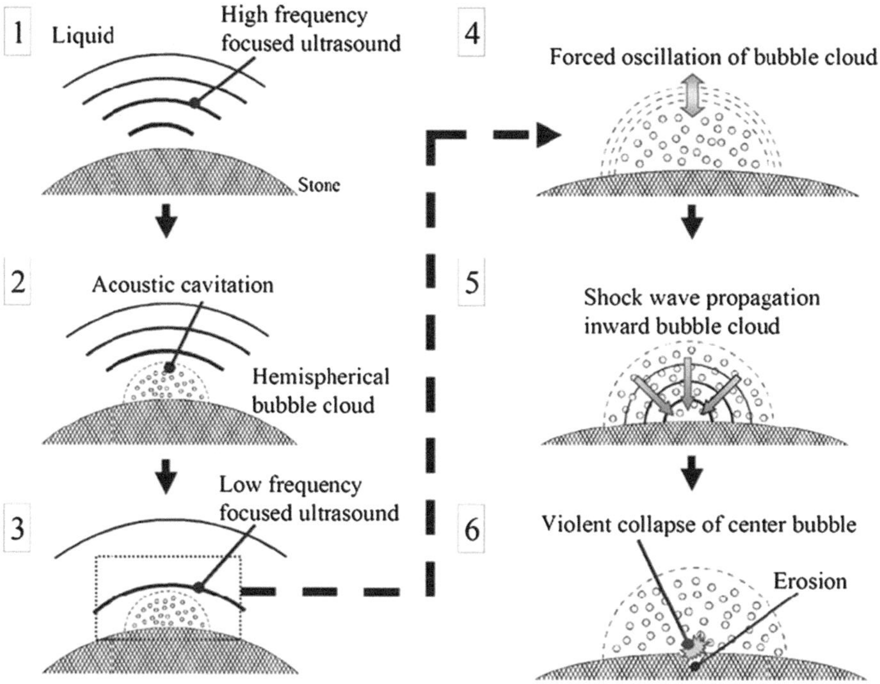

Fig. 5.29 Schematic of the method to control cavitation bubble clouds and the typical cavitation control waveform. Reproduced with permission from Yoshizawa et al. © Springer

cavitation at the solid surface can be well controlled in space (Matsumoto et al. 2005). From the shadowgraph photography of cloud collapse (Fig. 5.30b), we can see the shock wave propagation from bubble cloud that occurred immediately after the third frame of Fig. 5.30a. We can also see the shock wave signal of the cavitation cloud collapse that was captured by a hydrophone, as shown in Fig. 5.30c. The maximum pressure was about 3 MPa at 1.6 mm away from the focal point of the focused ultrasound.

This cavitation control method utilizing only cavitation erosion might produce less tissue damage and smaller stone fragments than conventional (SWL). This method aimed at exploiting an acoustic cavitation event produced by the wave is very different from traditional lithotripsy. Results indicate that application of HIFU lithotripsy to natural stones produces small stone fragments (<1 mm) as a result of localized cavitation on the stone surface. This characteristic of comminution via cavitation erosion is an attractive outcome with respect to minimizing the size of stone fragments produced during treatment.

Similar work by a group at the University of Michigan utilized very short (<50 cycles) high-intensity (peak negative pressure >6 MPa) ultrasound pulse sequences at low duty cycles (0.1–5 %) to initiate and maintain a cavitation bubble cloud in lithotripsy (Duryea et al. 2011). After performing ultrasound treatments at different

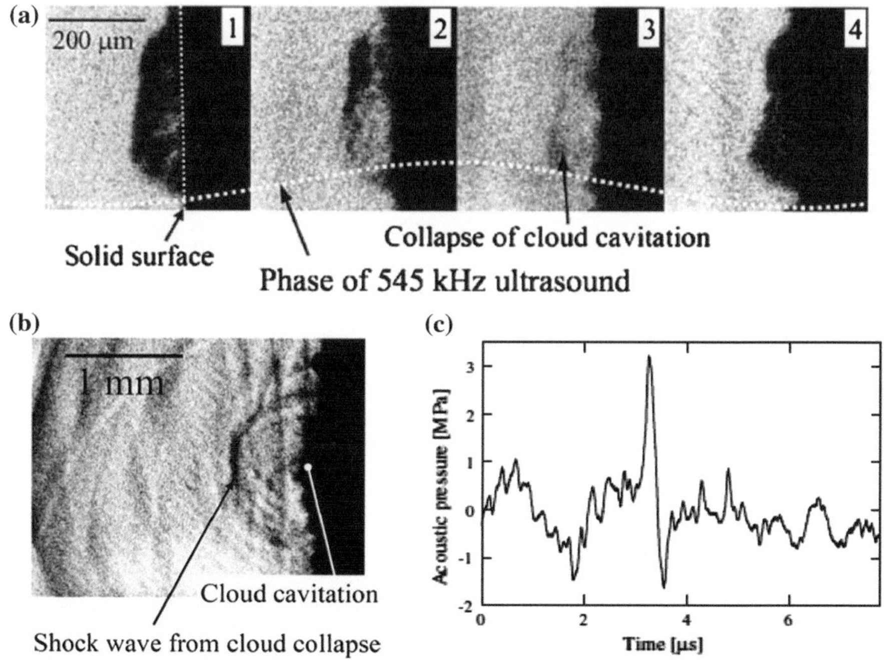

Fig. 5.30 **a** High-speed photography showing the forced collapse of a cavitation bubble cloud (camera interframe time, 275 ns; exposure, 100 ns). **b** Shadowgraph photography of the shock wave emitted from the collapse of the cavitation bubble cloud (camera exposure, 5 ns). **c** Shock wave signal of the cavitation cloud collapse. Reproduced with permission from Matsumoto et al. © Elsevier

pressure levels, they observed the stone damage morphology shown in Fig. 5.31. No visually apparent damage was observed when the peak negative pressure p_- was lower than 3 MPa. As p_- increased from 7 to 21 MPa, the typical area of damage on the face of the stone increased. At the maximum peak negative pressure of 21 MPa, the average erosion rate was approximately 26 mg/min. From the calculated spatial variability in the acoustic backscatter signal of passive detection during lithotripsy treatments (Fig. 5.32, lower panels), we can see that the bubble cloud action on the stone surface increased as a function of pressure. The high-speed photographs of the cavitation bubbles on the surface of the stone confirmed these results, showing larger bubble clouds at higher pressures (Fig. 5.32, top panels). Residual stone fragments generated by histotripsy treatment appeared to be smaller than 100 μm.

Moreover, Maxwell et al. developed a new method of lithotripsy that uses short, broadly focused bursts of ultrasound rather than shock waves to fragment urinary calculi, and this technique is called burst wave lithotripsy (BWL) (Maxwell et al. 2015). Their results demonstrate that stone comminution with BWL using broadly focused ultrasound bursts is feasible as a potential noninvasive treatment method for nephrolithiasis. They found that bursts produced fractures in artificial and

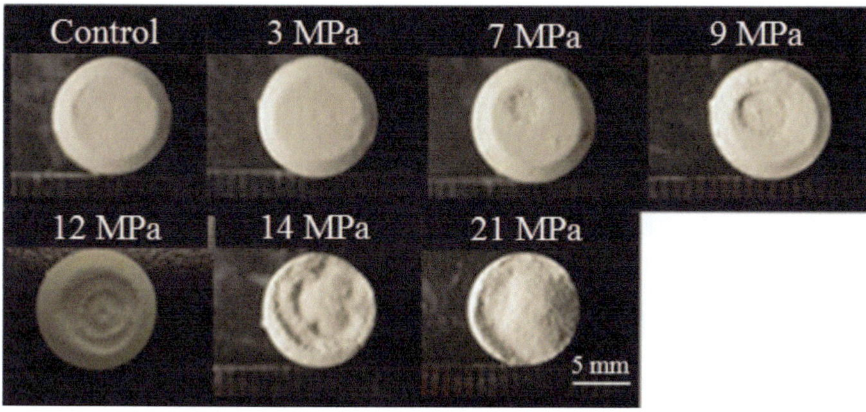

Fig. 5.31 The stone damage morphology. Reproduced with permission from Duryea et al. © IEEE

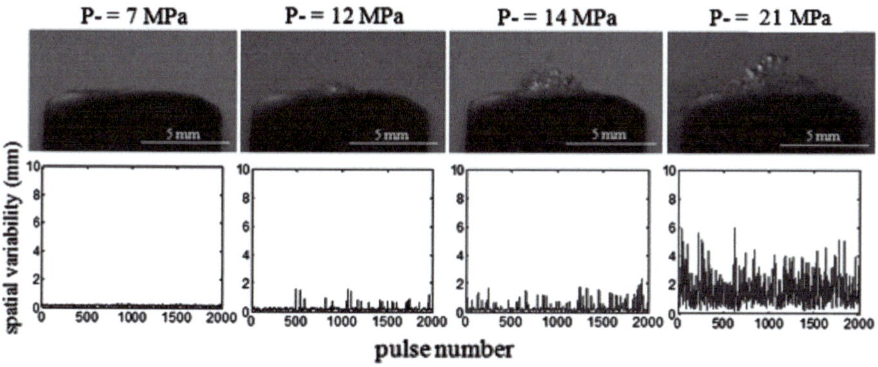

Fig. 5.32 Bubble activity at the stone surface. Reproduced with permission from Duryea et al. © IEEE

human calculi that led to the fragmentation of all stone types, but the duration of exposure required to reduce stones to passable fragments varied by stone type. Adjusting the fundamental ultrasound frequency allowed for stone fragment size to be controlled. These characteristics suggest that BWL may represent a potentially viable noninvasive alternative to SWL.

These results suggest that cavitation control with HIFU may offer a potential adjunct to SWL procedures, and it is possible that a synergistic combination of the two modalities could eventually lead to faster and more complete stone comminution.

5.4.2 Histotripsy

Short, HIFU pulses delivered at certain pulse repetition frequencies (PRFs) can achieve the most energy-efficient tissue erosion and form sharp, clear boundaries for holes of defined perforations, especially in soft tissue (Xu et al. 2004, 2005a, b). As shown in Fig. 5.33, at a spatial peak pulse average intensity (I_{SPPA}) of 3000 W/cm^2, tissue erosion occurred but did not perforate the tissue, while an I_{SPPA} >= 3500 W/cm^2 produced tissue perforation, and the perforation area increased with higher intensities. This technique can be considered a form of soft tissue lithotripsy ("histotripsy") (Roberts et al. 2006; Xu et al. 2007; Parsons et al. 2006).

Histotripsy-induced tissue erosion or tissue fractionation can be directly attributed to controlled acoustic cavitation. Inertial cavitation (or transient cavitation) is the main mechanism of the process, which is sustained by the high-amplitude and temporally changing acoustic backscatter observed during the erosion process. This signal—which probably occurs as sound reflects from a dynamically varying bubble flock caused by the high-intensity ultrasound pulses to the erosion area—is often regarded as a signature of cavitation. The acoustic backscatter signal observed from the erosion zone can be divided into three processes—initiation, maintenance, and extinction, as shown in Fig. 5.34 (Xu et al. 2005b).

The first change in the backscatter signal, called initiation, indicates the appearance of cavitation, as shown in Fig. 5.34; the backscattering signal is initiated at time "a," "c," and "e." The initiation of cavitation is closely related to the erosion process, such that without initiation, erosion is never observed. Therefore, the presence or absence of initiation can be used to predict erosion or lack of erosion. In

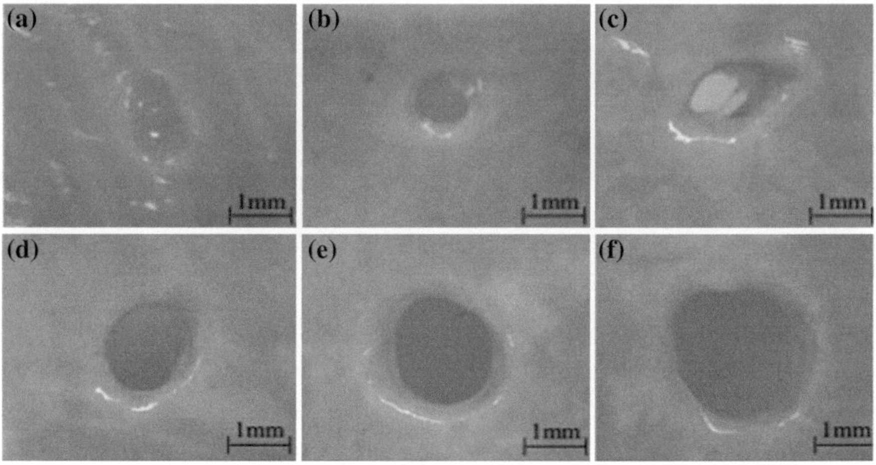

Fig. 5.33 Erosion and perforation in porcine atrial wall tissue generated by ultrasound pulses at different I_{SPPA} values. The I_{SPPA} values in (**a**), (**b**), (**c**), (**d**), (**e**), and (**f**) are 3000, 3500, 4000, 5000, 7000, and 9000 W/cm^2, respectively. Reproduced with permission from Xu et al. © Elesvier

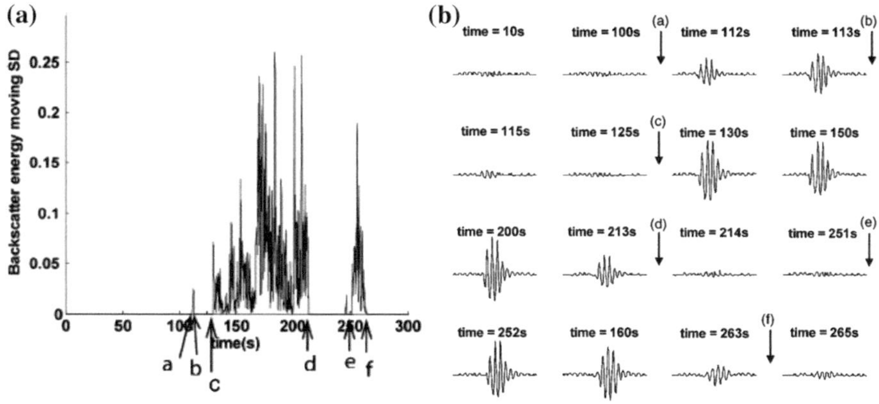

Fig. 5.34 The moving standard deviation (SD) of the backscatter energy (**a**) and corresponding waveforms (**b**). Ultrasound pulses with a PD of 3 cycles, a PRF of 20 kHz, an I_{SPPA} of 4000 W/cm^2, and gas saturation of 40 % were applied. Reproduced with permission from Xu et al. © Acoustical Society of America

Fig. 5.34, we can see that the backscattering signal does not appear immediately as the ultrasound pulses are applied to the tissue, there exists a period of time (from time 0 to "a") called the initiation delay time. The initiation delay time is highly correlated with the acoustic parameters, gas saturation, and probably other parameters. With higher pulse intensity and gas saturation, the initiation delay time is shorter, and histotripsy can be more efficient (Xu et al. 2004).

The initiation and the tissue erosion processes are highly correlated (as shown in Fig. 5.35); the initiation could be used as an excellent indicator of cavitation as well as histotripsy. However, even erosion does not occur without initiation (Fig. 5.35a), more than just initiation is required to produce erosion. If the initiating sequence is short and only contains several pulses (initiation "a" and extinction "b" were detected), it can just damage the membrane of the surface, not perforate the tissue (Fig. 5.35b). Several maintenance pulses, which have lower pulse intensity than pulses in the initiation phase, are needed to achieve tissue erosion or perforation (Fig. 5.35c). In Fig. 5.35c, initiation "c" was detected and tissue was perforated at "d." Some acoustic parameters such as pulse intensity, pulse repetition frequency (PRF), and pulse duration (PD) can control the interaction between the microbubbles and the acoustic field, and this interaction can affect the maintenance pulse. It is important to determine how to control and predict the maintenance phase of the high-backscatter environment so that it can be carried out at the lowest energy levels required.

The extinction is defined as the highly backscattering signal distinguished, as shown in Fig. 5.34, and the backscattering signal is extinguished at time "b" and "d," and the tissue is finally perforated at "f." Ideally, the extinction of the varying backscattering signal occurs at the same time as tissue erosion. Thus, the extinction of the enhanced backscatter signal can be used as a signature to monitor tissue

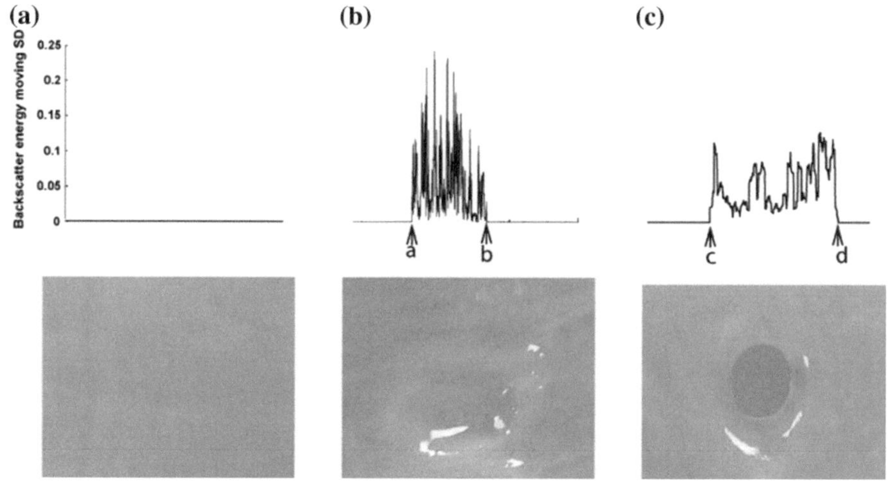

Fig. 5.35 The initiation and extinction phenomena and corresponding tissue effects generated. All three tissue samples were treated for 8 min by ultrasound pulses with a PD of 3 cycles, a PRF of 20 kHz, an I_{SPPA} of 3500 kHz, and gas saturation of 40–45 %. Reproduced with permission from Xu et al. © Acoustical Society of America

erosion. Similar to the initiation and maintenance phases, acoustic pulse parameters and gas saturation may also affect extinction.

As discussed above, the initiation, maintenance, and extinction phases are important for controlling the histotripsy process caused by cavitation. The three processes likely coincide with the acoustic pulse parameters, such as pulse intensity, pulse duration, and pulse repetition frequency. Therefore, we can use the appropriately designed series of acoustic pulses and change these parameters to control and optimize cavitation-induced histotripsy.

Lee et al. (2014) studied the controlled generation of single microbubbles (approximately 100 μm) at solid surfaces in water. They deterministically generated a single microbubble over the focal area by focusing a nanosecond pressure pulse (<100 ns). Such controlled bubble nucleation was realized by the strong negative pressure generated by pulsed optical excitation of a carbon–nanotube–polymer composite and its tight focusing (approximately 100 μm). The generated negative-pressure amplitude (>50 MPa), which was further increased by acoustic interference at an impedance-mismatched boundary, even exceeded the cavitation threshold of distilled water in free-field (boundary-free) conditions. Thus, this strong pressure can activate many submicron nuclei in both water and at surfaces, thus making the nucleation process controllable but irrelevant to a pit array (<10 μm in diameter). Overall, the ability to control bubble formation can lead to new applications in cell-level therapies such as ablation of individual cancerous cells (microhistotripsy).

5.4.3 Sonothrombolysis

Thrombosis-related diseases, such as myocardial infarction, stroke, pulmonary embolism, deep venous thrombosis, systemic embolism, and others, are by far the leading causes of mortality and morbidity worldwide. Therefore, various thrombolytic therapies have been developed, and some have progressed to clinical trials and shown effective results (Meairs et al. 2012). Current thrombolytic therapies used in the clinic include thrombolytic drugs (Kyrle and Eichinger 2005; Moll 2008), catheter-based intravascular techniques (Gutt et al. 2005; Kim et al. 2006), or a combination of the two (Alesh et al. 2007; Bjarnason et al. 1997). However, each of these methods has its own set of drawbacks. Drugs are not site-specific, resulting long treatment times (several hours), and they are associated with risk of bleeding. Catheters also carry a risk of bleeding and are invasive to the blood vessel wall.

The potential of using ultrasound for treating thrombolytic diseases has been extensively studied for several decades. Ultrasound is a completely novel, non-pharmacological approach that offers the potential to increase reperfusion and limit bleeding complications, either alone or in combination with contrast agents. Ultrasound can also be used as an adjunct to thrombolytic therapy (Rosenschein et al. 2000; Tachibana and Tachibana 1995; Shaw et al. 2009). EKOS Corp. is a commercial entity that utilizes catheter-based ultrasound for enhanced sonothrombolysis. However, treatments in which ultrasound is used in combination with thrombolytic drugs can still cause drug-associated hemorrhaging. In addition, treatments in which ultrasound is used in combination with contrast agents poses the risk of blood vessel damage due to the uncontrollable rupturing of UCAs (Miller et al. 1996; Qiao et al. 2013). Catheter-delivered external transducer ultrasound is another technique that has been researched by many groups, but it is considerably more invasive than externally delivered ultrasound and suffers from high technical demands and other potential risks.

Focused US (FUS) has been shown to be an effective thrombolytic method without using thrombolytic drugs, both in vitro and in vivo. It is noninvasive and requires only short treatment times (several minutes) (Westermark et al. 1999; Frenkel et al. 2006; Maxwell et al. 2009). However, to obtain proper focusing of the ultrasonic beam, the ultrasound frequency has to be carefully selected. Although some reports indicate that low-frequency ultrasound at the tens of kilohertz level could increase the disruptive effect of enzymatic thrombolysis and suggest that it seems more effective than at higher frequencies, many researchers have concentrated on the megahertz frequency range because of the difficulty in achieving adequate focusing at long wavelengths.

In studies where high-intensity FUS was used to break down blood clots, cavitation collapse was suggested to be the underlying cause of clot lysis (Rosenschein et al. 2000; Westermark et al. 1999). The collapse of bubbles in a fluid subjected to ultrasound gives rise to transient, yet powerful microjets that can disintegrate clots mechanically. Previous studies have shown that thrombolysis is more efficient using pulsed FUS rather than continuous FUS (Westermark et al. 1999), as the degree of acoustic shielding is minimized by the relatively long pauses between pulses, allowing intense cavitation to take place in the focus region (Rosenschein et al. 2000; Westermark et al. 1999). More specifically, our hypothesis on the mechanism of enhanced thrombolysis using pulsed ultrasound is that each pulse creates a cluster of microbubbles localized at the transducer focus. Collapse of the microbubbles causes local stress that removes a portion of the targeted clot. Residual microbubbles may also act as nuclei that can be excited by subsequent pulses, predisposing the clot in the focal region to further damage. Other studies have explored how changing acoustic parameters (particularly PRF) can enhance cavitation effects when using pulsed ultrasound (Tran et al. 2003; Sapozhnikov et al. 2002; Chang et al. 2001). Therefore, it may be possible to achieve a synergistic effect between residual cavitation nuclei by adjusting the timing between two consecutive pulses. It can be hypothesized then that in pulsed FUS-induced thrombolysis, if the size distribution of microbubbles is within an active size range in which the bubbles oscillate strongly in response to the applied acoustic pressure (Yasui et al. 2008; Hauptmann et al. 2013), the cavitation activity will be enhanced and result in optimal thrombolysis.

Figure 5.36 shows the number of clot debris particles induced by 1.2-MHz FUS at three duty cycles, as well as that for the control group (Xu et al. 2015). The acoustic power was 60 W and the PRF was 454 Hz, and the figure shows the mean debris distributions for particle diameters between 2 and 30 μm. The number of clot debris particles generated using a 9 % duty cycle was higher than at either of the other two duty cycles (Fig. 5.36b).

Figure 5.36a clearly shows that the number of small particles (<10 μm) and large particles (10 < D < 30 μm) was highest when the 9 % duty cycle was used. As the mass is proportional to the number of particles for a given range of diameters, the mass of both the small and large particles at a 9 % duty cycle was larger than the mass at the two other tested duty cycles (2.3 and 18 %), as illustrated in Fig. 5.37a.

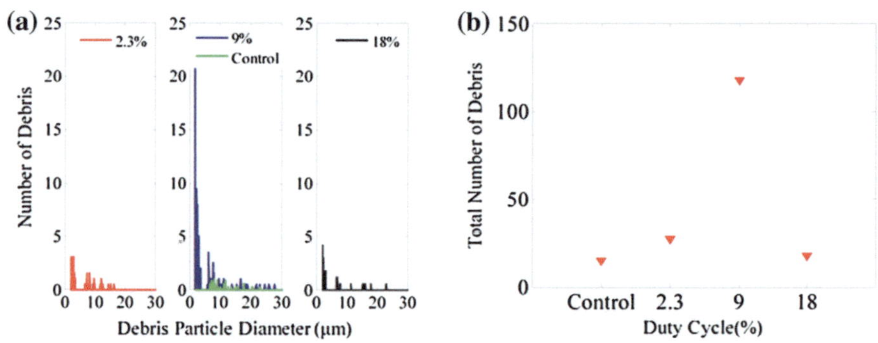

Fig. 5.36 **a** Histograms of the number of clot debris particles with sizes ranging from 2 to 30 μm. **b** The total number of clot debris particles as a function of duty cycle. Reproduced with permission from Xu et al. © Elsevier

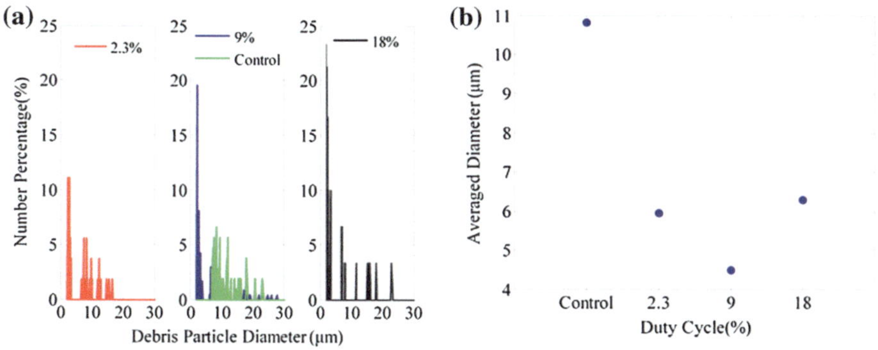

Fig. 5.37 **a** Histograms of the mass of clot debris particles. **b** The total mass of the clot debris particles as a function of duty cycle. Reproduced with permission from Xu et al. © Elsevier

Therefore, the total mass of the debris particles generated at 9 % duty cycle was the largest of the three test conditions (Fig. 5.37b).

Figure 5.38a shows the percentage of clot debris particles induced by FUS using the three test duty cycles as well for in the control group. FUS treatment performed at the 9 % duty cycle produced a higher percentage of smaller particles (<6 μm) than the two other duty cycles. Figure 5.38b clearly shows that the mean diameter of clot debris particles in the sample treated with 9 % duty cycle was smaller than the mean diameter of the clot debris produced using the other two duty cycles. These results indicate that when FUS is delivered at 9 % duty cycle, the clot is more likely to be finely abraded.

Looking more carefully at the control group, we can see that Fig. 5.36b shows that the total number of clot debris particles in the control group was lower than the number of clot debris particles generated at 2.3 % and the 18 % duty cycles, while the total mass of the clot debris particles in the control group was larger than the

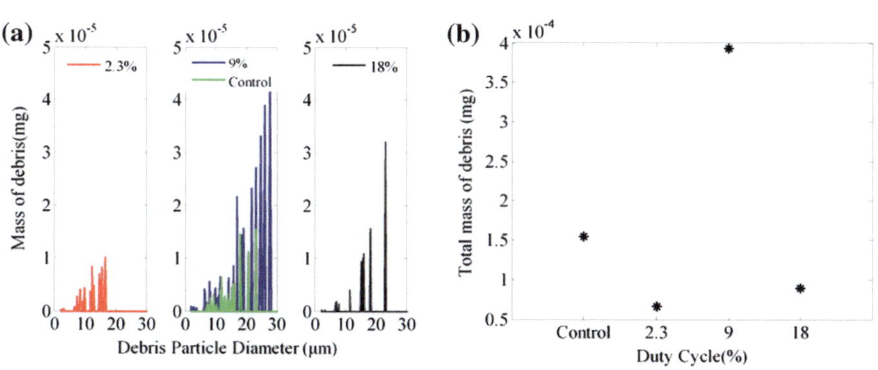

Fig. 5.38 **a** Histograms of the percentage of clot debris particles. **b** The average diameter of clot debris particles as a function of duty cycle. Reproduced with permission from Xu et al. © Elsevier

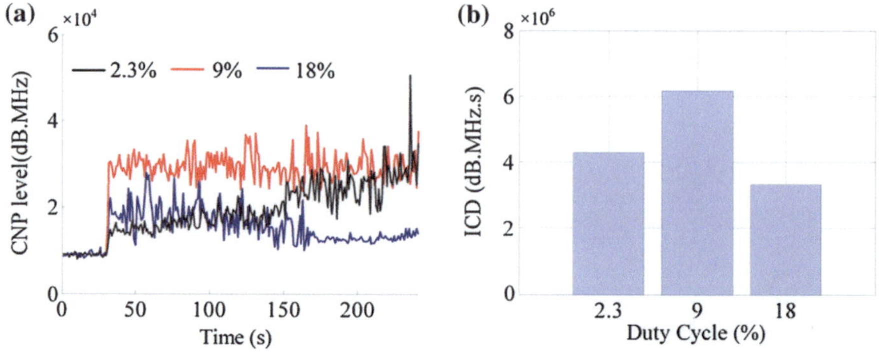

Fig. 5.39 **a** Passively detected broadband noise emission over time. **b** Inertial cavitation dose generated during the total treatment time as a function of duty cycle. Reproduced with permission from Xu et al. © Elsevier

total mass of clot debris particles generated at 2.3 and 18 % duty cycles (Fig. 5.37b). Also, clot debris particles in the control group were larger in size than in the other groups, shown in Fig. 5.38b. These findings suggest that although using pulsed ultrasound at 2.3 and 18 % duty cycles did not increase the total debris mass when compared to the control group, these settings can help fragment large clot debris particles into smaller ones.

Next, the inertial cavitation activity over time was examined for the same three duty cycles, and the results are shown in Fig. 5.39a. Inertial cavitation activity increased to its steady-state value immediately upon exposure to FUS when the duty cycle was set to 9 %. However, a delay in inertial cavitation was observed for the 2.3 % duty cycle, and a gradual decay in cavitation was observed over time for the 18 % duty cycle. To quantify the cavitation activity during the entire treatment time, inertial cavitation dose (ICD) was calculated by integrating the area under the cavitation noise power (CNP) value curve over time. The ICD from the 9 % duty cycle treatment was greater than the ICD measured using the other two duty cycles (Fig. 5.39b).

The evolution of the size distribution of the cavitation bubbles, shown in Fig. 5.40, was monitored for all three tested duty cycles. In Fig. 5.40a, when the duty cycle was 2.3 %, the size distribution of the cavitation bubble from the 1st, 103rd, and 105th pulse increased slightly, but it appeared to be smaller than the resonance radius of the emission frequency of the FUS transducer ($R_{res} \approx 3/f_0$, f_0 = 1.2 MHz, solid vertical line in Fig. 5.40). When the duty cycle was 9 %, the size distribution of the cavitation bubble for the 1st, 103rd, and 105th pulse remained nearly unchanged, right around the resonance radius (Fig. 5.40b). When the duty cycle was increased to 18 %, the mean size of the cavitation bubble for the 1st, 103rd, and 105th pulse increased to a size that was substantially larger than the resonance radius (Fig. 5.40c).

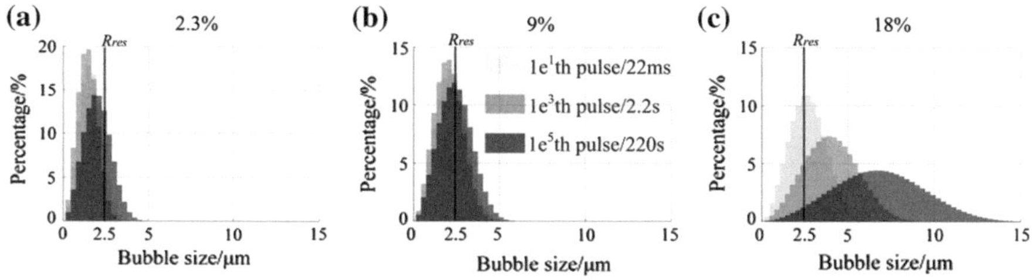

Fig. 5.40 Size distributions of the induced cavitation bubbles after the 1st, 103rd, and 105th pulses at a duty cycle of **a** 2.3 %, **b** 9 %, and **c** 18 %. Reproduced with permission from Xu et al. © Elsevier

The data presented here suggest that there is an optimal duty cycle at which the efficiency of thrombolysis and cavitation activity is strongest. This optimal duty cycle is reached when the size distribution of the cavitation bubbles is centered around the linear resonance size of the transmission frequency of the FUS transducer. On the one hand, if the duty cycle is too short, only a subset of the partially dissolved bubbles can be reactivated. On the other hand, if the duty cycle is too long, bubbles can grow beyond the resonance size due to rectified diffusion and coalescence, thereby diminishing their cavitation activity. The confinement of sonication to pulse wave trains imposes a limiting effect on bubble growth and coalescence and lowers the maximum bubble size. The rationale behind the pulsing strategy is to generate cavitation bubbles using high-intensity pulses that provide properly sized nuclei by adjusting the time between two consecutive pulses as a means to control the size of cavitation nuclei for subsequent pulses and to sustain optimal cavitation.

5.4.4 *Vascular Applications*

5.4.4.1 Tumor Angiogenesis Disruption

The term angiogenesis refers to the growth of new blood vessels, which is an essential process for tumor growth, progression, and metastasis; thus, preventing or stopping angiogenesis is widely considered an emerging technique in cancer therapy. If we can stop cancers from growing blood vessels, then we can slow the growth of the cancer or even shrink it. Anti-angiogenic drugs are treatments that stop tumors from growing their own blood vessels and inhibit angiogenesis, and they may provide a form of cancer therapy worthy of serious exploration.

Tumor angiogenesis is critical for tumor development, reproduction, and repair (growth and metastasis). This specific type angiogenesis, however, produces defective, leaky, and fragile vessels. Discontinuities or gaps that allow hemorrhage and facilitate permeability are common features, and they occur as a result of poor

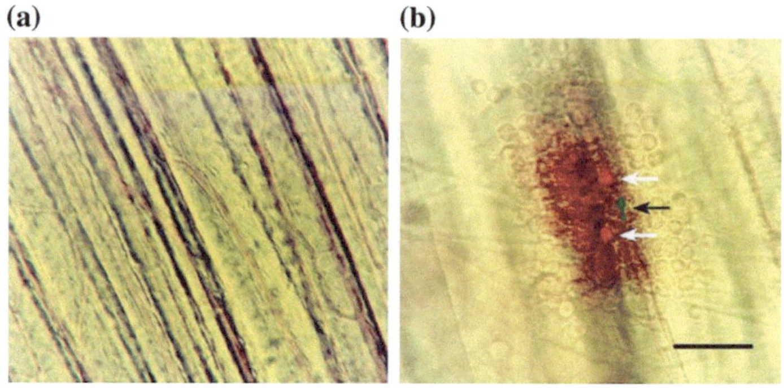

Fig. 5.41 Intravital images of microvessels in rat spinotrapezius muscle. **a** Normal muscle and intact capillaries under transillumination (×20). **b** Composite image created from transilluminated and epi-illuminated images after ultrasound exposure in the presence of microbubbles. Scale bar, 20 μm. Reproduced with permission from Skyba et al. © American Heart Association

cell–cell junctions and defects in endothelial cell (EC) barrier functions. The basement membrane that envelops the ECs and pericytes of tumor vessels may have extra layers that are loosely associated with ECs.

Ultrasound has been used to destroy such vessels in vivo by exciting microbubbles intravascularly (Skyba et al. 1998). In these studies, before microbubble injection no microvessel ruptures were seen after ultrasound exposure. Similarly, no ruptures were seen when microbubbles were injected in the absence of ultrasound (Fig. 5.41a). However, when microbubble injection and ultrasound exposure were performed simultaneously, the microbubbles destroyed microvessels ≤7 μm in diameter (mostly capillaries). The destruction was evidenced by the presence of microvessel rupture and non-viable cells, as shown in Fig. 5.41b, which shows a ruptured capillary with extravasation of red blood cells (RBCs). In the figure, the white arrows indicate propidium iodide (PI)-labeled non-viable cells while the black arrow indicates fragments of destroyed microbubble.

As shown above, microbubble-enhanced ultrasound cavitation could potentially disrupt tumor vasculatures or small blood vessels, depending on the effective acoustic pressure amplitude. This general hypothesis is schematically illustrated in Fig. 5.42, where acoustic cavitation combines high-pressure amplitude pulsed ultrasound and circulating microbubbles. Liu et al. (2012) demonstrated that tumor neovasculature (angiogenesis) can be physically disrupted with microbubble-enhanced high-pressure amplitudes but low-intensity (I_{SPTA}) pulsed ultrasound. Complete cessation of tumor circulation happened immediately after it was insonated percutaneously and received an intravenous injection of microbubbles, and the cessation persisted as long as 24 h, corresponding to a significant vascular disruption observed by histology. This study showed that tumor circulation could be completely blocked off for 24 h when treated with 4.8-MPa pulses. Also, the contrast-enhanced ultrasound (CEUS) gray scale value (GSV) indicated that there was significant GSV drop-off in both of the two experimental groups but none

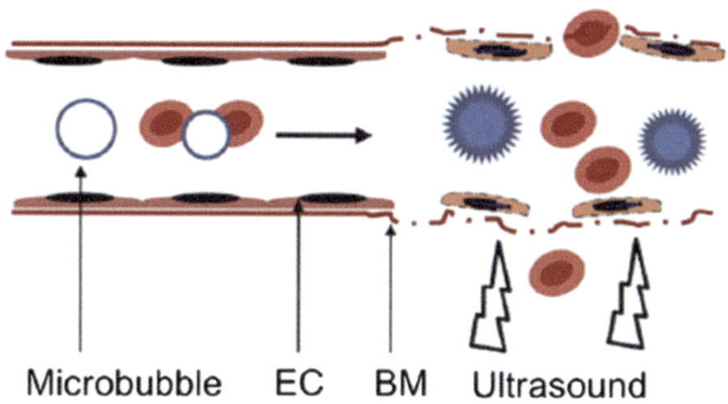

Fig. 5.42 Schematic representation of vascular disruption through the mechanical effects of microbubble-enhanced cavitation. Reproduced with permission from Liu et al. © Elsevier

in the controls. Histology showed that the tumor microvasculature was disrupted into diffuse hematomas accompanied by thrombosis, intercellular edema, and the formation of multiple cysts. Also, the 24 h of blocked tumor circulation resulted in massive necrosis of the tumor. Thus, microbubble-enhanced ultrasound provides a new, simple physical method for anti-angiogenic therapy and may have great potential for clinical applications.

5.4.4.2 Localized Cavitation Damage of the Vessel Wall

In animal models, the most common method used to induce atherosclerosis is balloon injury of the artery endothelium. According to the postulated mechanisms for atherogenesis, lesions in the artery's endothelium might be the main mechanism in rabbit models. Thus, another potential method that for damaging the arterial endothelium is focused ultrasound-induced cavitation, in particular, inertial cavitation.

Usually, gas-filled microbubbles that serve as UCAs can be used to enhance the diagnostic capabilities in ultrasound imaging. When microbubbles are exposed to higher amplitude ultrasound, microbubbles undergo a fast inertial cavitation and then collapse. When the inertial collapse occurs near a boundary, asymmetric contraction of microbubbles can generate strong shock waves and high-speed fluid microjets, leading to localized biological effects on nearby cells and structures. The application of ultrasound to microvessels containing microbubbles can produce tunneling in tissue, jet-induced EC damage, capillary rupture, and arterial injury. The results from these studies indicate that the interaction of ultrasound with UCAs in the vascular system potentially causes EC dysfunction and injury. But, no studies have yet shown if the endothelium of large vessels can be injured by ultrasound cavitation.

However, in for developing a rabbit model of atherosclerosis, the vessel that needs to be damaged is the aorta, which is considerably larger than a microvessel. Here, we describe experiments conducted to confirm whether focused ultrasound

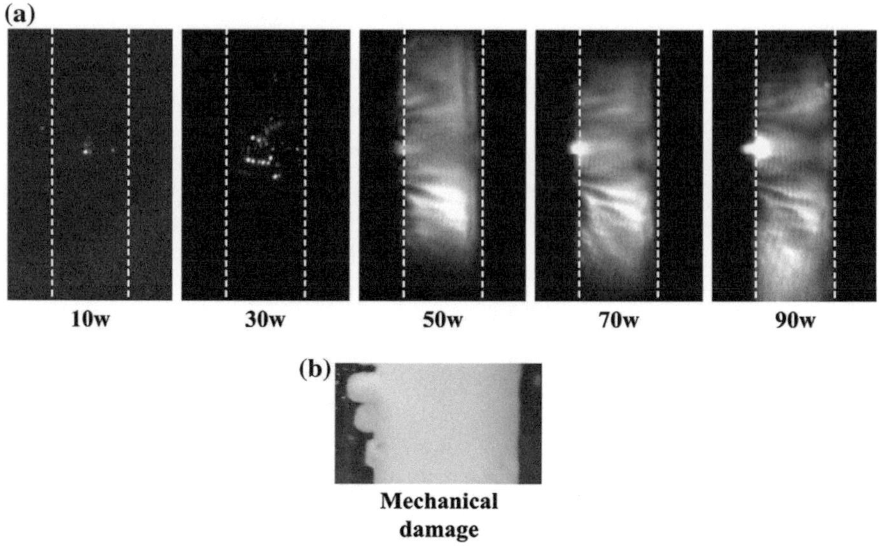

Fig. 5.43 **a** The SL images within the vessel at different input powers. **b** The cylindrical holes formed on the vessel wall, from the *bottom to top*, corresponding to input powers of 50, 70, and 90 W. The ultrasound wave was irradiated from *left to right*. The dashed lines indicate the position of the vessel wall. Reproduced with permission from Qiao et al. © Elsevier

(FU)-induced cavitation can cause endothelial lesions in rabbit abdominal aortas. If so, then FU-induced cavitation may be able to replace artery balloon injury, by inducing a localized, eccentric vulnerable plaque at a desired site.

For vessels larger than 2 mm in diameter, convective cooling can significantly weaken the thermal effects of FU. But, the continued presence of ultrasound will cause repetitive cavitation and acoustic microstreaming, making possible continuous wave-induced cavitation effects in large vessels. Via the sonoluminescence (SL) method, mechanical damage was observed and high-speed cameras were used to investigate the combination effect of UCAs and continuous FU in large phantom vessels with a diameter of 10 mm (without considering thermal effects) (Qiao et al. 2013). As shown in Fig. 5.43, when the focus was positioned at the proximal wall, a cylindrical hole along the acoustic axis opposite the ultrasound wave propagation direction was observed at an input power equal to or greater than 50 W. When the focus was located at the distal wall, only small tunnels were found. The place where the cylindrical hole formed corresponded to where bubbles gathered and emitted brilliant light near the wall. Without UCAs, neither the bright SL nor the cylindrical hole was found. However, the concentration of UCAs had little influence on the SL distribution and the length of the cylindrical hole. The SL intensity near the proximal vessel wall and the length of the cylindrical hole both increased with the input power. Thus, these findings should be considered for large-vessel therapy and the use of UCAs.

5.4.5 Ultrasound-Enhanced Delivery of Drugs and Genetic Materials

Another application of the mechanical effects associated with cavitation is ultrasound-enhanced drug delivery and gene transfection, which involves sonoporation/sonophoresis, disruption of the blood brain barrier (BBB), and cavitation-enhanced extravasation.

Cavitation is the primary mechanism of ultrasound-enhanced drug delivery and gene transfection (Bailey et al. 2003). Cavitation microjets puncture holes in cell membranes and convect drugs into cells, tissues, and blood clots. In addition, flow around the oscillating bubble creates compressive, tensile, and shear stress on the biological interface. The flow aids in diffusion through mixing and opens channels by straining membranes.

5.4.5.1 Sonoporation/Sonophoresis

The rapid collapse of cavitation bubbles leads to the formation of shock waves that can disrupt tissues and enhance drug transport (implicated in sonophoresis and sonoporation). Furthermore, collapsing bubbles near a surface experience non-uniformities in their surroundings that result in the formation of high-velocity microjets. These microjets can penetrate into the tissue or generate secondary stress waves in tissues (implicated in sonophoresis). Sonoporation is a transient increase in cell membrane permeability caused by ultrasound-driven microbubbles and is of potential use as a method for gene delivery and drug delivery (Miller et al. 2002, 1996; Bao et al. 1997; Ferrara et al. 2007). Liquid jets provide increased transport of heat and gas by streaming and have the capacity to puncture the cell membrane, producing openings that could allow for the transport of extracellular material into the cell (Ohl and Ikink 2003; Prosperetti 1997).

Sonoporation uses ultrasound energy and UCAs to facilitate cell permeabilization, and cavitation phenomena are critically dependent on the existence of nucleation sites. However, very few cavitation nuclei exist normally in tissue or blood. Introducing microbubbles can therefore dramatically increase the number of cavitation nuclei and reduce the cavitation threshold. Thus, cavitation-enhanced mechanical mechanisms can enhance drug transport across interfaces, facilitate the drug/gene uptake by cells, and locally enhance the therapeutic efficacy of pharmaceutical agents.

In SEM images, Zhong et al. (2011) directly observed membrane poration as a result of exposing a cell–microbubble mixture to pulsed ultrasound, as shown in Fig. 5.44. This figure shows HL-60 leukemia cells with randomly positioned pores on the membrane; these cells were immediately fixed after 3 s of ultrasound exposure at 0.5 MPa peak negative pressure. In this image, the majority of the holes have a diameter between 0.1 and 0.5 μm.

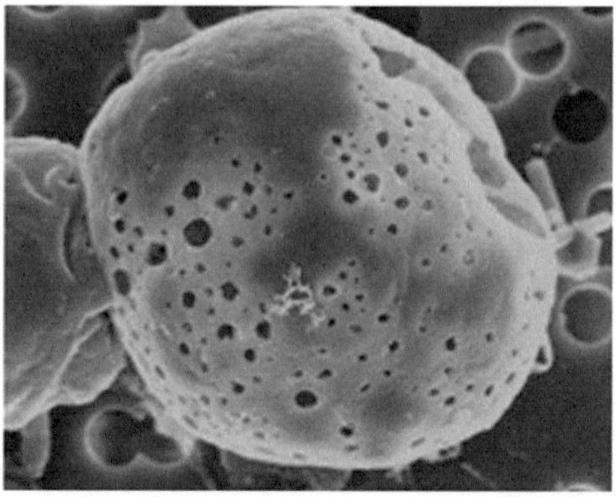

Fig. 5.44 Scanning electron microscopy (SEM) images of HL-60 leukemia cells that received a 3-s ultrasound exposure (1 MHz center frequency, 10 % duty cycle, 1 kHz pulse repetition frequency, 0.5 MPa peak negative pressure, 1 % microbubbles). Reproduced with permission from Zhong et al. © Elsevier

Sonoporation can also be affected by microstreaming, whereby different microstreaming patterns generate different shear stress and stretch/compression distributions in the vicinity of a bubble on a wall. Previous studies using micro-PIV have shown that ultrasound-driven microstreaming flows around bubbles can help mix therapeutic agents into the surrounding blood and assist in sonoporative delivery of molecules across cell membranes (Collis et al. 2010). However, microstreaming patterns show significant variations around bubbles, suggesting that sonoporation may be either enhanced or inhibited in different zones across a cellular surface. Thus, alternating these patterns may result in improved sonoporation. The clear and reproducible delineation of microstreaming patterns based on driving frequency makes frequency-based pattern alternation a feasible alternative to the clinically less desirable practice of increasing sound pressure for an equivalent sonoporative effect. Surface divergence is proposed as a measure relevant to sonoporation.

Applying ultrasonic field pulses while microbubbles are present and adhering to cells generates a critical transmembrane shear force, which leads to the formation of pores once the membrane breakdown force is exceeded. Rapidly oscillating microbubbles generate a fluid flow over the cell surface. This flow is termed microstreaming and is probably responsible for the disruption of cell membrane by tearing the lipid bilayer membrane open (Marmottant and Hilgenfeldt 2003).

One of the suggested mechanisms of sonoporation is linked to shear stress exerted on the cell membrane by acoustic microstreaming generated by a contrast microbubble pulsating nearby a cell. An illustration of this sonoporation mechanism is shown in Fig. 5.45. Figure 5.45a shows a microbubble in contact with an EC, whereby the microbubble volumetrically oscillates in the stable cavitation regime. In the compression phase of the microbubble (panel a, middle picture), the microbubble pulls on the cell membrane. The liquid surrounding the bubble–wall

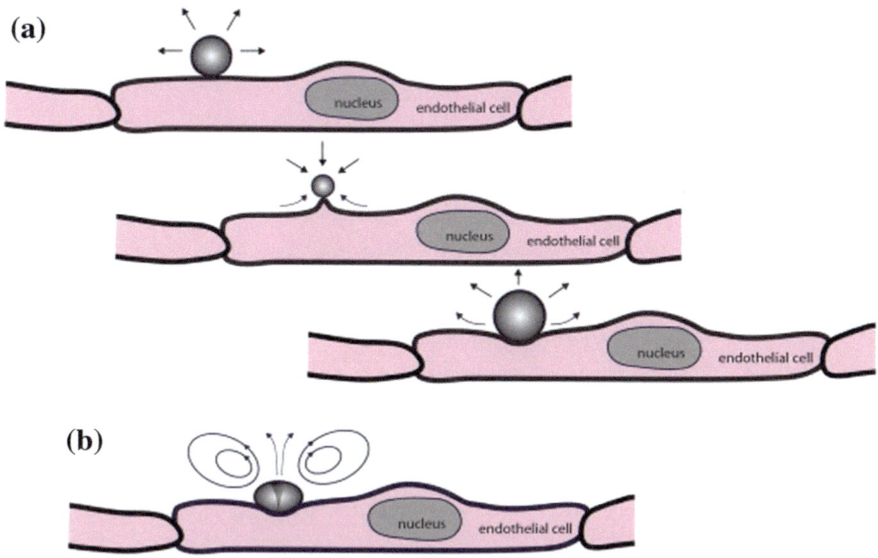

Fig. 5.45 Illustration of common microbubble phenomena nearby cells. **a** Stable cavitation, in which the bubble massages the cell membrane; **b** Transient cavitation, in which violent collapse and jets occur, and acoustic microstreaming around the bubble can become significant. Reproduced with permission from Kooiman et al. © Elsevier

interface shears along the cell membrane, as denoted by the arrows. In the expansion phase (panel a, bottom picture), the microbubble exerts a normal force on the cell membrane, and the shear motion along the cell membrane is pointed outward. Thus, it is implied that the liquid and cell around the microbubble are stretched and sheared at the frequency of the incoming ultrasound wave in the stable cavitation regime. Since the microbubble oscillations are mild, the oscillations can be sustained over a long duration, potentially setting up an acoustic streaming pattern as illustrated in Fig. 5.45b. With increasing pressure, the bubble also enters the inertial cavitation regime and will display more violent behavior, such as a collapse and jetting, as shown.

5.4.5.2 Disruption of the Blood Brain Barrier

The BBB is a unique, selective barrier formed by ECs that hampers transport and diffusion of molecules from the vasculature to the brain (Cecchelli et al. 2007), as shown in Fig. 5.46. Drug delivery to the brain remains challenging, because the BBB complicates the delivery of diagnostic and therapeutic agents to the brain. For example, the BBB prevents more than 98 % of all small-molecule drugs and approximately 100 % of large-molecule neurotherapeutics from entering the brain. Only small-molecule drugs with high lipid solubility and a low molecular mass

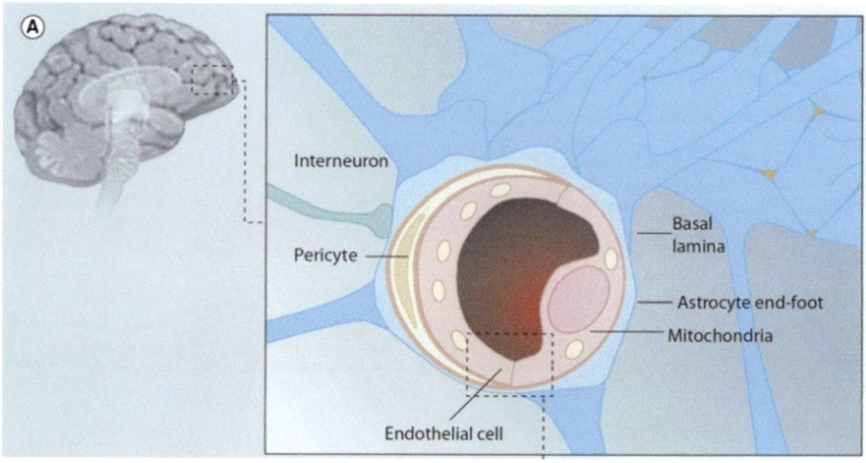

Fig. 5.46 The blood–brain barrier. Reproduced with permission from Cecchelli et al. © Nature Publishing Group

under 400–500 Da can cross the BBB in pharmacologically significant amounts (Pardridge 2001). However, to deliver the therapeutic agents to brain, the BBB needs to be disrupted or opened to increase the brain uptake of the drugs. A successful drug-delivery system to the brain should be efficient, safe, convenient, noninvasive, and localized to a specific volume in the brain. These capabilities ensure maximal therapeutic outcome and minimal unwanted side effects.

Due to the mechanical effects of acoustic cavitation, focused ultrasound combined with microbubbles can temporarily open the BBB, as shown in Fig. 5.47. Since the first report about BBB disruption using high power unfocused ultrasound (Ballantine 1960), several studies have demonstrated the feasibility of using focused ultrasound for the localized and reversible disruption of the BBB in animal models (Vykhodtseva et al. 2008; Hynynen 2008; Tung et al. 2011). Most studies have combined low-pressure amplitude and low time-averaged acoustic power burst sonications with intravascular injection of preformed microbubbles to produce BBB disruption without damage to neurons.

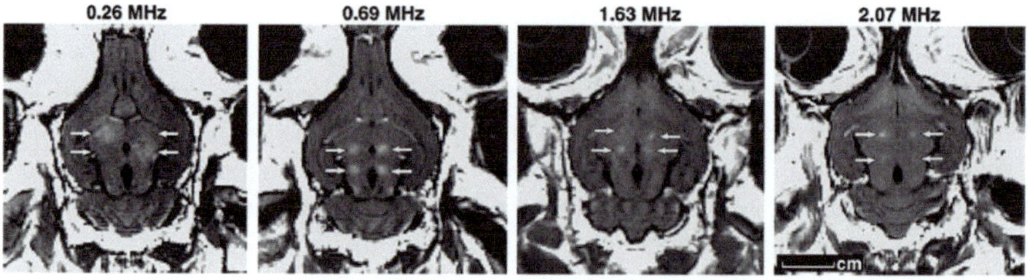

Fig. 5.47 Local BBB disruption in a rabbit brain produced by FUS at four ultrasound frequencies: 2.1, 1.63, 0.69, and 0.26 MHz. Reproduced with permission from Vykhodtseva et al. © Elsevier

Recent studies into BBB disruption—aimed at delivering large molecules to treat conditions such as Alzheimer's disease—have shown that sonoporation of the BBB can be achieved without requiring violent inertial cavitation. This important observation could be explained by multiple alternative theories describing how large molecules pass through the BBB. Two theories are based on molecular passage across the BBB through stretch-activated channels—pores in cell membranes that open based on their sensitivity to changes in mechanical stress. In one of these theories, the first-order oscillations of contrast agent bubbles are thought to directly cause activation through "cellular massage" (oscillation of the cell membrane). In the other theory, the second-order microstreaming flow is thought to generate flow fields that develop shear stresses over the cell membranes, resulting in tension and stretching of the membrane walls that cause channel activation. These two theories are plausible, given that in general, studies using ECs show that changes or disruptions in typical cell behavior can be caused by the application of shear stresses. Finally, a third theory on molecular passage across the BBB does not involve passage across stretch-activated channels into individual cells, but rather is based on the opening of gaps between otherwise tightly packed ECs of the BBB.

To determine how sonoporation of the BBB works, Tung et al. investigated the physical mechanism of BBB opening at different bubble sizes (Tung et al. 2011). They found that inertial cavitation was required for smaller bubbles to induce BBB opening. However, with larger bubbles that have diameters similar to the capillary diameter and thus may be in contact with the capillary wall, BBB opening was induced through nonlinear oscillation instead of inertial cavitation. No significant damage was detected at the BBB opening threshold for all bubble sizes. The BBB opening threshold, the (ICD), and the BBB opening volume all depended on bubble size; the larger (4–5 and 6–8 μm) diameter bubbles used with lower pressure amplitudes (0.20–0.30 MPa) were determined to be safe and consistent for opening the BBB.

Overall, sonoporation of the BBB has great potential to produce reversible opening of the BBB without brain tissue damage. Targeted BBB disruption could potentially aid in the delivery of chemotherapy agents, antibodies, and genes to the brain, as well as the delivery of molecules to other neural tissues.

5.4.5.3 Cavitation-enhanced Extravasation for Drug Delivery

To improve efficiency of anticancer drugs, they must be delivered efficiently through the tumor vasculature, cross the vessel wall, and traverse the tumor tissue to reach viable cells in the tumor. But, recent studies have shown that the unique physiological barriers presented by tumors eventually limit this process. As a result of tumor proliferation, some cells in tumors are located more than 100 mm away from blood vessels, forming hypoxic regions that contain aggressive viable cancer cells and neighboring necrotic regions located further away from vessels that contain dead cells, starved by lack of oxygen and nutrients.

One possible approach to reach viable tumor cells is to systemically inject pharmaceuticals simultaneously with an ultrasound contrast agent and to then apply ultrasound to the target region. The oscillations of the bubble or ultrasound-induced cavitation can alter the vessel walls and allow for the extravasation of the pharmaceutical agent (Ferrara et al. 2007; Arvanitis et al. 2011). Usually, increasing the permeability of the vessel wall is the first step in enhancing drug delivery to non-endothelial cells. Increasing penetration distance is the second step after extravasation of therapeutic agents to hypoxic regions and poorly vascularized tumor regions. After extravasation and penetration, the mediation of cellular uptake by ultrasound is another key process to be controlled. In this process, cavitation is the most likely mechanism to induce extravasation, while pore formation can be induced to mediate cell uptake.

5.5 Summary

The mechanical effects of cavitation play an important role in a wide range of biomedical ultrasound applications. This chapter provides an overview of the fundamentals of cavitation-enhanced mechanical mechanisms that are involved in many biomedical applications. Such mechanisms include the acoustic microstreaming that develops in liquids in the vicinity of vibrating gas bubbles in an ultrasound field at megahertz frequencies, the shear stress induced by microstreaming, and the formation of shock waves and microjets during symmetrical or asymmetrical bubble collapse. When an ultrasound intensity is applied at a low level, well below 100 mW/cm^2, the oscillation of microbubbles in sound fields can result in relatively modest and controllable acoustic microstreaming and stresses, and these events may assist in the mixing and delivery of some compounds and enhance cellular uptake. When the ultrasound intensity is higher, the bubbles will collapse after reaching their maximum expansion and generate high-pressure shock waves due to rebound. When bubbles oscillate near a boundary, the collapse will become non-spherical and induce a high-speed liquid jet toward or away from the boundary. In Sects. 5.2 and 5.3, we reviewed these phenomena, as they pertain to cavitation-enhanced mechanical effects, from both a theoretical and an experimental perspective. We then briefly introduced several practical and possible applications that involve these effects in Sect. 5.4. From these applications, we see that although acoustic microstreaming, jet formation and shock waves are introduced separately in this chapter, it is difficult to isolate their effects independently for each application; the bioeffects of these mechanical, non-thermal processes always occur in combination with each other.

References

Alesh I, Kayali F, Stein PD. Catheter-directed thrombolysis (intrathrombus injection) in treatment of deep venous thrombosis: a systematic review. Cathet Cardiovasc Interv. 2007;70(1):145–50.

Arvanitis CD, Bazan-Peregrino M, Rifai B, Seymour LW, Coussios CC. Cavitation-enhanced extravasation for drug delivery. Ultrasound Med Biol. 2011;37(11):1838–52.

Bailey M, Khokhlova V, Sapozhnikov O, Kargl S, Crum L. Physical mechanisms of the therapeutic effect of ultrasound (a review). Acoust Phys. 2003;49(4):369–88.

Bao S, Thrall BD, Miller DL. Transfection of a reporter plasmid into cultured cells by sonoporation in vitro. Ultrasound Med Biol 1997;23(6):953-959.

Barnett S. Nonthermal issues: cavitation—its nature, detection and measurement. Ultrasound Med Biol. 1998;24:S11–21.

Bjarnason H, Kruse JR, Asinger DA, Nazarian GK, Dietz CA Jr, Caldwell MD, Key NS, Hirsch AT, Hunter DW. Iliofemoral deep venous thrombosis: safety and efficacy outcome during 5 years of catheter-directed thrombolytic therapy. J Vasc Interv Radiol. 1997;8(3):405–18.

Brujan E-A. Shock wave emission from laser-induced cavitation bubbles in polymer solutions. Ultrasonics. 2008;48(5):423–6.

Brujan EA. Cavitation in non-newtonian fluids. Berlin: Springer-Verlag; 2011.

Brujan EA, Ikeda T, Matsumoto Y. Jet formation and shock wave emission during collapse of ultrasound-induced cavitation bubbles and their role in the therapeutic applications of high-intensity focused ultrasound. Phys Med Biol. 2005;50(20):4797–809.

Brujan EA, Nahen K, Schmidt P, Vogel A. Dynamics of laser-induced cavitation bubbles near elastic boundaries: influence of the elastic modulus. J Fluid Mech. 2001;433:283–314.

Brujan EA, Ohl CD, Lauterborn W, Philipp A. Dynamics of laser-induced cavitation bubbles in polymer solutions. Acustica. 1996;82:423–30.

Caskey CF, Stieger SM, Qin S, Dayton PA, Ferrara KW. Direct observations of ultrasound microbubble contrast agent interaction with the microvessel wall. J Acoust Soc Am. 2007;122 (2):1191–200.

Cecchelli R, Berezowski V, Lundquist S, Culot M, Renftel M, Dehouck M-P, Fenart L. Modelling of the blood–brain barrier in drug discovery and development. Nat Rev Drug Discov. 2007;6 (8):650–61.

Chang PP, Chen W-S, Mourad PD, Poliachik SL, Crum LA. Thresholds for inertial cavitation in albunex suspensions under pulsed ultrasound conditions. IEEE Trans Ultrason Ferroelectr Freq Control. 2001;48(1):161–70.

Chen H, Brayman AA, Evan AP, Matula TJ. Preliminary observations on the spatial correlation between short-burst microbubble oscillations and vascular bioeffects. Ultrasound Med Biol. 2012;38(12):2151–62.

Chen H, Brayman AA, Kreider W, Bailey MR, Matula TJ. Observations of translation and jetting of ultrasound-activated microbubbles in mesenteric microvessels. Ultrasound Med Biol. 2011a;37(12):2139–48.

Chen H, Kreider W, Brayman AA, Bailey MR, Matula TJ. Blood vessel deformations on microsecond time scales by ultrasonic cavitation. Phys Rev Lett. 2011b;106(3):034301.

Coakley WT, Nyborg W. Cavitation; dynamics of gas bubbles; applications. In: Fry FJ (ed) Ultrasound: its applications in medicine and biology. Amsterdam: Elsevier Scientific Publishing Co.; 1978. pp. 77–159.

Collis J, Manasseh R, Liovic P, Tho P, Ooi A, Petkovic-Duran K, Zhu Y. Cavitation microstreaming and stress fields created by microbubbles. Ultrasonics. 2010;50(2):273–9.

Davidson B, Riley N. Cavitation microstreaming. J Sound Vib 1971;15(2):217–233.

Dijkink R, Ohl C-D. Measurement of cavitation induced wall shear stress. Appl Phys Lett. 2008;93(25):254107.

Doinikov AA, Bouakaz A. Acoustic microstreaming around a gas bubble. J Acoust Soc Am. 2010a;127(2):703–9.

Doinikov AA, Bouakaz A. Acoustic microstreaming around an encapsulated particle. J Acoust Soc Am. 2010b;127(3):1218–27.
Doinikov AA, Bouakaz A. Theoretical investigation of shear stress generated by a contrast microbubble on the cell membrane as a mechanism for sonoporation. J Acoust Soc Am. 2010c;128(1):11–9.
Doinikov AA, Bouakaz A. Effect of a distant rigid wall on microstreaming generated by an acoustically driven gas bubble. J Fluid Mech. 2014;742:425–45.
Duryea AP, Maxwell AD, Roberts WW, Xu Z, Hall TL, Cain CA. In vitro comminution of model renal calculi using histotripsy. IEEE Trans Ultrason Ferroelectr Freq Control. 2011;58(5):971–80.
Eisenmenger W. The mechanisms of stone fragmentation in ESWL. Ultrasound Med Biol. 2001;27(5):683–93.
Elder SA. Cavitation microstreaming. J Acoust Soc Am. 1959;31:54–64.
Ferrara K, Pollard R, Borden M. Ultrasound microbubble contrast agents: fundamentals and application to gene and drug delivery. In: Yarmush ML, editor. Annual review of biomedical engineering, vol. 9. Palo Alto: Annual Reviews; 2007. p. 415–47.
Frenkel V, Oberoi J, Stone MJ, Park M, Deng C, Wood BJ, Neeman Z, Horne M III, Li KC. Pulsed high-intensity focused ultrasound enhances thrombolysis in an in vitro model 1. Radiology. 2006;239(1):86–93.
Gormley G, Wu J. Observation of acoustic streaming near Albunex® spheres. J Acoust Soc Am. 1998;104(5):3115–8.
Gutt CN, Oniu T, Wolkener F, Mehrabi A, Mistry S, Büchler MW. Prophylaxis and treatment of deep vein thrombosis in general surgery. Am J Surg. 2005;189(1):14–22.
HT. Ballantine Jr, Bell E, Manlapaz J. Progress and problems in the neurological applications of focused ultrasound. J Neurosurg. 1960;17:858–876.
Harper JD, Dunmire B, Wang Y-N, Simon JC, Liggitt D, Paun M, Cunitz BW, Starr F, Bailey MR, Penniston KL. Preclinical safety and effectiveness studies of ultrasonic propulsion of kidney stones. Urology. 2014;84(2):484–9.
Hauptmann M, Struyf H, De Gendt S, Glorieux C, Brems S. Evaluation and interpretation of bubble size distributions in pulsed megasonic fields. J Appl Phys. 2013;113(18):184902.
Hynynen K. Ultrasound for drug and gene delivery to the brain. Adv Drug Deliv Rev. 2008;60(10):1209–17.
Ikeda T, Yoshizawa S, Tosaki M, Allen JS, Takagi S, Ohta N, Kitamura T, Matsumoto Y. Cloud cavitation control for lithotripsy using high intensity focused ultrasound. Ultrasound Med Biol. 2006;32(9):1383–97.
Kim HS, Patra A, Paxton BE, Khan J, Streiff MB. Catheter-directed thrombolysis with percutaneous rheolytic thrombectomy versus thrombolysis alone in upper and lower extremity deep vein thrombosis. Cardiovasc Intervent Radiol. 2006;29(6):1003–7.
Kolb J, Nyborg WL. Small-scale acoustic streaming in liquids. J Acoust Soc Am. 1956;28(6):1237–42.
Kooiman K, Vos HJ, Versluis M, de Jong N. Acoustic behavior of microbubbles and implications for drug delivery. Adv Drug Deliv Rev. 2014;72:28–48.
Krasovitski B, Kimmel E. Shear stress induced by a gas bubble pulsating in an ultrasonic field near a wall. IEEE Trans Ultrason Ferroelectr Freq Control. 2004;51(8):973–9.
Kyrle PA, Eichinger S. Deep vein thrombosis. Lancet. 2005;365(9465):1163–74.
Lauterborn W. Numerical investigation of nonlinear oscillations of gas bubbles in liquids. J Acoust Soc Am. 1976;59(2):283–93.
Lauterborn W, Bolle H. Experimental investigations of cavitation-bubble collapse in the neighbourhood of a solid boundary. J Fluid Mech. 1975;72(02):391–9.
Lee T, Baac HW, Ok JG, Youn HS, Guo LJ. Controlled generation of single microbubble at solid surfaces by a nanosecond pressure pulse. Phys Rev Appl. 2014;2(2):024007.
Lewin PA, Bjørnø L. Acoustically induced shear stresses in the vicinity of microbubbles in tissue. J Acoust Soc Am. 1982;71(3):728–34.

Liu X, Wu J. Acoustic microstreaming around an isolated encapsulated microbubble. J Acoust Soc Am. 2009;125(3):1319–30.

Liu Z, Gao S, Zhao Y, Li P, Liu J, Li P, Tan K, Xie F. Disruption of tumor neovasculature by microbubble enhanced ultrasound: a potential new physical therapy of anti-angiogenesis. Ultrasound Med Biol. 2012;38(2):253–61.

Marmottant P, Hilgenfeldt S. Controlled vesicle deformation and lysis by single oscillating bubbles. Nature. 2003;423(6936):153–6.

Matsumoto Y, Allen JS, Yoshizawa S, Ikeda T, Kaneko Y. Medical ultrasound with microbubbles. Eur J Ultrasound. 2005;29(3):255–65.

Matula T, Chen H. Micro/Manoparticles in ultrasound imaging and therapy. In: Yang X, editor Nanotechnology in modern medical imaging and interventions. Hauppauge: Nova Science; 2013. pp 143–164.

Maxwell AD, Cain CA, Duryea AP, Yuan L, Gurm HS, Xu Z. Noninvasive thrombolysis using pulsed ultrasound cavitation therapy–histotripsy. Ultrasound Med Biol. 2009;35(12):1982–94.

Maxwell AD, Cunitz BW, Kreider W, Sapozhnikov OA, Hsi RS, Harper JD, Bailey MR, Sorensen MD. Fragmentation of urinary calculi in vitro by burst wave lithotripsy. J urol. 2015;193(1):338–44.

Meairs S, Alonso A, Hennerici MG. Progress in sonothrombolysis for the treatment of stroke. Stroke. 2012;43(6):1706–10.

Miller DL. Particle gathering and microstreaming near ultrasonically activated gas-filled micropores. J Acoust Soc Am 1988;84(4):1378–1387.

Miller MW, Miller DL, Brayman AA. A review of in vitro bioeffects of inertial ultrasonic cavitation from a mechanistic perspective. Ultrasound Med Biol. 1996;22(9):1131–54.

Miller DL, Pislaru SV, Greenleaf JF. Sonoporation: mechanical DNA delivery by ultrasonic cavitation. Som cell mole genet 2002; 27 (1-6):115-134

Minnaert M. XVI On musical air-bubbles and the sounds of running water. Lon Edinb Dublin Philos Mag J Sci 1933;16(104):235–248.

Mitragotri S. Healing sound: the use of ultrasound in drug delivery and other therapeutic applications. Nat Rev Drug Discov. 2005;4(3):255–60.

Moll S. A clinical perspective of venous thromboembolism. Arterioscler Thromb Vasc Biol. 2008;28(3):373–9.

NCRP. Exposure criteria for medical diagnostic ultrasound II: criteria based on all known mechanisms (NCRP report 140). National Council on Radiation Protection and Measurements, Maryland; 2002.

Nyborg WL. Acoustic streaming near a boundary. J Acoust Soc Am. 1958;30(4):329–39.

Nyborg WL. Ultrasonic microstreaming and related phenomena. Br J Cancer. 1982;45 Suppl V:156.

Ohl C, Ikink R. Shock-wave-induced jetting of micron-size bubbles. Phys Rev Lett. 2003;90 (21):214502.

Ohl CD, Kurz T, Geisler R, Lindau O, Lauterborn W. Bubble dynamics, shock waves and sonoluminescence. Philos Trans R Soc Lon Ser A: Math Phys Eng Sci. 1999;357(1751):269–94.

Pardridge WM. Brain drug targeting: the future of brain drug development. Cambridge, UK: Cambridge University Press; 2001.

Parsons JE, Cain CA, Abrams GD, Fowlkes JB. Pulsed cavitational ultrasound therapy for controlled tissue homogenization. Ultrasound Med Biol. 2006;32(1):115–29.

Pecha R, Gompf B. Microimplosions: cavitation collapse and shock wave emission on a nanosecond time scale. Phys Rev Lett. 2000;84(6):1328.

Philipp A, Lauterborn W. Cavitation erosion by single laser-produced bubbles. J Fluid Mech. 1998;361:75–116.

Plesset M, Chapman R. Collapse of an initially spherical vapour cavity in the neighbourhood of a solid boundary. J Fluid Mech. 1971;47:283–90.

Postema M, Van Wamel A, Lancée CT, De Jong N. Ultrasound-induced encapsulated microbubble phenomena. Ultrasound Med Biol. 2004;30(6):827–40.

Prentice P, Cuschieri A, Dholakia K, Prausnitz M, Campbell P. Membrane disruption by optically controlled microbubble cavitation. Nat Phys. 2005;1(2):107–10.

Prosperetti A. A new mechanism for sonoluminescence. J Acoust Soc Am 1997;101(4):2003–2007.

Qiao Y, Yin H, Li Z, Wan M. Cavitation distribution within large phantom vessel and mechanical damage formed on surrounding vessel wall. Ultrason Sonochem. 2013;20(6):1376–83.

Rassweiler JJ, Knoll T, Köhrmann K-U, McAteer JA, Lingeman JE, Cleveland RO, Bailey MR, Chaussy C. Shock wave technology and application: an update. Eur Urol. 2011;59(5):784–96.

Rayleigh L. VIII. On the pressure developed in a liquid during the collapse of a spherical cavity. Lond Edinb Dublin Philos Mag J Sci. 1917;34(200):94–8.

Roberts WW, Hall TL, Ives K, Wolf JS Jr, Fowlkes JB, Cain CA. Pulsed cavitational ultrasound: a noninvasive technology for controlled tissue ablation (histotripsy) in the rabbit kidney. J Urol. 2006;175(2):734–8.

Rooney JA. Hemolysis near an ultrasonically pulsating gas bubble. Science. 1970;169(3948):869–71.

Rosenschein U, Furman V, Kerner E, Fabian I, Bernheim J, Eshel Y. Ultrasound imaging-guided noninvasive ultrasound thrombolysis preclinical results. Circulation. 2000;102(2):238–45.

Sapozhnikov OA, Khokhlova VA, Bailey M, Williams JC Jr, McAteer JA, Cleveland RO, Crum LA. Effect of overpressure and pulse repetition frequency on cavitation in shock wave lithotripsy. J Acoust Soc Am. 2002;112(3 pt 1):1183–95.

Shaw GJ, Meunier JM, Huang S-L, Lindsell CJ, McPherson DD, Holland CK. Ultrasound-enhanced thrombolysis with tPA-loaded echogenic liposomes. Thromb Res. 2009;124(3):306–10.

Skyba DM, Price RJ, Linka AZ, Skalak TC, Kaul S. Direct in vivo visualization of intravascular destruction of microbubbles by ultrasound and its local effects on tissue. Circulation. 1998;98 (4):290–3.

Tachibana K, Tachibana S. Albumin microbubble echo-contrast material as an enhancer for ultrasound accelerated thrombolysis. Circulation. 1995;92(5):1148–50.

Tho P, Manasseh R, Ooi A. Cavitation microstreaming patterns in single and multiple bubble systems. J Fluid Mech. 2007;576:191–233.

Tran BC, Seo J, Hall TL, Xu Z, Ives K, Fowlkes JB, Cain CA. In vivo comparison of multiple pulse and CW strategies for microbubble-enhanced ultrasound therapy. In: IEEE symposium on ultrasonics. IEEE; 2003. pp. 909–912.

Tung Y-S, Vlachos F, Feshitan JA, Borden MA, Konofagou EE. The mechanism of interaction between focused ultrasound and microbubbles in blood-brain barrier opening in mice. J Acoust Soc Am. 2011;130(5):3059–67.

Vogel A, Lauterborn W, Timm R. Optical and acoustic investigations of the dynamics of laser-produced cavitation bubbles near a solid boundary. J Fluid Mech. 1989;206:299–338.

Vos HJ, Dollet B, Versluis M, De Jong N. Nonspherical shape oscillations of coated microbubbles in contact with a wall. Ultrasound Med Biol. 2011;37(6):935–48.

Vykhodtseva N, McDannold N, Hynynen K. Progress and problems in the application of focused ultrasound for blood–brain barrier disruption. Ultrasonics. 2008;48(4):279–96.

Westermark S, Wiksell H, Elmqvist H, Hultenby K, Berglund H. Effect of externally applied focused acoustic energy on clot disruption in vitro. Clin Sci. 1999;97:67–71.

Wu J. Theoretical study on shear stress generated by microstreaming surrounding contrast agents attached to living cells. Ultrasound Med Biol. 2002;28(1):125–9.

Wu J, Du G. Streaming generated by a bubble in an ultrasound field. J Acoust Soc Am. 1997;101 (4):1899–907.

Xu S, Zong Y, Feng Y, Liu R, Liu X, Hu Y, Han S, Wan M. Dependence of pulsed focused ultrasound induced thrombolysis on duty cycle and cavitation bubble size distribution. Ultrason Sonochem. 2015;22:160–6.

Xu Z, Fowlkes JB, Ludomirsky A, Cain CA. Investigation of intensity thresholds for ultrasound tissue erosion. Ultrasound Med Biol. 2005a;31(12):1673–82.

Xu Z, Fowlkes JB, Rothman ED, Levin AM, Cain CA. Controlled ultrasound tissue erosion: the role of dynamic interaction between insonation and microbubble activity. J Acoust Soc Am. 2005b;117(1):424–35.

Xu Z, Hall TL, Fowlkes JB, Cain CA. Optical and acoustic monitoring of bubble cloud dynamics at a tissue-fluid interface in ultrasound tissue erosion. J Acoust Soc Am. 2007;121(4):2421–30.

Xu Z, Ludomirsky A, Eun LY, Hall TL, Tran BC, Fowlkes JB, Cain CA. Controlled ultrasound tissue erosion. IEEE Trans Ultrason Ferroelectr Freq Control. 2004;51(6):726–36.

Yasui K, Tuziuti T, Lee J, Kozuka T, Towata A, Iida Y. The range of ambient radius for an active bubble in sonoluminescence and sonochemical reactions. J Chem Phys. 2008;128(18):184705.

Yoshizawa S, Ikeda T, Ito A, Ota R, Takagi S, Matsumoto Y. High intensity focused ultrasound lithotripsy with cavitating microbubbles. Med Biol Eng Compu. 2009;47(8):851–60.

Zhong W, Sit WH, Wan JM, Yu AC. Sonoporation induces apoptosis and cell cycle arrest in human promyelocytic leukemia cells. Ultrasound Med Biol. 2011;37(12):2149–59.

Zhou Y, Cocks FH, Preminger GM, Zhong P. Innovations in shock wave lithotripsy technology: updates in experimental studies. J Urol. 2004;172(5):1892–8.

Zhu S, Cocks FH, Preminger GM, Zhong P. The role of stress waves and cavitation in stone comminution in shock wave lithotripsy. Ultrasound Med Biol. 2002;28(5):661–71.

Chapter 6
Cavitation Control and Applications

Yangzi Qiao, Shanshan Xu, Nan Chang, Hui Yin and Mingxi Wan

Keywords Effect of acoustic parameters · Effect of ambient temperature and pressure · Constrained cavitation · Spatial and temporal control · Cavitation dosage control

6.1 Introduction

Cavitation is an important phenomenon that can occur both during diagnostic and therapeutic applications. It is not only an indicator of the transient high pressure and temperature induced by intense bubble collapse, but it also mediates mechanical and thermal effects. As far as mechanical effects, collapsing bubbles can emit shock waves and liquid jets, which are considered the main mechanisms that cause mechanical damage (Kodama and Tomita 2000). Regarding thermal effects, cavitation can enhance heating by confining the acoustic energy within the cavitation region, and it can also enhance viscous absorption at the bubble surface and generate high-frequency broadband harmonics, which have higher heat absorption coefficients. Therefore, cavitation has great potential to enhance therapeutic effi-

Y. Qiao · S. Xu · N. Chang · H. Yin · M. Wan (✉)
The Key Laboratory of Biomedical Information Engineering of Ministry of Education, Department of Biomedical Engineering, School of Life Science and Technology, Xi'an Jiaotong University, Xi'an 710049, People's Republic of China
e-mail: mxwan@mail.xjtu.edu.cn

Y. Qiao
e-mail: qiaoyangzi@gmail.com

S. Xu
e-mail: xushanshan@mail.xjtu.edu.cn

N. Chang
e-mail: chana0131@163.com

H. Yin
e-mail: yhlovegxm@gmail.com

M. Wan et al. (eds.), *Cavitation in Biomedicine*,
DOI 10.1007/978-94-017-7255-6_6

ciency. However, uncontrolled cavitation may dramatically alter acoustic energy, heat distributions, and cause migration of bubble cloud, all of which result in the generation of unpredictable lesions and poorly reproducible therapies. Thus, we need a better understanding of how to control various aspects of cavitation.

Recent studies on controlling cavitation can be categorized as either basic research mainly focused on the parameters that influence cavitation as well as its physical mechanisms, or they can be categorized as controlled application studies that focus on histotripsy, cavitation-controlled ablation, drug delivery, etc. For these approaches, the acoustic parameters (input pressure, pulse length, etc.) are changed according to different treatment needs. However, between these two research areas, there is an important gap: Very few studies investigate how to spatially and temporally control bubble distribution, which is the key factor that influences energy efficiency and determines therapeutic results.

In this chapter, we introduce findings from both types of studies—basic research and applications—and connect these findings with key considerations for controlling the spatial and temporal evolution of bubble distribution. Sections 6.2–6.4 discuss parameters that may influence cavitation: temperature, pressure, and frequency. For example, within media, the surrounding temperature and ambient pressure will not only affect the vaporization process that is closely related to the inception of cavitation nuclei, but they also influence the intensity of cavitation activities. Also, with respect to the transmission of ultrasound, the acoustic pressure and frequency heavily influence cavitation. Optimizing these acoustic parameters has thus been the focus of attempts to control cavitation. Furthermore, two special cases aimed at capitalizing on cavitation should also be examined. First, phase-shift nanoemulsions (PSNEs) which are sensitive to both temperature and acoustic pressure, and their response to environmental changes should be comprehensively studied. Second, systems that use a combination of multiple waves with different amplitudes have also been investigated to enhance cavitation effect (Takagi et al. 2010). We discuss these two cases in Sects. 6.2 and 6.4, respectively.

Section 6.5 mainly focuses on the spatial and temporal control of bubble distribution. Employing standing waves is the most common method used to trap bubbles. Early in 1976, Miller (1977) observed stable bubble arrays in a 1-MHz standing-wave field. To balance the radiation force, Yin et al. (2014) used an acoustic reflector to trap the bubbles where they were generated and avoid energy distribution distortion. In another study, Li et al. (2013) demonstrated that they could control the bubble cloud by changing the geometry of the transducer to be spherical. High-speed photography is a powerful tool to visualize the spatial distribution and dynamics of cavitation. However, because not all the bubbles contribute to intense cavitation, methods beyond high-speed photography have been developed. For example, considerable research has focused on sonoluminescence (SL) and sonochemiluminescence (SCL), which are direct indictors of inertial cavitation.

Finally, in Sects. 6.6–6.8 we discuss applications that are closely related to cavitation control, including histotripsy, cavitation-controlled ablation, and drug delivery. Appropriate tuning of acoustic parameters can break down and remove soft tissue which is termed as histotripsy. This technique can be divided, depending on the mechanism of action, into three different types: cavitation-based histotripsy, boiling histotripsy, and microtripsy beyond the intrinsic cavitation threshold. Another application of cavitation, ablation, calls for longer ultrasound exposure, which can lead to heat accumulation. However, cavitation is also very important during ablation, so two types of cavitation control systems have been constructed to enhance heating and avoid bubble distortion. First, the open-loop system is always based on the preliminary experiments, which elucidate the relationship between passive cavitation detection (PCD) signals and temperature, and this system assumes that controlled cavitation is closely related to sustained enhanced heating. Second, the closed-loop system contains real-time feedback circuits, which can adjust the treatment parameters according to the cavitation signals processed after collection. Hockham et al. (2010) have given out the schematic of a closed cavitation control system and verified its effectiveness.

These promising applications have sparked much research into how to take advantages of cavitation. However, only a few publications have systematically and thoroughly covered ways to control cavitation, which is of considerable importance and thus the subject of this chapter. While this research is still in its infancy, we are trying to thoroughly investigate these cavitation control methods and to give out a hint on how best to apply them to various applications.

6.2 Effect of Temperature on Cavitation

6.2.1 Effect of the Liquid Temperature

Liquid temperature is one of the fundamental thermodynamic parameters that determines the rate and nature of cavitation bubble collapse (Liu et al. 2013). Changes in the liquid temperature can affect the liquid's properties, such as viscosity, vapor pressure, surface tension, and density, and all of these parameters can in turn affect cavitation bubble dynamics. In addition, temperature variation can also change the dissolved gas content of the fluid, which can also influence cavitation (Auret et al. 1993). Hence, here we discuss various studies that determined how temperature affects cavitation.

Early in 1993, Auret et al. (1993) reviewed many previous works on how the liquid temperature affects cavitation erosion. From the studies they reviewed, they concluded that up to a maximum temperature, the erosion rate increases slightly with increasing temperature, and then it decreases considerably. The peak erosion rate generally occurred about midway between the freezing and the boiling points of the cavitating liquid. Then in 1997, Ahmed studied the influence of temperature on bubble pressure and cavitation erosion (Ahmed 1998). Later, by detecting the temperatures of various places in a cavitating flow field, Huang and Zhuang (2008) verified that cavitation is an endothermic process. Recently, Hattori et al. also reported how temperature affected cavitation erosion of 9Cr steel in liquid metal, and they found that the cavitation erosion rate increased with increasing temperature and reached a peak at temperature of about 70 °C (shown in Fig. 6.1, where the mean depth of erosion rate (MDER) equals the mass loss divided by the material density) (Hattori et al. 2013).

In another study, Liu et al. (2011) explored how temperature influences the oscillation properties of a laser-induced cavitation bubble near a rigid boundary in water. They recorded variations in the bubble radius with time at different ambient temperatures and compared them to the simulation result obtained from a modified Rayleigh–Plesset equation (Fig. 6.2). The results show that temperature heavily influences both the maximum bubble radius and the bubble lifetime.

In another report by the same group (Liu et al. 2013), the maximum and minimum values of the bubble radius and the lifetime of the bubble were reported at different temperatures, while the bubble energy was calculated according to Rayleigh's theory (Rayleigh 1917) for spherical cavitation bubble collapse (Table 6.1). From Table 6.1, it can be concluded that the lifetime and the energy of the cavitation bubble increase with the liquid temperature, which means that the bubble will collapse more violently in liquid at higher temperatures. As such, the impact force induced by the cavitation bubble collapse was stronger in liquids of higher temperatures, but this force decreased after reaching a peak at about 50 °C (Fig. 6.3).

The authors explained this phenomenon by treating the gas content in water as a cushion. When the water temperature increased, the cushion effect was reduced because there was less gas in the ambient media, thus increasing the bubble-induced impact force. Further increasing the temperature, the increasing saturation vapor

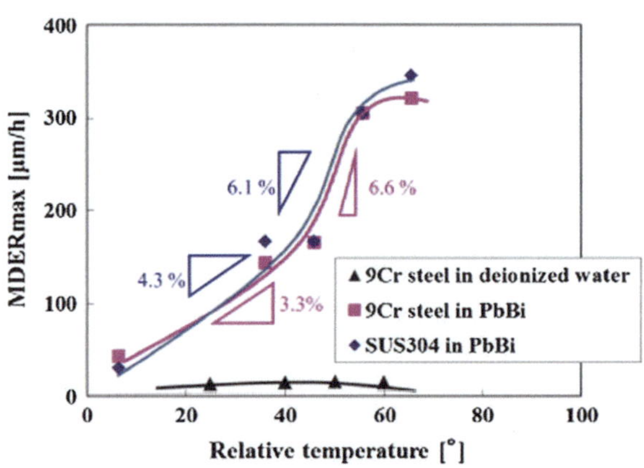

Fig. 6.1 Relationship between temperature and the maximum mean depth erosion rate (MDERmax). MDER was defined as the mass loss divided by the material density, the eroded area, and the exposure time interval. Reproduced with permission from Hattori et al. (2013). © 2013 Elsevier B.V.

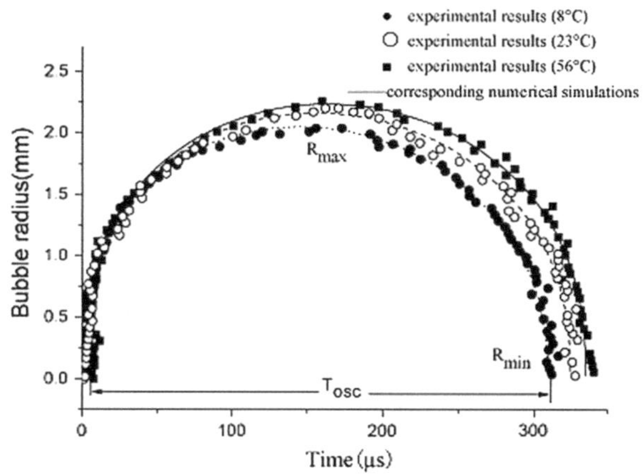

Fig. 6.2 Time evolution of the bubble radius $R(t)$ in distilled water at different ambient temperatures. Reproduced with permission from Liu et al. (2011). © 2010 Elsevier GmbH

pressure inside the bubble became a more dominant factor and reduced the impact force and thus the erosion rate (Liu et al. 2013).

Investigations of how temperature affects SL in water have also been conducted, one specifically by Hiller et al. (1992). In this study, the authors speculated that lower water temperatures would cause the water viscosity to increase, which would thereby increase the bubble's stability, thus permitting the forcing acoustic pressure to become higher before bubble collapse occurred. A higher forcing pressure can ensure a smaller minimum radius, a higher internal temperature, and finally stronger emission of light (Fig. 6.4).

In another study on SL, Germano et al. (2010) studied the temperature dependence (in water) of a single bubble's SL threshold, which was defined as the minimum acoustic pressure needed for light emission (Fig. 6.5). As seen here, the measured values suggest an approximately linear dependence of the threshold up to

Table 6.1 Bubble parameters in distilled water at different ambient temperatures (Liu et al. 2011)

Temperature (°C)	R_{max} (mm)	R_{min} (mm)	T (μs)	E (mJ)
0	2.02	0.063	302.0	3.45
8	2.06	0.061	306.6	3.66
15	2.11	0.050	315.4	3.93
23	2.14	0.030	321.0	4.10
28	2.17	0.041	325.0	4.28
33	2.19	0.030	327.8	4.40
39	2.20	0.040	330.0	4.46
45	2.21	0.050	336.0	4.52
51	2.23	0.045	338.8	4.64
56	2.24	0.046	339.2	4.71
63	2.26	0.040	340.6	4.83
69	2.27	0.030	341.0	4.90

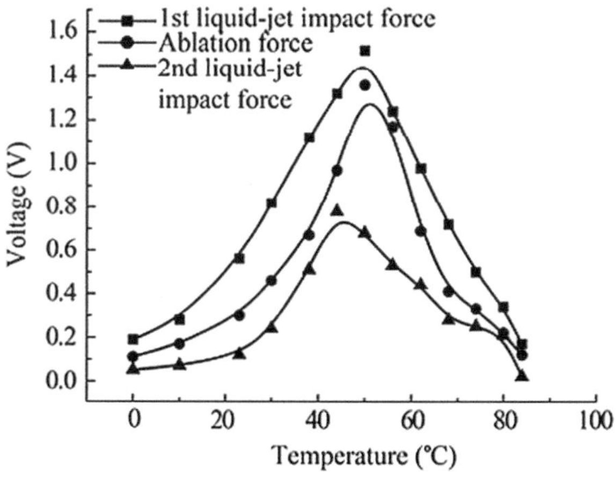

Fig. 6.3 Relationship between the liquid jet impact force and ambient temperature. Reproduced with permission from Liu et al. (2013). © Tianjin University of Technology

about 14 °C and an almost independent behavior above 20 °C. The abrupt changes in the threshold slope at given values of water temperature suggest different mechanisms of interaction between physical parameters, and these mechanisms depend on temperature and the onset of SL.

The temperature dependence of single-bubble sonoluminescence (SBSL) has also been examined in sulfuric acid, as reported by Moshaii et al. (2011). In this study, maximum SL emissions were acquired by calculating the phase diagrams of a single SL bubble at different ambient temperatures. An opposite trend was obtained in sulfuric acid compared to what was seen in water (Fig. 6.6). This difference originates from different instability mechanisms that determine the ultimate phase parameters of SL in water and sulfuric acid. In water, the phase parameters are determined by shape instability of the cavitation bubble due to low viscosity; in sulfuric acid, the liquid viscosity is high and phase parameters are mainly restricted by positional instability.

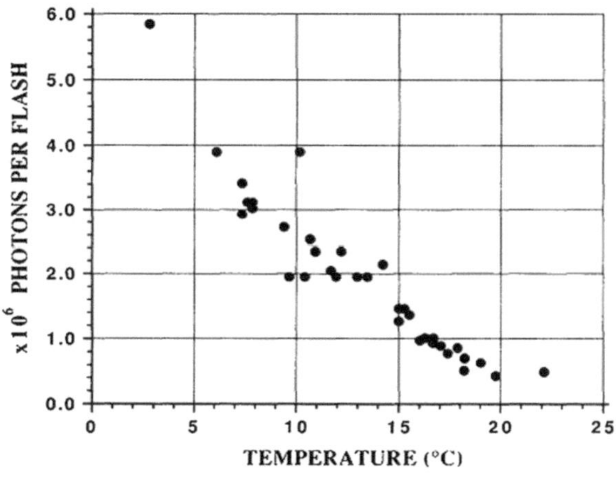

Fig. 6.4 Number of photons emitted per flash of SL as a function of temperature. Reproduced with permission from Hiller et al. (1992). © 1992 The American Physical Society

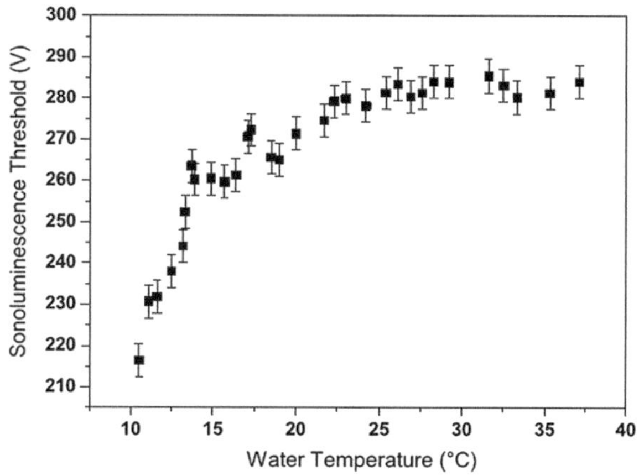

Fig. 6.5 Single-bubble sonoluminescence (SBSL) threshold versus water temperature. Reproduced with permission from Germano et al. (2010). © 2009 Elsevier B.V.

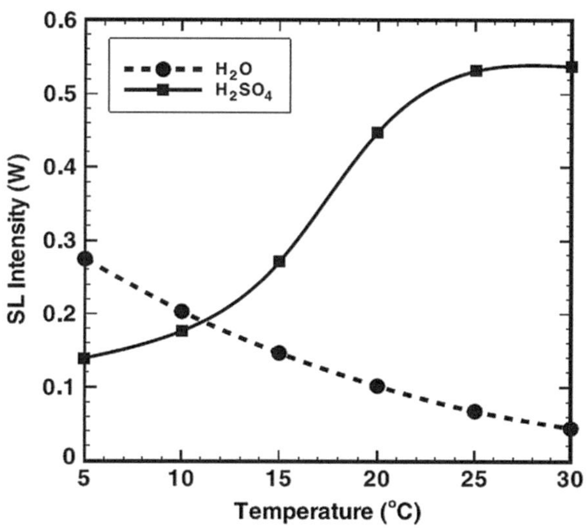

Fig. 6.6 Dependence of the maximum SL intensity on the temperature of water and of sulfuric acid (85 %). Reproduced with permission from Moshaii et al. (2011). © 2011 American Physical Society

6.2.2 *Temperature Dependence of PSNE-Induced Cavitation*

6.2.2.1 Effect of Ambient Temperature on PSNE-Induced Cavitation

PSNEs are relatively new ultrasound contrast agents, having potential applications in vessel occlusion, drug delivery, and high-intensity focused ultrasound (HIFU) thermal ablation (Kawabata et al. 2008; Fabiilli et al. 2010; Kripfgans et al. 2002). They have a small (nanoscale) initial diameter and a long lifetime in vivo. At the desired location, PSNEs can undergo a phase-shift process and change into gas

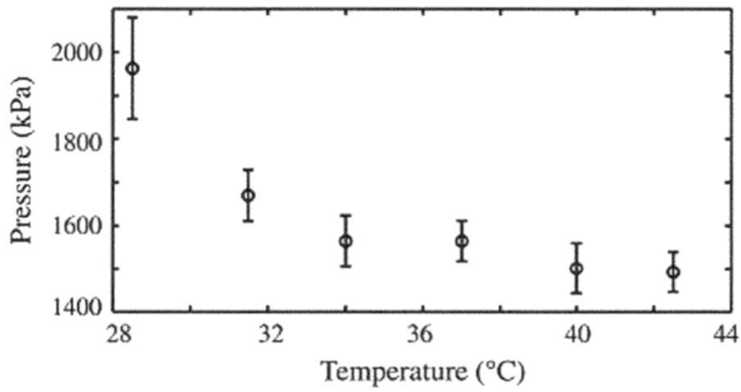

Fig. 6.7 Vaporization pressure threshold measurements as a function of droplet sample temperature. Each point represents the mean value ± the standard error from 10 trials. Reproduced with permission from Reznik et al. (2011). © 2011 World Federation for Ultrasound in Medicine & Biology

bubbles under intense acoustic exposure, a process called acoustic droplet vaporization (ADV). Because they are stabilized by their shell materials, PSNEs are not easily heated to vaporization, and they are prone to staying in a superheated state. The phase transition occurs when the surrounding pressure decreases below the saturated vapor pressure of the nanodroplets. Hence, this change in local pressure drives the phase transformation (Lin and Pitt 2013). But, changes in temperature can also influence the droplet's degree of superheating, which is related to the threshold of the local pressure for phase transition.

Reznik et al. (2011) investigated the relationship between droplet temperature and the local pressure threshold (Fig. 6.7). From Fig. 6.7, we can conclude that an increase in droplet temperature causes the pressure threshold for phase transition to drop. This happens because increasing the temperature causes an increase in internal energy of the droplet liquid, which leads to an increase in the vapor pressure of the perfluorocarbons (PFCs) that mostly often used core materials of PSNEs. As a result, a lower negative pressure is required to induce an energetically favorable phase transition.

A similar trend was also reported by Zhang and Porter (2010). They found an inversely proportional relationship between the ADV threshold and the ambient temperature, as shown in Fig. 6.8.

6.2.2.2 Effect of Different Phase-Shift Temperatures on PSNE-Induced Cavitation

Recently, Pitt et al. (2014) studied the phase transition process of PSNEs induced by ultrasound. The perfluorocarbons they used were PFC_5 and PFC_6, whose phase change temperatures are 29 and 56 °C, respectively. Figure 6.9 shows the simplified model they used, whereby P_∞ is the sum of the atmospheric pressure, hydrostatic pressure, and acoustic pressure. P_g is the total pressure inside the gas phase, including the vapor pressure of the perfluorocarbon liquid plus the vapor pressure of

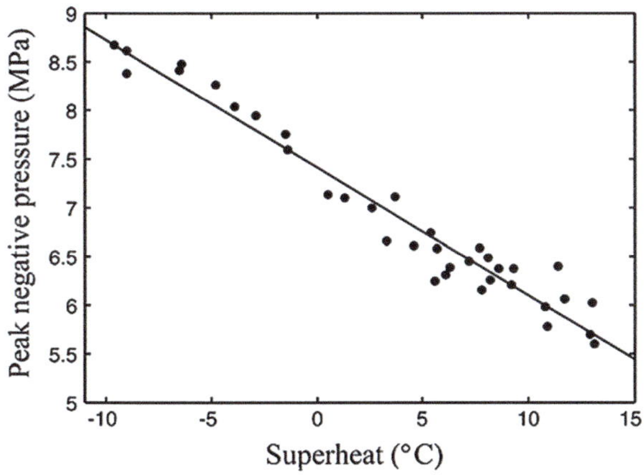

Fig. 6.8 ADV threshold as a function of the degree of superheating. The data were measured in the phantom with a dodecafluoropentane volume fraction of 0.30 %. Reproduced with permission from Zhang and Porter (2010). © 2010 World Federation for Ultrasound in Medicine & Biology

Fig. 6.9 Schematic diagram of the emulsion droplet (*upper left*) and gas phase surrounding the droplet (*upper right*). Reproduced with permission from Pitt et al. (2014). © 2013 Elsevier B.V.

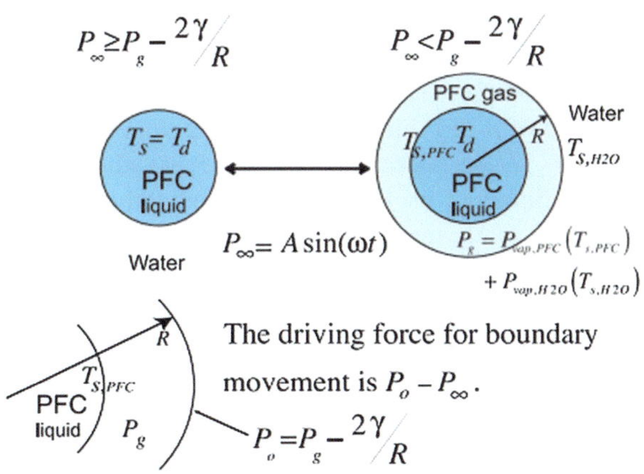

the water. P_0 is the gas-phase pressure minus the Laplace pressure $2\gamma/R(t)$. The modified bubble dynamics equation is as follows:

$$\frac{Pg(T,t) - P_\infty(t)}{\rho_L} = R\frac{d^2R}{dt^2} + \frac{3}{2}\left(\frac{dR}{dt}\right)^2 + \frac{4\upsilon_L}{R}\frac{dR}{dt} + \frac{2\gamma}{R\rho_L} \tag{6.1}$$

The vapor pressure generated by different PFCs can be obtained from the following equation:

$$P_{\mathrm{vap}}(T_s) = \exp\left(A + B/T_s + C\ln T_s + DT_s^E\right) \tag{6.2}$$

Table 6.2 Vapor pressure coefficients of PFC_5 and PFC_6 (Pitt et al. 2014)

Compound	*A*	*B*	*C*	*D*	*E*
PFC_5	72.801	5.5390×10^3	7.5271	6.1648×10^{-17}	6.0000
PFC_6	63.705	5.6667×10^3	6.0445	2.7396×10^{-17}	6.0000

where *A*, *B*, *C*, *D,* and *E* are constants. The coefficients for PFC_5 and PFC_6 are listed in Table 6.2.

Compared to PFC_6, the vapor pressure generated by PFC_5 is higher, forming larger bubble, and this would lead to a higher collapse velocity and more intense cavitation activities.

Phase shifting was also investigated by Kawabata et al. (2005), who explored the feasibility of control the induction of phase shifting by changing the ratio of 2*H*, 3*H*-perfluoropentane (boiling point: 53.5 °C) to perfluoropentane (*n*-PFP, boiling point: 29.3 °C). The acoustic intensity threshold for the phase shift of the emulsions was determined using the B-mode images before and after exposure to pulsed ultrasound. They studied the brightness changes in the echographic image as a function of *n*-PFP ratio to the total volume of perfluorocarbon (Fig. 6.10, left) and the concentration of *n*-PFP in the emulsion when the ratio of the volume of *n*-PFP was 100 % (Fig. 6.10, right).

In this study, the authors found that the acoustic intensity threshold decreased as the ratio of *n*-PFP increased, until the emulsion contained equal volumes of *n*-PFP and 2*H*, 3*H*-perfluoropentane. However, when the ratio of the *n*-PFP was 100 %, the brightness of the echography image did not change with the concentration of *n*-PFP. It is likely that the interaction of *n*-PFP and 2*H*, 3*H*-perfluoropentane has an impact on the threshold. The authors also gave an equation for the energy (E_H)

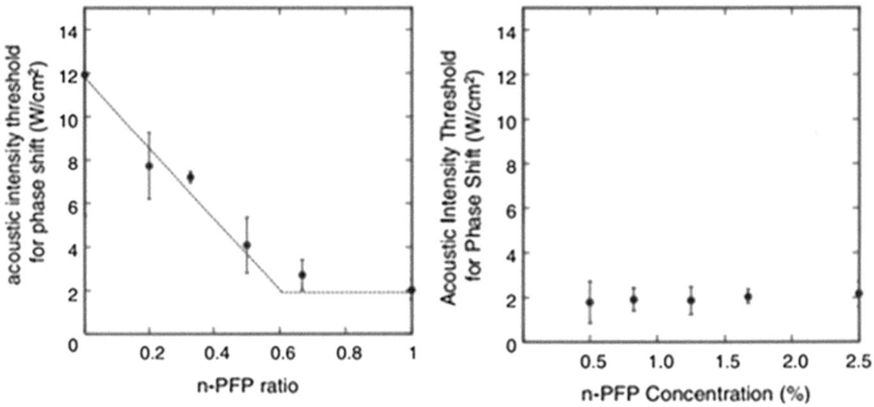

Fig. 6.10 The acoustic intensity threshold for a phase shift depends on the ratio (vol./vol.) of *n*-PFP to total perfluorocarbon (*left*) and on the concentration of *n*-PFP in the emulsion (*right*). Reproduced with permission from Kawabata et al. (2005). © 2005 The Japan Society of Applied Physics

needed to change the phase of $2H$, $3H$-perfluoropentane: $E_H = K_0 * c_H - K_1 * c_L$ $(E_H > 0)$, whereby K_0 and K_1 are constants and c_H and c_L are the concentrations of $2H$, $3H$-perfluoropentane and n-PFP, respectively.

However, while this particular study did not involve cavitation, many other studies have focused on cavitation of PSNEs, which can enhance the mechanical and thermal effects of HIFU. For example, we have explored (unpublished) cavitation intensity, spatial distribution, and phase transition induced by PSNEs with different shifting temperatures (PFC_5 and PFC_6). In our studies, the working frequency of HIFU was 1.2 MHz, and the pulse length and the duty cycle were 10 μs and 1:20, respectively. The phase transition process was also detected by B-mode imaging. PFC_5 can be evaporated into bubbles with just a single pulse (10 μs) at an electric input power of 1 W, which equals a spatially averaged intensity (I_{SAL}) of about 22 W/cm^2. The acquired image is shown in Fig. 6.11. However, for PFC_6, the B-mode image showed no evident changes after a single-pulse sonication. After increasing the pulse number, we found that some bubbles could be detected when the pulse number was 20. When the pulse number was 50, an apparent phase transition was captured (Fig. 6.12). As seen in this figure, the size of bubble cloud was larger than that formed with only one pulse (Fig. 6.11).

In addition to B-mode images, SL and SCL methods were also used to investigate PSNE-induced cavitation activities and their spatial distribution during pulsed HIFU exposure (Figs. 6.13 and 6.14). SL and SCL images were captured by an EMCCD camera, and, because of the time resolution limitation, the total exposure time of EMCCD was 20 s, equaling 2 s of HIFU exposure in total.

Determining the threshold of cavitation remains a key area of research for controlling cavitation activities. In these figures, we can see that the SL threshold of PFC_6 was 5 W, while that of PFC_5 was <4 W. The SL distribution did not appear to be different in these two groups. When luminal was added for SCL, the brightness of the light was enhanced, as shown in Fig. 6.14. The SCL threshold of PFC_5 was 2 W, while that of PFC_6 was 3 W. Comparing these two figures, because SL originates only from the extreme conditions within the core of the cavitation bubble,

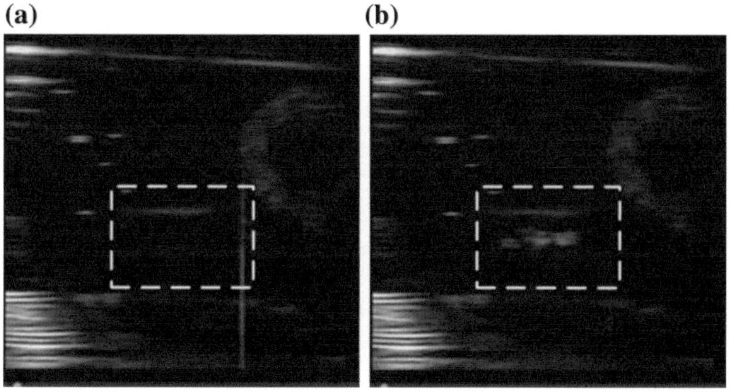

Fig. 6.11 B-mode images of PFC_5 **a** before and **b** after HIFU sonication (single pulse)

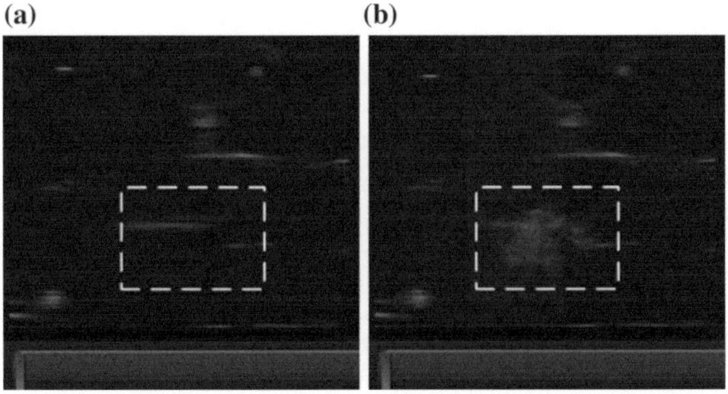

Fig. 6.12 B-mode images of PFC_6 **a** before and **b** after HIFU sonication (50 pulses)

it only occurs in a very small region. However, SCL is produced by the secondary chemical reaction between luminal molecules and $HO^{\cdot}$ radicals, and its active region and time period are both considerably broader. From Fig. 6.14, we can see that the SCL intensity and light-emitting region grew with increasing power for both PFC_5 and PFC_6 groups. However, the PFC_5 group showed a more evident

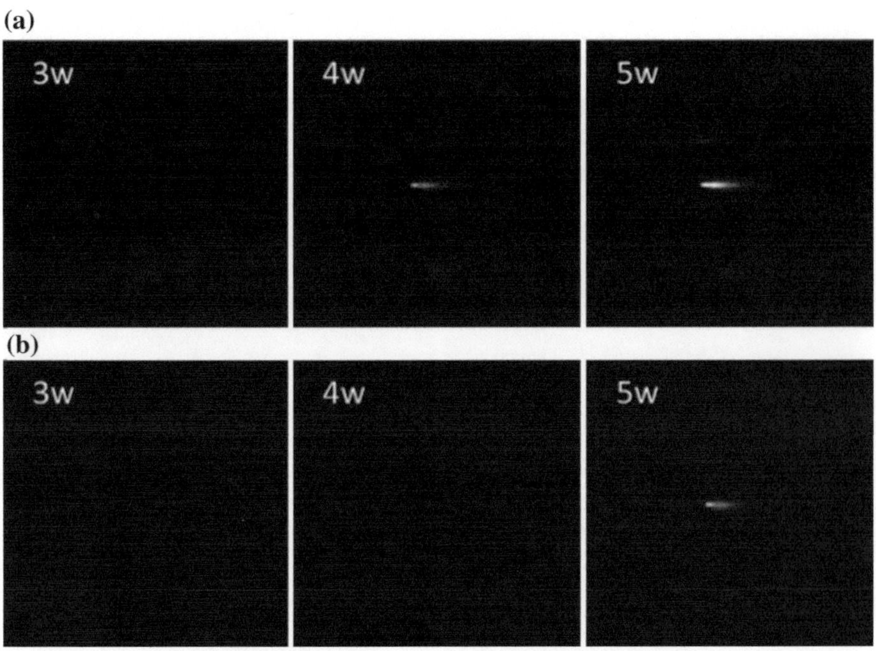

Fig. 6.13 The sonoluminescence (SL) images of PFC_5 (**a**) and PFC_6 (**b**) at different input powers: 3, 4, and 5 W (from left to right)

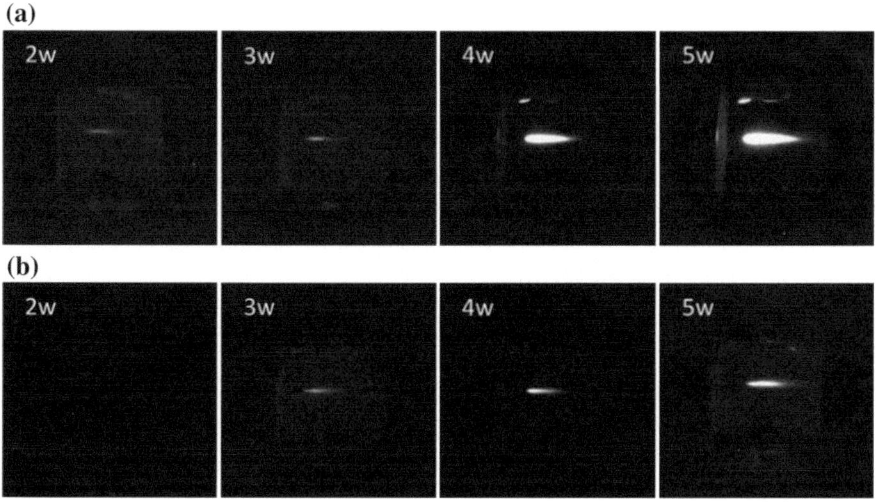

Fig. 6.14 The sonochemiluminescence (SCL) images of PFC_5 (**a**) and PFC_6 (**b**) at different input powers

transformation, especially when the input power increased from 3 to 4 W. Correspondingly, Singh et al. (2012) presented an illustration of the local pressure variation as an ultrasound wave propagates through the medium and of the vaporization threshold of PFC_5 and PFC_6 (Fig. 6.15). Here, we can see that PFC_5

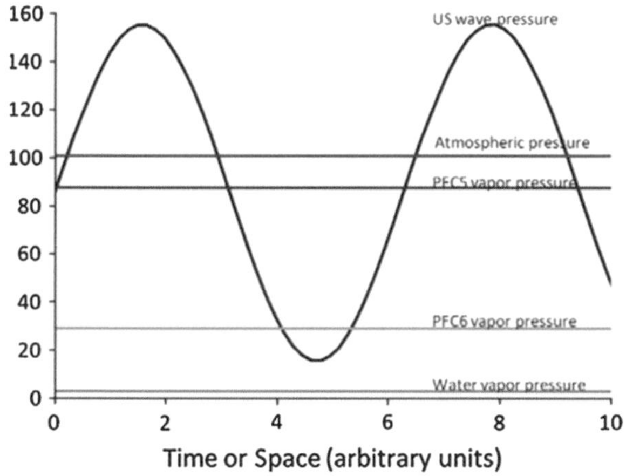

Fig. 6.15 The local pressure (kPa) in a medium oscillates as an ultrasound wave propagates through the medium. The *horizontal lines* represent the vapor pressures of PFC_6, PFC_5, and water at 25 °C. When the local pressure falls below the vapor pressure of a liquid, vaporization can occur. Reproduced with permission from Singh et al. (2012). © 2012 Elsevier B.V.

may commence its phase transition much earlier and for a longer time during the acoustic cycle.

According to the B-mode results, the input powers used in these SL and SCL studies are sufficient to evaporate both PFC_5 and PFC_6. However, the degree to which these bubbles—filled with different core materials—were able to enhance the cavitation intensity and affect the distribution did vary, especially according to the SCL results. This may be closely related to the physical and chemical properties of the core materials, such as their boiling temperatures, superheated states, and different energy conservation efficiencies.

6.3 Effect of Pressure on Cavitation

6.3.1 *Effect of Static Pressure*

Cavitation can be roughly defined as a process of nucleation in a liquid when the ambient pressure falls below the vapor pressure (P_v). It should be distinguished from another similar process: boiling, which happens when the temperature is raised above the saturated vapor/liquid temperature (Brennen 2013). These two processes in water can be clearly depicted by Fig. 6.16. The curve from the triple point T_r to the critical point C separates the liquid and vapor domains.

However, it is worth noting that the curve $P_v(T)$ shown in Fig. 6.16 is not an absolute boundary separating liquid from vapor. Deviations from this curve can exist when a rapid phase change occurs. Even in almost static conditions, a phase change may not necessarily occur at a pressure lower than P_v. For example, if we consider the so-called Andrews isotherms in the p-ϑ diagram, where the specific volume is $\vartheta = 1/\rho$ and ρ is the density (Fig. 6.17), such curves can be approximated in the liquid and vapor domains by the van der Waals equation of state. Along the path from A to M in this diagram, the transformation from liquid to vapor can be avoided. Along the AM path, the liquid is in metastable equilibrium and can even withstand negative absolute pressures, i.e., tensions, without any phase transformation. Hence, when the local absolute pressure drops below the vapor pressure, this does not necessarily ensure that the inception of cavitation will occur. However, one related cavitation parameter is the static delay, which is defined as the difference between the actual pressure and the vapor pressure when cavitation occurs. In some case, there is also a dynamic delay, which is due to inertial phenomena associated with the time necessary for vapor cavities to be observable (Franc and Michel 2006).

In another study, Caupin and Herbert (2006) regarded nucleation as an energy accumulation process, where nucleation is proportional to the bubble's volume and is damped by the energy cost of the liquid–vapor interface. They also calculated the cavitation pressure threshold, which is based on vapor pressure. Thus, for a bubble of radius R at pressure P, which is lower than the vapor pressure P_v, the damping energy for bubble growth can be written as follows:

$$E(R) = \frac{4}{3}\pi R^3(P - P_v) + 4\pi R^2\sigma \tag{6.3}$$

where P_v represents the saturated vapor pressure and σ stands for the liquid–vapor surface tension. Hence, for a bubble of critical radius $R_C = 2\sigma/(P_v - P)$, the damping energy becomes $E_b = (16\pi\sigma^3)/\left[3(P_v - P)^2\right]$, which is the energy barrier for a bubble's spontaneous growth. The rate of nucleation triggered by thermal fluctuations is $\Gamma = \Gamma_0 \exp[-E_b/(k_B T)]$, where k_B is the Boltzmann constant, T is the absolute temperature and Γ_0 is the product of the number density of nucleation sites and an attempt frequency for nucleation. An approximation for this can be written as $\Gamma_0 \approx \left(\frac{4}{3}\pi R_C^3\right)^{-1}\frac{k_B T}{h}$, where h is Planck's constant. For an experiment conducted in a volume V during a time τ, the cavitation inception probability is $\sum = 1 - \exp(-\Gamma V \tau)$. When $\sum = 1/2$, the corresponding pressure can be defined as the cavitation threshold P_{cav}, which can be written as follows:

$$P_{cav} = P_v - \left(\frac{16\pi\sigma^3}{3k_B T}\frac{1}{\ln(\Gamma_0 V\tau/\ln 2)}\right)^{1/2} \tag{6.4}$$

This classical nucleation theory assumes that the thickness of the bubble wall can be neglected compared to the bubble radius, making it a crude approximation when the critical radius becomes on the order of the liquid–vapor interfacial width. Furthermore, this theory possesses a spinodal limit, which means that at a spinodal

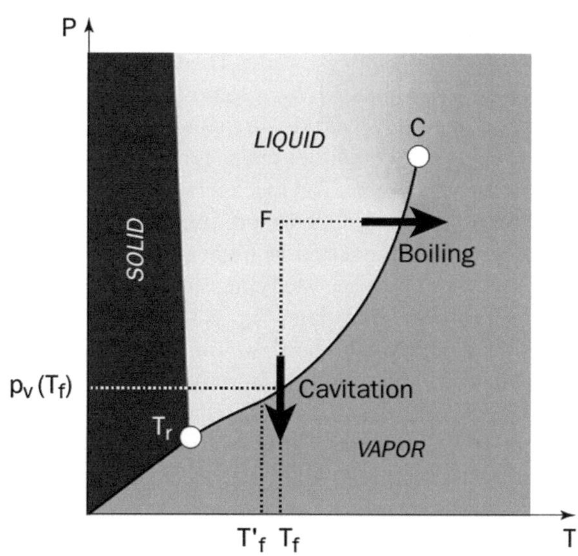

Fig. 6.16 Phase diagram. Reproduced with permission from Franc and Michel (2006). © 2004 Kluwer Academic Publishers

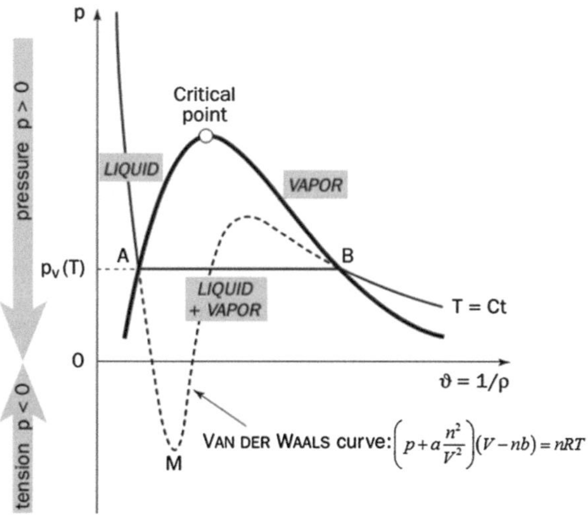

Fig. 6.17 Andrews-isotherms. Reproduced with permission from Franc and Michel (2006). © 2004 Kluwer Academic Publishers

pressure, it ceases to be metastable and becomes macroscopically unstable. However, regardless of this limitation, nucleation theory as described here is a generic feature of all liquids.

6.3.2 *Effect of Overpressure on Cavitation Suppression*

According to the general definition above, cavitation occurs when the local pressure drops below the vapor pressure. Therefore, when the local static pressure rises, this should be accompanied by suppression of cavitation. Khokhlova et al. (2006) described one such overpressure situation. They examined cavitation during lesion formation in a gel phantom under pressure conditions of atmospheric pressure plus 100 bars overpressure. The mechanism by which cavitation enhanced the heating of the tissue was through the absorption of waves scattered or emitted by the bubbles, which are often converted to higher and more readily absorbed frequencies (Holt and Roy 2001). In their study, an aluminum pressure chamber was used to create the overpressure condition (Fig. 6.18), and a CCD camera placed in front of the pressure chamber was used to film lesion formation. The acoustic power was adjusted to 32 W with 100 % duty cycle to eliminate the nonlinear propagation effect (Fig. 6.19b), which the authors showed was more effective at enhancing heating and accelerating lesion formation (Khokhlova et al. 2006). The level of overpressure was set to 100 bars, exceeding the maximum negative pressure of the HIFU waveform.

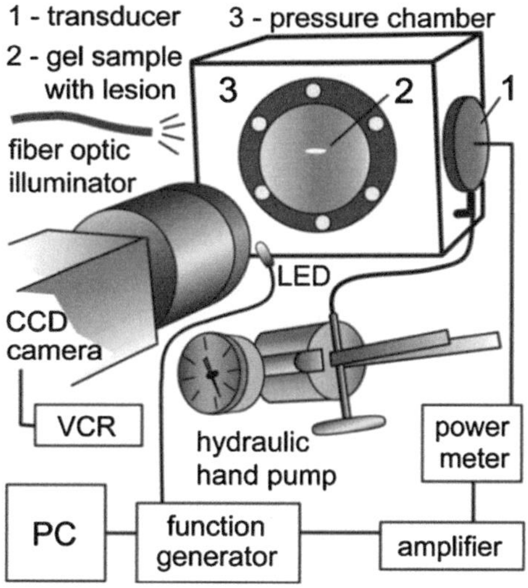

Fig. 6.18 Experimental arrangement for video imaging of HIFU-induced single lesions in gel under elevated static pressure. Reproduced with permission from Khokhlova et al. (2006). © 2006 Acoustical Society of America

The volume of the lesion formed in the degassed gel phantom under atmospheric pressure was much bigger than that which formed in the overpressure condition (Fig. 6.19a, circled by the red dotted line). Also, the lesion inception time (15 s) under the atmospheric pressure condition was earlier than the inception time in the overpressure condition (24 s), indicating that cavitation was effectively suppressed by the high local static pressure.

In another study, the overpressure effect on cavitation bubbles was investigated in a shock wave lithotripsy field and recorded by a high-speed digital camera at a rate of 40,500 frames per second (Sapozhnikov et al. 2002). A polyethylene terephthalate (PETE) bottle filled with non-degassed water was positioned in the lithotripter field with a charging potential of 18 kV corresponding to a 350 ± 80 bar

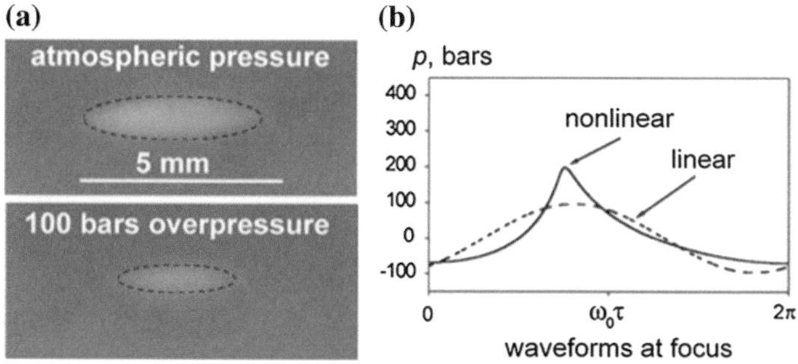

Fig. 6.19 **a** Final lesions formed in degassed gel under standard atmospheric static pressure (1 bar) and under overpressure (100 bars); **b** Corresponding waveforms at the focus, with and without accounting for the effects of nonlinear propagation. Reproduced with permission from Khokhlova et al. (2006). © 2006 Acoustical Society of America

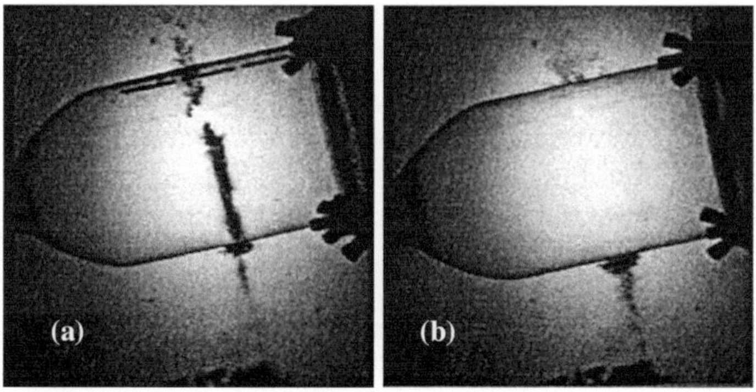

Fig. 6.20 High-speed images in and around a polyethylene terephthalate bottle. Pressure in the bottle was **a** 1 bar and **b** 4 bars. Reproduced with permission from Sapozhnikov et al. (2002). © 2002 Acoustical Society of America

pressure (Fig. 6.20). The pressure outside the bottle was 1 bar, while inside the bottle the local static pressures were either 1 bar or 4 bars. From the figure, we can see that for the pressurized bottle under high static pressure, cavitation bubbles were dramatically suppressed.

In the same study, the passive cavitation detection (PCD) signals were simultaneously recorded with video images under static pressures of either 1 bar or 1.5 bars (Fig. 6.21). According to the PCD signal at the static pressure of 1 bar, there

Fig. 6.21 Simultaneous high-speed camera images and passive cavitation detection (PCD) for $p_0 = 1$ bar and 1.5 bars. Reproduced with permission from Sapozhnikov et al. (2002). © 2002 Acoustical Society of America

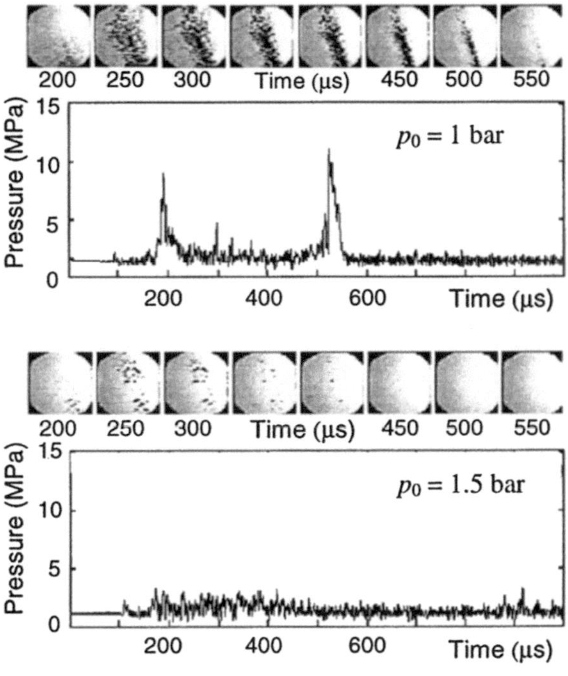

was a strong spike at 180 μs corresponding to the time when the cavitation bubble was first seen, and there was another one ~350 μs later when the bubbles collapsed. While at a static pressure of 1.5 bars, only very sparse and short-lived bubbles appeared, and very little signal was detected by the PCD.

6.3.2.1 Effect of Acoustic Pressure

In their investigations on acoustic cavitation bubbles' behavior in water, Huang et al. (2006) designed an experiment setup to study a single bubble's dynamic under different driving sound pressures (Fig. 6.22). A pair of piezoelectric transducers was tightly attached at two opposite sites on the equator of a spherical 100 mL flask to produce a standing wave. A single bubble can be trapped in the antinode of a standing wave, and the authors imaged the oscillation process using a long-distance microscope (Hirox KH-3000) through a quartz window opened on the surface of the flask. The illuminating system, used to detect and record the SL phenomenon, consisted of four parts: a 532-nm Nd YAG laser generator, an acousto-optic modulator (AA sa AA. MT. 350), a digital delay/pulse generator (SRS DG535), and a photomultiplier tube (PMT, Hamamtsu R212).

In the experiment conducted by Herbert et al. (2006) to study the cavitation pressure in ultrapure water, a hemispherical shaped piezoelectric transducer was used to reach higher negative pressures and to keep cavitation within the middle of bulk liquid, to avoid wall-induced heterogeneous cavitation. The transducer was held in a stainless steel cell and was controlled by the pressure gauge and the bellow. The cell was able to produce high pressures, and it was calibrated by the static pressure method. The cell was immersed in a thermostated bath (operated between −10 and 80 °C) (Fig. 6.23). The "echo method" (Imre et al. 2002) was used because it is simple to implement and offers a wide range of applicability to detect cavitation bubbles. These methods also required only one transducer for producing and detecting the bubbles.

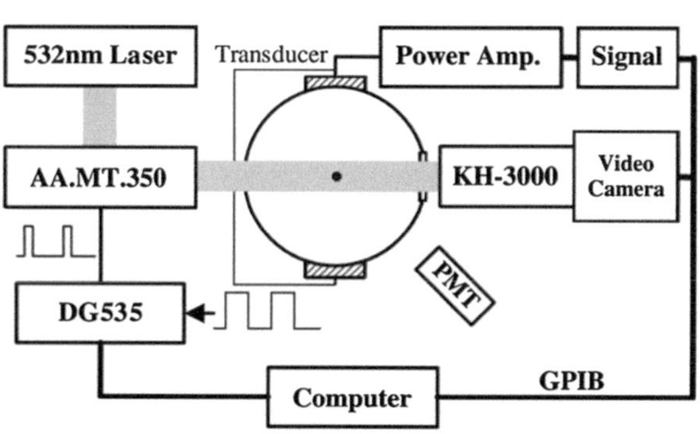

Fig. 6.22 Sketch of the experimental setup. Reproduced with permission from Huang et al. (2006). © 2006 Elsevier B.V.

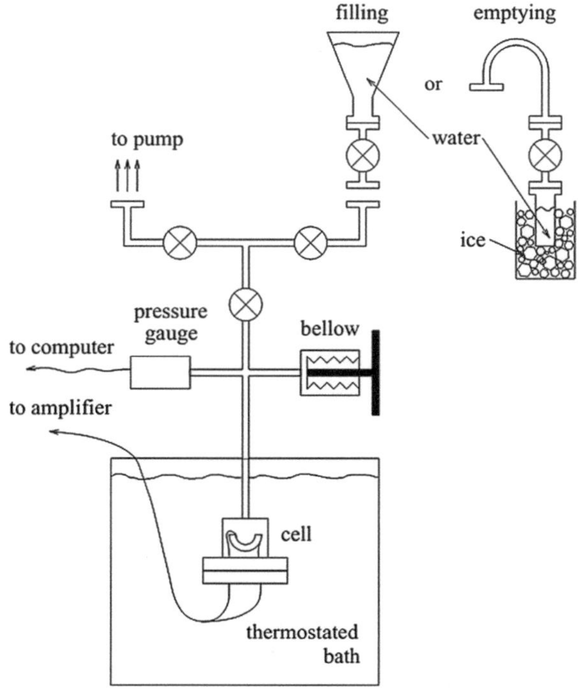

Fig. 6.23 Sketch of the experimental setup. Reproduced with permission from Herbert et al. (2006). © 2006 The American Physical Society

According to the Rayleigh–Plesset equation, which is used to describe an uncoated spherical bubble's behavior in an incompressible infinite liquid, the sound pressure is one of the main influencing factors on bubble dynamics:

$$R\frac{\mathrm{d}^2R}{\mathrm{d}t^2}+\frac{3}{2}\left(\frac{\mathrm{d}R}{\mathrm{d}t}\right)^2=\frac{1}{\rho}\left[P\mathrm{g}(R,t)-P_0-P(t)\right]+\frac{R}{\rho c}\frac{\mathrm{d}}{\mathrm{d}t}\left[P\mathrm{g}(R,t)-P(t)\right] -\frac{4v}{R}\frac{\mathrm{d}R}{\mathrm{d}t}-\frac{2\sigma}{\rho R} \tag{6.5}$$

where R is the radius of the spherical bubble, $P_\mathrm{g}(R,t)$ and P_0 are the gas pressure inside the bubble and the ambient pressure, respectively, and $P_\mathrm{a}(t)=P_\mathrm{a}\sin\omega t$, P_a is the amplitude of the driving pressure, and ω is the angular frequency. Parameters ρ, c, σ, and v stand for the density, speed of sound, surface tension, and shear viscosity of the liquid, respectively.

Huang et al. (2006) described bubble dynamics under various sound pressures using the first experimental setup introduced in this section. They documented the maximum radius R_m of the oscillating bubble and then obtained the ambient radius R_0 by fitting a numerical calculation of $R(t)$ based on the Rayleigh–Plesset equation to the experimental data. They found that the same variation tendency is possessed by R_m and R_0. As the sound pressure P_a increased, the maximum radius R_m and the

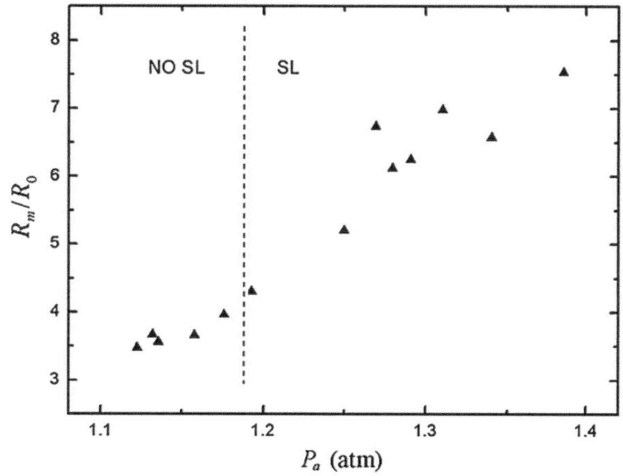

Fig. 6.24 The compression ratio of the maximum bubble radius to the ambient radius (R_m/R_0) as a function of the driving sound pressure. Reproduced with permission from Huang et al. (2006). © 2006 Elsevier B.V.

ambient radius R_0 decreased to minimum values at $P_a \approx 1.2$ atm. With a further increase in P_a, R_m, and R_0 increased again. The turning point was also the point where the bubble began to emit light, i.e., producing SL. A monotonously increasing relationship was found between the sound pressure and the ratio of R_m to R_0, as shown in Fig. 6.24. From their study, we can conclude that enhancing the irradiation power can increase the compression ratio of the bubble despite that the mass inside the bubble also increases.

In the second experiment designed by Herbert et al. (2006), a few cycles of a 1-MHz traveling wave were generated by a hemispherical piezoelectric transducer and tightly focused at the center of the transducer, which was immersed in liquid. The experimental volume was $V = 2.1 \times 10^{-4}$ mm^3, and the effective experiment time was 45 ns. When enough amplitude was applied, cavitation occurred. The occurrence probability of cavitation Σ (seen in Sect. 6.3.1) was measured by repeating the sound bursts for a given amplitude. The experimental setup had high

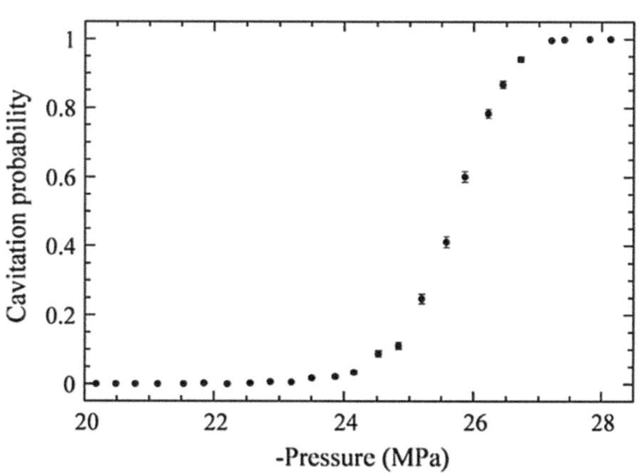

Fig. 6.25 Cavitation probability as a function of pressure at 4 °C. Reproduced with permission from Caupin and Herbert (2006). © 2006 Académie des sciences

reproducibility for obtaining a Σ-pressure curve (Fig. 6.25), where Σ ranged from 0 to 1 in a narrow pressure range. This is reasonable because of the strong exponential dependence of Σ on E_b (seen in Sect. 6.3.1). The corresponding pressure at the point when Σ equals 1/2 can be defined as the threshold for cavitation P_{cav}.

6.4 Effect of Frequency on Cavitation

6.4.1 Frequency Dependence of Ultrasonic Cavitation

Frequency is also one of the most important ultrasound parameters because it can influence cavitation bubble properties such as size distribution, lifetime, and collapse velocity. At higher acoustic frequencies, bubbles have less time to expand, leading to smaller bubble sizes. Brotchie et al. (2009) detected bubble size distribution by changing the pulse separation time at different ultrasound powers and frequencies. With this increase in pulse separation time, the bubbles dissolved, which led to a decrease in SCL intensity. Thus, they calculated bubble size distribution according to Eq. 6.6. The result is shown in Fig. 6.26, where we can see that higher frequencies generated smaller and more narrowly distributed bubble clouds.

$$\left(\frac{\mathrm{DC}_s}{\rho_g R_0^2}\right)t = \frac{1}{3}\left(\frac{RT\rho_g R_0}{2M\gamma} + 1\right) \tag{6.6}$$

In addition to bubble size distribution, cavitation is also closely linked with the intense collapse of bubbles, which generates SL and high-energy free radicals. To explore this further, Yasui (2002) studied the influence of ultrasonic frequency on

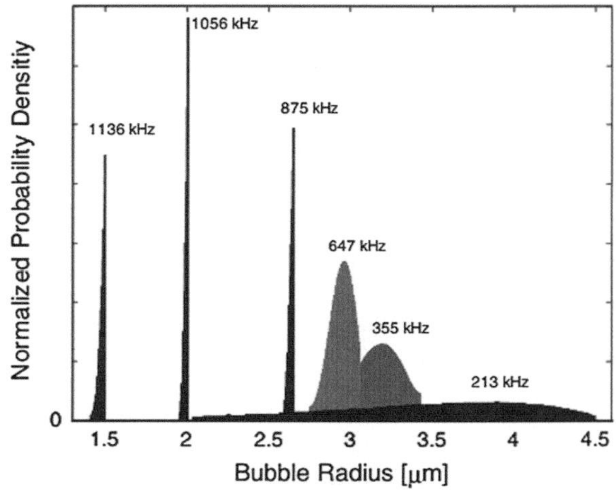

Fig. 6.26 The bubble size distributions for different frequencies. Reproduced with permission from Brotchie et al. (2009). © 2009 The American Physical Society

multi-bubble sonoluminescence (MBSL). They used frequencies of 20, 140 kHz, and 1 MHz. At these frequencies, the shorter period available for bubble expansion restrained the bubbles into a certain volume, making the bubble collapse milder. At 1 MHz, they considered that the light originated mainly in plasma emissions, while at 20 kHz it originated both from the chemiluminescence of OH radicals and in the plasma emissions. Their results showed that the range of bubble radii capable of producing SL narrowed as the ultrasound frequency increased, which was in agreement with results reported by Brotchie et al. (2009). In another similar study, Beckett and Hua (2001) carried out a comprehensive investigation of ultrasonic frequency and its role in sonochemical activities and SL, where they measured the SL intensity at 205, 358, 618, and 1071 kHz (Fig. 6.27). At the optimal frequency of 358 kHz, the broadest light spectrum was generated and the greatest intensity was found for bubbles with a radius of 300 nm. However, when the frequency was as high as 1071 kHz, the light was diminished relative to the other frequencies. The diagram in Fig. 6.28 illustrates how frequency influenced acoustic cavitation events.

6.4.2 *Enhancement of Cavitation by Multiple Frequencies*

Determining a single optimal frequency is important for cavitation control, but multiple frequencies can also be applied to enhance cavitation. By combining frequencies f_1 (high) and f_2 (low), new $f_1 + f_2$ and $f_1 - f_2$ frequencies can be

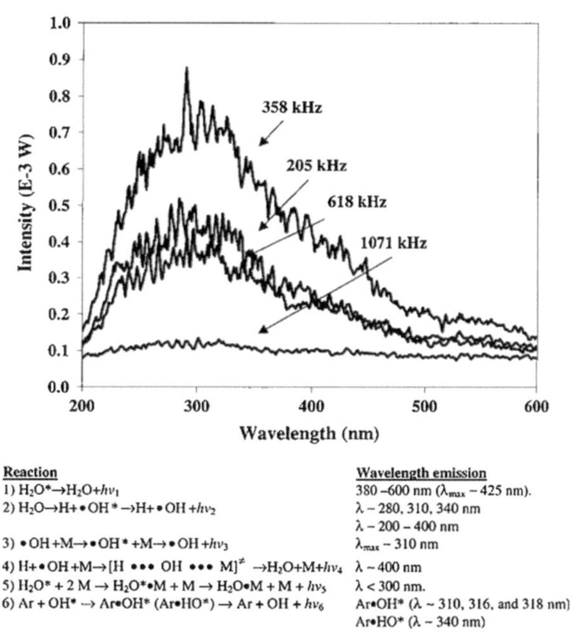

Fig. 6.27 SL spectra at four discrete ultrasonic frequencies. Reproduced with permission from Beckett and Hua (2001). © 2001 American Chemical Society

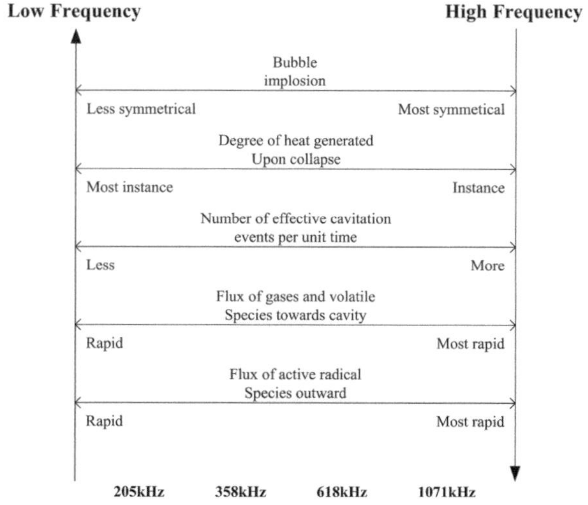

Fig. 6.28 Scheme of acoustic cavitation events as a function of frequency during continuous ultrasonic irradiation. Reproduced with permission from Beckett and Hua (2001). © 2001 American Chemical Society

obtained. Many studies have showed that this frequency combination can broaden the size of bubbles involved in cavitation and can enhance cavitation effects (Ciuti et al. 2003; Prabhu et al. 2004; Moholkar 2009). Figure 6.29 shows a typical setup for a frequency combination chamber.

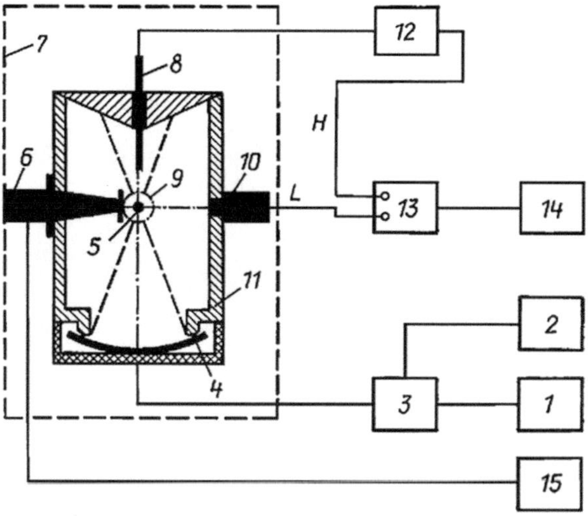

Fig. 6.29 Experimental schematic diagram. (*1*) High-frequency (HF) generator, (*2*) pulse generator, (*3*) mixer, (*4*) HF transducer, (*5*) focal spot of the HF transducer, (*6*) low-frequency transducer, (*7*) light-tight box, (*8*) hydrophone with bandwidth from 10 kHz to 5 MHz, (*9*) intensive cavitation zone, (*10*) photomultiplier with bandwidth from 200 to 700 nm, (*11*) preamplifier, (*12*) spectrum analyzer, (*13*) oscilloscope, (*14*) PC, and (*15*) LF generator. Reproduced with permission from Ciuti et al. (2003). © 2003 Elsevier B.V.

In one such study, the SL method was used to detect the intensity and spatial distribution of inertial cavitation after multiple frequencies was applied. The output signal H of a hydrophone was recorded simultaneously (Ciuti et al. 2003). Figure 6.30 shows the wave sequences and the corresponding hydrophone output *H* and of the photomultiplier *L*. The upper and lower bars indicate the time sequence of high-frequency (HF: 880 kHz) and low-frequency (LF: 27.2 kHz) waves, respectively. The SL intensities underwent their second increase when the LF field was turned off (around 25 s), while the output signals detected by the hydrophone decreased. This occurred because LF can generate more effective bubble nuclei than the naturally existing nuclei in the liquid. However, when both the LF and HF fields were applied simultaneously, the bubble concentration was so high that interactions between bubbles and shielding effects diminished the SL efficiency. After removing the LF field, the bubble density quickly decreased to an optimal value, leading to the sharp increase of SL intensity and a decrease in the hydrophone output. Thus, changing the time sequence of the HF and LF fields might be an effective way to control cavitation.

As cavitation bubble clouds form, negative and positive pressures play important, but different roles. Positive pressure is a key factor in the formation of a cavitation bubble could, while negative pressure is the primary factor that initiates cavitation to begin with (Maxwell et al. 2011). Based on this information, Yasuda et al. (2013) studied the controlled generation of cavitation bubbles inside gel phantom using a dual-frequency ultrasound exposure method. They introduced four kinds of waveforms shown in Fig. 6.31, which represent the four possible combinations of two elements: a positive peak pressure-emphasized wave and a negative peak pressure-emphasized wave. The bubble clouds generated by these different ultrasound waveforms are shown in Fig. 6.32. Comparing the PP and NN results, we can see that NN is more effective in generating sparsely distributed bubbles. Looking at the NP result, some cavitation bubbles were first generated in the focal region, and then when the waveform became positive, the bubble cloud grew faster. Here, the cavitation bubbles were generated by the N (negative) waves and stimulated to grow by the P (positive) waves. By combining wave sequences, the distribution of cavitation bubbles could be controlled. For example, in situations where cavitation should be prevented, the NN sequence should be used.

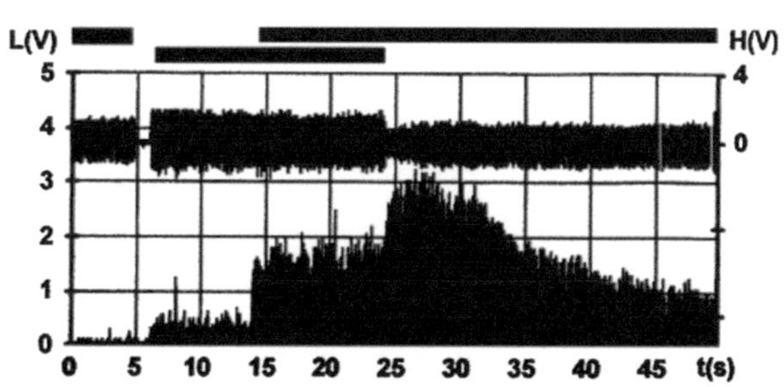

Fig. 6.30 Typical wave sequences used to enhance the cavitation efficiency. Reproduced with permission from Ciuti et al. (2003). © 2003 Elsevier B.V.

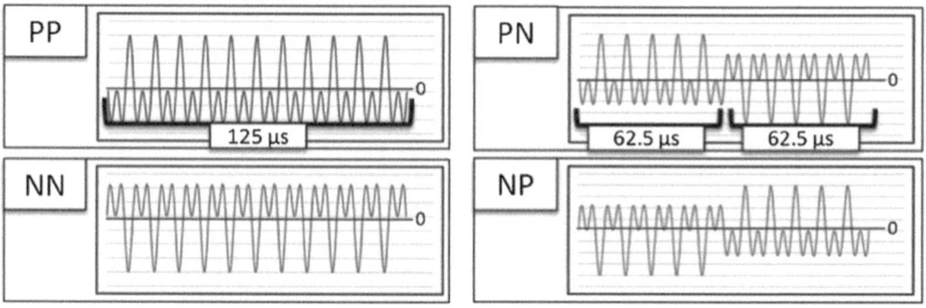

Fig. 6.31 Four kinds of ultrasound sequences to control bubble generation. Reproduced with permission from Yasuda et al. (2013). © 2013 The Japan Society of Applied Physics

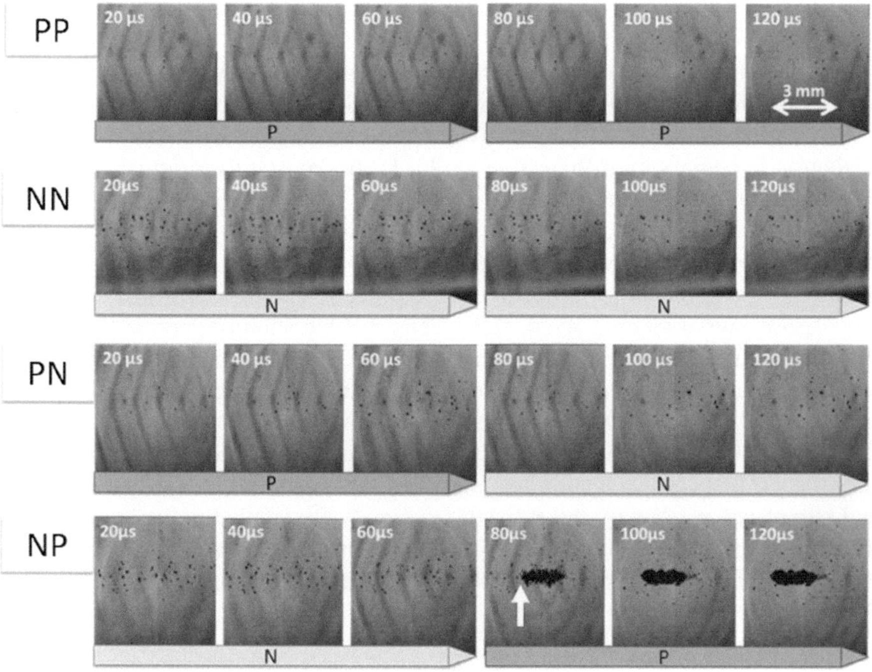

Fig. 6.32 High-speed images of generated cavitation bubbles in gel. Reproduced with permission from Yasuda et al. (2013). © 2013 The Japan Society of Applied Physics

6.5 Radiation Force-Constrained Cavitation

All forms of wave motion, including electromagnetic waves, transverse waves on an elastic string, surface waves in liquid and longitudinal sound waves, exert a kind of unidirectional force on absorbing and reflecting obstacles in their path. This force is widely called the radiation pressure or radiation force. In order to constrain cavitation bubbles in an ultrasonic field, this acoustic radiation force needs to be balanced. One way to balance this force is to artificially build an acoustic standing wave, which can significantly help in constraining cavitation.

In 1977, Miller described a phenomenon whereby bubbles formed stable arrays in a 1-MHz standing-wave field (Miller 1977). They employed a "Harvey chamber," which is an acoustic resonator consisting of an ultrasonic transducer and a disk-shaped quartz crystal. The curvilinear-distributed bubbles appeared as bright spots, as shown in Fig. 6.33. These bubbles in the array were coincident with pressure-null surfaces of the standing-wave field. In this array, there was an irregular round bubble that was considered cavitationally active, and because this bubble was trapped in position, we can see that constraining cavitation is possible.

Miller also observed a two-dimensional (2D) array using the same apparatus. As we can see in this 2D array in Fig. 6.34, the bubble movement and associated fluid flow was somewhat chaotic. Also, when the acoustic pressure was below the threshold at which stable arrays break down, we can see that the distance between neighbor bubbles in this array was maintained at a similar spacing to that of the linear array.

Another standing-wave experiment was performed by Nomura et al. (2005), who used schlieren photography to observe cavitation bubble streaming generated in standing-wave field (Fig. 6.35). In this experiment, the ultrasonic transducer was mounted on the bottom and the acoustic reflector was on the top. Due to the ultrasonic wave irradiation, compression and rarefaction waves were formed inside the resonator. The spacing between the bands was equal to half of the wavelength of the standing wave. Also, we can see that acoustic cavitation formed between the bands.

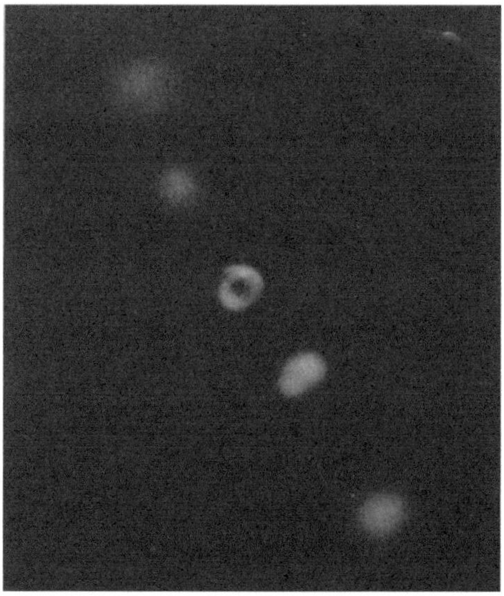

Fig. 6.33 A linear array in a Harvey chamber, viewed with a small angle from its axis. One bubble in the array (diameter of ~20 μm) is in focus. Reproduced with permission from Miller (1977). © 1977 Acoustic Society of America

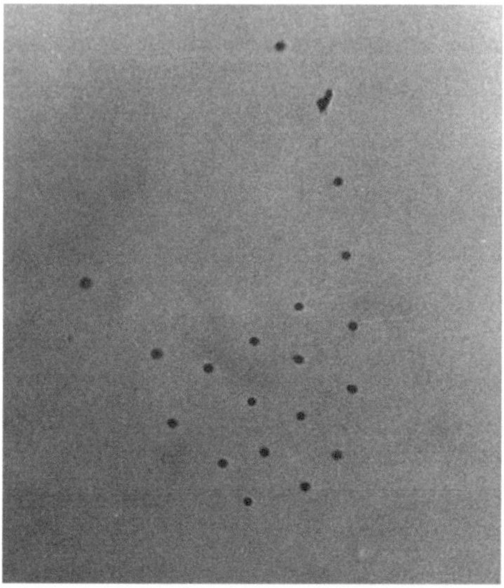

Fig. 6.34 A two-dimensional array captured within a 3-μs flash of light. Reproduced with permission from Miller (1977). © 1977 Acoustic Society of America

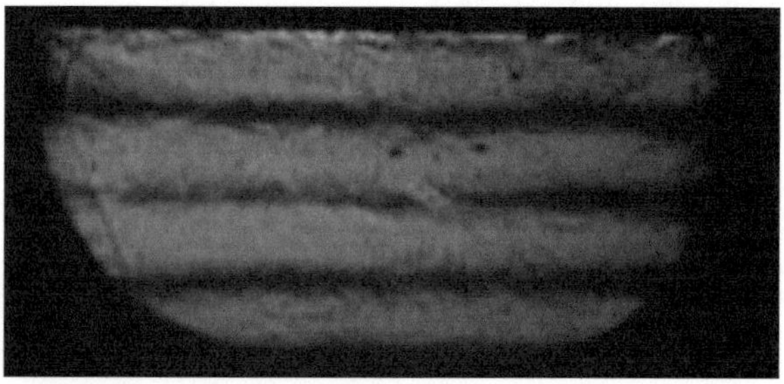

Fig. 6.35 Visualization of an acoustic field by schlieren photography. The dark bands correspond to antinodes of sound pressure. Reproduced with permission from Nomura et al. (2005). © 2005 The Japan Society of Applied Physics

As an important therapeutic ultrasound source, HIFU fields can also be used to control and constrain cavitation. Normally, cavitation bubbles created by HIFU would not remain stable because HIFU imparts a strong radiation force. However, numerous of HIFU setups consisting of a transducer and a reflector have been used to manipulate cavitation. For example, Yin et al. (2014) built a standing-wave field by such a method (Fig. 6.36). When applied acoustic intensity is lower than breakdown threshold for a stable array, cavitation bubbles can be generated in the focal region, and then, the bubbles can be constrained by the two opposing radiation forces. Figure 6.37 (unpublished) shows bubbles being trapped in focal region of a HIFU standing-wave field, where the ultrasonic transducer and acoustic reflector

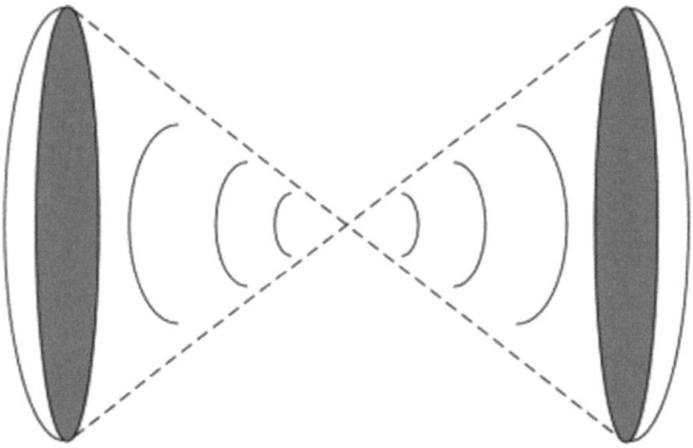

Fig. 6.36 Standing-wave field based on a transducer and a reflector

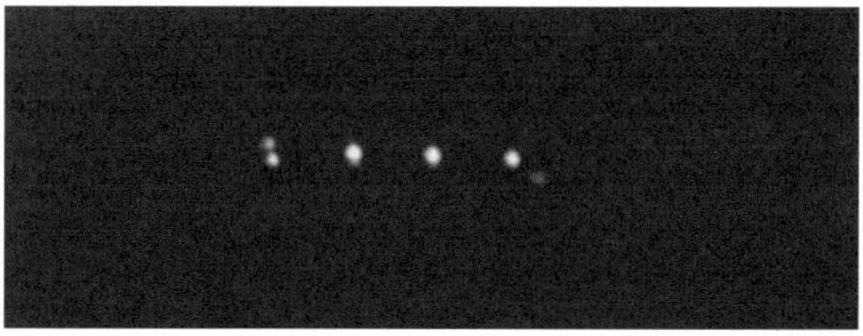

Fig. 6.37 Trapped bubbles in the focal region of an HIFU standing-wave field

were placed horizontally. In addition, SCL is also employed to map the cavitation in this study because SCL can reflect the distribution of most active cavitation bubbles (Fig. 6.37). The reflector could avoid bubble directional displacement induced by acoustic radiation force under certain electric power (≤50 W), however, when the applied electric power equal to or exceed 70 W, the bubbles were moving in acoustic streaming (Fig. 6.38).

In another experiment, Li et al. (2013) developed a spherical cavity transducer with open ends. The acoustic field inside this transducer is intensively compressed to a subwavelength level, while the intensity of sound pressure in the focal region significantly increases. The transducer is shown in Fig. 6.39. Compared to a normal HIFU transducer, this spherical transducer keeps cavitation bubbles in the center of the transducer. The resulting SCL images using different HIFU transducers are shown in Fig. 6.40 (unpublished). Because the spherical transducer can focus the bubbles so precisely, it gives perfect spherical light emission in the focal region, while a normal HIFU transducer produces elongated light emission. The author suggested if this new form of transducer could be applied in HIFU treatment in the future, the focusing precision would be enhanced a lot and the subwavelength therapy could be achieved to improve the treatment effects, while minimizing undesired thermal injury to normal tissues.

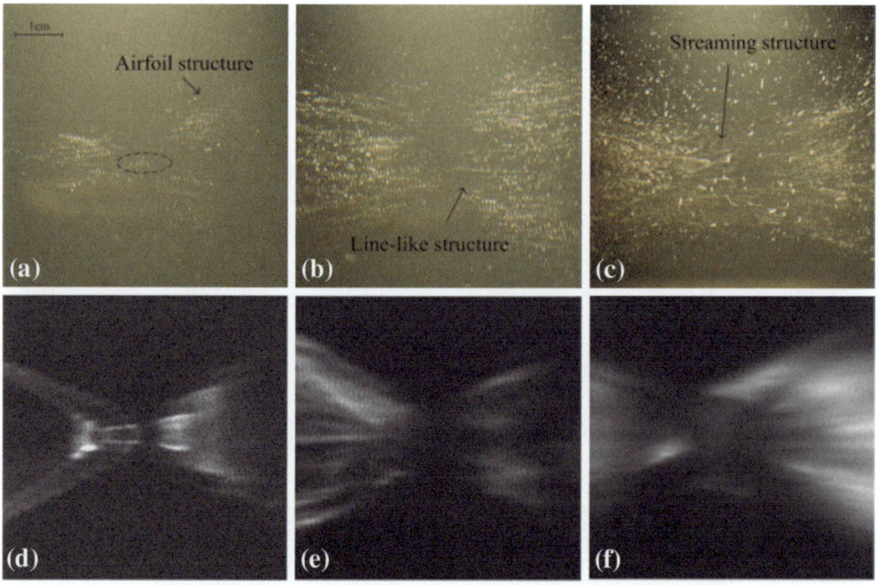

Fig. 6.38 Spatial distribution of bubbles and SCL at 10 W (**a**, **d**), 50 W (**b**, **e**), and 90 W (**c**, **f**) in counterbalanced group (with reflector). Exposure time of SCL images was 10 s. Reproduced with permission from Yin et al. (2014). © Elsevier

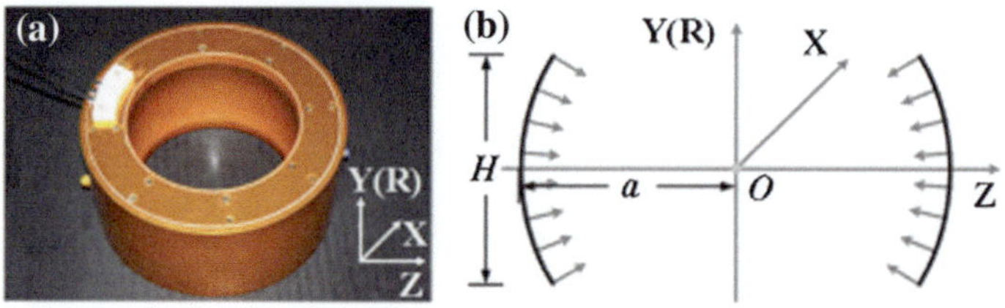

Fig. 6.39 Spherical cavity transducer with open ends. Reproduced with permission from Li et al. (2013). © AIP Publishing

Another area where cavitation bubbles are usually constrained is near boundaries, including fluid–rigid boundaries, soft tissue surfaces, and fluid–gas boundaries. These situations affect single-bubble shapes more than they do multi-bubble distributions, and they are discussed in Chap. 2.

In most cases, to constrain bubbles in a cavitation field, a standing-wave field is necessarily to build. By controlling the amplitude, frequency, and phase delay of the transducers, bubbles can be trapped and their movement can be controlled. In addition, acoustic tweezers can also be used to manipulate bubble movement. As opposed to optical tweezers, acoustic tweezers have the unique advantage of being able to propagate through non-transparent media. Therefore, developments into additional research instrument that can generate, control, and monitor cavitation will likely become a focus of cavitation research in the near future.

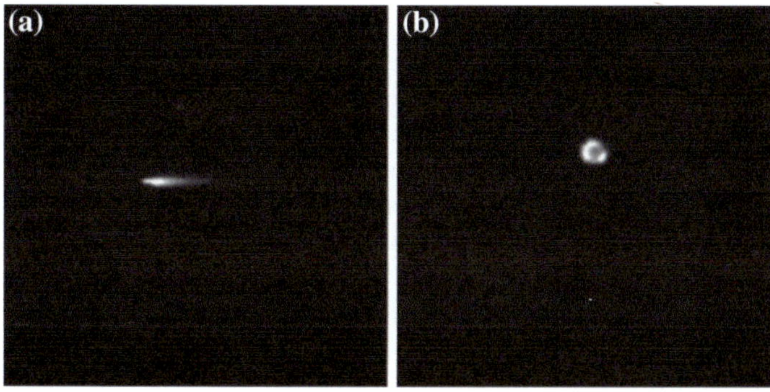

Fig. 6.40 SCL images in focal regions of **a** a normal HIFU transducer and **b** a spherical HIFU transducer

6.6 Cavitation-Controlled Tissue Histotripsy

6.6.1 The Principle

Histotripsy is a combination of "histo" and "tripsy," which, respectively, mean "soft tissue" and "breakdown." Thus, the term indicates that tissue can be removed in a localized manner. This technique was developed at the University of Michigan over 10 years ago. Histotripsy is one of the bioeffects of focused ultrasound, but it is not the only outcome. Focused ultrasound applied at different intensity levels can produce a variety of bioeffects, which are summarized nicely in Fig. 6.41 (Maxwell et al. 2012). At extreme ultrasound intensities, shock waves can form at the focus

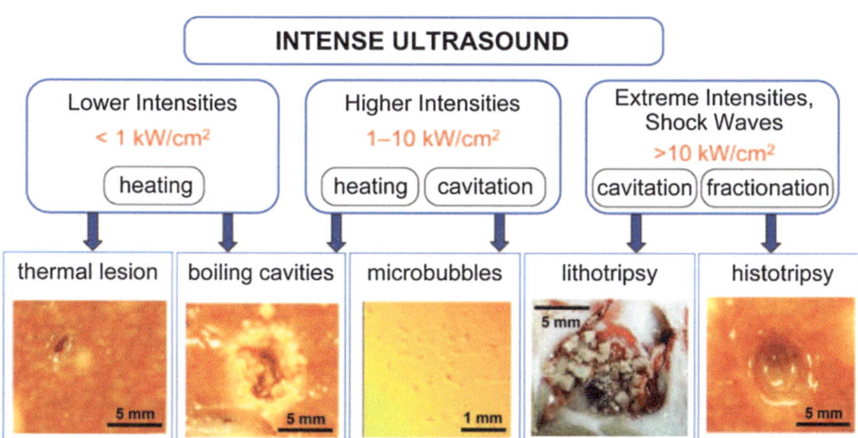

Fig. 6.41 Bioeffects of focused ultrasound at different focal intensity levels. Reproduced with permission from Maxwell et al. and Acoustics Today (2012). © 2011 Acoustical Society of America. On-line article can be found on http://acousticstoday.org/issues/2012AT/Oct2012/index.html#?page=28

and cause significant mechanical damage with little heat effects. This is called cavitation-based histotripsy. Early studies have shown that short, intense pulses delivered at a certain pulse repetition frequency can achieve energy-efficient histotripsy and sharp, clearly defined perforations (Cooper et al. 2004). As precise tissue histotripsy is highly related to the activities of cavitation bubbles, many studies have focused on controlling cavitation during this process. Besides, a group at University of Washington developed another approach to controlling called boiling-based histotripsy, in which in situ shock amplitudes and the time to boil can be estimated. Thus, when the pulse length is longer than the time to boil and there is enough ultrasound off time to allow cooling, mechanical lesions can be produced that lack thermal damage. The lesions formed by boiling-based histotripsy are very reproducible in terms of their location, shape, size, and the lesion content. In addition Lin et al. (2014) recently proposed a "microtripsy" technique, which can generate a dense bubble cloud with only a single negative half-cycle beyond the intrinsic cavitation threshold, and it can achieve precise tissue removal.

There are several methods available to study the evolution of cavitation activities, such as high-speed photography, SCL imaging, and the PCD method. In addition, various acoustic parameters, discussed below, have been studied and optimized for controlling tissue histotripsy.

6.6.2 Effect of Pulse Length

6.6.2.1 Cavitation-Based Histotripsy

When the negative pressure is high enough, bubble clouds can be generated within just 3–10 cycles. Maxwell et al. (2011) observed bubble clouds induced by only a 3-cycle waveform of a 1-MHz spherically focused transducer. The waveform they used is shown in Fig. 6.42. Here, we can see that the original sinusoidal waves were changed into shock waves with high peak positive pressures and low negative pressures. This very high peak positive shock can be released by the bubbles' surface. In Figs. 6.43 and 6.44, we can see, respectively, individual bubbles as they formed at the focus and the formation of bubble clouds. In the images of individual bubble development, we see that the bubble was formed within the first cycle and grew larger in the next two cycles. The subsequent shockwaves acting on these single bubbles led to bubble clouds that stemmed from the center of the single bubble and grew toward the transducer, as shown in Fig. 6.44. This occurs mainly because the bubble has a soft surface, which can invert the pressure and result in a strong tensile wave toward the transducer. At $t = 2.75$ μs, a spherical wave reflected by the individual bubble became visible. It then interacted with the incident negative portion of the cycle and created dense bubble clouds. This is depicted in Fig. 6.45, which shows both a sketch of the shock scattering from a bubble and the pressure distribution on the acoustic axis. The interaction between the bubble

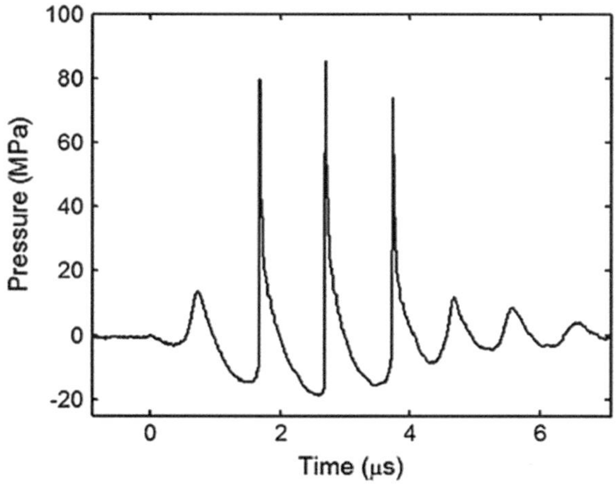

Fig. 6.42 A three-cycle focal pressure waveform of the histotripsy transducer used to generate bubble clouds. Reproduced with permission from Maxwell et al. (2011). © 2011 Acoustical Society of America

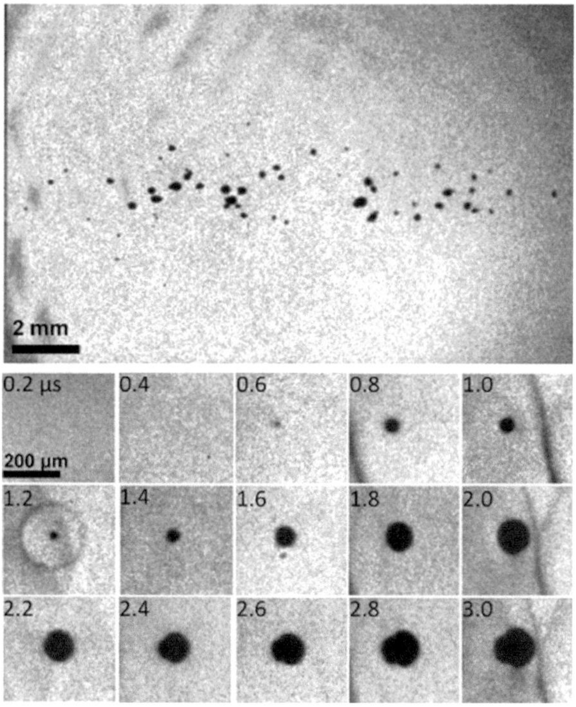

Fig. 6.43 Single bubbles generated at the focus of a histotripsy transducer (*top*) and temporal behavior of an individual bubble during the first three cycles of a pulse (*bottom*). Reproduced with permission from Maxwell et al. (2011). © 2011 Acoustical Society of America

surface and the shock waves can generate a very large transient rarefaction wave toward the transducer, inducing the growth of bubble clouds.

Later, Maxwell et al. (2013) also studied the probability of generating inertial cavitation in just a single cycle or in two cycles of a 1.1-MHz focused ultrasound

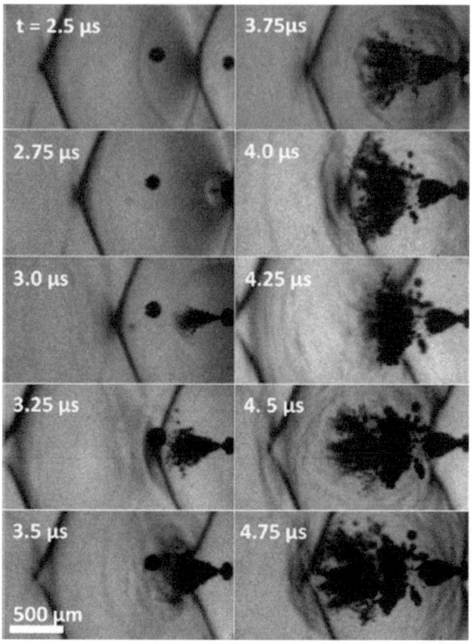

Fig. 6.44 Initiation of a bubble cloud. Reproduced with permission from Maxwell et al. (2011). © 2011 Acoustical Society of America

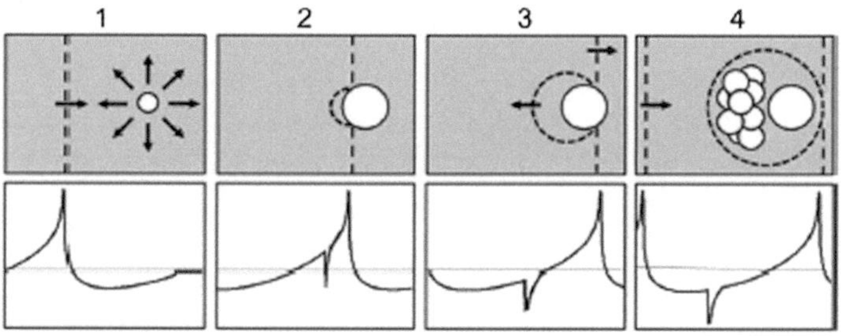

Fig. 6.45 Conceptual sketch of shock scattering from a bubble (*top*) and pressure distribution on the acoustic axis (*bottom*). Reproduced with permission from Maxwell et al. (2011). © 2011 Acoustical Society of America

pulse. They also investigated the corresponding pressure threshold that can consistently generate inertial cavitation. This work represents preliminary investigations into microtripsy, which will be introduced in the next section. In the study, inertial cavitation was identified by both high-speed photography and PCD. Figure 6.46 shows examples of the PCD temporal signals, the corresponding frequency-domain amplitude spectra, and high-speed photographs. According to their statistical results, a negative pressure of 27 MPa has the highest chance of

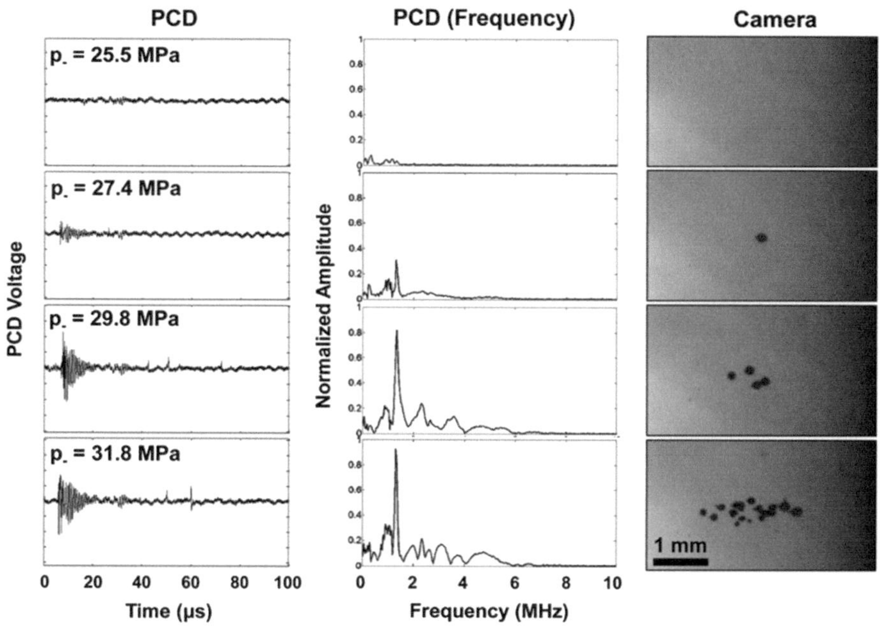

Fig. 6.46 Sample PCD temporal signals (*left*), corresponding windowed and normalized frequency-domain amplitude spectra (*center*), and corresponding camera photographs (*right*) of cavitation at the transducer focus. Reproduced with permission from Maxwell et al. (2013). © 2013 World Federation for Ultrasound in Medicine & Biology

generating cavitation in water and other water-based media. However, in the non-water-based media, the threshold for cavitation differs a lot, between ~15 MPa for adipose tissue to >36 MPa for olive oil. Importantly though, these data can be used to enhance or suppress cavitation during therapeutic applications.

In a different study, Wang et al. (2010) provided details on cavitation activities within and after a therapeutic pulse by capturing high-speed results within 5 cycles and 20 cycles of 1-MHz pulses with P_{-}/P_{+} greater than 19/125 MPa, respectively. Under both cycle conditions, the evolution of bubble clouds was generally the same, such that they grew with the input therapy pulse and dissolved gradually after the pulse. However, a shorter pulse produced smaller overall bubble clouds (2010). From Figs. 6.47 and 6.48, we can observe that in the first 10–15 cycles, the bubble clouds grew to their maximum size, but then when more acoustic energy was delivered, the size of the bubble cloud remained the same. This occurred because the bubble cloud shielded the acoustic energy from being transmitted into the focal region, thus preventing the violent growth and collapse of the bubbles, which is needed for efficient tissue removal.

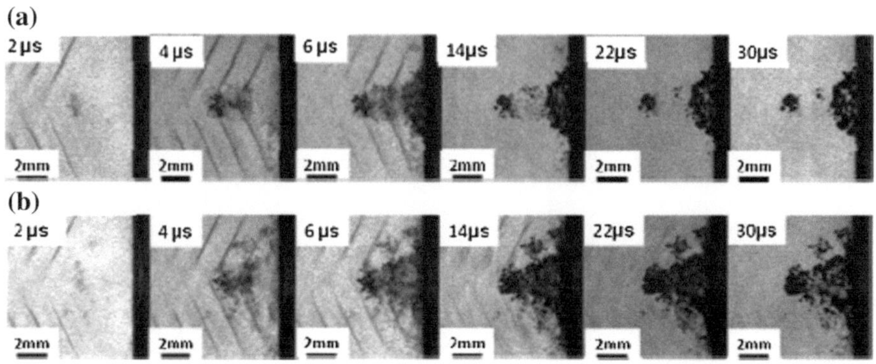

Fig. 6.47 Bubble cloud evolution within a therapeutic pulse. Reproduced with permission from Wang et al. (2010). © 2010 American Institute of Physics

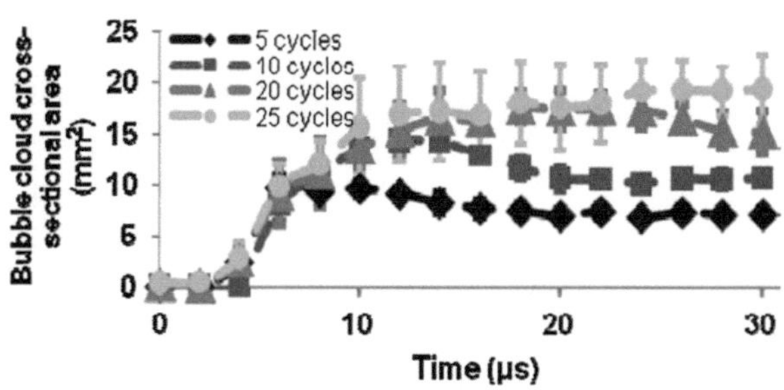

Fig. 6.48 Evolution of the overall bubble clouds within 30 μs after the arrival of the therapeutic pulse. Reproduced with permission from Wang et al. (2010). © 2010 American Institute of Physics

6.6.2.2 Boiling Histotripsy

Boiling histotripsy can also fractionate and remove tissue, but its mechanism is quite different from cavitation-based histotripsy. It involves pulses, each of which is normally longer than the time to boil, and it also requires a longer off time between pulses so that the thermal injury is negligible compared to the mechanical damage. It is believed that shocks at the focus are the main contributors to heat absorption, and the higher harmonics generated by high-intensity ultrasound have higher heat absorption coefficients. Using the weak shock theory, Khokhlova et al. (2011) estimated the time to boil as follows:

$$t_b = \frac{\Delta T c_v}{H} \tag{6.7}$$

where ΔT is the change from the ambient temperature to 100 °C, c_v is the heat capacity per unit volume, and H is the heating rate induced by the shocks. Acoustic parameters such as pulse length and duty factor were defined according to the

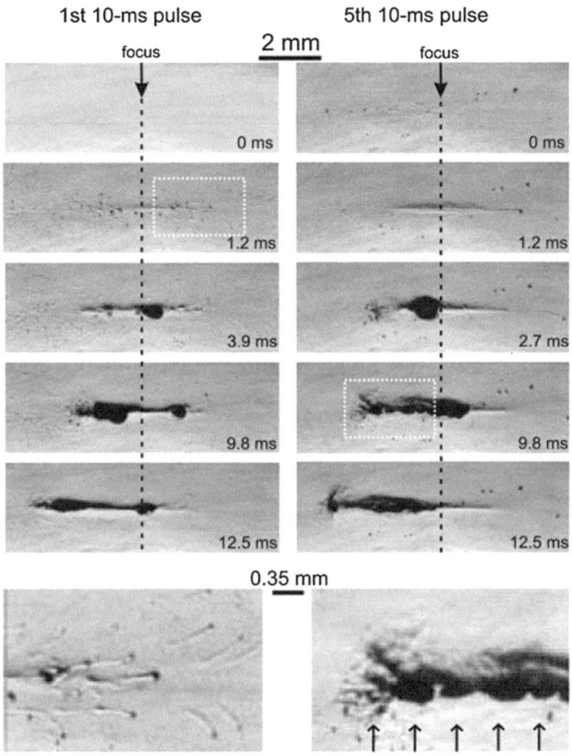

Fig. 6.49 High-speed images of HIFU pulses in a transparent 7 % bovine serum albumin (BSA) polyacrylamide gel during the first (*left*) and fifth (*right*) pulses. Reproduced with permission from Khokhlova et al. (2011). © 2011 Acoustical Society of America

calculation results. Figure 6.49 shows the high-speed images in the gel during the first and fifth pulses. The HIFU was working at 2.158 MHz with a pulse length of 10 ms and a pulse repetition frequency (PRF) of 1 Hz. During the first pulse, a large boiling bubble was observed at 3.9 ms. However, on the fifth pulse it was observed a little earlier, at 2.7 ms, which may have been due to heat accumulation. For the rest of the pulses, boiling occurred at nearly the same time of 2–3 ms, indicating that the process was quite predictable and repeatable. Figure 6.50 shows the lesions formed in the gel. Here, we can see that large boiling bubbles contributed to tearing of the gel, and there was no evident thermal lesion. Compared to cavitation bubbles, which dissolve quickly after HIFU is turned off, boiling bubbles may be able to provide longer echogenicity on B-mode images. The structure shown in B-mode images of boiling bubbles coincided well with this study's lesion created in the gel, indicating that boiling histotripsy can be monitored within the tissue by B-mode imaging.

B-mode images, record of the transducer voltage, and the lesions produced in excised bovine liver were shown in (Fig. 6.51). In the B-mode images captured after the 1st, 12th, and 33rd HIFU pulses, we can see highly echogenic regions, indicating boiling bubbles. Also, as indicated by the transducer voltage recording, the onset of boiling bubbles caused the transducer voltage to fluctuate, which

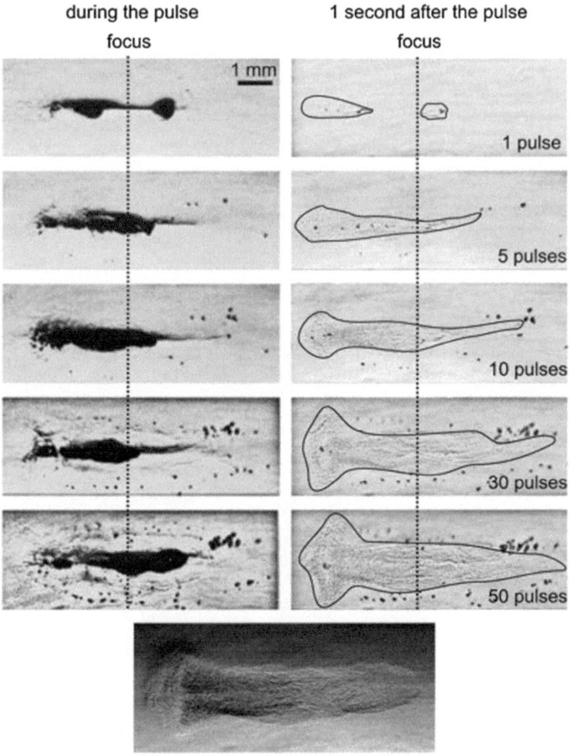

Fig. 6.50 Lesions formed in the gel during pulsed HIFU exposure. Reproduced with permission from Khokhlova et al. (2011). © 2011 Acoustical Society of America

corresponded well with the theoretical prediction. Furthermore, all the lesions had the same locations, sizes, and tadpole shapes, agreeing well with results from the gel experiments.

To determine how a boiling bubble creates tissue fragmentation, Simon et al. (2012) bored a cylindrical hole through tissue and captured the acoustic fountain and tissue atomization at the tissue–air interface during histotripsy pulses. Figure 6.52 shows their proposed mechanism of tissue fractionation. As the focus of HIFU possesses the highest energy, they easily created a vapor cavity with a pressure-release surface. The interaction between shock waves and these vapor cavities led to the formation of an acoustic fountain and tissue atomization. In addition, they also captured disconnected droplets in the chain and attributed tissue fractionation to unstable droplet oscillation and explosion, and they also found that the tissue wetness affected the atomization and histotripsy rate. A higher portion of water lowered the cavitation threshold and made the tissue removal easier.

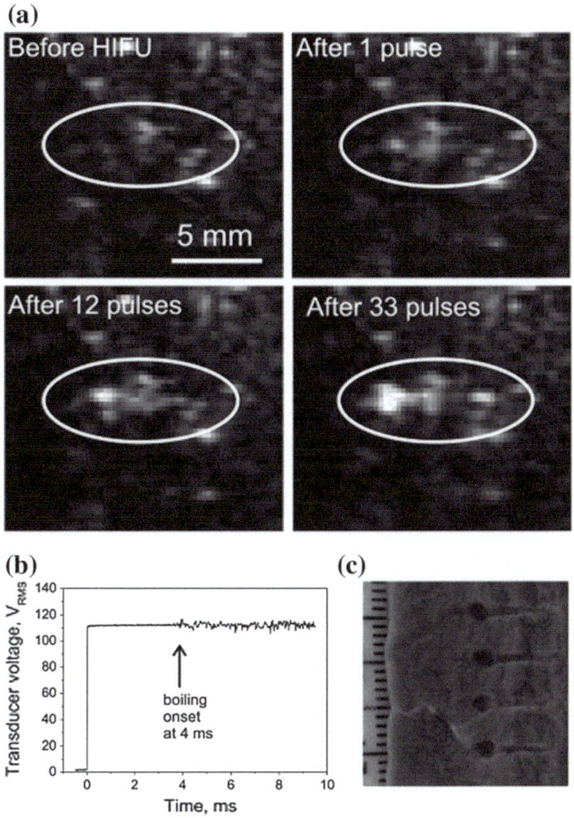

Fig. 6.51 **a** B-mode images, **b** the transducer voltage recording, and **c** the lesions produced by four independent exposures using a 2.158-MHz HIFU transducer. Reproduced with permission from Khokhlova et al. (2011). © 2011 Acoustical Society of America

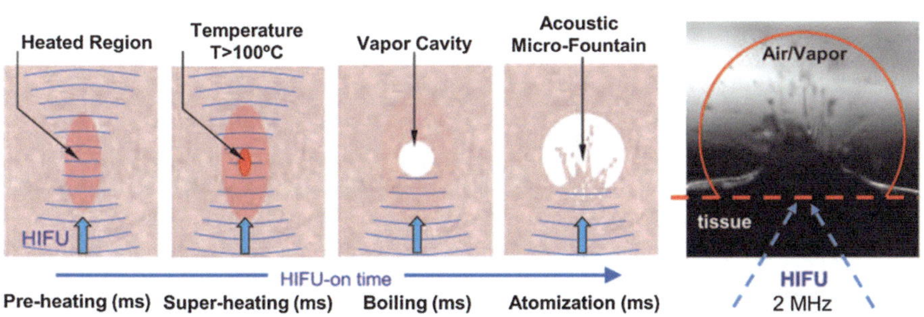

Fig. 6.52 Proposed boiling histotripsy mechanism (*four left panels*), and tissue atomization and fountain at the free tissue–air interface (*right panel*). Reproduced with permission from Simon et al. (2012). © 2012 Institute of Physics and Engineering in Medicine

6.6.2.3 Microtripsy

Recently, Lin et al. (2014) proposed another kind of histotripsy: microtripsy, which can directly generate dense bubble clouds with only a single negative half-cycle beyond the intrinsic cavitation threshold. The functionality of this method is mainly related to the peak negative pressure rather than positive pressure, tissue character, or dissolved gases. By carefully adjusting the applied peak negative pressure, the lesion can be precisely controlled. Figure 6.53 shows a representative 1.5-cycle acoustic waveform measured by a fiber optic probe hydrophone (FOPH). The lesions generated by microtripsy were evaluated using a red blood cell (RBC) tissue-mimicking phantom. The samples were treated with 500 microtripsy pulses at a PRF of 1 Hz. The intrinsic cavitation threshold was calculated according to the following equation:

$$P_{\text{cav}}(p_{-}) = \frac{1}{2}\left[1 + \text{erf}\left(\frac{p_{-} - p_t}{\sqrt{2\sigma^2}}\right)\right] \tag{6.8}$$

where P_{cav} is the probability of generating a dense bubble cloud. The term erf is the error function, p_t is the pressure that makes P_{cav} equal to 0.5, and σ is the variable related to the transition width in the cavitation probability curve. Figure 6.54 illustrates the lesion size estimation using the lateral direction as an example. The lesion size estimation is based on a scaled profile where p_{-} values exceed the P_{LSE} (which was used to estimate the size of the lesion that would be produced in RBC phantom) threshold, defined as the p_{-} that gave a 50 % probability of observing at least one dense bubble cloud generation event over the course of the 500 delivered pulses.

The related images in Fig. 6.55 show the lesions and bubble clouds in the red blood cell phantoms at different p_{-} values. The size of the lesions increased linearly with the peak negative pressure, and the size of lesions was actually a little larger than the areas the bubble clouds filled. The average size of lesion widths ranged

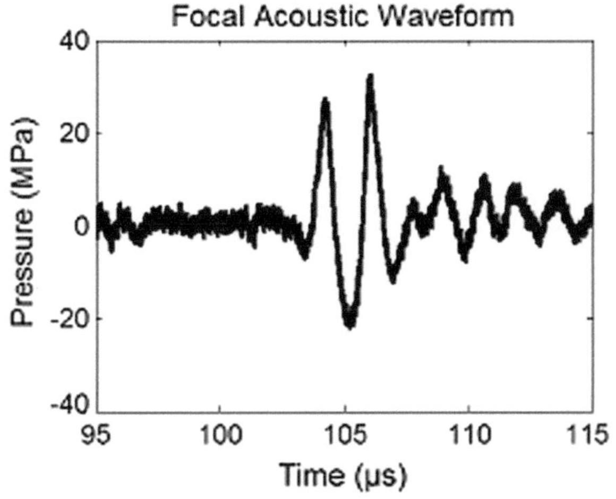

Fig. 6.53 A representative free-field 1.5-cycle acoustic waveform. Reproduced with permission from Lin et al. (2014). © 2014 IEEE

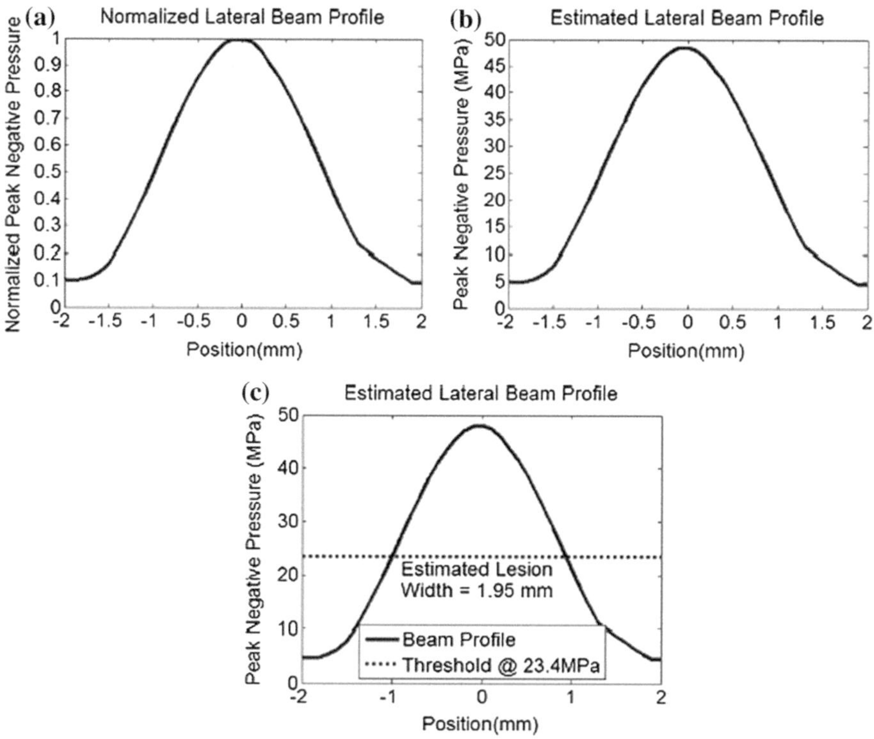

Fig. 6.54 The lesion size estimation using the lateral direction as an example: **a** a normalized lateral beam profile, **b** the estimated lateral beam profile and **c** the estimated lesion size. Reproduced with permission from Lin et al. (2014). © 2014 IEEE

from 0.9 mm (p_{-} = 26.2 MPa) to 3.91 mm (p_{-} = 80.7 MPa, which was not included in this figure), while the −6 dB beamwidth of the transducer was 1.8 × 4.0 mm and the estimated P_{LSE} was 23.4 MPa. The lesions formed in ex vivo canine kidney were also examined by B-mode imaging and histology. Their sizes were also close to the estimation.

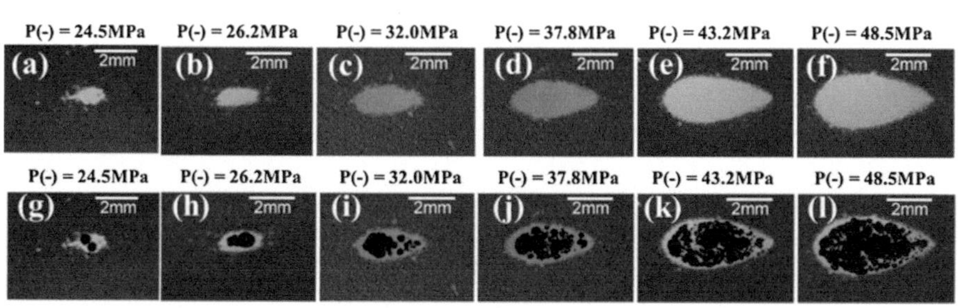

Fig. 6.55 **a–f** Representative lesion and **g–l** bubble cloud images in red blood cell phantoms after 500 histotripsy pulses had been applied. Reproduced with permission from Lin et al. (2014). © 2014 IEEE

Compared to other kinds of histotripsy, this method has some advantages. First, the lesion and bubble cloud generated by microtripsy can be precisely controlled and well-confined within the transducer focal zone. Second, the size of the lesion can be very small—on the subwavelength level—by allowing only a small fraction of the focal region to exceed the P_{LSE}. Third, this method is insensitive to the center frequency of the transducer, which opens the application to low-frequency transducers and allows deep penetration of the ultrasound wave.

6.6.3 *Effect of Duty Cycle*

Duty cycle mainly affects the decay time of a bubble cloud. If the duty cycle is too high, there is not enough time for bubble clouds to dissolve and they overlap with later pulses, scattering the ultrasound waves and shielding the target. However, if the duty cycle is too low, the residual nuclei from previous pulses cannot be used again. For shorter pulse durations, there seems to be an optimal PRF. However, for longer pulse durations, bubble clouds may need more time to dissolve. Figure 6.56 shows the relationship between the histotripsy rate and PRF for a fixed pulse length (Xu et al. 2004). The working frequency of the transducer was 788 kHz. The p_-/p_+ = 11.6/36.0, 11.1/33.4, 10.6/31.5, 10.6/31.5 MPa for pulse lengths of 3, 6, 12, and 24 cycles, respectively. For PD = 3, 6, and 12 cycles, the maximum erosion rates appeared at 19.6, 6.90 and 4.95 kHz, respectively. For PD = 24 cycles, the erosion rate increased with the PRF.

In another study, Cao et al. (2013) studied the effect of duty cycle on inertial cavitation dose (ICD) and stable cavitation dose (SCD). For PD = 10 μs, both the ICD and SCD decreased evidently with the duty cycle. However, increasing the PD up to 80 μs, the differences of ICD and SCD between different duty cycles almost diminished (Fig. 6.57). The roles, that both inertial and stable cavitation play in tissue erosion, are also investigated. ICD amplitude was much larger than SCD amplitude. Thus, inertial cavitation can lead to production of shock waves, high-velocity liquid jets, and strong shear forces that can penetrate the tissue and cause mechanical damage, while stable cavitation is less likely to contribute to tissue erosion.

Figure 6.58 shows their results of tissue erosion when the duty cycle was set to 1:10. Figure 6.58a shows the tissue erosion created with a pulse length of 10 μs, which produced a well-defined hole that was larger than the erosion produced at pulse lengths of 20 μs (Fig. 6.58b) and 40 μs (Fig. 6.58c). With longer pulse lengths, the bubble cloud formed during earlier pulses shielded the target, causing the ultrasound energy to be scattered and absorbed. However, the bioeffects that occur during tissue erosion are still not clear and must be studied further.

Khokhlova et al. (2011) also investigated the influence of pulse duration and duty factor during boiling histotripsy. They identified three types of lesions and

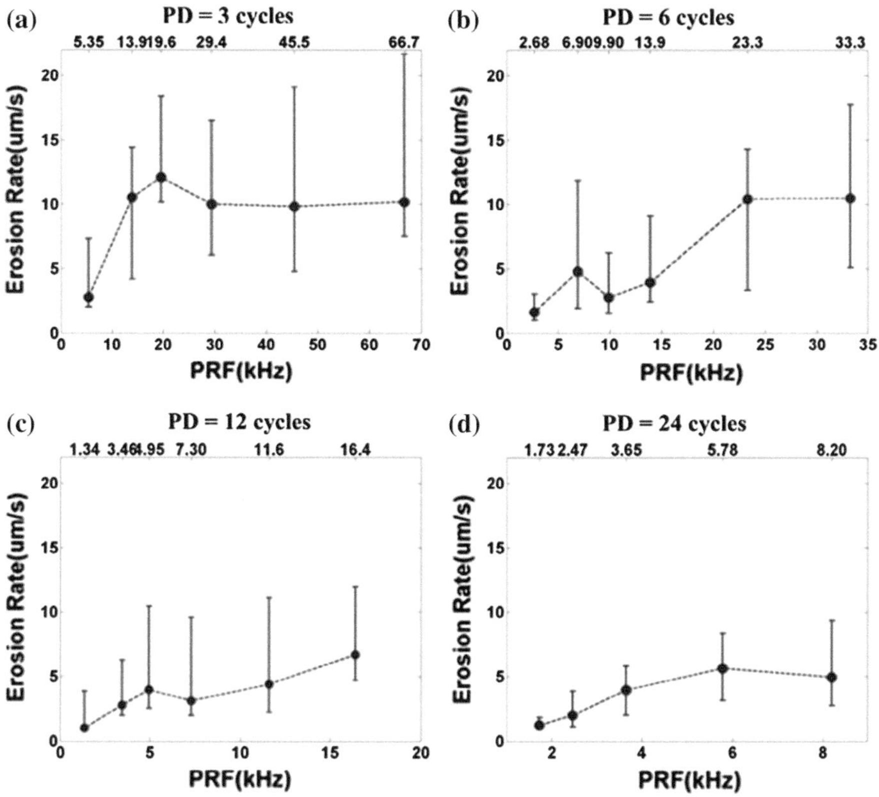

Fig. 6.56 The relationship between the erosion rate and duty cycle for a fixed pulse length. Reproduced with permission from Xu et al. (2004). © 2004 IEEE

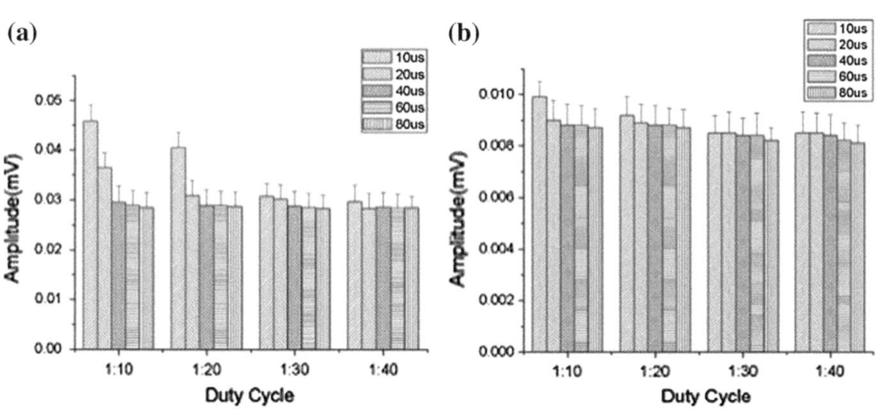

Fig. 6.57 PCD results: **a** broadband noise, ICD and **b** subharmonic, SCD. Reproduced with permission from Cao et al. (2013). © 2013 Elsevier B.V.

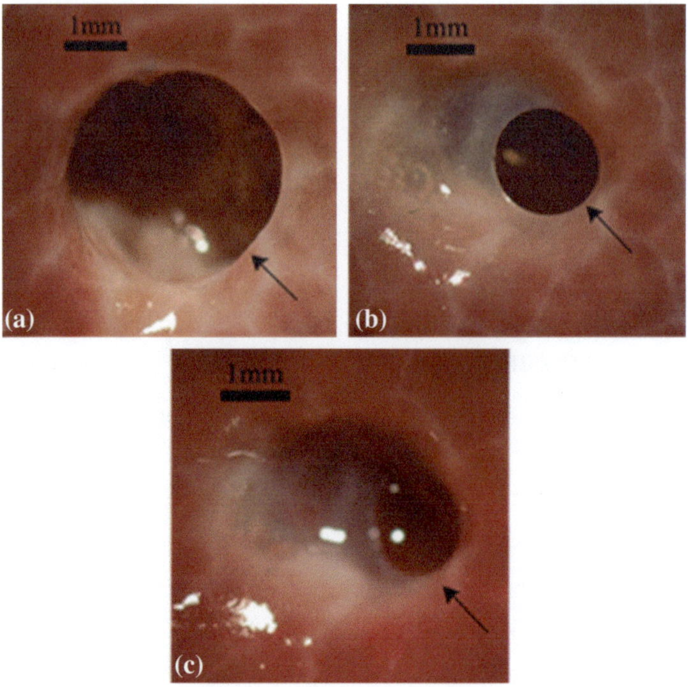

Fig. 6.58 **a** A well-defined hole was created with a pulse length of 10 μs and a duty cycle of 1:10. **b**, **c** Erosion result under the same experimental condition, except with pulse lengths of **b** 20 μs and **c** 40 μs. Reproduced with permission from Cao et al. (2013). © 2013 Elsevier B.V.

scored them from 1 to 3, where a type 1 lesion was a cavity filled with liquid that could be easily removed; a type 2 lesion had mixed mechanical and thermal damage; and a type 3 lesion was a solid thermal lesion with a vaporized core. Table 6.3 shows the effects of pulse length and duty factor on the lesion type. Higher duty cycles generated more evident thermal lesions.

Table 6.3 Effect of pulse duration and duty factor on the type of lesion produced in ex vivo heart samples (Khokhlova et al. 2011)

Pulse duration (ms)		Duty factor			
		0.005	0.01–0.02	0.03–0.2	0.2–1
4–10	(1–3t_b)	1	1	2	3
10–30	(3–10t_b)	1	2	2	3
30–100	(10–30t_b)	2	2	3	3
100–500	(> 30t_b)	3	3	3	3

6.6.4 Histotripsy Monitoring by Imaging Feedback

Histotripsy can cause tissue to transform from a soft solid to a fluid-like homogenate, changing its elasticity and backscatter intensity. Wang et al. (2012) monitored histotripsy using image-based feedback information, taking into account this tissue transition. For example, by quantifying the decrease of backscatter intensity, they estimated the degree of tissue fractionation using B-mode imaging. However, the sensitivity of this technique was not very high. Figure 6.59a shows the B-mode images of lesions produced in the phantom. When 100 therapeutic pulses were

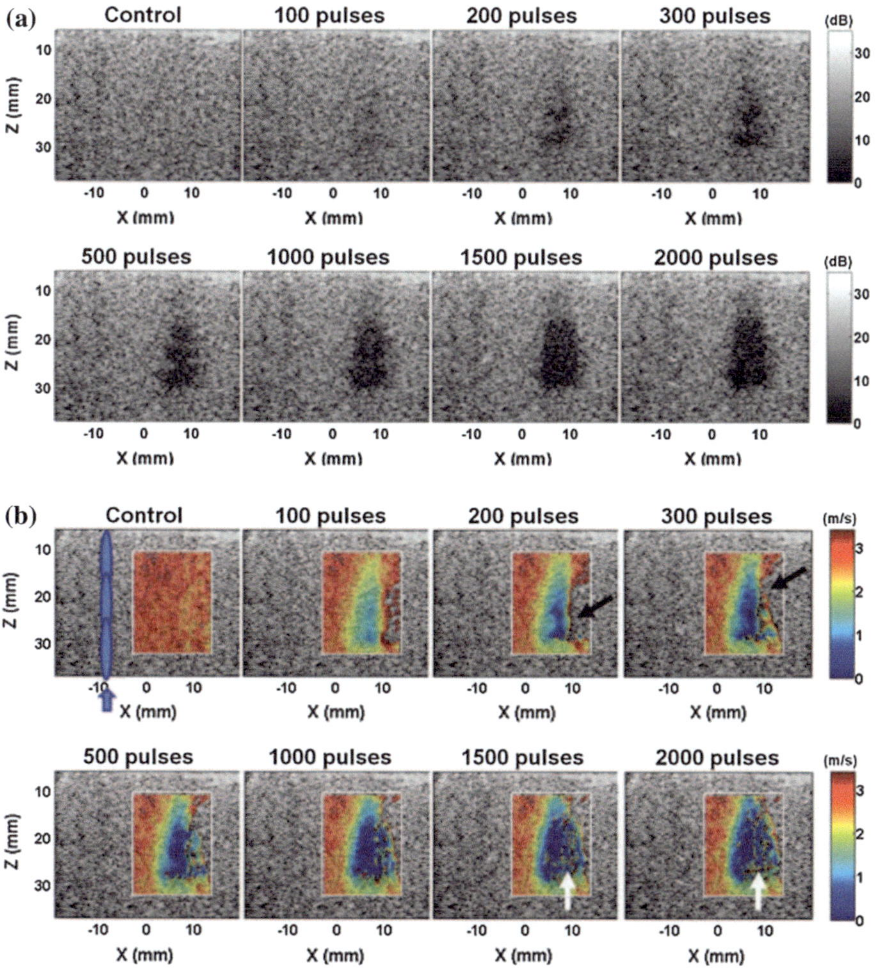

Fig. 6.59 **a** B-mode and **b** shear wave velocity images of a representative lesion produced in agar–graphite phantoms with increasing numbers of therapy pulses per treatment location. Reproduced with permission from Wang et al. (2012). © 2012 IEEE

applied, the treated area was not easily identified. On the other hand, when the elasticity of the lesions was measured with ultrasound shear wave elastography, there was higher specificity, sensitivity, and better correlation with tissue elasticity to the degree of tissue fractionation. Figure 6.59b shows the shear wave velocity images of the lesions. The shear wave velocity decreased in the treated volume as the pulse number increased, indicating a softer area was produced. Compared to B-mode images, this technique allowed for easier identification of the treated region, especially at lower numbers of applied pulses, as in Fig. 6.59b; a clear difference can be seen at 100 pulses. As there was strong correlation between the formed lesion and this imaging feedback method, it represents a promising way to monitor non-invasive histotripsy therapy.

6.7 Cavitation Dosage Control During Ablation

6.7.1 *Effect of Cavitation on Ablation Process*

Cavitation plays a key role in enhancing HIFU ablation, which can cause protein denaturation and cell death. Cavitation bubbles affect HIFU ablation in three main ways: first, the oscillating cavitation bubbles can trap acoustic energy and generate high-frequency broadband noises that have higher heat absorption capacity (Coussios et al. 2007). Second, shearing stress can convert acoustic energy to heat, increasing the opportunity for viscous absorption at the bubble surface (Thomas et al. 2005). Finally, when cavitation bubbles reflect acoustic waves, this can force the lesion shape to grow unpredictably and cause irreversible damage to healthy tissue (Chen et al. 2003). Thus, to make use of cavitation-enhanced heat absorption and reduce the lesion distortion, many researchers have focused on controlling cavitation during tissue ablation.

6.7.2 *Methods for Controlling Cavitation During Ablation*

6.7.2.1 Open-Loop Cavitation Control System

Thomas et al. (2005) monitored the temperature near the focus of HIFU using an E-type thermocouple, and they collected cavitation data using a PCD transducer during and after a 10-s continuous HIFU irradiation. A data acquisition (DAQ) computer was used to collect data and control the function generator which produces sinusoidal signal fed to the HIFU transducer. When the input power was lower than the cavitation threshold, there was no detectable heating. When cavitation threshold was exceeded, however, the temperature did change, as shown in Fig. 6.60. From 0 to 5 s, the temperature kept rising. From 5 to 10 s, even though

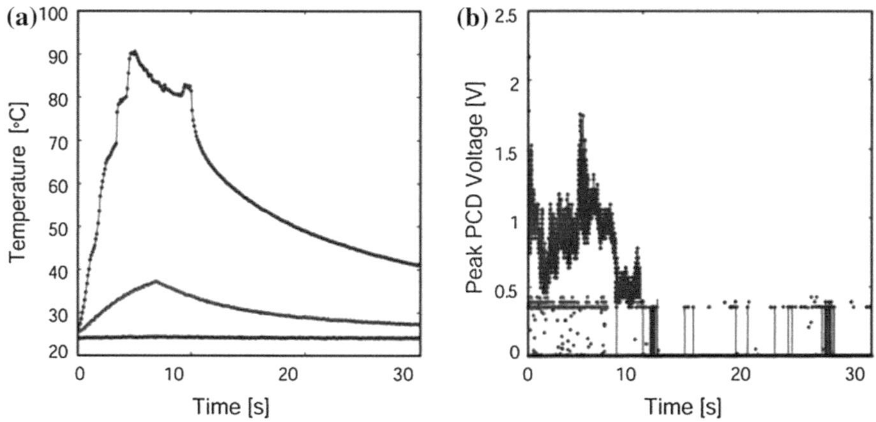

Fig. 6.60 Plot showing **a** temperature and **b** peak PCD signal as a function of time during and after insonation for three different focal pressures. Reproduced with permission from Thomas et al. (2005). © 2005 Acoustical Society of America

the HIFU source continued to deliver constant acoustic intensity, both the temperature and the PCD signals decreased sharply. This was because the cavitation bubbles transitioned from enhancing heating to forming a bubble shield. Cavitation bubbles that formed in front of the HIFU focus shielded the initial target and caused a decline in both temperature and PCD signals. As the dramatic increase of temperature always coincides with the onset of inertial cavitation activities, cavitation should be control to maintain bubble-enhanced heating and avoid bubble shield effects.

In another study, open-loop cavitation control was used to measure PCD signals instead of temperature, whereby the goal of sustaining the bubble-enhanced heating was assumed to be controlling cavitation (Coussios et al. 2007). A schematic of the cavitation control loop during HIFU exposure is shown in Fig. 6.61, consisting of an HIFU generation loop (in black) and a passive cavitation detection loop (in gray).

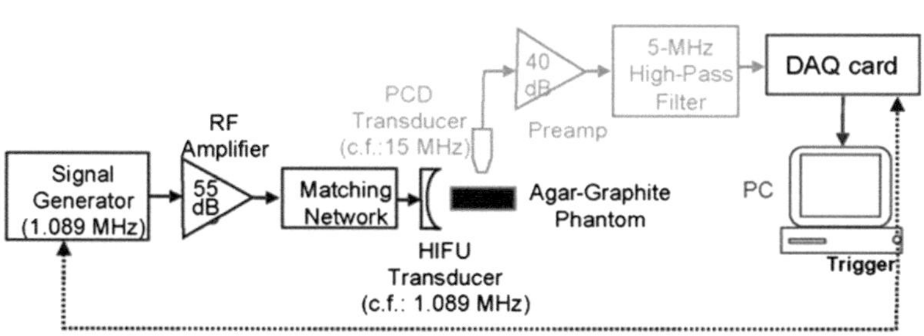

Fig. 6.61 Schematic of the cavitation control loop. Reproduced with permission from Coussios et al. (2007). © 2007 American Institute of Physics

In this study, pulsed HIFU excitation was used to form a population of preexisting bubbles which were used for the next pulses. On the other hand, the bubble clouds were confined to the focal region and would not foster the development of the cavitation field. A series of pulse lengths and duty cycles were tested for cavitation control, and the results are shown in Fig. 6.62. The results revealed optimal acoustic parameters of a 10 cycles pulse length and 20 % duty cycle for sustaining a high level of cavitation activity.

6.7.2.2 Closed-Loop Cavitation Control System

The open-loop cavitation control systems, like those discussed above, usually employ low-duty cycles, which facilitate heat diffusion and prevent temperature increases. Also, studies with these open-loop systems are often conducted in tissue-mimicking phantoms that have homogeneous acoustic parameters. However, cavitation in vivo is highly sensitive to local tissue characteristics and is difficult to predict, requiring a real-time control system for sustaining thermally relevant

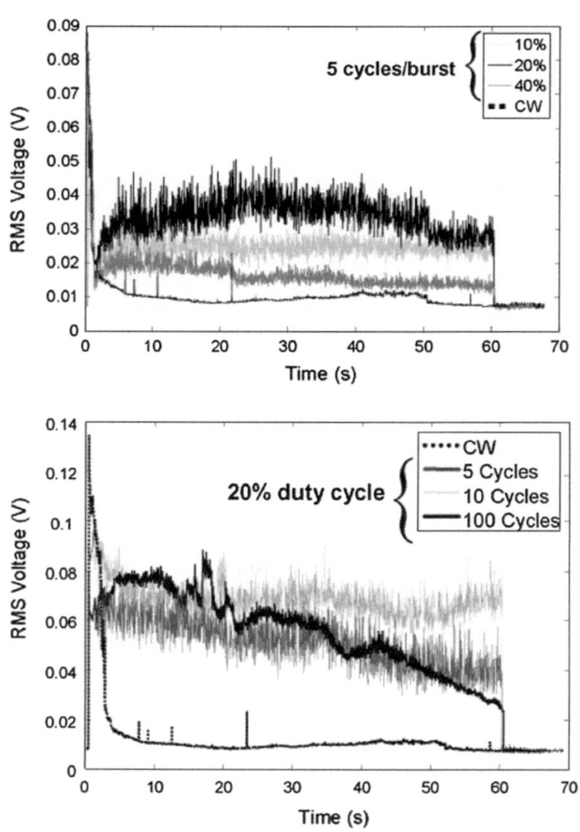

Fig. 6.62 The ability to maintain a constant level of cavitation activity by varying the duty cycle and the pulse duration following 1 s of continuous-wave HIFU exposure at a peak rarefaction pressure of 2.86 MPa. Reproduced with permission from Coussios et al. (2007). © 2007 American Institute of Physics

acoustic cavitation. Regarding a real-time system, the detected signals should be analyzed and compared periodically to preset range, within which the amount of energy was sufficient to produce desired effects, while at the same time, cavitation was restricted within the safety limits to prevent undesired effects.

Hockham et al. (2010) offered a schematic showing the main features of their sustained thermally relevant cavitation control system, shown in Fig. 6.63. In their studies, they found an optimal setting of K_1 and K_2 to control the function generator. Many values were used as set point: if the maximum was exceeded, K_1 was applied to the error signal; if the variance fell below the minimum, a different gain K_2 was applied to the error signal.

Their results of focal pressure amplitude, cavitation activity, focal temperature rise, and frontal position of the bubble cloud are shown in Fig. 6.64, and they are compared to results without the closed-loop feedback control. Without control, the temperature rose sharply at first and then dropped below the desired level for ablation. With control though, cavitation was sustained for a period of time, which ensured sufficient energy for tissue ablation. Also, the position of the bubble cloud was kept nearly stable within the HIFU focus, ensuring safe operation for the entire HIFU exposure process.

6.7.2.3 Multiple-Amplitude Enhanced Ablation

A combination of low- and high-amplitude ultrasound can also enhance the heating effect of HIFU. Takagi et al. (2010) and Nakamura et al. (2013) used a

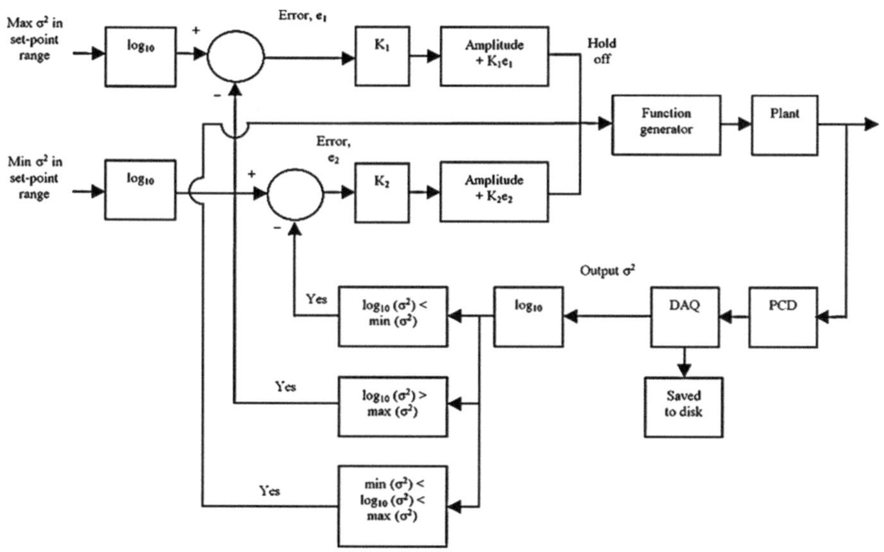

Fig. 6.63 Schematic diagram of the control system implanted for sustaining thermally relevant cavitation in the focal region of a HIFU transducer. Reproduced with permission from Hockham et al. (2010). © 2010 IEEE

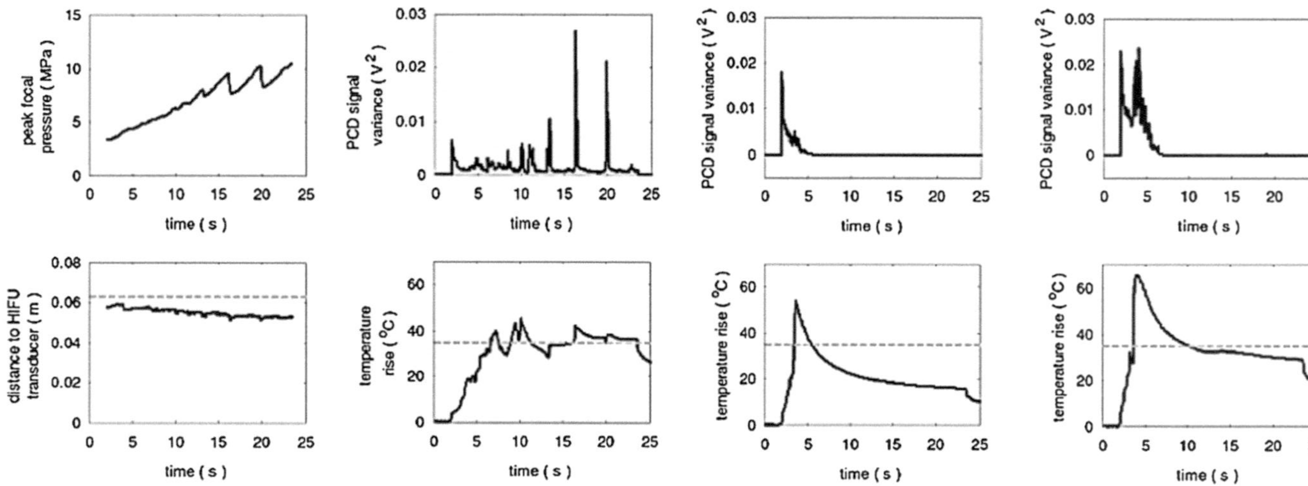

Fig. 6.64 *Columns 1 and 2*, results with closed-loop control; *Columns 3 and 4*, results without control. Clockwise, focal pressure amplitude, cavitation activity, focal temperature, and frontal position of the bubble cloud. Reproduced with permission from Hockham et al. (2010). © 2010 IEEE

high-intensity burst above the cavitation threshold to generate cavitation bubbles in chicken fillets, immediately followed by application of a low-intensity continuous-wave ultrasound at a certain intensity level and duration to sustain the triggered cavitation and to heat the tissue (Fig. 6.65). The subharmonics from the cavitation bubbles were detected, and the relationship between the half-subharmonic signal amplitude and the coagulation volume after the HIFU ablating was investigated.

The results indicated that tissue coagulation was more effective with a multiple-amplitude HIFU sequence than conventional HIFU exposure (Fig. 6.66). The authors proposed that this was due to the cavitation bubbles generated by the triggering pulse at the focus spot, which enhanced the heating locally and efficiently. With the triggered HIFU sequence, the shape of tissue coagulation was spherical, but with conventional HIFU exposure it was ellipsoidal, which is consistent with the known shape for the focal region of HIFU.

In follow-up work, Umemura et al. (2013) proposed that the second harmonic superimposed short pulses of focused ultrasound with extremely high intensity could be used as the trigger pulses (Fig. 6.67). These trigger pulses could again be immediately followed by focused ultrasound for heating, but using an intensity similar to or less than that of the normal HIFU treatment that sustained the triggered cavitation in the earlier study in excised chicken breast tissue. First, they used the negative peak pressure-emphasized waves to initiate cavitation, and then they used the positive peak pressure-emphasized waves to enlarge the generated microbubble clouds. Finally, they used the heating waves to heat the microbubbles in the clouds.

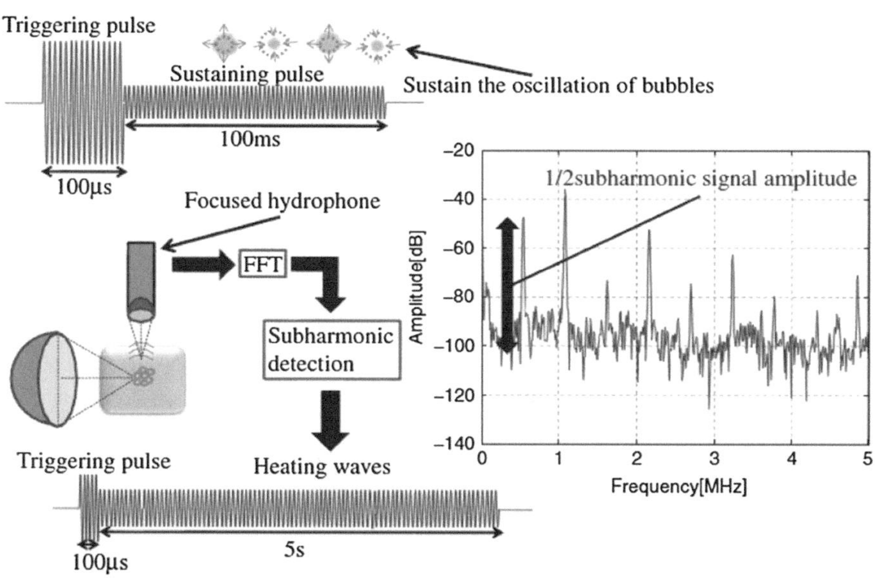

Fig. 6.65 Schematic of subharmonic detection method and experimental procedure. Reproduced with permission from Takagi et al. (2010). © 2005 The Japan Society of Applied Physics

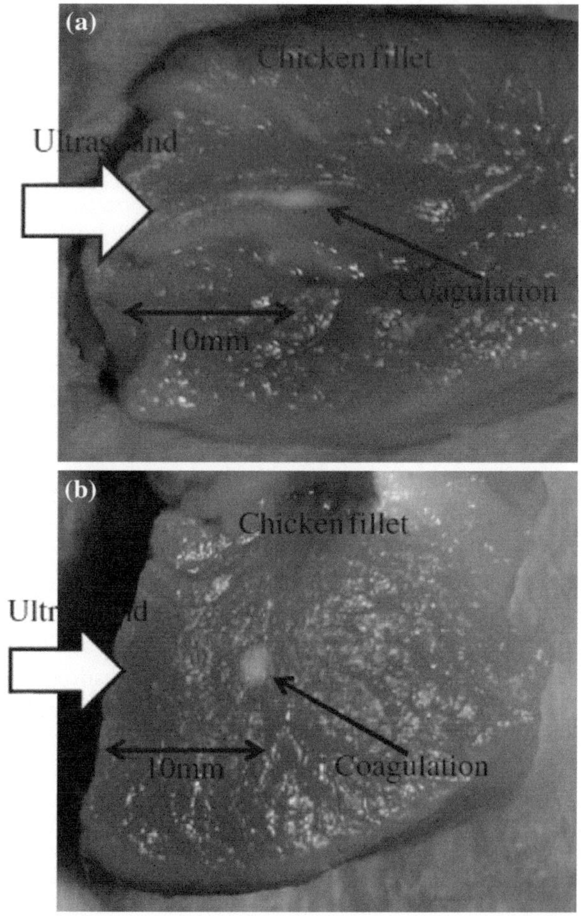

Fig. 6.66 Difference in partially coagulated region between **a** conventional HIFU exposure (heating waves alone) and **b** triggered HIFU exposure. Reproduced with permission from Takagi et al. (2010). © 2005 The Japan Society of Applied Physics

The effectiveness of the generated clouds in accelerating ultrasonically induced thermal coagulation was confirmed (Fig. 6.68), and they suggested that the combined ultrasonic exposure sequence might be ideal for the acoustic generation of microbubbles to enhance HIFU treatment.

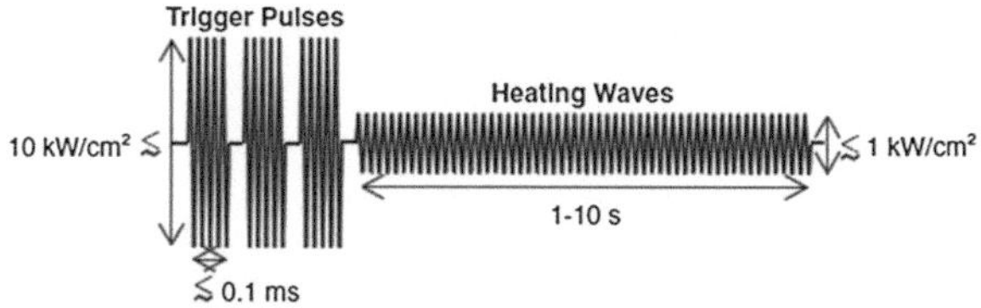

Fig. 6.67 Triggered HIFU exposure sequence consisting of extremely high-intensity trigger pulses followed by moderately high-intensity heating waves. Reproduced with permission from Umemura et al. (2013). © 2013 The Japan Society of Applied Physics

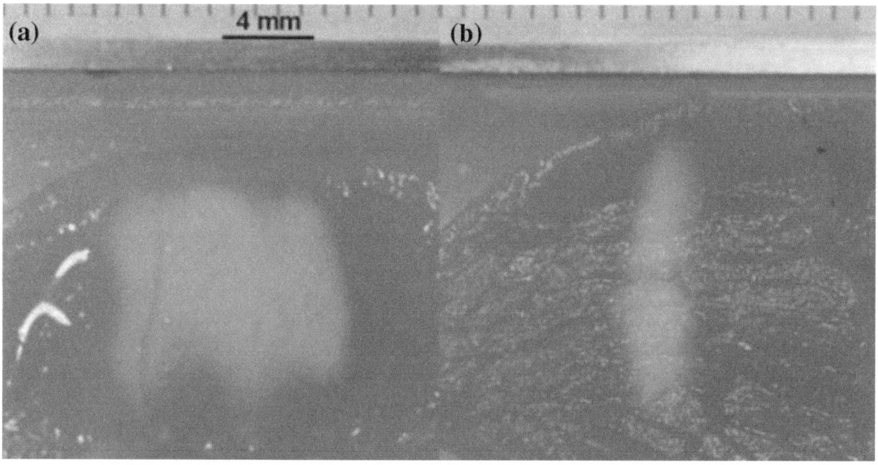

Fig. 6.68 Gross cross section of excised chicken breast tissue blocks after they were exposed to heating waves **a** with and **b** without prior exposure to the trigger pulse. Reproduced with permission from Umemura et al. (2013). © 2013 The Japan Society of Applied Physics

6.7.2.4 Laser-Enhanced Ablation

Cavitation bubbles can be formed by direct heating and subsequent explosive vaporization of liquid through optical breakdown, which was the result of irradiating an electrolyte solution with a focused laser. Laser-induced cavitation has been used for a broad range of applications (Ibrahimkutty et al. 2012; Lohse 1998; Frenz et al. 1996). Recently, Cui and Yang (2011) reported a heat enhancement effect during photoacoustic imaging (PAI)-guided HIFU therapy. PAI is an emerging non-ionizing and noninvasive imaging technology based on the photoacoustic effect, and it has high optical contrast and excellent spatial resolution (Xu and Wang 2006). This technique has been applied for early detection of cancers and monitoring thermal lesions generated by HIFU or other means (Funke et al. 2009; Chitnis et al. 2010; Gusev et al. 1993; Esenaliev et al. 1999; Xu and Wang 2006). Earlier reports have suggested that this heat enhancement is caused by the nucleation of cavitation that occurs when the diagnostic laser light beam illuminates the sample concurrently with HIFU radiation. In one such study that applied a laser dose below the safety limit recommended by American National Standards Institute, Cui et al. (2013) used a tunable optical parameter oscillator (OPO) laser with a pulse repetition rate of 10 Hz as the irradiation source to induce cavitation. In addition, a condenser lens was used to make the laser light confocal with the 5-MHz HIFU transducer (Fig. 6.69). Cavitation activity was investigated by PCD. The temperature increase was obtained from a T-type thermocouple that was inserted into the tissue sample through a needle and placed 0.5 mm away from the focal point of the HIFU transducer.

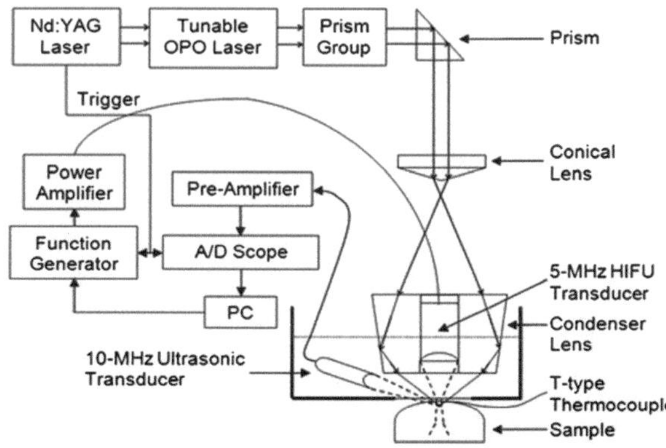

Fig. 6.69 System schematic. Reproduced with permission from Cui and Yang (2011). © 2011 American Institute of Physics

During ablation, there was a clear temperature enhancement when the laser system was operating, indicating that HIFU heating can be significantly enhanced by the assistance of laser light (Fig. 6.70). After HIFU exposure, the tissue sample was sliced, revealing an axial section through the lesion (Fig. 6.71). Lesions were larger when the laser light was on, compared with lesions formed without laser light. Also, empty cavities were observed in the central region of the lesions, suggesting that cavitation might have occurred during the heating process when the laser concurrently illuminated the sample.

These observations demonstrate that PAI-guided HIFU is advantageous. This technique can facilitate the HIFU treatment process by enhancing heating at a relatively low HIFU intensity, while PAI monitors the treatment process, without requiring any micro- or nanoparticles to be present in the bloodstream.

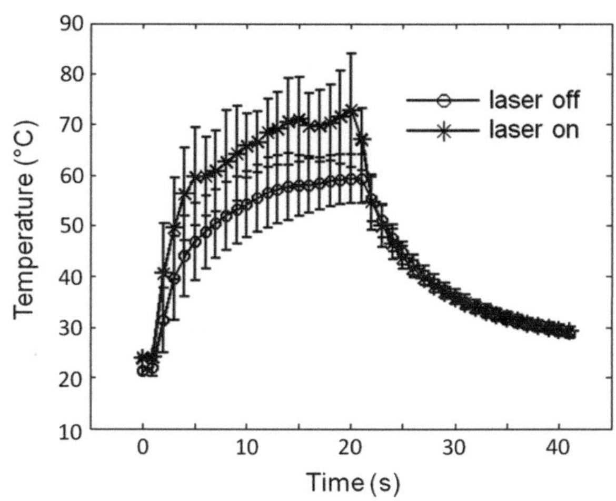

Fig. 6.70 Temperature enhancement during PAI-guided HIFU. Reproduced with permission from Cui and Yang (2011). © 2011 American Institute of Physics

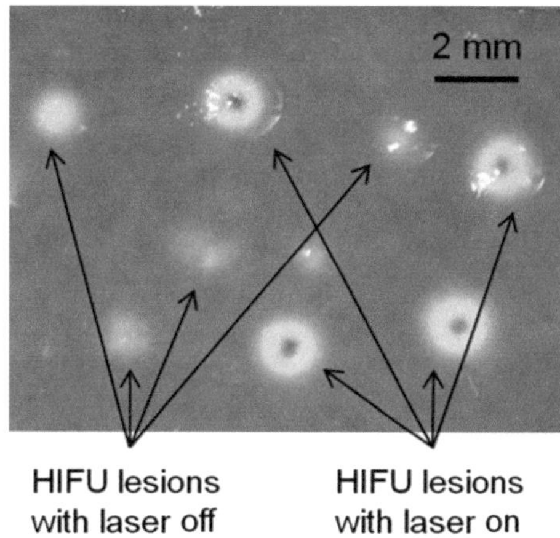

Fig. 6.71 Photograph of the HIFU lesions inside a tissue sample after HIFU exposure. Reproduced with permission from Cui and Yang (2011). © 2011 American Institute of Physics

6.8 Drug Delivery Controlled by Low-Intensity Focused Ultrasound

Drug delivery using therapeutic ultrasound is a developing technique, whereby drug molecules are driven into the target area using ultrasound waves. This technique requires proper drug carriers, of which the most commonly used are ultrasound contrast agents (UCAs). More conventionally, UCAs are used as cavitation nuclei to enhance cavitation efficiency. Cavitation appears to play two roles during drug delivery: first, it disrupts the structure of the carrier and releases the drug. Second, it changes the permeability of cell membranes and capillaries. The challenges researchers face are how to control the amount of drug released and how to release at the right location. These challenges are related with the characteristics of the ultrasound transducer and to acoustic parameters.

To date, most focused ultrasound-induced therapeutic applications are performed using a separate single-element transducer with a fixed focus size, which cannot vary as needed. A linear array capable of synthesizing the focus to have variable size will be competitive for drug delivery applications, since microbubbles can be destroyed in a predefined region according to the treatment plan. In terms of cavitation detection, standard PCD using a focused or unfocused single-element transducer has been employed to detect cavitation arising from ultrasound–UCA interactions. However, standard PCD limits the spatial sensitivity as well as clinical applicability due to the confocal setting between devices. Recently, multi-element ultrasound array detectors have been used to passively image cavitation during continuous-wave HIFU ablation or other pulsed-wave therapies such as drug delivery under the development of customized beamforming methods. These multi-element transducers can detect cavitation at a lower exposure amplitude compared to single-element PCD transducers. In one such study, Xu et al. (2014) proposed a combined approach based on a programmable diagnostic ultrasound scanner, in which a variable-sized focal region where the acoustic pressure was above the UCA fragmentation threshold was synthesized by reasonably matching the excitation voltage and the transmit aperture. They monitored the temporal and spatial distribution of the UCAs during the destruction progress using ultrafast 2D transmission and reception at 400 Hz. Simultaneously during destruction, they extracted the broadband noise emission from the backscattering of the focused release bursts (i.e., destruction pulses). The authors hypothesized that the microbubbles could be destroyed in this variable region, and the destruction efficiency as well as the corresponding ICD could be regulated by manipulating the transmission parameters. We discuss their experiments and results in this section.

6.8.1 Composite Sequence and Focal Pattern of Low-Intensity Focused Ultrasound

6.8.1.1 Composite Sequence of Imaging and Destruction Pulse

In their experiments, S. Xu et al. used a programmable Ultrasonix research platform scanner (Ultrasonix, Richmond, BC, Canada). A diagram of proposed strategy is illustrated in Fig. 6.72. A composite sequence was used, consisting of an ultrafast 2D imaging frame (Fig. 6.72a) and a focused destruction pulse (Fig. 6.72c), which were delivered and received along the same sector line, to destroy and image

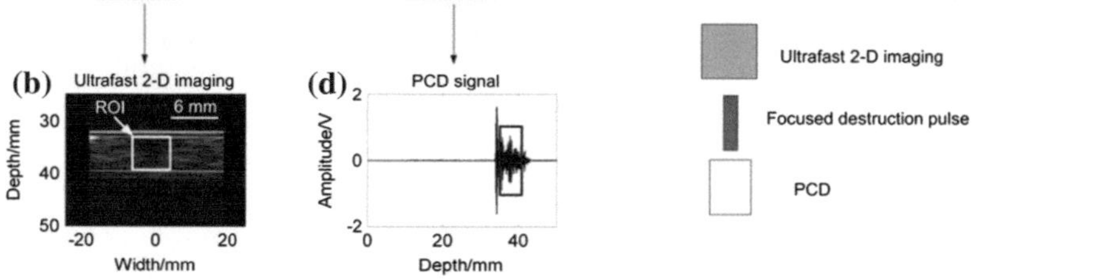

Fig. 6.72 **a** Beam pattern of transmitted wide beam used for monitoring microbubble destruction. **b** An ultrafast 2D ultrasound image with the region of interest (ROI) indicated by the *white box*. **c** Beam pattern of focused destruction pulse used for microbubble destruction. **d** Backscattering of focused destruction pulse with the signals in the *black box* for subsequent calculation in the frequency domain. **e** Composite sequence for microbubble destruction, ultrafast 2D ultrasound monitoring imaging before and after microbubble destruction and array-based PCD during destruction. Reproduced with permission from Xu et al. (2014). © American Institute of Ultrasound in Medicine

microbubbles (Fig. 6.72e). The composite sequence was performed on a 38-mm aperture by a 128-element linear array (L14-5/38, center frequency: 5 MHz, −3 dB bandwidth: 70 %; Ultrasonix, Richmond, BC, Canada) at a sampling frequency of 20 M/s. Beamformed radio frequency (RF) data of the backscattering signal from the focused destruction pulse were stored in cine memory, and prebeamformed channel RF data of the backscattering signal from ultrafast 2D imaging were stored in Sonix DAQ (Ultrasonix, Richmond, BC, Canada).

6.8.1.2 Focal Pattern of Low-Intensity Focused Ultrasound

The focal region patterns excited with four transmit apertures are displayed in Fig. 6.73. As expected, the width of the destruction pulse was enlarged as the transmit aperture decreased. From Fig. 6.72e, we can see that the maximum

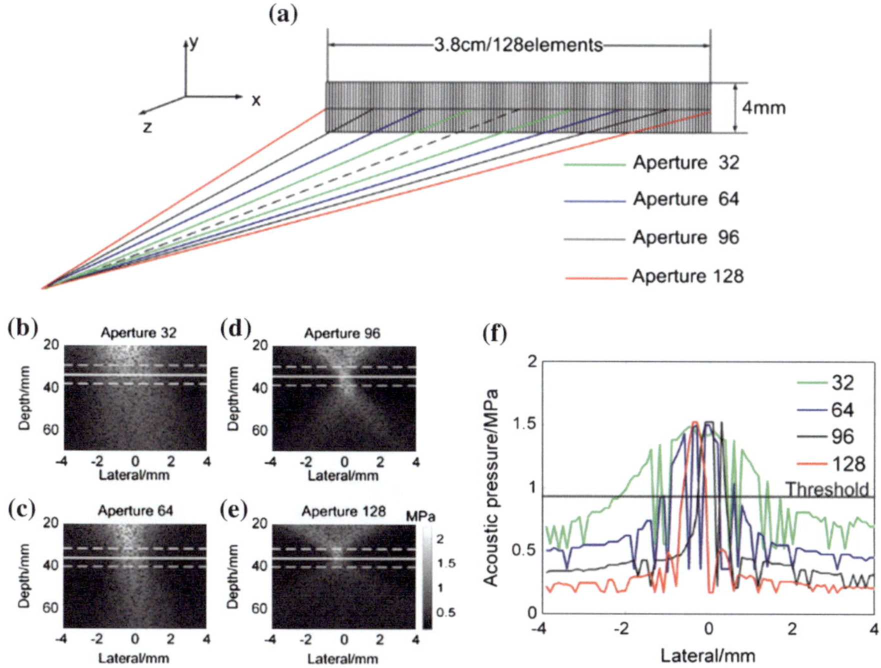

Fig. 6.73 **a** The linear array dimension and the active elements for each given aperture of the destruction pulse. **b–e** The axial–transverse profiles of a single-cycle 5-MHz destruction pulse excited with 32, 64, 96, and 128 elements. **f** Acoustic pressure versus lateral position for the focal region synthesized with four transmit apertures. The *dashed white lines* in panels (**b–e**) indicate the position of latex tube, and the *solid white line* represents the axial position where the acoustic pressure in (**f**) was extracted. The up-and-down patterns on each line in (**f**) were caused by the asynchronous working of ultrasound transmission and the receiver during the acoustic field calibration. Reproduced with permission from Xu et al. (2014). © American Institute of Ultrasound in Medicine

acoustic pressure in each focused ultrasound–UCA interaction region was approximately 1.52 MPa.

6.8.2 *Spatially Controlled Microbubble Destruction*

6.8.2.1 Monitoring Imaging

Figure 6.74 shows the ultrafast 2D images at specific time instants during the sonication, excited with four different apertures. By direct observation, the destruction region when the smallest aperture (32 elements) was used continued to enlarge until the exposure time reached ~1 s and then the area remained almost unchanged for the rest of the exposure. In cases where the destruction pulse was excited with larger apertures (64, 96, or 128 elements), the destruction region continued to grow over the whole sonication period. We can see from the figure that at a specific time instant, the destruction area in the small-aperture sonication experiment was larger than that excited with the larger apertures.

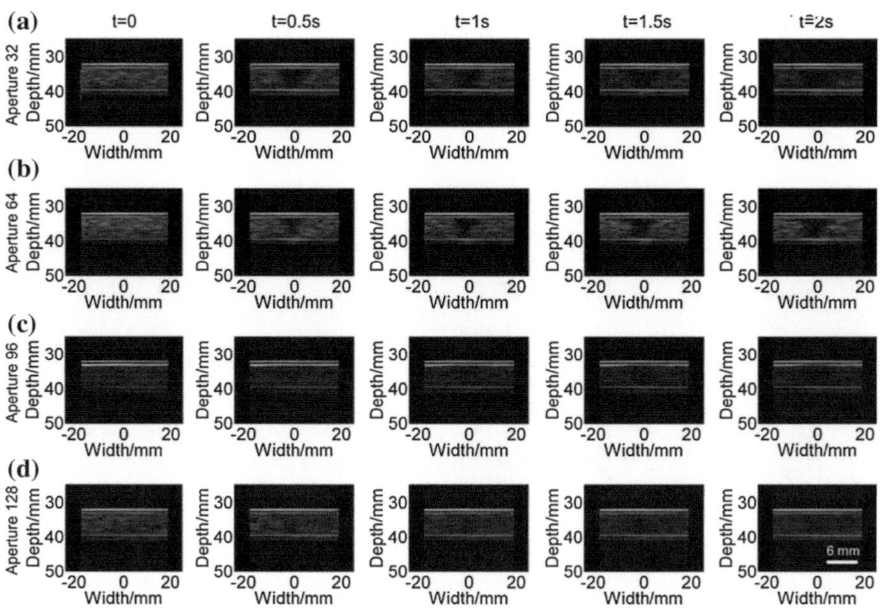

Fig. 6.74 Ultrafast 2D images acquired at specific time instants during microbubble destruction in a static microbubble suspension. **a–d** Images correspond to experiments performed at the transmit aperture of 32, 64, 96, and 128 elements. The destruction pulse was a single-cycle pulse centered at 4 MHz. Reproduced with permission from Xu et al. (2014). © American Institute of Ultrasound in Medicine

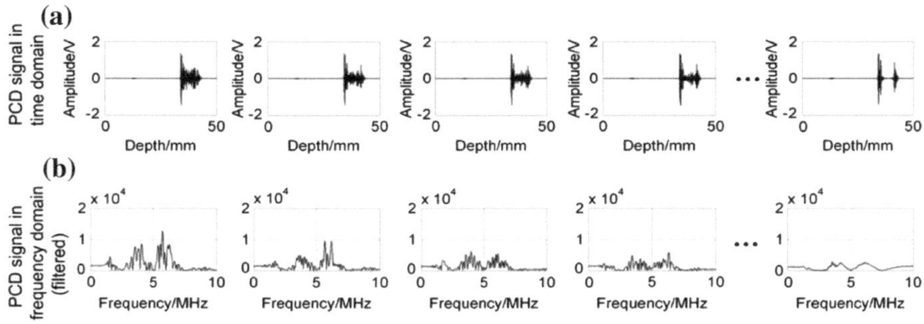

Fig. 6.75 Typical sequence examples of **a** the destruction pulse backscattering signal in the time domain and **b** the destruction pulse backscattering signal filtered in the frequency domain. Reproduced with permission from Xu et al. (2014). © American Institute of Ultrasound in Medicine

6.8.2.2 Simultaneous Cavitation Activity

Figure 6.75a, b is the time-varying backscatter signals of destruction pulse in the time and frequency domains, respectively. As shown, the backscattering of the destruction pulse included broadband noise emissions, which faded away with time due to the exhaustion of microbubbles. Based on these results, the destruction efficiency as well as the corresponding inertial cavitation dose could be regulated by manipulating the transmission parameters.

6.8.3 Temporally Controlled Microbubble Destruction

6.8.3.1 Influence of Aperture

Figure 6.76 shows the root-mean-square (RMS) curves calculated from the back-scattering traces of the destruction pulse and corresponding time-intensity curves

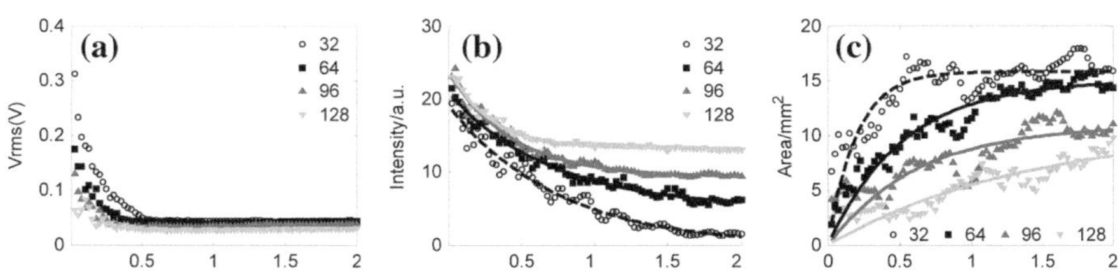

Fig. 6.76 The measured **a** root-mean-square (RMS) curves, **b** TICs, and **c** TACs as a function of time for different transmit apertures of the focused destruction pulse (32, 64, 96, and 128 elements). The least-squares fits of the TIC and TAC to the exponential function are shown as the smooth lines in (**b**) and (**c**). One dot in every 10 is shown for visibility of the graph (a.u. means arbitrary units). Reproduced with permission from Xu et al. (2014). © American Institute of Ultrasound in Medicine

(TICs) and time-area curves (TACs) from the ultrafast 2D monitoring at four transmit apertures (32, 64, 96, 128 elements). The destruction pulse was generated with a single-cycle pulse at 4 MHz. From Fig. 6.76a, we can see that the detected broadband noise emission was higher at smaller aperture and that the RMS curve for this transmit aperture had a larger area under curve, viz. ICD. Furthermore, the RMS curves showed a slight difference between different apertures beyond 0.5 s. From Fig. 6.76b, it is apparent that intensity of the surviving microbubbles in the fixed ROI decreased more rapidly as the transmit aperture decreased, implying that microbubbles were destructed more efficiently in a focal region synthesized with a smaller transmit aperture. The difference in the final intensity levels achieved for each sonication due to the remaining microbubbles in the ROI indicated that in the fixed ROI, fewer microbubbles were destructed as the transmit aperture increased. Figure 6.76c illustrates the temporal evolution of the destruction area at each aperture. At the smallest aperture (32 elements), the destruction region expanded rapidly for the first 0.5 s because of the initiation by the composite sequence, and then, it slowly reached a relatively stabilized value (15.88 mm^2); nevertheless, at the largest aperture (128 elements), the destruction region enlarged roughly linearly with time and then reached a non-stabilized destruction area of <8 mm^2 by the end of 2 s.

6.8.3.2 Influence of Pulse Length

Figure 6.77 shows the RMS curves from the backscattering of the destruction pulse and the corresponding TICs and TACs calculated from the ultrafast 2D monitoring

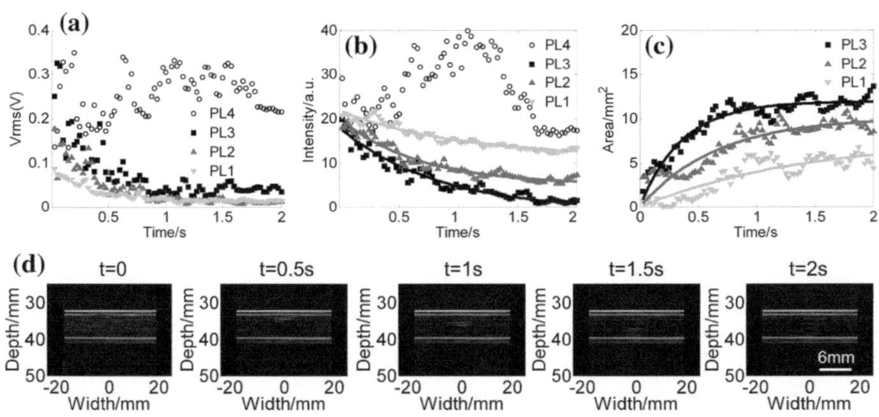

Fig. 6.77 The measured **a** RMS curves, **b** TICs, and **c** TACs as a function of time with different pulse lengths of the focused destruction pulse (1, 2, 3, and 4 cycles). The least-squares fits of the TIC and TAC to the exponential function detailed in the text are shown as the smooth lines in (**b**) and (**c**). One dot in every 10 is shown for visibility of the graph. **d** The ultrafast 2D images acquired at specific time instants during the sonication where a "cascade" effect was observed using a 4-cycle destruction pulse (a.u. means arbitrary units). Reproduced with permission from Xu et al. (2014). © American Institute of Ultrasound in Medicine

at four pulse lengths (1, 2, 3, and 4 cycles). The destruction pulse was excited with an aperture of 128 elements at 5 MHz. In Fig. 6.77a, it is apparent that the detected broadband noise emission was higher at longer pulse lengths and that the RMS curve for longer pulse lengths had larger ICD values. Figure 6.77b shows that as pulse lengths became longer, the intensity decreased more rapidly and the final intensity in the fixed ROI was lower. Figure 6.77c illustrates that in terms of the area of destruction region, the rise rate (β) and the plateau (P) of the exponential fitting curve both increased with the pulse length. Specifically, at a pulse length of 3 cycles, the destruction area expanded rapidly from the beginning of the composite sequence to ~1 s, and then, it expanded slowly to a relatively stable size (12 mm^2); nevertheless, at a pulse length of 1 cycle, the destruction area enlarged roughly linearly with time and reached a non-stabilized destruction area of <6 mm^2 by the end of 2 s. Further, it should be noted that when the destruction pulse was excited with a 4-cycle pulse, both the TIC and the RMS curve demonstrated an increase and a subsequent fall. From the ultrafast 2D images shown in Fig. 6.77d, we can see a "cascade" phenomenon whereby old bubbles were rapidly destroyed and then new bubbles were produced and pushed toward the distal wall of the tube. Hence, the destruction area in this sonication condition was not calculated.

6.8.3.3 Influence of Frequency

Figure 6.78 shows the RMS curves calculated from the backscattering traces of destruction pulse and corresponding TICs and TACs from the ultrafast 2D monitoring at four transmit frequencies (3, 4, 5, and 6 MHz). The destruction pulse was excited with a single-cycle pulse at an aperture of 96 elements. It should be noted that the acoustic pressure of the destruction pulse excited with four frequencies was calibrated to be 1.9 MPa by tuning the excitation voltage. From Fig. 6.78a, it is shown that the detected broadband noise emission was higher at the lower frequency and that the RMS curve associated with the lower frequency had a larger

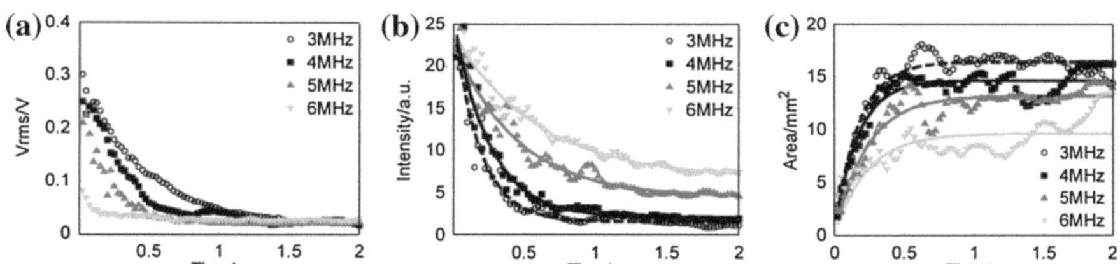

Fig. 6.78 The measured **a** RMS curves, **b** TICs, and **c** TACs as a function of the transmit frequency of the destruction pulse (3, 4, 5, and 6 MHz). The least-squares fits of the TIC and TAC to the exponential function detailed in the text are shown as the smooth lines in (**b**) and (**c**) The destruction pulse was excited with a single-cycle pulse at an aperture of 96 elements. One dot in every 10 is shown for readability of the graph (a.u. means arbitrary units). Reproduced with permission from Xu et al. (2014). © American Institute of Ultrasound in Medicine

ICD. From Fig. 6.78b, we can see that the intensity of the surviving microbubbles in the fixed ROI decreased more rapidly with decreasing frequency, implying that the microbubbles were destroyed more efficiently in a focal region synthesized with a lower frequency. The difference of the final intensity levels achieved in each sonication due to the remaining microbubbles in the ROI indicated that in the fixed ROI, fewer microbubbles were destroyed as the frequency increased. Figure 6.78c illustrates the temporal evolution of the destruction area at each frequency. At 3 MHz, the destruction area expanded rapidly from the beginning of the composite sequence to <0.5 s and achieved a destruction area of 16.4 mm^2. However, at 6 MHz, the destruction area only reached 9.6 mm^2 after 0.5 s.

In this section, we discussed a new strategy to achieve microbubble destruction in a spatially and temporally controlled manner. Overall, this technique is a potentially helpful tool for monitoring and UCA-enhanced therapeutic applications in vivo.

6.9 Summary

Cavitation and bubble dynamics have been studied for many years. Early on, cavitation occurrence was often specifically avoided because it is random, difficult to control and caused mechanical damage to industrial devices. However, it is very useful in biomedical applications, where it can act as a mediator of both the mechanical and thermal effects of certain therapies. But whether cavitation can be properly used to enhance therapeutic efficiency depends on controlling the cavitation process, both spatially and temporally.

When discussing cavitation process control, there are two scopes of approach. One is fundamentally based on acoustic parameters or adjusting the ambient environmental conditions, through which we can regulate the inception, amplitude, intensity of bubble collapse, etc. Therefore, we first reviewed the effects of temperature, pressure, and frequency on cavitation progress (Sects. 6.2–6.4). These kinds of studies always involve single-bubble dynamics, which pave the way to help us understand cavitation mechanisms and are the basis of controlling cavitation and using it in further applications. The other approach for discussing cavitation control involves the biomedical applications of cavitation, which depend on the activities of bubble clouds. The dynamics of bubble clouds are much more complex than those of single bubbles: in clouds, the region where cavitation takes place is determined by the ultrasound field and the energy distribution. Conversely, the generated cavitation bubbles can distort the pressure field and shield the target. Thus, the acoustic parameters are not the only important factor; the structure of the clouds and their interactions also affect the results. Therefore, the cavitation process should be carefully adjusted according to different biomedical needs. In Sects. 6.6–6.8, we covered some specific cavitation control-related applications, such as precise removal of soft tissue, enhancement of HIFU ablation, and temporally and spatially controlled drug delivery. To connect these two approaches, Sect. 6.5

focused on radiation force-constrained cavitation. By building a standing field or using an acoustic counterbalance appliance, bubble clouds can be trapped where they are generated or needed, regardless of the acoustic radiation force. As such, some fundamental research about bubble clouds can also be carried out in free fields.

While good progress has been made in cavitation control and associated applications, challenges remain. Cavitation can induce many physical, chemical, and biological effects, so investigations must still be performed to find out how to take advantage of each effect and prevent undesired effects. For wider therapeutic applications and studies, optimizing acoustics and designing new ultrasound devices is also highly important.

References

Ahmed S. Investigation of the temperature effects on induced impact pressure and cavitation erosion. Wear. 1998;218:119–27.

Auret J, Damm O, Wright G, Robinson F. Cavitation erosion of copper and aluminium in water at elevated temperature. Tribol Int. 1993;26:421–9.

Beckett MA, Hua I. Impact of ultrasonic frequency on aqueous sonoluminescence and sonochemistry. J Phys Chem A. 2001;105:3796–802.

Brennen CE. Cavitation and bubble dynamics. New York: Cambridge University Press; 2013.

Brotchie A, Grieser F, Ashokkumar M. Effect of power and frequency on bubble-size distributions in acoustic cavitation. Phys Rev Lett. 2009;102:084302.

Cao H, Yin H, Qiao Y, Zhang S, Wan M. Sonochemiluminescence observation and acoustic detection of cavitation induced by pulsed HIFU at a tissue–fluid interface. Ultrason Sonochem. 2013;20:1370–5.

Caupin F, Herbert E. Cavitation in water: a review. CR Phys. 2006;7:1000–17.

Chen WS, Lafon C, Matula TJ, Vaezy S, Crum LA. Mechanisms of lesion formation in high intensity focused ultrasound therapy. Acoust Res Lett Online. 2003;4:41–6.

Chitnis PV, Brecht H-P, Su R, Oraevsky AA. Feasibility of optoacoustic visualization of high-intensity focused ultrasound-induced thermal lesions in live tissue. J Biomed Opt. 2010;15:021313–5.

Ciuti P, Dezhkunov N, Francescutto A, Calligaris F, Sturman F. Study into mechanisms of the enhancement of multibubble sonoluminescence emission in interacting fields of different frequencies. Ultrason Sonochem. 2003;10:337–41.

Cooper M, Xu Z, Rothman ED, Levin AM, Advincula AP. Fowlkes JB, Cain CA. Controlled ultrasound tissue erosion: the effects of tissue type, exposure parameters and the role of dynamic microbubble activity. In: Ultrasonics symposium. IEEE; 2004. pp. 1808–1811.

Coussios CC, Collin JR, Muckle AP. Non-invasive monitoring and control of inertial cavitation dynamics during HIFU exposure in vitro. In: 6th International symposium on therapeutic ultrasound, vol 1. AIP Publishing; 2007. pp. 164–70.

Cui H, Yang X. Enhanced-heating effect during photoacoustic imaging-guided high-intensity focused ultrasound. Appl Phys Lett. 2011;99:231113.

Cui H, Zhang T, Yang X. Laser-enhanced cavitation during high intensity focused ultrasound: An in vivo study. Appl Phys Lett. 2013;102:133702.

Esenaliev RO, Karabutov AA, Oraevsky AA. Sensitivity of laser opto-acoustic imaging in detection of small deeply embedded tumors. IEEE J Sel Top Quantum Electron. 1999;5:981–8.

Fabiilli ML, Lee JA, Kripfgans OD, Carson PL, Fowlkes JB. Delivery of water-soluble drugs using acoustically triggered perfluorocarbon double emulsions. Pharm Res. 2010;27:2753–65.

Franc JP, Michel JM. Fundamentals of cavitation, vol. 76. Dordrecht: Springer; 2006.

Frenz M, Paltauf G, Schmidt-Kloiber H. Laser-generated cavitation in absorbing liquid induced by acoustic diffraction. Phys Rev Lett. 1996;76:3546.

Funke AR, Aubry J-F, Fink M, Boccara A-C, Bossy E. Photoacoustic guidance of high intensity focused ultrasound with selective optical contrasts and time-reversal. Appl Phys Lett. 2009;94:054102.

Germano M, Alippi A, Bettucci A, Brizi F, Passeri D. Water temperature dependence of single bubble sonoluminescence threshold. Ultrasonics. 2010;50:81–3.

Gusev VE, Karabutov AA, Hendzel K. Laser optoacoustics. New York: American Institute of Physics; 1993.

Hattori S, Nimura A, Yada H. Effect of temperature on cavitation erosion of 9Cr steel in liquid metal. Wear. 2013;303:269–75.

Herbert E, Balibar S, Caupin F. Cavitation pressure in water. Phys Rev E. 2006;74:041603.

Hiller R, Putterman S, Barber BP. Spectrum of synchronous picosecond sonoluminescence. J Acoust Soc Am. 1992;92:2454.

Hockham N, Coussios CC, Arora M. A real-time controller for sustaining thermally relevant acoustic cavitation during ultrasound therapy. IEEE Trans Ultrason Ferroelectr Freq Control. 2010;57:2685–94.

Holt RG, Roy RA. Measurements of bubble-enhanced heating from focused, MHz-frequency ultrasound in a tissue-mimicking material. Ultrasound Med Biol. 2001;27:1399–412.

Huang D, Zhuang Y. Temperature and cavitation. Part C: J Mech Eng Sci. 2008;222:207–11.

Huang W, Chen W, Liu Yn, Gao X. The evolution of the cavitation bubble driven by different sound pressure. Ultrasonics. 2006;44:e407–10.

Ibrahimkutty S, Wagener P, Menzel A, Plech A, Barcikowski S. Nanoparticle formation in a cavitation bubble after pulsed laser ablation in liquid studied with high time resolution small angle x-ray scattering. Appl Phys Lett. 2012;101:103104.

Imre AR, Maris HJ, Williams PR. Liquids under negative pressure. In: Proceedings of the NATO advanced research workshop of liquids under negative pressure Budapest, Hungary, 23–25 Feb 2002, vol 84. Dordrecht: Springer.

Kawabata KI, Asami R, Azuma T, Yoshikawa H, Umemura SI. Cavitation assisted HIFU with phase-change nano droplet. In: International ultrasonics symposium. IEEE; 2008. pp. 780–783.

Kawabata KI, Sugita N, Yoshikawa H, Azuma T, Umemura SI. Nanoparticles with multiple perfluorocarbons for controllable ultrasonically induced phase shifting. Jpn J Appl Phys. 2005;44:4548.

Khokhlova TD, Canney MS, Khokhlova VA, Sapozhnikov OA, Crum LA, Bailey MR. Controlled tissue emulsification produced by high intensity focused ultrasound shock waves and millisecond boiling. J Acoust Soc Am. 2011;130:3498–510.

Khokhlova VA, Bailey MR, Reed JA, Cunitz BW, Kaczkowski PJ, Crum LA. Effects of nonlinear propagation, cavitation, and boiling in lesion formation by high intensity focused ultrasound in a gel phantom. J Acoust Soc Am. 2006;119:1834–48.

Kodama T, Tomita Y. Cavitation bubble behavior and bubble–shock wave interaction near a gelatin surface as a study of in vivo bubble dynamics. Appl Phys B. 2000;70:139–49.

Kripfgans OD, Fowlkes JB, Woydt M, Eldevik OP, Carson PL. In vivo droplet vaporization for occlusion therapy and phase aberration correction. IEEE Trans Ultrason Ferroelectr Freq Control. 2002;49:726–38.

Li F, Wang H, Zeng D, Fan T, Geng H, Tu J, Guo X, Gong X, Zhao C, Wang Z. Sub-wavelength ultrasonic therapy using a spherical cavity transducer with open ends. Appl Phys Lett. 2013;102:204102.

Lin CY, Pitt WG. Acoustic droplet vaporization in biology and medicine. BioMed Res Int. 2013.

Lin KW, Kim Y, Maxwell AD, Wang TZ, Hall TL, Xu Z, Fowlkes B, Cain CA. Histotripsy beyond the "intrinsic" cavitation threshold using very short ultrasound pulses:"Microtripsy". IEEE Trans Ultrason Ferroelectr Freq Control 2014;61:251–265.

Liu X, Hou Y, Liu X, He J, Lu J, Ni X. Oscillation characteristics of a laser-induced cavitation bubble in water at different temperatures. Optik—Int J Light Electron Opt. 2011;122:1254–7.
Liu X, Long Z, He J, Li B, Liu X, Zhao J, Lu J, Ni X. Experimental study of temperature effect on the growth and collapse of cavitation bubbles near a rigid boundary. Optoelectron Lett. 2013;9:317–20.
Lohse D. Cavitation physics: lasers blow a bigger bubble. Nature. 1998;392:21.
Maxwell A, Sapozhnikov O, Bailey M, Crum L, Xu Z, Fowlkes B, Cain C, Khokhlova V. Disintegration of tissue using high intensity focused ultrasound: two approaches that utilize shock waves. Acoustic Today. 2012;8:24–37.
Maxwell AD, Cain CA, Hall TL, Fowlkes JB, Xu Z. Probability of cavitation for single ultrasound pulses applied to tissues and tissue-mimicking materials. Ultrasound Med Biol. 2013;39:449–65.
Maxwell AD, Wang TY, Cain CA, Fowlkes JB, Sapozhnikov OA, Bailey MR, Xu Z. Cavitation clouds created by shock scattering from bubbles during histotripsy. J Acoust Soc Am. 2011;130:1888–98.
Miller DL. Stable arrays of resonant bubbles in a 1-MHz standing-wave acoustic field. J Acoust Soc Am. 1977;62:12–9.
Moholkar VS. Mechanistic optimization of a dual frequency sonochemical reactor. Chem Eng Sci. 2009;64:5255–67.
Moshaii A, Tajik-Nezhad S, Faraji M. Temperature dependency of single-bubble sonoluminescence in sulfuric acid. Phys Rev E. 2011;84:046301.
Nakamura K, Asai A, Sasaki H, Yoshizawa S, Umemura S. Large volume coagulation utilizing multiple cavitation clouds generated by array transducer driven by 32 channel drive circuits. Jpn J Appl Phys. 2013; 52:07HF10.
Nomura S, Mukasa S, Kuroiwa M, Okada Y, Murakami K. Cavitation bubble streaming in ultrasonic-standing-wave field. Jpn J Appl Phys. 2005;44:3161.
Pitt WG, Singh RN, Perez KX, Husseini GA, Jack DR. Phase transitions of perfluorocarbon nanoemulsion induced with ultrasound: a mathematical model. Ultrason Sonochem. 2014;21:879–91.
Prabhu AV, Gogate PR, Pandit AB. Optimization of multiple-frequency sonochemical reactors. Chem Eng Sci. 2004;59:4991–8.
Rayleigh L. VIII. On the pressure developed in a liquid during the collapse of a spherical cavity. Philos Mag J Sci. 1917;34:94–8.
Reznik N, Williams R, Burns PN. Investigation of vaporized submicron perfluorocarbon droplets as an ultrasound contrast agent. Ultrasound Med Biol. 2011;37:1271–9.
Sapozhnikov OA, Khokhlova VA, Bailey M, Williams JC Jr, McAteer JA, Cleveland RO, Crum LA. Effect of overpressure and pulse repetition frequency on cavitation in shock wave lithotripsy. J Acoust Soc Am. 2002;112:1183–95.
Simon JC, Sapozhnikov OA, Khokhlova VA, Wang Y-N, Crum LA, Bailey MR. Ultrasonic atomization of tissue and its role in tissue fractionation by high intensity focused ultrasound. Phys Med Biol. 2012;57:8061.
Singh R, Husseini GA, Pitt WG. Phase transitions of nanoemulsions using ultrasound: Experimental observations. Ultrason Sonochem. 2012;19:1120–5.
Takagi R, Yoshizawa S, Umemura S. Enhancement of localized heating by ultrasonically induced cavitation in high intensity focused ultrasound treatment. Jpn J Appl Phys. 2010; 49:07HF21.
Thomas CR, Farny CH, Coussios CC, Roy RA, Holt RG. Dynamics and control of cavitation during high-intensity focused ultrasound application. Acoust Res Lett Online. 2005;6:182–7.
Umemura SI, Yoshizawa S, Takagi R, Inaba Y, Yasuda J. Enhancement of focused ultrasound treatment by acoustically generated microbubbles. Jpn J Appl Phys. 2013;52:07HA02.
Wang TY, Hall TL, Xu Z, Fowlkes JB, Cain CA. Imaging feedback of histotripsy treatments using ultrasound shear wave elastography. IEEE Trans Ultrason Ferroelectr Freq Control. 2012;59:1167–81.

Wang TY, Maxwell AD, Park S, Xu Z, Fowlkes JB, Cain CA. Why are short pulses more efficient in tissue erosion using pulsed cavitational ultrasound therapy (histotripsy)? In: 9th International symposium on therapeutic ultrasound (ISTU), vol 1. AIP Publishing; 2009. pp. 40–3.

Xu M, Wang LV. Photoacoustic imaging in biomedicine. Rev Sci Instrum. 2006;77:041101.

Xu S, Hu H, Jiang H, Xu Za, Wan M. Ultrafast 2-dimensional image monitoring and array-based passive cavitation detection for ultrasound contrast agent destruction in a variably sized region. J Ultrasound Med. 2014;33:1957–70.

Xu Z, Ludomirsky A, Eun LY, Hall TL, Tran BC, Fowlkes JB, Cain CA. Controlled ultrasound tissue erosion. IEEE Trans Ultrason Ferroelectr Freq Control. 2004;51:726–36.

Yasuda J, Asai A, Yoshizawa S, Umemura S-i. Efficient generation of cavitation bubbles in gel phantom by ultrasound exposure with negative-followed by positive-peak-pressure-emphasized waves. Jpn J Appl Phys. 2013;52:07HF11.

Yasui K. Influence of ultrasonic frequency on multibubble sonoluminescence. J Acoust Soc Am. 2002;112:1405–13.

Yin H, Qiao Y, Cao H, Li Z, Wan M. Cavitation mapping by sonochemiluminescence with less bubble displacement induced by acoustic radiation force in a 1.2 MHz HIFU. Ultrason Sonochem. 2014;21:559–65.

Zhang P, Porter T. An in vitro study of a phase-shift nanoemulsion: a potential nucleation agent for bubble-enhanced HIFU tumor ablation. Ultrasound Med Biol. 2010;36:1856–66.

Chapter 7
Cavitation Imaging in Tissues

Hui Zhong, Hong Hu, Chen Bai, Runna Liu, Jinjin Wan and Mingxi Wan

Keywords Active cavitation imaging · Passive cavitation imaging · Transient differential imaging · Subharmonic cavitation imaging · Bubble wavelet transform · Super-resolution reconstruction · 3D cavitation imaging

7.1 Introduction

The two predominant mechanisms by which therapeutic ultrasound works are through heating and mechanical damage, especially for high-intensity focused ultrasound (HIFU) therapies such as HIFU ablation and pulsed HIFU (pHIFU) (Kennedy 2005; ter Haar and Coussios 2007). Cavitation plays important roles in both mechanisms, as cavitation bubbles oscillate in nonlinear manner to increase acoustic energy absorption at high harmonics and violently collapse to generate high temperatures and mechanical forces. Cavitation is considered to be a secondary mechanism for HIFU ablation (Kennedy 2005) but the primary mechanism

H. Zhong · H. Hu · C. Bai · R. Liu · J. Wan · M. Wan (✉)
The Key Laboratory of Biomedical Information Engineering of Ministry of Education, Department of Biomedical Engineering, School of Life Science and Technology, Xi'an Jiaotong University, Xi'an 710049, People's Republic of China
e-mail: mxwan@mail.xjtu.edu.cn

H. Zhong
e-mail: bmezhonghui@mail.xjtu.edu.cn

H. Hu
e-mail: huhong1020@gmail.com

C. Bai
e-mail: baichen1986@gmail.com

R. Liu
e-mail: lrn_xjtu@163.com

J. Wan
e-mail: sufayer@163.com

M. Wan et al. (eds.), *Cavitation in Biomedicine*,
DOI 10.1007/978-94-017-7255-6_7

for pHIFU, especially for histotripsy (Xu et al. 2004, 2005b). Some other applications also utilize cavitation, such as sonoporation, as well as drug and gene delivery. For these therapy techniques to be developed, however, they must be adequately monitored for safety and efficiency using imaging techniques. Monitoring the dynamic changes of cavitation activity is essential for understanding the cavitation mechanisms, controlling and optimizing treatment processes, and improving treatment efficiency.

We described methods for detecting and mapping cavitation in water in Chap. 2, where we outlined the basic tools used to investigate cavitation, mainly acoustic and optical methods. However, for HIFU and pHIFU applications, cavitation usually occurs in tissues, not water. Tissues have different viscoelastic properties than water, and bubbles in tissues exhibit different oscillation behaviors than those in water. In tissues, as opposed to in water, there is a higher threshold to generate inertial cavitation because of the presence of elasticity in tissues (Yang and Church 2005). In addition, most tissues are not transparent or heterogeneous, and they have higher acoustic attenuation than water does. All these differences mean that monitoring cavitation in tissues must be done differently than it is in water. We can define a parameter called cavitation-to-tissue ratio (CTR), which is the average signal intensity in a cavitation area divided by the average signal intensity in an untreated tissue area. This ratio can be considered an index to evaluate how well images of cavitation regions contrast with the surrounding tissue.

Because acoustic cavitation imaging methods are suitable for observing opaque tissues while optical methods are not, this chapter focuses on acoustic methods. These types of cavitation imaging methods mainly include active cavitation imaging (ACI) and passive cavitation imaging (PCI). ACI can detect residual cavitation bubbles by transmitting the interrogating pulses and receiving the echoes of the cavitation bubbles between HIFU exposures. Conversely, PCI obtains cavitation images by receiving and analyzing the emission signals from cavitation bubbles. Unlike with ACI, PCI can obtain stable and transient cavitation images when HIFU is on, while ACI can obtain higher spatial resolution than PCI. And in ACI, the most important thing is how to distinguish cavitation bubbles from tissues. Therefore, in this chapter, we propose several ACI-based methods for improving the CTR. First, we propose differential imaging, including integral and transient differential imaging methods, which can be used to eliminate the signals of the normal tissue outside the HIFU-irradiated area and also exclude signals from the thermal lesions induced in the treated area. Subsequently, we introduce the pulse-inversion-based second-/subharmonic imaging method and the cavitation bubble wavelet transform (CBWT) imaging method, which can be used to further improve the CTR. In addition, we also present a Doppler-based ACI method called bubble Doppler. In Chap. 2, we described in detail the PCI (or PCM) methods in the time domain and the Fourier domain, so in this chapter, we present the corresponding results in tissues. We also introduce some additional techniques for improving the resolution of PCI images.

Normal line-by-line ACI has a relatively low frame rate, and it cannot depict the transient behavior of cavitation very well. However, cavitation images with high frame rates can be obtained using ultrafast ACI (UACI), which is based on the plane wave transmission technique, but this greatly decreases the signal-to-noise ratio (SNR). In this chapter, we introduce two different UACI methods, based on coherent compounding and adaptive beamforming, respectively, for improving the SNR and the CTR.

Cavitation activity actually occurs in three-dimensional (3D) space; thus, 3D cavitation imaging is essential for providing spatial distribution information about cavitation. In this chapter, we present some basic principles and methods for 3D cavitation imaging using two-dimensional (2D) arrays.

Some studies have reported using magnetic resonance imaging (MRI) to monitor cavitation during HIFU treatment, and we present them at the end of this chapter. However, although MRI has good precision in temperature mapping and superior contrast between necrotic and normal tissues, there is currently no proof that it has considerable advantages for monitoring cavitation activity.

7.2 Active Cavitation Imaging

7.2.1 Conventional ACI and Differential Imaging Methods

Conventional ACI mainly refers to B-mode ultrasound cavitation imaging for monitoring cavitation activity that arises specifically in HIFU therapy. Some researchers deem that in conventional B-mode ultrasound images, the formation of a hyperechoic region at the HIFU treatment site results from bubble activity induced by HIFU (ter Haar et al. 1989; Rabkin et al. 2005). However, some normal tissues display hyperechogenicity in the absence of cavitation, so in these tissues, the cavitation area is difficult to identify and the CTR can be rather low. But, differential imaging methods can solve this problem. With this method, the backscattered signals from the surrounding tissue can be eliminated by subtracting the signals acquired before HIFU exposure from the signals acquired after HIFU exposure, thus improving the CTR of the obtained differential image.

However, the hyperechoic region might reflect integrated information, including not only cavitation bubbles but also the tissue thermal lesion, since some studies have found that echogenicity enhancement was still observed long after HIFU was performed (Seo et al. 2002; Zhang et al. 2009; Zhong et al. 2007). Thus, differential images reflect the integral changes of a treated area, and such images can be called integral differential images. Unfortunately, integral differential images produce obvious movement artifacts due to radiation forces induced by HIFU.

Thus, we describe a transient differential imaging method here, which uses two consecutive data frames acquired between HIFU exposures. It can monitor the cavitation bubble activity and exclude the information associated with normal tissue and the tissue thermal lesion, and also greatly reduce movement artifacts.

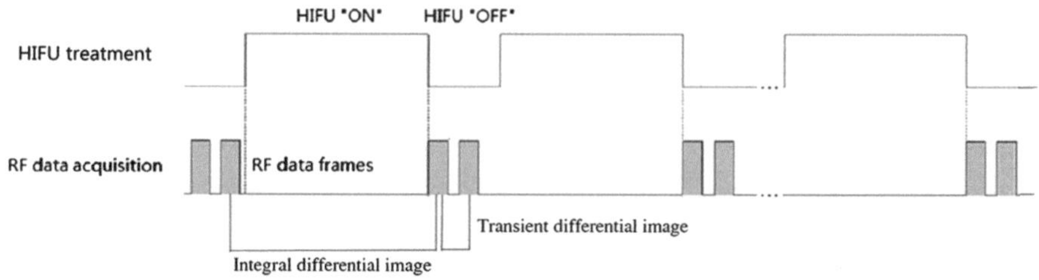

Fig. 7.1 Time sequence of HIFU treatment and RF data acquisition

To obtain the integral and transient differential images at the same time, HIFU treatment and data acquisition can be performed as follows: Two radio frequency (RF) data frames are acquired before HIFU treatment as the reference data frames. Then, HIFU delivery is briefly interrupted for a period of time to acquire two consecutive RF data frames. HIFU is completely off during the data acquisition to avoid interference from the HIFU beam. The time sequences of HIFU treatment and RF data acquisition are shown in Fig. 7.1.

The reference frame can consist of one RF data frame acquired before HIFU treatment. The first of each couple of RF data frames acquired at different times during HIFU treatment could then be selected, upon which the reference frame is subtracted, giving the integral differential images. In this way, the changes between the integral differential images reflect the "slow" changes of the treated area, which may include the formation of a thermal lesion and the production of cavitation bubbles during HIFU treatment, because the time between the frames is long enough (usually >0.5 s) to form a thermal lesion and simultaneously produce cavitation bubbles.

Similarly, we can then use two consecutive data frames obtained between HIFU exposures to monitor the "fast" changes within a treated area, where the data frame rate determines how quickly the changes are monitored. One frame usually needs several tens of milliseconds or even several hundreds of microseconds if using plane wave transmission, and the corresponding frame rate is several tens of hertz or even several thousands of hertz. Since HIFU is completely off and no more acoustic energy is absorbed by samples during the period of RF data acquisition, changes in the thermal lesion during this period should only depend on thermal conduction. The bioheat equation is as follows (Pennes 1948):

$$\frac{\partial T}{\partial t} = k\Delta T + \frac{q_v}{c_p} \tag{7.1}$$

where k is the temperature conductivity, such that $k = K/c_p$ and K is the thermal conductivity; c_p is the heat capacity per unit volume at a constant pressure; and q_v is the heat source induced by absorption of ultrasound (it is equal to zero here due to no HIFU). ΔT is the temperature difference between the temperature at the focus site and the equilibrium temperature (room temperature). Since the time between two consecutive RF frames is only several tens of milliseconds or even several hundreds of microseconds, from this equation, we can calculate and obtain that the thermal conduction area and the increase in temperature will not be large enough to form a new thermal lesion. Thus, the pure thermal lesion should change little between two consecutive RF frames. Therefore, the change in the tissue thermal lesion can be ignored between two consecutive RF frames, and the obtained transient differential image would not include information from the thermal lesion. Conversely, if cavitation bubbles are produced during HIFU treatment, there would be obvious changes in the backscattered signals between two consecutive RF frames due to the rapid changes of bubble activity, including bubble movement, dissolution, and/or collapse. Therefore, these transient differential images can reflect bubble activity only. Moreover, since the transient differential images are obtained with two consecutive RF data frames that are acquired when HIFU is off, no radiation force is applied to the tissue, greatly reducing the movement artifacts.

Differential images can be obtained by using several algorithms, including absolute difference (AD), the sum of absolute differences (SAD), and the sum of squared differences (SSD). The SAD and SSD algorithms can be adopted along each scan line of the RF frames. The algorithms can be calculated as follows (Zhong et al. 2006):

$$\mathrm{AD}(z,x) = |S_1(z,x) - S_2(z,x)| \tag{7.2}$$

$$\mathrm{SAD}(z,x) = \sum_{i=-\frac{n-1}{2}}^{\frac{n-1}{2}} |S_1(z+i,x) - S_2(z+i,x)| \tag{7.3}$$

$$\mathrm{SSD}(z,x) = \sum_{i=-\frac{n-1}{2}}^{\frac{n-1}{2}} (S_1(z+i,x) - S_2(z+i,x))^2 \tag{7.4}$$

where $S_1(z,x)$ and $S_2(z,x)$ refer to the corresponding A-line RF echo of two data frames acquired before and after HIFU treatment, respectively, or two consecutive data frames; z is the depth, x is the transverse direction of the image, and n is the calculation window width.

Typical experiment results are displayed in Fig. 7.2. In this experiment, HIFU delivery was briefly interrupted for 100 ms every 500 ms to acquire two consecutive RF data frames (one frame needs ~40 ms). The total integrated HIFU treatment time was 36 s at an electrical power of 80 W. Figure 7.2 shows the dynamic changes of different differential monitoring images with time. Figure 7.2a shows the B-mode images. The contrast of the B-mode images is poor, and the treated area is not easy to identify, especially at the initial stage of treatment due to the presence of hyperechogenic tissue. In Fig. 7.2b, the series of integral differential images at different exposure times are shown. In these images, the size of the white area increases with time corresponding to the increase of the tissue lesion size and the area of bubble activity. The integral differential images have good contrast at the

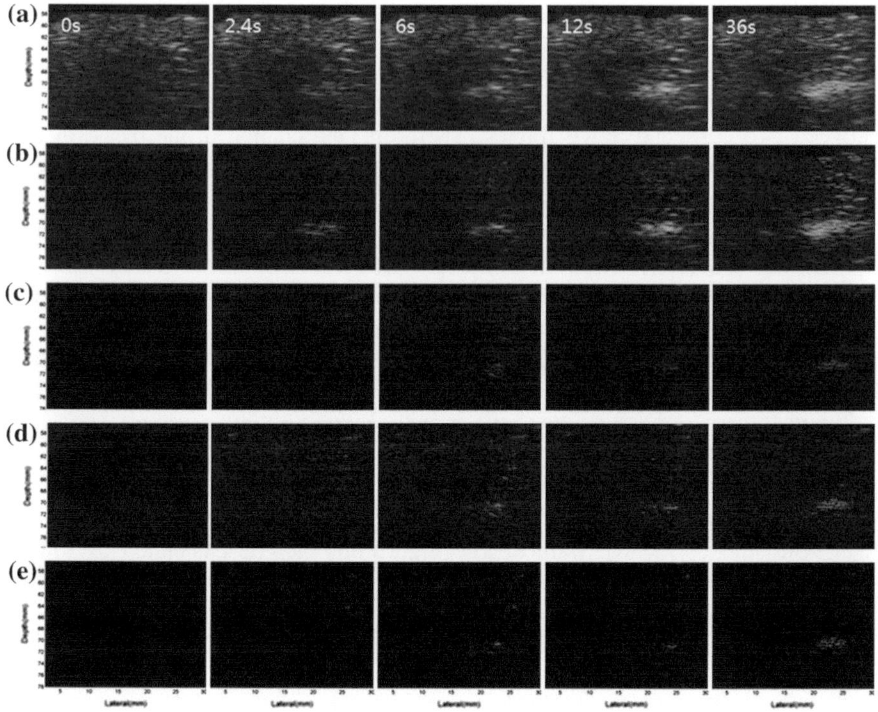

Fig. 7.2 Dynamic monitoring images captured at 0, 2.4, 6, 12, and 36 s during HIFU therapy with a power of 80 W: **a** B-mode images, **b** integral differential images, and transient differential images with **c** AD, **d** SAD, and **e** SSD algorithms

initial stage of treatment, but some artifacts are produced at later stages due to tissue movement induced by HIFU radiation forces. In Fig. 7.2c–e, the transient differential images obtained from two consecutive RF frames are shown. In the transient differential images, there is little information at the initial stage of treatment due to weak cavitation bubble activity, but they have good contrast at later stages of treatment. In addition, the white areas seem to be smaller than in the integral differential images. Moreover, in Fig. 7.2c–e, the movement artifacts are greatly reduced. We can also see that the SSD images have the highest contrast, while the AD and SAD images have lower, but similar contrast. Comparing the AD and SAD with the SSD images, we can see that SSD images have lower noise than the AD or SAD images at the cost of losing some small signals.

Figure 7.3a represents selections within a region of interest (ROI) of treated tissue and normal tissue for calculating the CTR. Figure 7.3b shows the changes in the CTR for each of the differential monitoring images with time. Here, we can see that B-mode images have the lowest CTR, SAD and AD images have slightly higher CTR values, and the SSD images have the highest CTR at all times, except for at ~2.4 s, when the integral differential image had the highest CTR. This may have arisen because at this time, the thermal lesion had formed, but cavitation was

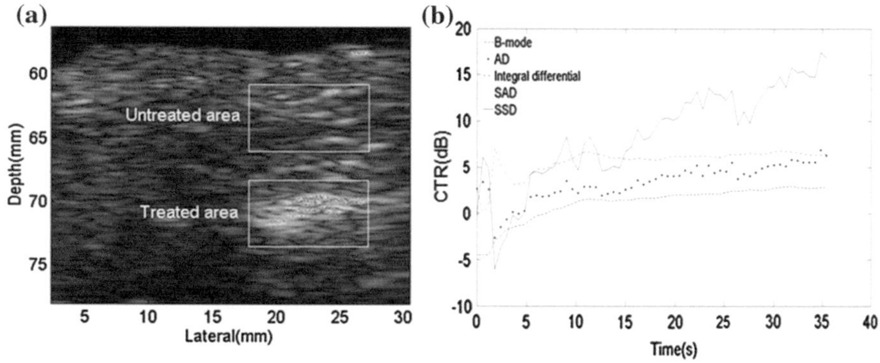

Fig. 7.3 CTR of differential monitoring images. **a** Illustration of a selection of the region of interest (ROI). **b** CTR of different differential monitoring images with time

not yet occurring, which may have caused the CTR of integral differential image to increase, while the CTR of the transient differential images was still low. After that, cavitation began, which caused the CTR of transient differential images to increase, while that of integral differential image changed not large due to the production of movement artifacts reducing the CTR.

7.2.2 Ultrafast ACI

By using differential imaging methods, we can suppress signals from normal tissues and thermal lesions in conventional B-mode ACI, allowing for high-CTR cavitation images. However, conventional B-mode imaging has a low frame rate (tens of hertz) due to its line-by-line scan mode, and it thus cannot monitor the rapid process of cavitation activity very well. To capture the transient behavior of cavitation bubbles, cavitation images with high frame rates should be acquired, requiring the need for an appropriate ultrafast active cavitation imaging (UACI) technique. Ultrafast ultrasonic imaging was first used by Fink's group to image the transient shear wave, and this technique is based on plane wave transmission and the delay-and-sum algorithm (PWDAS) (Sandrin et al. 1999). Ultrafast ultrasonic imaging can obtain a full ultrasound image from one single transmission, leading to imaging frame rates up to 10 kHz, only depending on the image depth. Nevertheless, it has a poor SNR and low resolution because of the low acoustic pressure of the plane wave due to lack of focusing.

To address these issues, Fink's group later introduced a coherent compounding method to PWDAS-based UACI to detect and locate the cavitation events both in vitro and in vivo with high sensitivity (Gateau et al. 2011a, b). Recently, Hu et al. (2015) developed an adaptive beamforming-based UACI with high spatial–temporal resolution by combining plane wave transmission with minimum variance (MV) beamforming and coherence factor (CF) weighting. In this section, we first described the coherent compounding-based UACI in concise words, then introduced the adaptive beamforming-based UACI, and focused on typical experiment results when using this method to assess in vitro bubble behavior for monitoring HIFU treatment.

7.2.2.1 Coherent Compounding-Based UACI

Montaldo et al. (2009) developed an ultrafast imaging technique based on plane wave transmission, in which a coherent plane wave compounding method was introduced to improve the poor SNR and resolution of plane wave ultrasonography. Gateau et al. (2011a, b) employed this method as a UACI approach and combined it with passive cavitation detection to detect single nucleation events and monitor the individual behavior of induced bubbles both in vitro and in vivo.

The basic principle of this coherent compounding method is PWDAS-based UACI, as illustrated in Chap. 2 (Sect. 2.5). Here though, the plane wave is transmitted with inclination α, as shown in Fig. 7.4a. The new delays can be calculated as follows (Montaldo et al. 2009):

$$\tau(\alpha, x_1, x, z) = \left(z\cos\alpha + x\sin\alpha + \sqrt{z^2 + (x - x_1)^2}\right)/c \tag{7.5}$$

We can obtain the image with plane wave emission of a certain angle using the same method as described in Sect. 2.5, but with the new delays given above. By coherently compounding the images obtained with the emission of several tilted plane waves of different angles, the final image quality can be improved. Figure 7.4b–e shows the schematic representations of the synthetic plane waves. The compound method sends individual plane waves of different angles (as shown in Fig. 7.4b). The plane waves interfere constructively at a fixed focal point when a proper time delay is used (shown in Fig. 7.4c). Using the same set of waves, the

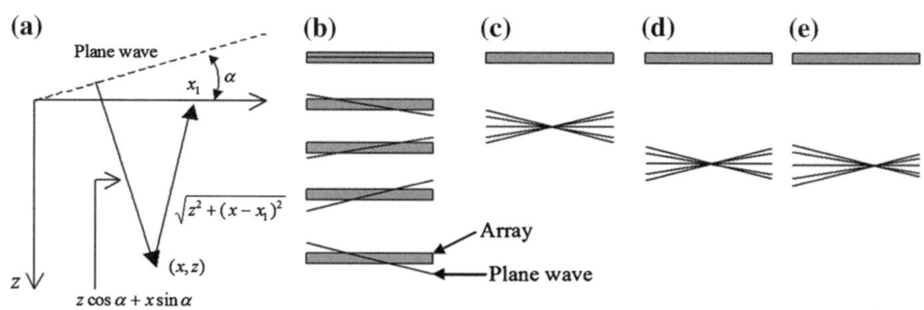

Fig. 7.4 Schematic diagram of coherent compounding-based UACI. **a** Time delays for a plane wave of angle α. **b** Individual plane waves sent by the compound method. **c**–**e** The addition of the plane waves with the adequate delays enables focus at different depths and lateral positions. Reproduced with permission from Montaldo et al. (2009). © 2009 IEEE

plane wave can be focused at different depths (Fig. 7.4d) or different lateral positions (Fig. 7.4e) (Montaldo et al. 2009).

Using coherent compounding-based UACI, Gateau et al. (2011a, b) were able to experimentally record the life of a cavitation bubble, from its nucleation to its disappearance or next excitation. In these experiments, a focused transducer (660 kHz, f-number = 1) driven by a high-power electric burst of one or two cycles was used to induce a cavitation event, and the resulting bubbles were detected by confocally aligning a linear array (4–7 MHz) with the transducer. A spatially compounded image was obtained by coherent recombination and beamforming of the backscattered echoes from pairs of successive plane wave illuminations with opposite tilts (Montaldo et al. 2009). Moreover, change detection (CD) images, which suppress the medium echoes and extract the bubble echoes, were obtained by an additional coherent subtraction of reference echoes before beamforming. In their study, plane waves with −2° and +2° tilts were used to form the images after coherent synthetic recombination. For each plane wave emission, a single-cycle pulse at 6 MHz was transmitted with each element of the array with the appropriate delay (Gateau et al. 2011b).

Figures 7.5 and 7.6 show bubble nucleation and bubble evolution in muscle tissues. The therapy transducer is excited by one-cycle pulses delivered at a pulse repetition frequency (PRF) of 1010 Hz and with peak negative pressures of −7.6 MPa. Images of the different detection acquisitions for the first four high-amplitude exposures are presented in Fig. 7.5, from which no hyperechogenic regions could be seen on the raw RF active images, while some scatterers appeared on the CD images. A single cavitating region was detected after the first shot, revealing bubble 1; two others were generated during the second shot: bubble 2, on the upper left-hand side, and bubble 4, on the lower left-hand side. Two more bubbles were induced by the third shot: bubble 3, just above bubble 2, and bubble 5, just below bubble 4. Bubble 2 disappeared after the fourth shot. The evolution of the local maximum at each bubble location is plotted in Fig. 7.6, from which we can observe the dramatic increase in the scattering amplitude that occurs when bubbles appear (Gateau et al. 2011b).

7.2.2.2 Adaptive Beamforming-Based UACI

As mentioned above, Fink's group proposed a tilted plane wave compounding method to improve the sensitivity and imaging quality of cavitation bubbles, but this came at the cost of decreasing the frame rate. To improve this, Hu et al. (2015) introduced an adaptive beamforming method using PWDAS-based UACI, in which plane wave transmission is combined with MV beamforming and CF weighting. This technique, called PWMVCF-based UACI, can be used to image the cavitation bubbles with good SNR and high spatial–temporal resolution. The principle and the algorithm process were illustrated in Chap. 2 (Sect. 2.6). To improve the CTR of cavitation images in tissue, the transient differential imaging method is used with all cavitation images.

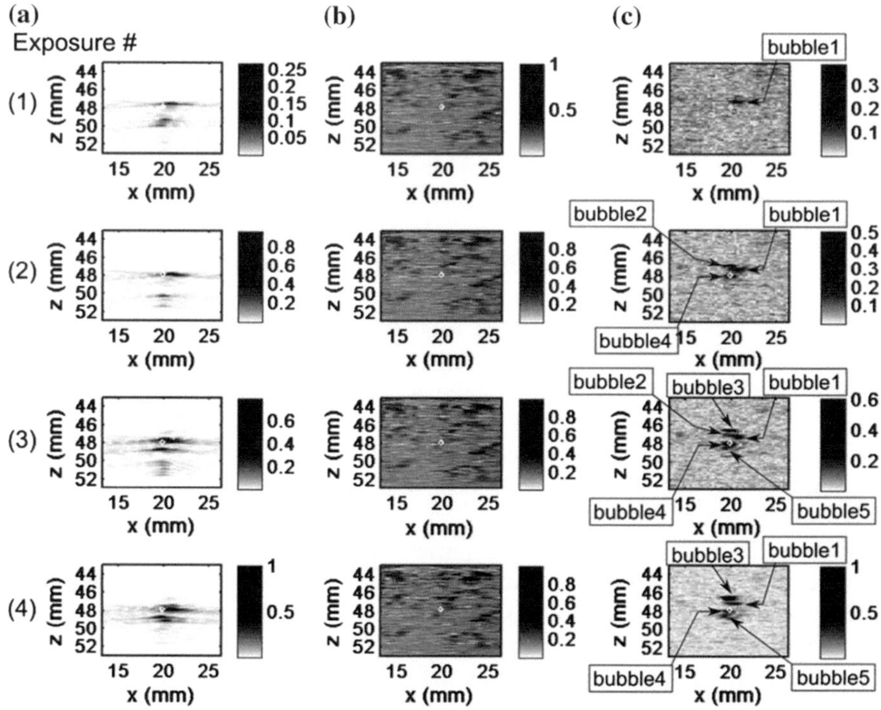

Fig. 7.5 Images of the different detection acquisitions for the first four high-amplitude exposures. Each row corresponds to a different shot: **a** the beamformed image corresponding to the high-pass-filtered passive RF data, **b** second active raw RF image for each shot, and **c** the corresponding change detection (CD) images. The *x*-axis origin is set to the first element of the array. For columns (**b**) and (**c**), the focus point of the single-element transducer is marked with a white rhombus ($z = 47.9$ mm, $x = 19.8$ mm). The beamformed data are normalized to the brightest point of the series for (**b**) and (**c**). Reproduced with permission from Gateau et al. (2011b). © 2011 IEEE

Then, we can evaluate the performances of conventional B-mode ACI, PWDAS-based UACI, and PWMVCF-based UACI in tissue in terms of CTR and the resolution of the images, as shown in Fig. 7.7. Taking the signals in the upper rectangular box as the tissue signals and those in the lower box as the cavitation signals, we calculated the CTRs to be 32.31, 7.35, and 22.81 dB for B-mode ACI, PWDAS-based UACI, and PWMVCF-based UACI, respectively. Compared to B-mode ACI, the CTR of PWDAS-based UACI was 24.97 dB lower because of the low power of plane wave transmission; however, the CTR of PWMVCF-based UACI was only 9.50 dB lower than that of B-mode ACI. To quantify the resolution, we can use the normalized −6 dB width of the autocorrelation function of the enveloped RF signal. The resolution gain is defined as the ratio of the −6 dB autocorrelation lengths of the enveloped RF data before beamforming to those after beamforming (Abeyratne et al. 1995). Compared to B-mode ACI, the axial and lateral resolution gains of PWDAS-based UACI were 0.82 and 0.69, respectively,

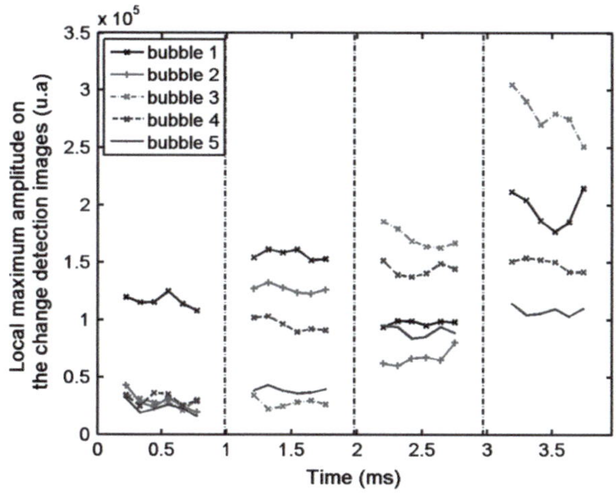

Fig. 7.6 Evolution of the scattering amplitude of the cavitation regions identified in Fig. 7.5. The *vertical dash-dotted lines* indicate the high-amplitude shots. Reproduced with permission from Gateau et al. (2011b). © 2011 IEEE

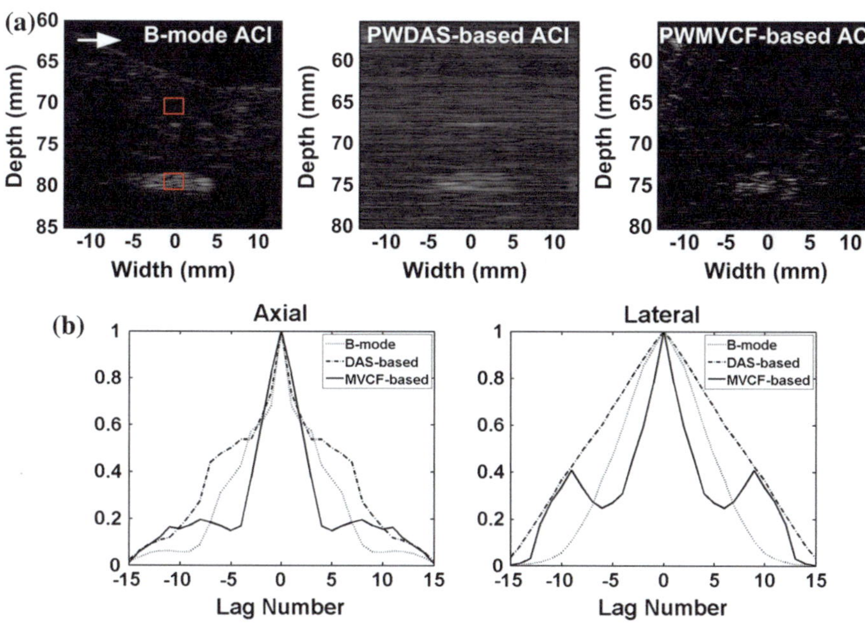

Fig. 7.7 Comparison of conventional B-mode ACI, PWDAS-based UACI, and PWMVCF-based UACI in an in vitro pig muscle. **a** Cavitation bubble images with B-mode ACI, PWDAS-based UACI, and PWMVCF-based UACI. **b** The corresponding autocorrelation functions of the enveloped RF data in (**a**). The dynamic range is 40 dB for all images. The *arrow* in the first image indicates the HIFU transmission direction, and the *rectangular boxes* indicate the ROI of tissue (*upper*) and cavitation (*below*) areas to calculate the CTR. The electrical power of HIFU was 60 W, and exposure time was 5 ms

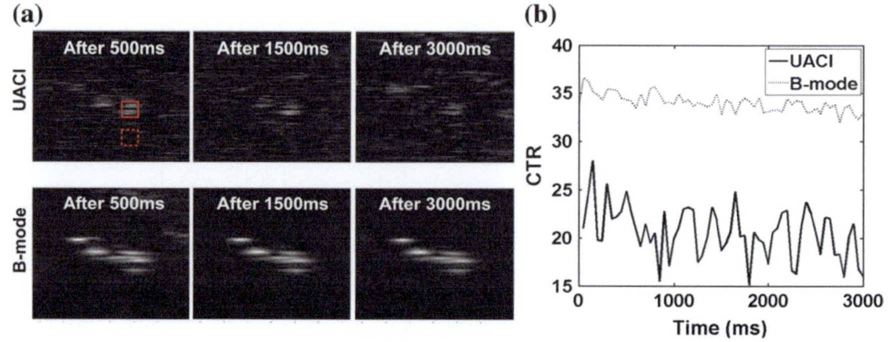

Fig. 7.8 Comparison of CTRs from B-mode ACI and PWMVCF-based UACI in in vitro pig muscle. **a** Cavitation bubble images for PWMVCF-based UACI and B-mode ACI after HIFU exposures of 500, 1500, and 3000 ms; **b** CTR curves of B-mode ACI and PWMVCF-based UACI. The electrical power of HIFU was 60 W, and exposure time was 5 ms

while for PWMVCF-based UACI, the axial and lateral resolution gains were 1.45 and 1.75, respectively.

Figure 7.8 shows CTR curves for B-mode ACI and PWMVCF-based UACI in tissue. From this figure, we see that the CTRs of PWMVCF-based UACI are lower than those of B-mode ACI, and the CTR curve for PWMVCF-based UACI is less smooth (more susceptible to noise interference) because of the low image SNR of plane wave.

Plane wave transmission allows the frame rate of UACI to be very high, but the signal energy is much lower than that in conventional B-mode imaging because in UACI, the ultrasound beams are not focused in the transmit mode. As a consequence, the CTRs and the resolution of the PWDAS-based UAC images are lower than those achieved with B-mode imaging. These features can both be improved with PWMVCF-based UACI, which combines plane wave transmission with MV beamforming and CF weighting. Although the CTR of PWMVCF-based UACI is still lower than that achieved with B-mode ACI, this technique greatly increases the frame rate and destroys far fewer bubbles than with B-mode imaging. Therefore, PWMVCF-based UACI might depict transient bubble behavior more accurately and may be better suited to imaging bubble distribution and monitoring bubble evolution.

To highlight this, Fig. 7.9 shows images of cavitation bubble distributions in tissue acquired with PWMVCF-based UACI method under different HIFU exposures. The HIFU exposures were set as follows: In the first row of Fig. 7.9, HIFU pulse durations of 0.5, 5, and 50 ms were accompanied with an electrical power of 100 W. In the second row of Fig. 7.9, an HIFU pulse duration of 5 ms was used at different electrical powers of 60, 80, and 100 W. In this figure, both the cloud area and the maximum intensity of cavitation bubbles increased with an increase of pulse duration. We can also see that higher HIFU excitation powers produced larger cloud areas of cavitation bubbles, but the overall maximum intensities were similar.

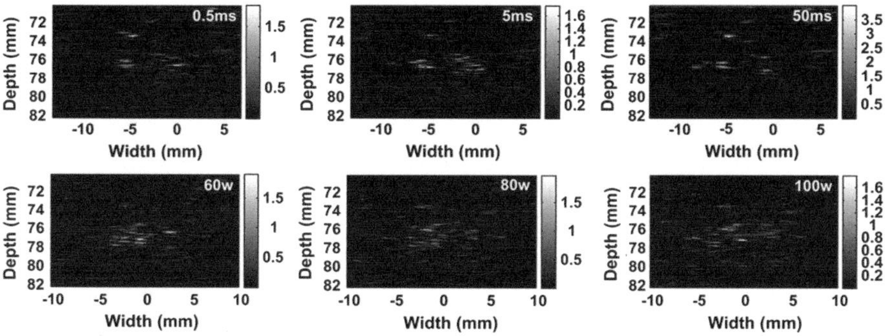

Fig. 7.9 PWMVCF-based UAC images of cavitation bubble clouds induced by HIFU with different exposure parameters in in vitro pig muscle. *Upper row* pulse durations for 0.5, 5, and 50 ms (excitation power was 100 W). *Lower row* excitation power for 60, 80, and 100 W (pulse duration was 5 ms)

7.2.3 Typical UACI Results for HIFU Therapy

Cavitation behaviors, including generation, movement, dissolution, and collapse, are mostly rapidly changing processes, especially in HIFU therapy. The PWMVCF-based UACI method introduced above can be used to monitor the spatial–temporal evolution of cavitation bubbles not only during ablation in a bulk tissue but also during tissue erosion at a fluid–tissue interface.

7.2.3.1 Evolution of Cavitation Bubbles in Tissue Ablation

Tissue ablation with HIFU is a thermal-based technique, which can make tissue temperature rise very rapidly to 56 °C (ter Haar et al. 1989). Cavitation plays an important role in this heating and almost cannot be avoided in tissue ablation because of the high energy of HIFU. However, the distribution of cavitation is difficult to predict, leading to unpredictable HIFU lesion shapes (Coussios et al. 2007). In addition, an excess of cavitation bubbles can shield the focus from acoustic exposure, causing prefocal deformation of the lesion. However, we can monitor the evolution of cavitation bubbles during tissue ablation using PWMVCF-based UACI, and this can help us better understand the relationship between cavitation and distorted ablation lesions, allowing for better control over the treatment.

In an HIFU ablation experiment, cavitation bubbles can be imaged in the intervals of HIFU exposures when the HIFU transducer is completely off, in order to avoid interference. Figure 7.10 shows the timing sequence used in HIFU ablation. A pair of T_2-delayed UAC images (UACIs) is obtained in every interval, and a subtraction operation is performed with them to exclude the slow-changing RF signals scattered by normal tissue and tissue denaturation. First, an initial UAC

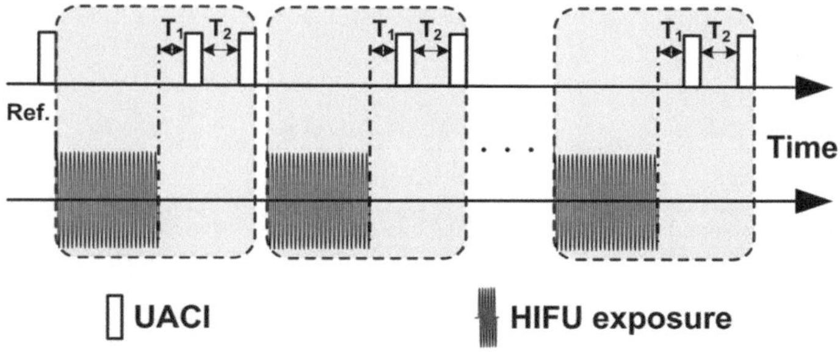

Fig. 7.10 Timing sequence for HIFU ablation and UACI

image is acquired as a reference signal of the backscattered echo from the tissue without cavitation bubbles. Then, a HIFU excitation is performed, and a pair of delayed UACIs is taken. The delay time T_1 is chosen to take into account the propagation path from the HIFU transducer to the ultrasound scanner, in order to avoid the interference by HIFU; T_2 is the time for the evolution of cavitation. During time T_2, the backscattered echo from the tissue can be regarded as constant, but that from the cavitation is varying. The process is then repeated in succession, and the data are stored and transferred to a computer for post-processing. The data acquisition parameters used in the experiments presented here are as follows: The time delay T_1 was 1 ms; the time delay T_2 varied with the exposure parameters; the acquisition time of one UAC image was 500 μs; and the sampling rate was 40 MHz.

Figure 7.11 shows a series of cavitation bubble images taken during ablation of bulk tissue with different exposure parameters. The pulse duration (PD) of 600 ms and 800 ms corresponded to duty cycle (DC) of 60 and 80 % and total treatment times of 80 and 60 s, respectively. The electrical power was 100 W, corresponding to a calculated total in situ spatial average intensity of about 2200 W/cm^2. The

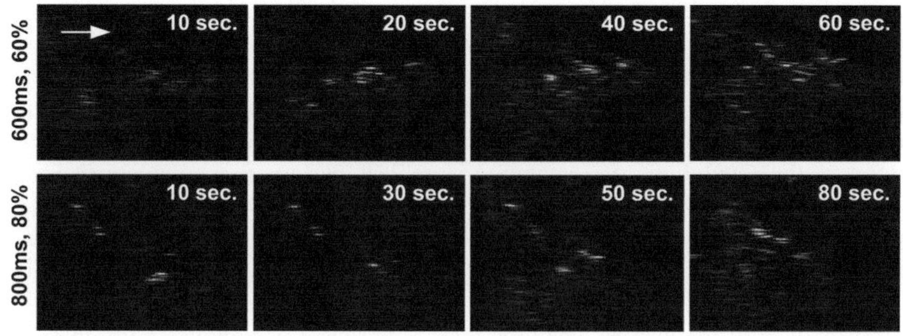

Fig. 7.11 Dynamic monitoring images of cavitation bubbles (20 mm × 15 mm) during ablation in bulk in vitro pig muscle under different exposure parameters. *Upper row* images for a PD of 600 ms and DC of 60 %. *Lower row* images for a PD of 800 ms and DC of 80 %. The *arrow* in the first image indicates the HIFU transmission direction, and the dynamic range is 40 dB for all images

region of cavitation bubbles increased with increasing treatment time. For a PD of 600 ms and DC of 60 %, the center of the bubble cloud changed little throughout the treatment, indicating that no significant shielding is being caused by cavitation. However, for a PD of 800 ms and DC of 80 %, the center of the bubble cloud moved toward the HIFU transducer, indicating that acoustic shielding occurred.

Thus, as shown in Fig. 7.11, monitoring the evolution of cavitation bubbles by PWMVCF-based UACI in bulk tissue ablation can help us observe whether or not acoustic shielding happens and we can then use this information to avoid prefocus lesions. Although the CTRs of UAC images are lower than those of B-mode images, this imaging method can obtain high frame rate images and has fewer effects on cavitation bubbles. Consequently, UACI methods are more suitable for investigating the transient behaviors of cavitation bubbles.

7.2.3.2 Evolution of Cavitation Bubbles in Tissue Erosion

Tissue erosion is a cavitation-based technique that uses pulsed high-intensity ultrasound to achieve mechanical disruption of soft tissue rather than thermal coagulation, and this approach can be regarded as histotripsy (Xu et al. 2004). Some investigations have shown that the tissue erosion process is highly correlated with the generation and maintenance of cavitation bubble clouds. Very short pulses with very high amplitudes successively delivered at certain pulse repetition frequencies (PRFs) can produce high treatment efficiencies (Xu et al. 2004, 2005a, b, 2007a, b, 2008). During tissue erosion, the cavitation bubbles present in the intervals between HIFU pulses survive for a long time (on the order of milliseconds) (Xu et al. 2007b) and they can influence the erosion pulses in two ways: They can serve as nuclei for easier generation of cavitation events, and they can act as strong scattering particles in sound propagation, resulting in energy "shadowing" (Xu et al. 2007b). The optimal exposure parameters should be selected to ensure that enough bubbles are produced by the previous pulse and dissipate to appropriate levels such that they do not cause heat deposition or energy shadowing, but there must be enough residual nuclei, so subsequent cavitation can easily occur. Therefore, we can use PWMVCF-based UACI to monitor the spatial–temporal evolution of cavitation bubbles during tissue erosion in order to better understand these cavitation–tissue erosion relationships and to better control the treatment (Hu et al. 2015).

In tissue erosion experiments, the time sequence is similar to that in tissue ablation experiments, except that HIFU is emitted in very short pulses instead of quasi-continuous waves. Limited by the buffer size of SonixDAQ and high-speed digitizer, the UACI was performed every 1000 HIFU pulses in the experiments presented here. Figure 7.12 shows the timing sequence used in these tissue erosion experiments, and the data acquisition parameters were as follows: The time delay T_1 between HIFU excitation and the first UAC image was 1 ms; the time delay T_2 between the two UAC images was 300 μs; the acquisition time of one UAC image was 150 μs; and the sampling rate was 40 MHz. The electrical power was adjusted

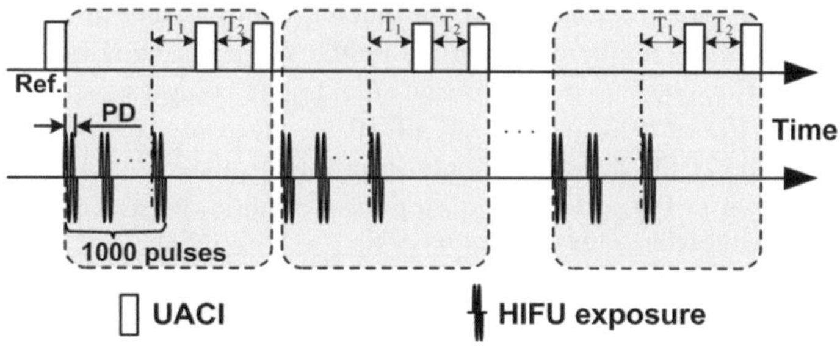

Fig. 7.12 Time sequence for tissue erosion and UACI

to 272 W, corresponding to the calculated total in situ spatial average intensity of about 6000 W/cm^2 in all experiments.

Figure 7.13 shows the typical results for monitoring cavitation bubbles during tissue erosion at the tissue–water interface. In Fig. 7.13a, we see that cavitation

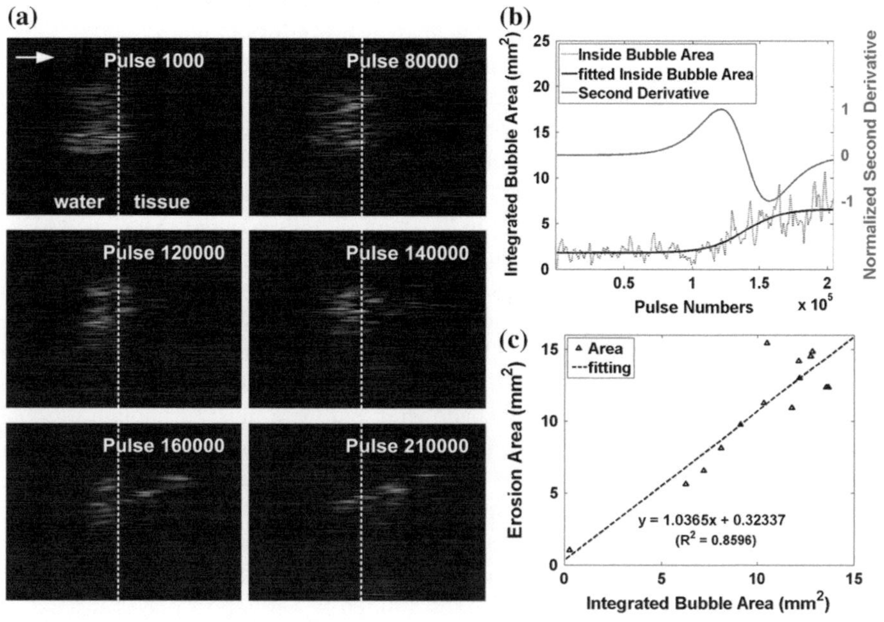

Fig. 7.13 Monitoring cavitation bubbles arising in tissue erosion at the tissue–water interface. **a** Dynamic cavitation bubble images (20 mm × 15 mm) with PD = 40 μs, PRF = 0.67 kHz, and total HIFU pulses = 210,000. The *arrow* in the first image indicates the HIFU transmission direction. The *dotted white lines* indicate the interface, of which the *left side* is water and the *right side* is tissue. The dynamic range is 40 dB for all images. **b** Changes in the integrated bubble areas with pulse number, corresponding to bubble images in (**a**). A sigmoid fit was used to approximate the tendency of the curve, and the second derivative of the fitted curve was calculated to locate the turning point. **c** Relationship between erosion area and integrated bubble area inside the tissue. Reproduced with permission from Hu et al. (2015). © Acoustical Society of America

bubbles initially distribute at the surface of the tissue and then gradually expand to the internal areas of the tissue with increasing HIFU pulses. Finally, almost the entire bubble cloud becomes distributed within the tissue. We then calculated the integrated bubble area, to use in the further quantitative analysis of the cavitation bubble evolution, as follows: After choosing an appropriate intensity threshold, the UAC image was converted into a binary image (1s and 0s); the pixels whose intensities were higher than the threshold were set to 1, and others were set to 0; the integrated bubble area was quantified according to the number of pixels labeled as 1s. Figure 7.13b shows the curves of the integrated bubble areas inside the tissue with different pulse numbers. The sigmoid fit was used to simplify the tendency of the curve, and the second derivative of the fitted curve was calculated to locate the turning point. Under tissue erosion treatments, the tissue is homogenized and the homogenate exchanges with water once the surface is removed by HIFU pulses, and this occurs until the erosion region is inundated with water. Therefore, while the bubble area within the tissue remains low, tissue damage has not yet developed. When this area begins to increase at the first turning point (maximum point of the second derivative of the curve), tissue damage starts developing. Then, as the area continues to increase, the area of tissue damage also continues to grow. When the area reaches saturation at the second turning point (minimum point of the second derivative of the curve), the damaged area does not grow further, and the treatment at this location should be finished. In the experiments presented here, tissue samples were sectioned along the imaging plane for observation of the damaged tissue morphology and then calculating the areas of damage after tissue erosion. The relationship between erosion area and internal bubble area was analyzed using fifteen independent experiments. The resulting correlation curve between internal bubble area and erosion area is shown in Fig. 7.13c, depicting a proportional relationship with a coefficient of 1.0365. This confirms that the area of damage produced during tissue erosion can be estimated from the integrated bubble area (Hu et al. 2015).

Using the PWMVCF-based UACI method to monitor the evolution of cavitation bubbles at the fluid–tissue interface during tissue erosion can actually give us considerable information about the formation of tissue erosion, whereby the erosion area can be estimated from the UACIs. Thus, we can potentially use these images to help us optimize tissue erosion treatment.

7.3 Second-Harmonic and Subharmonic Cavitation Imaging

Generally, conventional B-mode imaging can be used to monitor cavitation activity. However, the hyperechoic regions that often appear in B-mode monitoring images are thought to include the integrated information associated with both the tissue thermal lesion and cavitation bubbles (Zhang et al. 2009). Although transient differential imaging method introduced in Sect. 7.2.1 can eliminate the signals arising from normal tissue and thermal lesions in the treated tissue, the resulting image actually reflects the changes associated with cavitation bubbles—including movement, dissolution, and/or collapse—rather than cavitation itself. To acquire information about the pure cavitation bubbles and suppress the signals arising from normal tissue and the thermal tissue lesion, the proper signals that reflect the greatest differences between cavitation bubbles and the tissue must be extracted. One type of signal to investigate might be harmonic signals: Because cavitation bubbles are known to have much stronger nonlinearity than tissues, they may produce stronger harmonic signals.

Previous studies have shown that gas microbubbles can be induced to exhibit nonlinear motion in an ultrasound field. Because of this nonlinear motion, bubbles can generate harmonics, including the second harmonic, super-harmonics, ultra-harmonics, and subharmonics (Goldberg et al. 2001). Harmonic imaging has long been used in contrast-enhanced imaging. Thus, it seems reasonable that second-harmonic imaging can be also used in cavitation imaging during HIFU therapy, since cavitation bubbles have similar properties to contrast microbubbles. Cavitation bubbles can produce much stronger second-harmonic signals than tissues; therefore, second-harmonic imaging can theoretically obtain cavitation images with a higher CTR than fundamental imaging (B-mode imaging) can. However, there are two major limitations in second-harmonic cavitation imaging: First, the tissue, especially the necrotic tissue induced by HIFU, can also produce second-harmonic signals because of nonlinear propagation, which would cause the CTR to decrease; second, the second-harmonic signal undergoes greater attenuation than the fundamental signal, which would cause the SNR to decrease. In addition, the bandwidth limitation of the imaging probe would reduce the sensitivity of second-harmonic cavitation imaging.

Since bubbles alone can produce subharmonic signals, while tissues cannot (Goldberg et al. 2001), subharmonic imaging might be advantageous for cavitation bubble imaging. Subharmonic imaging likely provides more pure cavitation bubble information, producing high-CTR values, and undergoes less attenuation than second-harmonic imaging. However, if the transmission pulses have a wide bandwidth (one-cycle pulse), then the subharmonic signals induced by the cavitation bubbles would usually mix with the fundamental signals from the tissue, thus reducing the CTR (Zhong et al. 2014).

The pulse-inversion (PI) technique, which transmits two pulses with inverted phases, has been used in second-harmonic imaging for contrast-enhanced imaging (PI harmonic imaging). Since the PI technique can eliminate the fundamental signals, it can also be used in subharmonic imaging for the same purpose. In this section, typical cavitation imaging results are presented, in which PI-based second-harmonic and subharmonic methods were used.

In these experiments, HIFU delivery was briefly interrupted for 100 ms every 500 ms to acquire RF data frames. The total integrated HIFU treatment time was 36 s. The pulse-inversion transmission mode was selected in the ultrasound imaging system, and RF data frames were acquired between HIFU exposures with the HIFU completely off during data acquisition to avoid interference from the HIFU beam. Between every two HIFU exposures, one 2D RF data frame was acquired, which had 512 RF lines. The odd and the even RF lines were the echoes of positive and negative transmitted pulses, respectively. The 256 positive lines were selected to form a positive frame and the 256 negative lines to form a negative frame. Thus, one monitoring image included 256 scan lines. The summation frame was obtained by summing the positive frame and the negative frame. In the experiments, a linear array ultrasound imaging probe with a center frequency of 5.0 MHz and bandwidth of 80 % was used. Two types of imaging pulses were selected in the experiments. One was a one-cycle pulse with a 5.0 MHz frequency, and the other was a two-cycle pulse with a 3.0 MHz frequency. By using different algorithms, several monitoring images were obtained as follows:

1. The positive frame was used to form B-mode images.
2. Normal subharmonic images were acquired from the positive frame by using a digital band-pass filter with a center frequency of 2.5/1.5 MHz (for 5.0-/3.0-MHz transmission pulses, respectively) and a bandwidth of 60 %.
3. PI harmonic images were obtained from the summation frame by using a second-harmonic filter with a center frequency of 10.0/6.0 MHz (for 5.0-/3.0-MHz transmission pulses, respectively) and a bandwidth of 40 %.
4. PI subharmonic images were obtained from the summation frame by using the subharmonic band-pass filter (2.5/1.5 MHz center frequency, 60 % bandwidth).

Figure 7.14 shows the dynamic monitoring images during HIFU therapy (100 W) at times of 0, 12, 18, 24, and 36 s with different imaging algorithms. The imaging pulses transmitted were one-cycle pulses with a frequency of 5.0 MHz. From Fig. 7.14c, we can see that the PI harmonic images only poorly distinguish cavitation bubbles from the tissues with high nonlinearity. Furthermore, the high attenuation of the second harmonic and the limited bandwidth of the imaging probe (3–7 MHz) mean that the second-harmonic signals of 10.0 MHz in the treated area are too weak to detect. However, in Fig. 7.14b, d, we see that the normal subharmonic and PI subharmonic images eliminate signals from normal tissue and thus display cavitation bubble activity well. Especially in the second column of images, we can see that the PI subharmonic image shows some cavitation bubbles that are not visible in the normal subharmonic image. This suggests that the PI subharmonic images have higher sensitivity and contrast, and they are thus able to show cavitation bubbles better than normal subharmonic images.

Figure 7.15b shows how the CTR of different monitoring images changed with time. The selected ROIs of a treated area and a normal tissue are illustrated in Fig. 7.15a. In this figure, the PI subharmonic images have the highest CTR after about a 5-s HIFU exposure, at which time cavitation activity becomes obvious. The maximum value of the CTR in PI subharmonic images was 12 dB greater than that of B-mode images. However, the CTRs of PI harmonic images were lower than those of B-mode images, probably because the imaging pulses were chosen improperly.

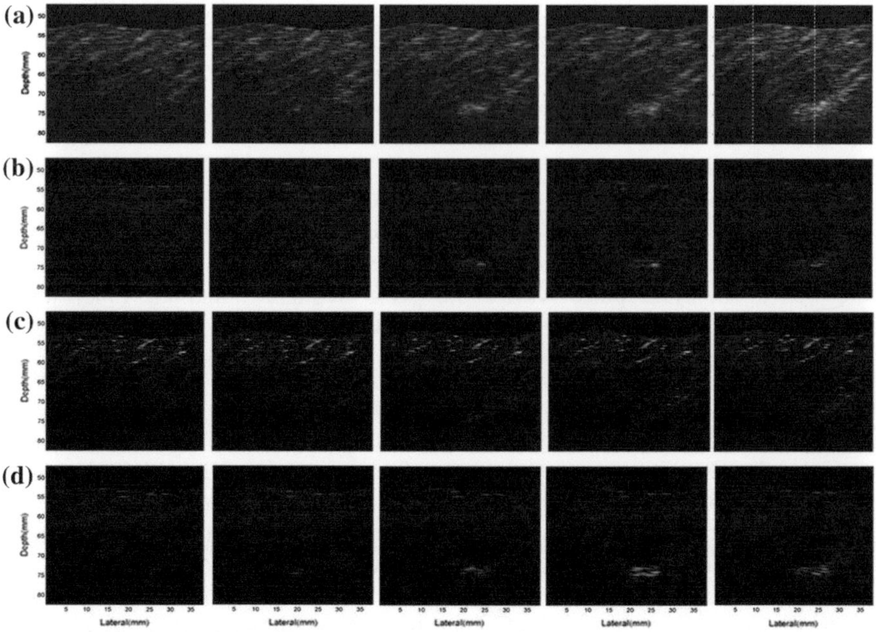

Fig. 7.14 Dynamic monitoring images with one-cycle 5-MHz imaging pulses during HIFU therapy at a power of 100 W at times 0, 12, 18, 24, and 36 s: **a** B-mode images, **b** normal subharmonic images, **c** PI harmonic images, and **d** PI subharmonic images. Reproduced with permission from Zhong et al. (2014). © 2014 American Scientific Publishers

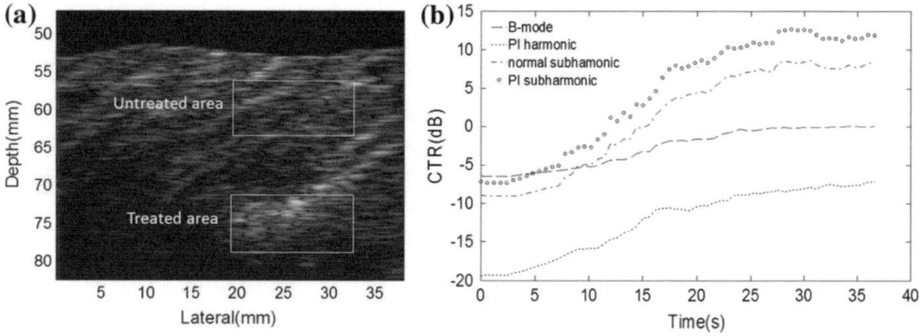

Fig. 7.15 CTR of different monitoring images with one-cycle 5-MHz imaging pulses: **a** illustration of a selected ROI; **b** CTR of different monitoring images with time

Figure 7.16 shows the dynamic monitoring images during HIFU therapy (100 W) at times 0, 12, 18, 24, and 36 s, where images were processed with different imaging algorithms. Here, the imaging pulses transmitted were two-cycle pulses with a frequency of 3.0 MHz. Different from Fig. 7.14c, we can see in

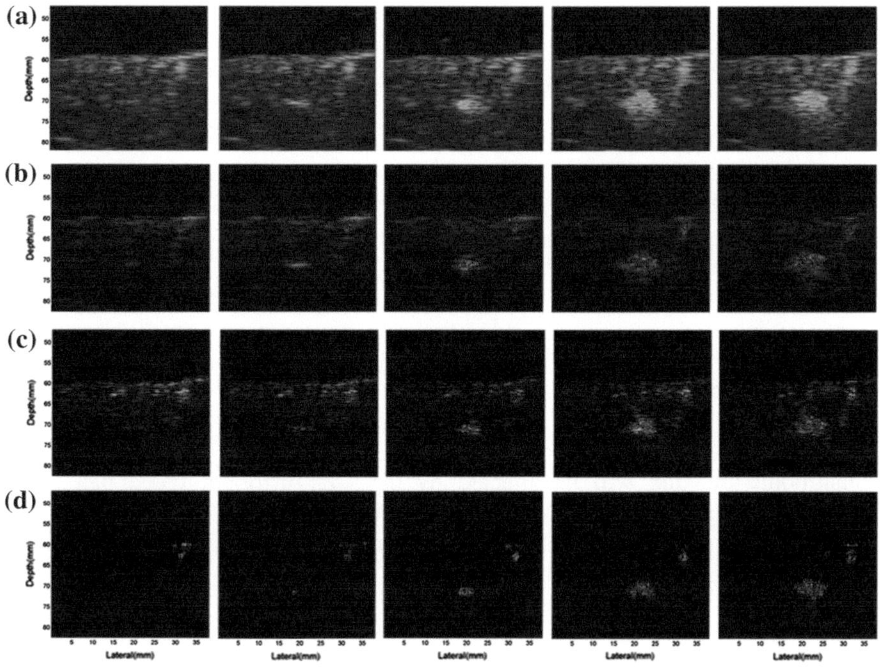

Fig. 7.16 Dynamic monitoring images with a two-cycle 3-MHz imaging pulse during HIFU therapy (100 W) at times of 0, 12, 18, 24, and 36 s: **a** B-mode images, **b** normal subharmonic images, **c** PI harmonic images, and **d** PI subharmonic images

Fig. 7.16c that the PI harmonic images adequately displayed cavitation bubbles in the treated area, although some tissues with high nonlinearity could not be eliminated. This is because the second harmonic of 6.0 MHz is within the bandwidth of the imaging probe, and the attenuation is not very high at this frequency. From Fig. 7.16d, we see that the PI subharmonic images eliminated the tissue information very well, but the resolution of the images was comparatively low because of the low frequency of 1.5 MHz.

In Fig. 7.17, we can see that the CTR of the PI subharmonic images is higher than that of the other images all times. The maximum value of the CTR in the PI subharmonic images was 17 dB greater than that in B-mode images. Also, the PI harmonic images had higher CTR values than those in B-mode images after a 5-s HIFU exposure. These results suggest that the PI subharmonic imaging method can capture information from pure cavitation bubbles very well, and the PI harmonic imaging method can also obtain good cavitation images when proper imaging pulses are chosen.

It should be noted, however, that the acoustic energy for generation of subharmonics must be beyond some certain threshold and the 1/2 subharmonic imaging is sensitive for bubbles with some certain size. Therefore, the sensitivity of

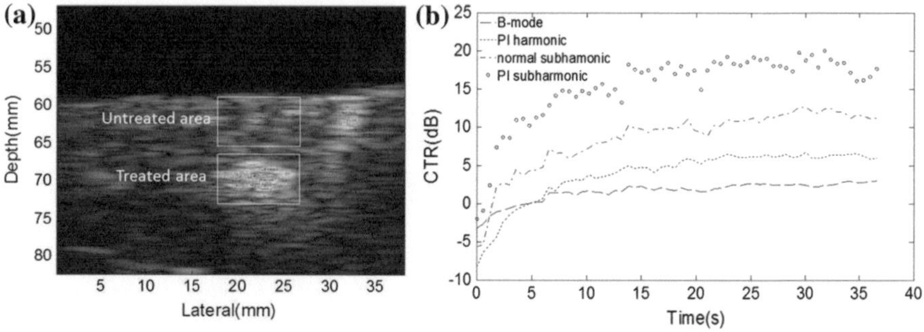

Fig. 7.17 CTR of different monitoring images with a two-cycle 3-MHz imaging pulse: **a** illustration of a selected ROI; **b** CTR of different monitoring images with time

subharmonic imaging will be affected by the size distribution and dissolution process of cavitation bubbles. More research should explore this area further.

7.4 Cavitation Imaging with the Bubble Wavelet Transform Technique

The detection of cavitation bubbles depends on both the properties of the bubble itself and those of surrounding tissues. To improve the CTR of cavitation images, the special properties of cavitation bubbles that distinguish them from tissues should be stimulated as much as possible, so the differences detected are as great as possible.

Second-harmonic cavitation imaging and subharmonic cavitation imaging utilize the fact that the nonlinearity of cavitation bubbles is much stronger than that of tissues to detect the cavitation bubbles. However, these two methods only make use of the signals of a certain frequency band (the second harmonic or subharmonics). The CTR could be potentially improved further if all the nonlinear components were utilized.

Therefore, a "cavitation bubble wavelet" could be constructed as a mother wavelet that is well matched with the echoes of cavitation bubbles. By using a wavelet transform technique, the cavitation bubbles can be identified from surrounding tissues. With this cavitation bubble wavelet transform (CBWT) technique, the mother wavelet should be constructed based on the dynamic oscillation model of gas bubbles in tissues. For simplicity, the RPNNP equation for free bubbles oscillating in water is used for mother wavelet construction in this section, which can be applied in circumstances where cavitation occurs in vessels. The method proposed in this section is different from the method proposed by Zhao et al. (2011), in which the bubble wavelet was constructed according to the modified Herring equation for encapsulated bubbles in ultrasound contrast imaging.

7.4.1 CBWT Imaging Method

Generally, the echoes backscattered from a cavitation bubble area contain two components if the cavitation is produced in tissues. The equation can be written as follows:

$$E(t) = E_C(t) + E_T(t) \tag{7.6}$$

where $E(t)$ represents the echo received by the imaging probe. $E_C(t)$ and $E_T(t)$ represent the echoes from the cavitation bubbles and those from the surrounding tissues, respectively. In the case of linear propagation, $E_T(t)$ is supposed to be zero in the harmonic imaging method. However, $E_T(t)$ cannot realistically be totally eliminated because of the nonlinear propagation of sound in tissues. Thus, the CTR is limited with the harmonic imaging method.

Continuous wavelet transform (CWT) is a mathematical tool, whereby a convolution of a signal is analyzed with a set of functions generated by the mother wavelet. The CWT of a signal $s(t)$ is expressed by the following integral:

$$W_s(a,b) = \frac{1}{\sqrt{a}} \int_{-\infty}^{+\infty} s(t) \cdot \psi^*_{a,b}(t)\mathrm{d}t \tag{7.7}$$

with

$$\psi_{a,b}(t) = \frac{1}{\sqrt{a}} \psi\left(\frac{t-b}{a}\right) \tag{7.8}$$

where $\psi(t)$ is the mother wavelet, a continuous function in both the time domain and the frequency domain. Then, b is the translation factor corresponding to the time, and a is the scale factor. The results of the CWT are a set of wavelet coefficients, which is the function of scale and time. The coefficients represent how closely the corresponding mother wavelet is correlated with the signal $s(t)$. The higher the coefficient, the more similar to the signal $s(t)$ the corresponding mother wavelet. The performance of CWT can be improved by utilizing more adapted mother wavelets. The choice of mother wavelet is the first issue that must be addressed, since different mother wavelets can draw out different characteristics of the signal.

In this section, CWT is utilized to analyze the correlation between the mother wavelet and the received echoes:

$$\mathrm{CWT}[E(t)] = \mathrm{CWT}[E_C(t) + E_T(t)] \tag{7.9}$$

Because of the linearity of CWT (Daubechies 1992), the following can be obtained:

$$\mathrm{CWT}[E(t)] = \mathrm{CWT}[E_C(t)] + \mathrm{CWT}[E_T(t)]$$
$$= \frac{1}{\sqrt{a}} \int_{-\infty}^{+\infty} E_C(t)\psi^*_{a,b}(t)\mathrm{d}t + \frac{1}{\sqrt{a}} \int_{-\infty}^{+\infty} E_T(t)\psi^*_{a,b}(t)\mathrm{d}t \tag{7.10}$$

From the equation, we see that $E_C(t)$ can be enhanced, while $E_T(t)$ can be simultaneously suppressed by CWT operation if the power spectrum of the selected mother wavelet is well matched with the echoes of cavitation bubbles.

Based on the principle above, the cavitation imaging algorithm can be implemented in three steps:

1. Select an appropriate mother wavelet based on the principle that the power spectrum of the selected mother wavelet should be well matched with the echoes from cavitation bubbles.
2. Use the CWT to analyze the correlation between the mother wavelet and the received echoes. The maximal correlation coefficients can be obtained at appropriate scales.
3. Select wavelet correlation coefficients at this scale to replace the data from the originally received echoes.

As mentioned above, distinguishing the cavitation bubble signals from surrounding tissue signals depends on the power spectrum of the selected mother wavelets and the echoes of the cavitation bubbles. Many theoretical models on the dynamics of a single oscillating microbubble have been established, and these can provide information about the desired echo of a single gas bubble. In this section, a new mother wavelet named the "cavitation bubble wavelet" is constructed according to the RPNNP equation. With the forth-order Runge–Kutta method, the RPNNP equation can be solved (Leighton 1994), which describes the response of a spherical bubble to a time-varying pressure field in an incompressible liquid:

$$\rho R\ddot{R} + \frac{3}{2}\rho\dot{R}^2 = \left(p_0 + \frac{2\sigma}{R_0} - p_V\right)\left(\frac{R_0}{R}\right)^{3\gamma}$$
$$- \left(p_0 + \frac{2\sigma}{R} - p_V\right) - 4\mu\frac{\dot{R}}{R_0} - p_{\mathrm{ext}}(t) \tag{7.11}$$

where p_V is the vapor pressure, ρ is the density of the surrounding liquid, σ is the surface tension, and μ is the viscosity. Here, $p_{\mathrm{ext}}(t)$ is the time-varying acoustic pressure. R_0, R, $\dot{R}$, and $\ddot{R}$ represent the initial radius, instantaneous radius, first time derivative of the radius, and second time derivative of the radius of the bubble, respectively. Other constants were chosen as follows: the specific heat ratio $\gamma = 1.4$ and the static pressure $p_0 = 10^5$ Pa. It should be noted that the RPNNP model does not consider the elasticity of the medium, so it can be used when cavitation is produced in water or blood, but a more appropriate model should be used when cavitation occurs in soft tissues.

The pressure of the bubble echo can be calculated as follows:

$$P = \frac{\rho}{d}\left(R^2\ddot{R} + R\dot{R}^2\right). \tag{7.12}$$

where d is the distance from the bubble to the receiver. After filtering and normalization, the predicted echoes can be used to construct the cavitation bubble wavelet.

There are two main differences between the cavitation bubble wavelet transform (CBWT) and the bubble wavelet transform (BWT) which was proposed for ultrasound contrast imaging (Zhao et al. 2011). First, the vibration properties and acoustic characteristics of contrast microbubbles depend on their physical properties and ambient environment, and many theoretical models for them have been established (Leighton 1994; Hoff 2001; Hoff et al. 2000; Morgan et al. 2000; de Jong et al. 1992; de Jong and Hoff 1993; Chatterjee and Sarkar 2003; Church 1995; Allen et al. 2002). For a free gas bubble, the RPNNP equation is used, while a modified Herring equation is solved for the time-dependent bubble radius and wall velocity; these outputs are used to formulate the predicted echo from a single encapsulated bubble (Morgan et al. 2000). Second, the size distribution of an ultrasound contrast agent (UCA) can be obtained; standard particle size analyzers can be used to measure the size distribution in vitro (Alter et al. 2009), while some acoustic sizing methods have been developed to estimate the size distribution of a type of UCA in vivo (Coussios et al. 2004; Sijl et al. 2008; Maresca et al. 2010). Therefore, a clear size distribution can be known for a given UCA. Conversely, the size of acoustic cavitation bubbles is difficult to measure, due to the chaotic nature and transient process of acoustic cavitation.

Experimental results of cavitation imaging using the CBWT method were obtained in water beneath tissue, as shown in Fig. 7.18. The physical constants for tap water at 20 °C are as follows: vapor pressure $p_V = 2330\ \mathrm{N/m^2}$, density $\rho = 998\ \mathrm{kg/m^3}$, surface tension $\sigma = 0.072\ \mathrm{N/m}$, and viscosity $\mu = 0.001\ \mathrm{Ns/m^2}$. With this technique, we must take into account different initial bubble radii. In the experiment, the initial radius (IR) changed from 0.5 to 2 μm with a step of 0.1 μm. The selected ROIs of the tissue and the cavitation area are shown in the left panel of Fig. 7.18j, while Fig. 7.18a, c, e, g represents the CTR varying with the scale factor at different IRs. The corresponding CBWT images at different IRs with the scale of the highest CTRs are shown in the right panels of Fig. 7.18b, d, f, h. From these images, we see that the CBWT images display the cavitation bubbles better than the B-mode images, which are shown in the left panels of Fig. 7.18b, d, f, h, suggesting the superiority of this technique. The optimal CBWT image is shown in the right panel of Fig. 7.18j, in which the IR corresponds to that of the highest CTR shown in Fig. 7.18i. The CTR of the optimal CBWT image is ~2.9 dB, better than that of the conventional B-mode image, whose CTR is ~2.2 dB. Figure 7.19 shows the results for monitoring cavitation activity using the CBWT method.

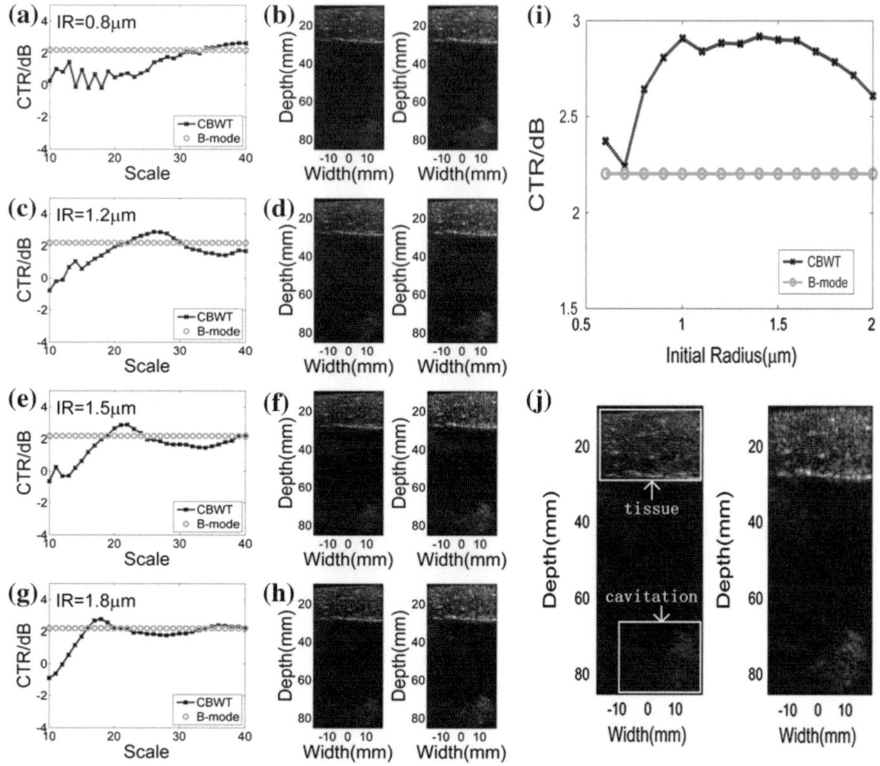

Fig. 7.18 Typical results with the CBWT method: **a**, **c**, **e**, and **g** are CTR versus scale with IRs of 0.8, 1.2, 1.5, and 1.8 μm, respectively; **b**, **d**, **f**, and **h** are the CBWT images (*right panel*) at the scales of 40, 26, 22, and 18, respectively, according to the results of (**a**), (**c**), (**e**), and (**g**). **i** is CTR versus IR, and **j** is the optimal CBWT image (*right panel*) at the scale of 32 and an IR of 0.9 μm

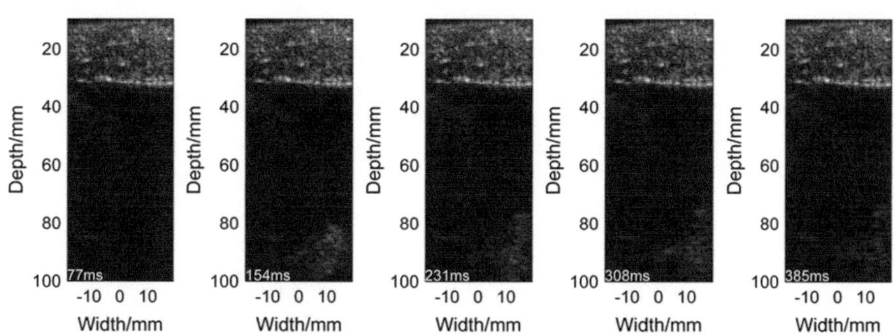

Fig. 7.19 Using CBWT images with time to monitor cavitation activity

7.4.2 CBWT Imaging Method with the Pulse Inversion

Pulse inversion is a useful technique to use with ultrasound harmonic imaging to extract the nonlinear contrast signal (Simpson et al. 1999; Shen and Li 2003; Liao et al. 2010; Pasovic et al. 2011). It can also stimulate the special characteristics of cavitation bubbles that differ from tissue characteristics, because cavitation bubbles produce very different echoes between two transmission pulses with inverted phases, while tissues do not.

Theoretically, the contrast between bubbles and tissues would be enhanced by using the CBWT method, whether or not the PI is employed. However, the maximum available CTR is limited because, to a certain extent, the signals from the tissue are correlated with the selected mother wavelet. By combining the PI technique with CBWT, the contrast between cavitation bubbles and the surrounding tissues can be improved by eliminating the tissue signals that include the fundamental signal. In the same way as described in Sect. 7.4.1, the PI-based CBWT can

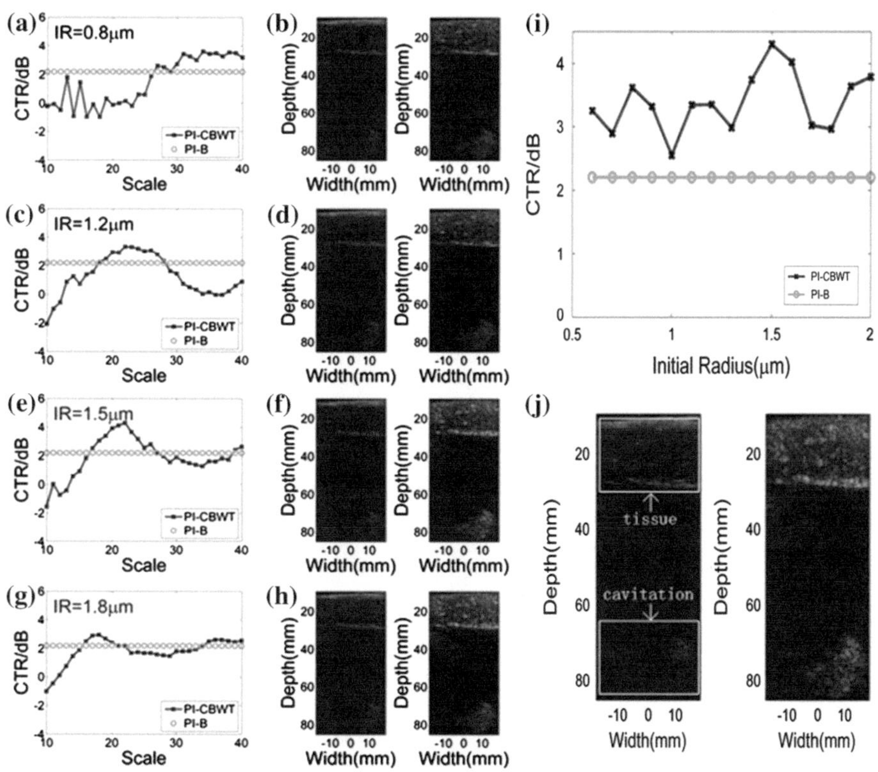

Fig. 7.20 Typical results with the PI-CBWT method: **a**, **c**, **e**, and **g** are CTR versus scale with IRs of 0.8, 1.2, 1.5, and 1.8 μm, respectively. **b**, **d**, **f**, and **h** are CBWT images (*right panel*) at the scales of 38, 24, 22, and 17, respectively. **i** is CTR versus IR, and **j** is the optimal CBWT image (*right panel*) at the scale of 22 and an IR of 1.5 μm

be constructed with a pair of pulses with inverted phases of 0° and 180°. The echoes replaced by the wavelet correlation coefficients from the two inverted pulses are summed to obtain one scan line of the final PI-CBWT. Figure 7.20 shows the process for how to obtain the optimal cavitation image based on the PI-CBWT method, which is similar to that of the CBWT method shown in Fig. 7.18. The optimal PI-CBWT image is shown in the right panel of Fig. 7.20j, in which the IR corresponds to that of the highest CTR shown in Fig. 7.20i. The CTR of the optimal PI-CBWT image is $\sim$4.3 dB, much better than that of the PI-based B-mode image, which is $\sim$1.1 dB.

In this section, the RPNNP equation for a free bubble was used, and the experiments were conducted in water beneath the tissue. The cavitation bubbles produced in the experiments may have relatively large sizes (submillimeter), whose resonance frequency was perhaps not in the range of the imaging probe's bandwidth. Thus, the CTR found in the CBWT images may have been limited. However, this might produce relatively small cavitation bubbles in tissues, which would perhaps be several to tens of micrometers in size. Therefore, the CBWT imaging method has the potential to be used for cavitation imaging in tissues with better CTR.

7.5 Bubble Doppler Technique for Cavitation Imaging

Recently, a new active cavitation imaging method based on the Doppler technique was proposed (Li et al. 2014). Ultrasound Doppler techniques are normally used to image moving scatterers in blood vessels, such as red blood cells and UCAs, and they can also detect bubbles trapped in the cracks and crevices of solid concretions in soft tissues (Lu et al. 2013). Therefore, Li et al. (2014) hypothesized that Doppler techniques could be used to detect cavitation bubbles induced by pulsed HIFU, and they obtained preliminary results in tissue-mimicking gel phantoms.

Three Doppler techniques—color Doppler, pulse-inversion Doppler, and decorrelation Doppler—can be fused and used for cavitation imaging, and this is called the bubble Doppler technique (Li et al. 2014). These individual Doppler techniques can provide different information about bubble distribution and be complementary to each other. For example, color Doppler can detect the change in the size of cavitation bubbles that dissolve after HIFU is turned off. Pulse-inversion Doppler is an indicator of nonlinearity of cavitation bubbles. Although it has similar sensitivity to color Doppler for detecting cavitation bubbles, pulse-inversion Doppler has an important advantage over color Doppler for in vivo applications: It can suppress the linear signal component induced by body motion or blood circulation. Finally, the interleaving Doppler technique, which is conceptually similar to the decorrelation Doppler technique, has the most sensitivity of all three Doppler techniques. However, it can only be used to detect the presence of bubbles, but it is not suitable for monitoring bubble activity (Li et al. 2014).

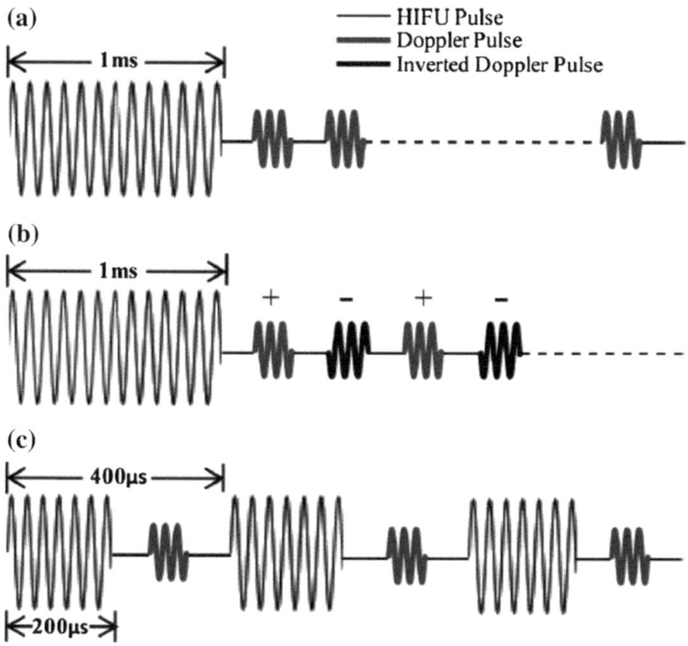

Fig. 7.21 Pulse sequences of three modes of bubble Doppler. **a** Color Doppler: One 1-ms HIFU pulse (*thin line*) was followed by fourteen 3-cycle Doppler pulses (*thick lines*). **b** Pulse-inversion Doppler: One 1-ms HIFU pulse was followed by fourteen Doppler pulses, and every other pulse was inverted (*thick lines*). **c** Interleaving Doppler: Each of the fourteen 200-μs HIFU pulses was followed by a single Doppler pulse. The HIFU Doppler pulse pair was repeated with a period of 400 μs. Reproduced with permission from Li (2014). © 2014 IEEE

Figure 7.21 shows the pulse sequences of the three modes of bubble Doppler that are used to synchronize HIFU pulses with the Doppler ensemble. In all cases, the Doppler ensemble pulses are transmitted in flash mode, in which all array elements are excited simultaneously to emit a quasi-plane wave. Figure 7.21a is color Doppler sequence, in which HIFU pulses were delivered for 1 ms at a PRF of 1 Hz followed by fourteen 3-cycle Doppler ensemble pulses with a PRF of 3 kHz. The total number of HIFU pulses was twenty, and each Doppler ensemble was transmitted with a 700-μs delay after the corresponding HIFU pulse to avoid interference. Figure 7.21b is a pulse-inversion Doppler sequence, which is similar to color Doppler with one modification: Every other pulse in the Doppler ensemble was inverted. Figure 7.21c shows interleaving Doppler sequences, in which fourteen 200-μs HIFU pulses were interleaved with fourteen Doppler ensemble pulses. Each HIFU pulse was followed by a single Doppler pulse with a 150-μs delay. The HIFU Doppler pulse pair was repeated with a period of 400 μs (Li et al. 2014).

For signal processing, the most common velocity and Doppler power estimation method (i.e., the autocorrelation algorithm) was used to obtain cavitation bubble

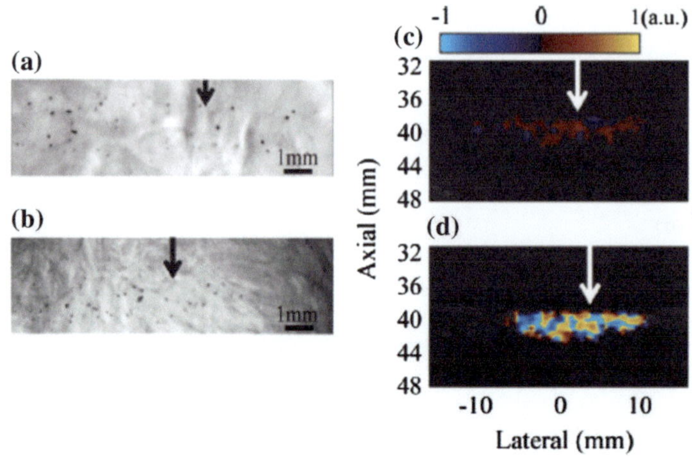

Fig. 7.22 Cavitation bubble imaging using bubble Doppler for two pulse sequences of color Doppler and interleaving Doppler: **a**, **b** are high-speed camera images; **c** color Doppler regime corresponding to (**a**); **d** interleaving Doppler regime corresponding to (**b**). Reproduced with permission from Li (2014). © 2014 IEEE

images. Figure 7.22 shows the typical results of cavitation bubble imaging using the bubble Doppler technique. High-speed camera images were also obtained simultaneously as shown in Fig. 7.22a, b, in which cavitation bubbles were observed as small dark dots. Figure 7.22c, d represents cavitation images from the color Doppler mode and the interleaving Doppler mode, respectively. In both cases, the size and position of the colored region in the bubble Doppler images corresponded well to those of the bubble distribution in the high-speed camera images, providing confirmation that the color region originated from the cavitation bubbles. In Fig. 7.22c, the displayed color contains alternating dark red and blue regions, indicating the very slowly, chaotically moving scatterers for the color Doppler pulse sequence. For the interleaving Doppler pulse sequence, the color region contains a color mosaic that is characteristic of the twinkling artifact (TA), as shown in Fig. 7.22d, indicating that the scatterers have a variety of speeds and move in all directions (Li et al. 2014).

In this study, it was also found that these three bubble Doppler techniques all had higher sensitivity than passive cavitation detection (PCD) and high-speed photography. The interleaving Doppler mode has the lowest threshold of all the Doppler techniques and could image cavitation bubbles when the HIFU focal peak negative pressure was above 1.9 MPa. Conversely, the color Doppler and pulse-inversion Doppler methods did not display cavitation bubbles until 3.1 MPa (Li et al. 2014).

7.6 Passive Cavitation Imaging

Active cavitation imaging can provide information about cavitation bubbles between HIFU exposures, but it cannot monitor cavitation activity when HIFU is on. Passive cavitation imaging (PCI) can solve this problem. However, unlike in ACI, PCI has poor axial resolution.

PCI was developed based on passive cavitation detection (PCD), which has many advantages in monitoring cavitation activity because it senses the acoustic emissions of cavitation bubbles. However, because of the passive receiver, the transducer is usually a single-element focused transducer with little capacity to use the source location to separate the mixture of emission traces it receives; it can, however, provide some information about spatially dependent sensitivity. However, if multielement arrays are used as passive receivers, they could provide the capacity for PCI and offer higher spatial and temporal resolution. As mentioned in Chap. 2, in order to generate cavitation images, the ultrasound probe needs to receive data from every element synchronously. However, most diagnostic ultrasound devices only offer the beamformed signal, preventing the original signals from being processed as a user-defined type. Fortunately, some ultrasound platforms currently undergoing research can provide users access to the original RF data coming from individual channels before the beamforming algorithm is applied, which enables the user to apply different and novel algorithms or techniques in order to achieve high spatial and temporal resolution. By turning off the pulse transmission, the dynamic beamforming algorithms of ultrasound scanners can be used to resolve cavitation mainly along the transverse direction of the linear array (Farny 2007; Farny et al. 2009) and to a small extent in the axial direction (Salgaonkar et al. 2009b; Leighton 1994).

7.6.1 *Time-Domain PCI*

According to the theory of passive cavitation mapping, mentioned in Sects. 2.8.1 and 2.8.2, the PCI method presented by Jensen et al. (2012) has been used to localize the position of tissue heating and ablation during HIFU exposure experiments in vitro. Thus, PCI can accurately realize spatial localization of HIFU lesion (Jensen et al. 2012). In this experiment, the HIFU transducer was always fixed in position, so its geometric focus remained at the midplane of the sample (at a depth of 20 mm in bovine liver). The 128-element linear array transducer with an aperture of 38 mm was arranged coaxially with the HIFU transducer. Each element had a nominal frequency response of 5–10 MHz. In order to compare the passive cavitation imaging results with B-mode images, a diagnostic ultrasound system was connected to the array and its mode could be switched between conventional B-mode and passive mode (no active pulse). In passive mode, it could record 64 channels of in-phase and quadrature (I/Q) data simultaneously during HIFU exposure. By using a digital comb filter with a bandwidth of 300 kHz for each comb, each channel was filtered in the frequency domain to obtain the broadband signal and the harmonics of the 1.06-MHz HIFU signal corresponding to inertial and stable cavitation, respectively (Gyöngy 2010).

Jensen et al. (2013) used a linear array (5–10 MHz) which is placed coaxially with 1.06-MHz HIFU transducer to passively receive emissions from bubbles in the ex vivo ox liver samples. As shown in Fig. 7.23, PCI shows spatial distribution of cavitation bubble in the HIFU focus and HIFU lesion presents elliptic shape. It should be noticed the problem that prediction of lesion seemed to deviate from cumulative passive imaging, which may be solved by estimation of tissue temperature. Moreover, when the amplitude of HIFU increased, such that the lesion region became overexposed and the tissue boiled, Jensen showed that both the broadband and harmonic maps shifted toward the HIFU transducer, which indicates a prefocal movement of cavitation activity (Bailey et al. 2001). Jensen et al. (2013) deemed that this experiment, in which the signals were processed straightly in the time domain, verified the effectiveness of the PCI method in tissue, whereby it nondestructively showed the location of the HIFU focus and gave a quantitative prediction of the lesion occurrence during HIFU treatment. The distribution of HIFU lesion region can be calculated by cumulative intensity of acoustic emissions from cavitation bubble; therefore, PCI can be used to dynamically predict HIFU focus and determine the damage position.

In the most recent development regarding the PCI method, 2D imaging of spatial–temporal cavitation was achieved in passive mode and then applied in drug release and gene therapy. This type of imaging allows us to predict the therapeutic result in human tissue, providing an impetus for using ultrasonic therapy. This more efficient and measurable method also allows us to map cavitation activity to drug delivery, monitor in vivo cavitation dynamics, and also assess the in vivo capabilities of PCI (Choi et al. 2014).

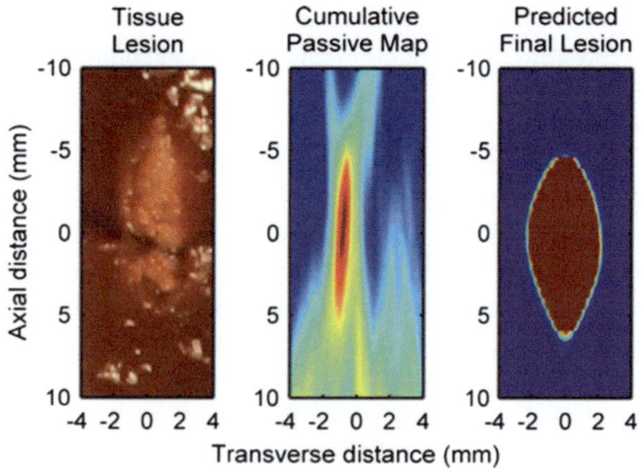

Fig. 7.23 Cumulative passive imaging of ox liver and prediction of final HIFU lesion. Reproduced with permission from Jensen et al. (2013). © IOP Science, 2013

7.6.2 Fourier Domain PCI

With the algorithm discussed in Sect. 2.8.3, the brightness value for a passive image point (Y, Z) is defined as the total beamformed energy from the acoustic sources sensed by linear array which is focused at the point. Hence, the image I can be expressed as in Eq. 2.26:

$$I(\omega, Y, Z) = \sum_{\forall\omega} \left| \sum_{\forall n} e^{j\omega T_n} S_n(\omega) \right|^2,$$

where $S(\omega)$ refers to the received emission signal and ω is the angular frequency of the radiation frequency component. Specifically, these angular frequencies correspond to the subharmonics and ultraharmonics of the sonication frequency for bubbles undergoing stable cavitation, whereas for inertial cavitation, these frequencies contain broadband frequencies (Salgaonkar et al. 2009a, b).

When we consider the analytical expression of PCI, by examining Parseval's theorem, we can see that the analytical expression of PCI is expressed in essentially the same way as the general PCI method because they are both described by the energy distribution of certain frequencies. Thus, for the analytical expression of PCI, the mathematical framework was designed and developed to optimize the parameters of passive imaging. We can use this analytical expression to then examine the effects and consequences for cavitation imaging when using different frequency bandwidths of cavitation emissions, beamforming configurations, and array dimensions.

Using the algorithm discussed in Sect. 2.8.3, Salgaonkar et al. (2009a, b) utilized a 192-element linear array whose center frequency was 7.5 MHz to capture the cavitation signals induced by a continuous-wave source. The method is mainly to receive data using a subaperture with a constant width (64 elements) and a constant f-number to beamform 16 receiving focal zones. Two ultrasound sources were used for sonication, with frequencies of 520 kHz and 2.2 MHz, respectively. The passively received signals were finally beamformed by the conventional delay-and-sum method. This experiment was performed in ex vivo bovine liver, and in order to passively image cavitation in the tissue, the sample was exposed to a 2.2-MHz continuous-wave focused ultrasound. With the propagation direction orthogonal to the image plane, the source was aligned and focused at a 20 mm depth on the imaging axis. The sonication amplitude was increased from 0 to 1.96 MPa peak-to-peak pressure.

After filtering the fourth harmonic, broadband emission energy was detected and spatially expressed as shown in Fig. 7.24a. This imaging modality, with the information in the frequency domain, can also show the spatial distribution of the acoustic emissions from the cavitation bubbles. According to Fig. 7.24a, the azimuthal position of the source focus may correspond to the maximum brightness value (at a 20 mm depth). Figure 7.24b illustrates that the broadband emission

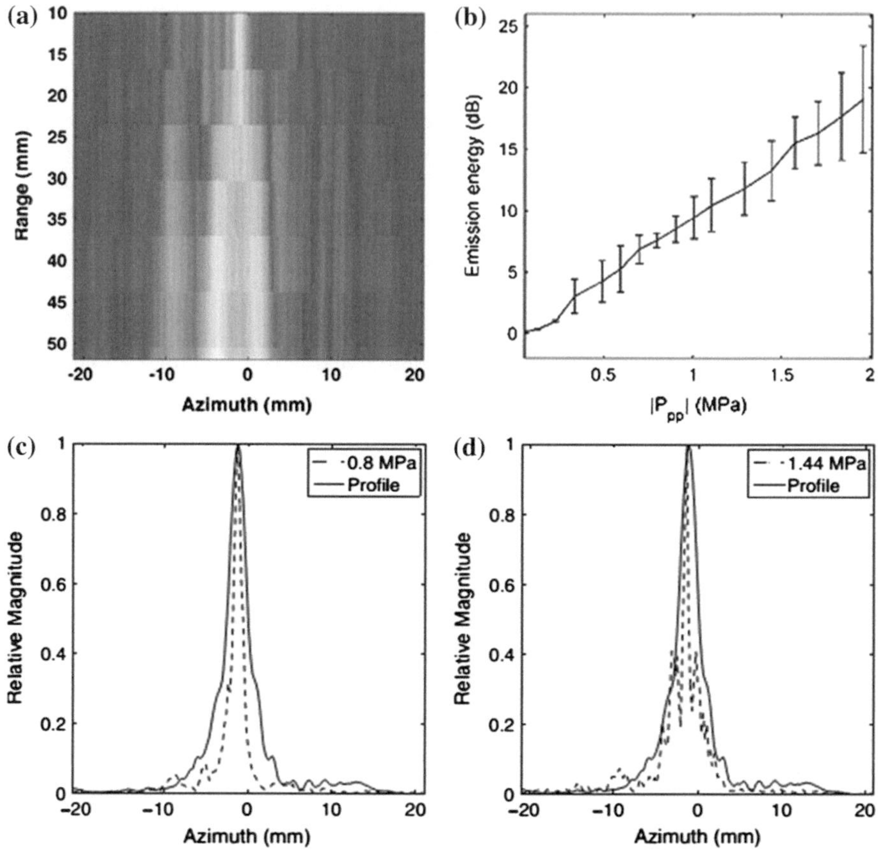

Fig. 7.24 Passive cavitation imaging in bovine liver tissue using broadband emissions. Reproduced with permission from Salgaonkar et al. (2009a, b). © 2009 Acoustical Society of America

energy increased monotonically with the sonication amplitude, which might be related to an increase in the inertial cavitation activity. Then, Fig. 7.24c, d shows comparisons between the relative magnitudes of the source pressure and the broadband emission energy with respect to the azimuth at the same positions but different sonication pressures. As the pressure increased, the broadband emission range broadened azimuthally.

Interestingly, the passive image in Fig. 7.24a is very similar to the simulated image of a point source at a 20 mm depth in water. However, in tissue, the threshold for inertial cavitation could be 3–40× higher than in water (Church and Yang 2006), because the broadband emissions should be influenced by the viscoelastic properties of the medium, the preexisting nuclei, and the dissolved gas content, among other factors (Salgaonkar et al. 2009b). Additionally, in tissues, inertial cavitation may take place at lower pressures because of the growth of preexisting nuclei in the continuous-wave condition. This situation may have applied to the liver sample used in the experiment, which was surrounded by blood and saline and thus may have had several preexisting cavitation nucleation sites.

From Fig. 7.24, we can also see that both the broadband emissions and the beam profile were able to detect the azimuthal position of the cavitation source. In other words, the passive image brightness pattern at the source focus depth corresponded to the peak of the beam profile along the array azimuth, as is clearly shown in Fig. 7.24. Moreover, this experiment highlighted two key points: (1) The passive image brightness pattern likely broadens with the increase in the sonication amplitude, possibly because the sidelobes disturb the transducer beamforming, and (2) the emission pattern broadens and is closer in width to the source beam pattern when the pressure increases.

7.6.3 Techniques for Improving the Resolution of PCI

7.6.3.1 Sparse Deconvolution Technique

Generally, cavitation can be imaged by conventional beamforming, in which the signals are appropriately delayed and summed together. However, this technique is unable to overcome the Rayleigh resolution limit, and the performance of imaging algorithm will degrade when there are interfering sources (Gyöngy 2010). Although adaptive beamforming and parametric techniques, such as maximum-likelihood and subspace-based methods, which are based on the assumption that cavitation bubbles can be considered as point sources, have been developed for overcoming the interference problem (Coviello et al. 2010; Krim and Viberg 1996), these methods fail when there is an extended bubble cloud.

In cavitation imaging, a vital problem that both beamforming and parametric techniques share is the bandlimited nature of cavitation signals. But, cavitation signals can be modeled as a series of emissions from temporally sparse inertial cavitation convolved by the impulse response of the transducer (Gyöngy and Coviello 2011). Spatial sparsity of cavitation emissions in the source field should be possessed. Therefore, the signal bandwidth can be increased by deconvolving the cavitation signals with sparsity preserving techniques, and then, the resolution of beamformed images can be improved.

In this section, two common and robust recovery methods for implementing sparse deconvolution are introduced to improve the resolution of cavitation images. They are the matching pursuit and the basis pursuit methods (Chen et al. 1998).

Cavitation bubbles are often considered to be an unknown source field $s(r,t)$ depending on position r and time t because of the nonlinear and chaotic response of the bubbles (Gyöngy and Coviello 2011). As a result, the pressure fluctuations in the emission from these sources are sampled by the receivers as $y_m = [y_{m,1}\cdots y_{m,N}]^T$ at time $t_n = n/f_s$, $n \in (1,\ldots,N)$, where f_S is the sampling rate. Assuming that the signal received is bandlimited, the cavitation recording should be as follows (Gyöngy and Coviello 2011):

$$\hat{y}_m = \int_v \frac{h(t) * s(r, t_n - |r - r_m|/c)}{4\pi|r - r_m|} \mathrm{d}r \tag{7.13}$$

where $h(t)$ is the impulse response of the receiver, * denotes temporal convolution, c is the speed of sound, and v is a volume enclosing the source region. Here, we suppose that $\hat{y}_m$ can be produced from the cavitation emission signals x_m:

$$\begin{aligned} \hat{y}_m &= Hx_m \\ H_{n,n'} &= \begin{cases} h_{n-n'} & |n - n'| \leq L \\ 0 & \text{otherwise} \end{cases} \\ x_{m,n} &= \int_v \frac{s(r, t_n - |r - r_m|/c)}{4\pi|r - r_m|} \mathrm{d}r \end{aligned} \tag{7.14}$$

where the impulse response $h = [h_{-L}, \ldots, h_L]$ has been discretized to times $[-L, \ldots, L]/f_S$. The $\hat{y}_m$ should be unique when the number of point sources is less than the number of receivers (Krim and Viberg 1996; Kim and Nelson 2003), and by choosing the minimum "energy" solution over space and time, the source field has a δ-function spatiotemporal autocorrelation. Hence, conventional beamforming is a simple but robust method for estimating the source field from the back-propagation of the recordings (Kim and Nelson 2003).

However, because of the bandlimited response of the cavitation signals, the image produced by conventional beamforming would become degraded, compromising its accuracy. With conventional beamforming specifically, the width of the point-spread function should be inversely proportional to the bandwidth of the signal that is beamformed (Marengo et al. 2000). Thus, sparse deconvolution provides a way to simultaneously better reconstruct cavitation emission signals with a broader bandwidth and reduce interference (Gyöngy and Coviello 2011).

As mentioned above, assuming that cavitation events are discrete and sparse, the intensity of cavitation event can be described quantificationally. Consequently, the received recordings x_m, from the elements $m \in (1, \ldots, M)$, are expected to consist of discrete peaks to describe the emission of each cavitation. To estimate x_m from y_m, the residual that describes the difference between the measured and modeled outputs must first be set (Gyöngy and Coviello 2011):

$$r_m = y_m - \hat{y}_m \tag{7.15}$$

Then, x_m is estimated by turning a linear system of equations into a problem involving the minimization of a weighted sum of two norms:

$$x_m = \arg\min\left\{\|\mathrm{r}\|_p + \lambda\|x_m\|_q\right\} \tag{7.16}$$

where λ represents a weighting parameter and $\|\cdot\|_p$ and $\|\cdot\|_q$ are p and q norms, respectively, for p, $q > 0$. Then, these solutions, rather than those from the classic approach, are expected to embody the sparsity. Although there are a number of variants based on sparse deconvolution, including the matching pursuit and the basis pursuit methods (Chen et al. 1998), the simplest situation was presented by Gyöngy (2010).

The matching pursuit method involves a simple algorithm when $p = 0$ and $\lambda = 0$, and it attempts to minimize the number of residual events (Mallat and Zhang 1993). Specifically, the convolutions should be performed using a dictionary of waveforms, and then, the results should subtract the waveform from the signal where the maximum absolute convolution is found. In reality, it is not possible to reduce the residual r_m in the l_0 norm because of noise, so this algorithm is to be performed with some tolerant level in the l_1 norm.

In contrast, the basis pursuit method is an optimization principle for the solution where $q = 1$, which estimates x_m with a low l_1 norm. Considering the noise measurement, if the impulse response matrix H satisfies the conditions of the restricted isometry property (RIP), the optimized estimation from all, at most K-sparse signals, should be as follows (Becker et al. 2011):

$$\mu\sqrt{1-\delta}\|x\|_2 \le \|Hx\|_2 \le \mu\sqrt{1+\delta}\|x\|_2 \tag{7.17}$$

Many algorithms have been introduced to solve this problem, such as the Douglas–Rachford (DR) algorithm (Becker et al. 2011).

The theory used to estimate the receiver impulse response suggests that although frequency-dependent attenuation will attenuate the higher-frequency components of the emissions, the 5–10-MHz receiver impulse response is assumed to dominate for propagation distances up to 100 mm. Considering the combined effects of the 5–10-MHz response of the transducer, the signal should pass a 4.6-MHz high-pass filter first (Gyöngy and Coviello 2011). Figure 7.25 shows the result of the impulse response estimation h'. This result depicts that the array response dominates over frequent-dependent attenuation when comparing the power spectrum of h' with the frequency calibration of the array.

As shown in Fig. 7.26, matching pursuit and basis pursuit deconvolution were applied to the data before beamforming and three maps produced by the same RF recording were compared. Figure 7.26 shows passive cavitation maps produced with non-deconvolved, matching pursuit with deconvolved, and basis pursuit with deconvolved data, respectively, at the peak rarefactional focal pressure (PRFP) of 1.3 MPa. Comparing these three passive maps from cavitation emissions, the original data depict a single cavitation event every two-second cycle of the HIFU insonation, which indicates a single bubble or a small bubble cluster. Conversely, the matching pursuit algorithm is able to express this bubble or cluster more clearly. Moreover, simulations showed that the basis pursuit algorithm is more suitable for

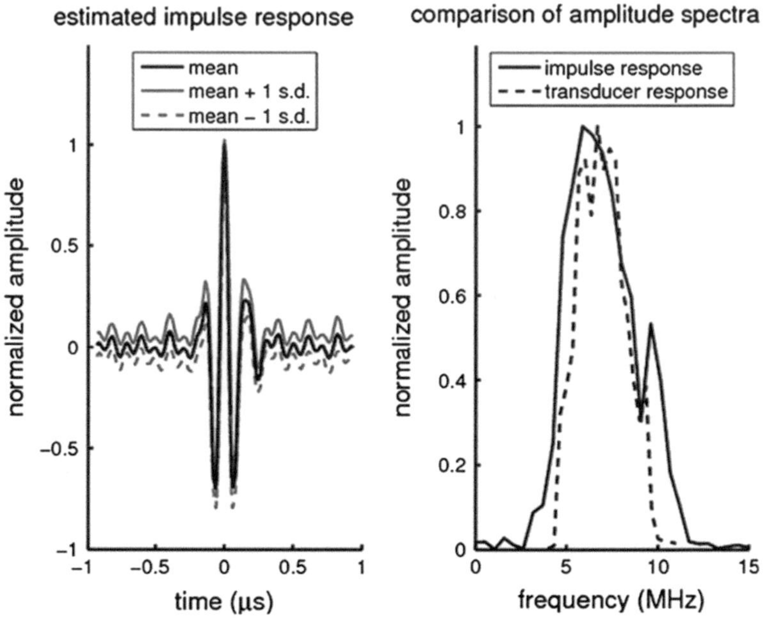

Fig. 7.25 Estimation of the receiver impulse response. Reproduced with permission from Gyöngy and Coviello (2011). © 2011 Acoustical Society of America

resolving several cavitation sources or a large cluster, as this method should possess the ability to show every local peak, while the others only indicate a single maximum. Both of sparse deconvolution algorithms were superior to the original method of passive cavitation imaging.

7.6.3.2 Adaptive Beamforming Technique

The conventional PCI algorithm based on time exposure acoustic (TEA) is to process the receiving data using short-time energy accumulation, but some limitations with this method often manifest as imaging artifacts that may be caused by high rarefactional pressure and the randomness of cavitation. Higher pressures can produce more cavitation bubbles that may interact with each other and develop further gather. General PCI cannot realize high resolution and low interference, which has been confirmed by a simulated broadband source (Coviello et al. 2013). Therefore, to overcome these deficiencies, here we consider using PCI with the robust Capon beamformer (RCB) that is extended by minimum variance distortionless response (MVDR).

In Fig. 7.27, we see that PCI combined with RCB removed imaging artifacts from the conventional images, providing a novel approach for obtaining PC images with improved resolution, good interference suppression, and robustness against steering vector errors (Coviello et al. 2013).

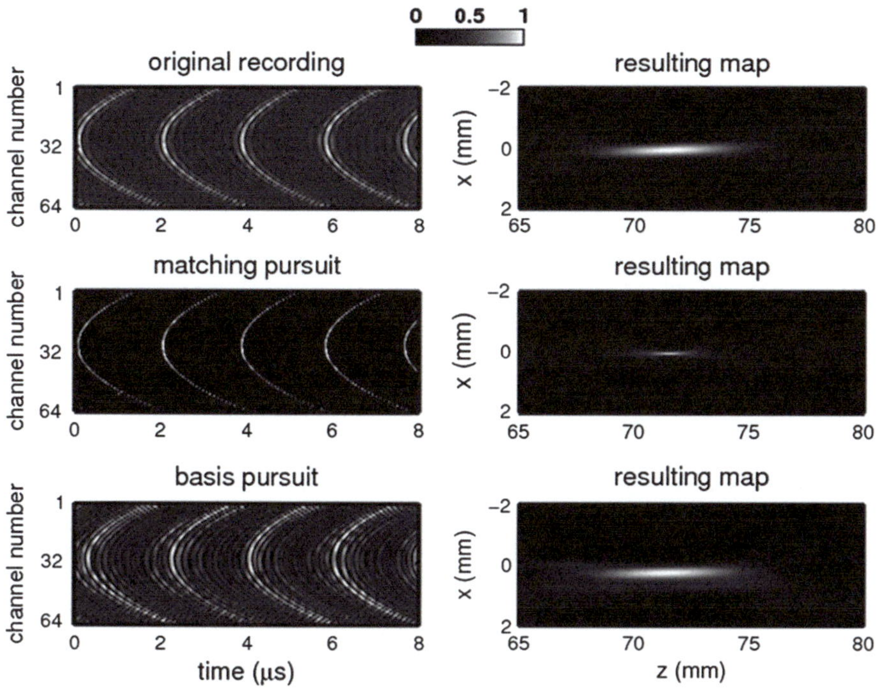

Fig. 7.26 Comparison of (*top*) no deconvolution, (*middle*) matching pursuit deconvolution, and (*bottom*) basis pursuit deconvolution on the generation of passive maps from cavitation emissions recorded at 1.3 MPa peak rarefactional focal pressure (PRFP). Reproduced with permission from Gyongy (2011). © 2011 Acoustical Society of America

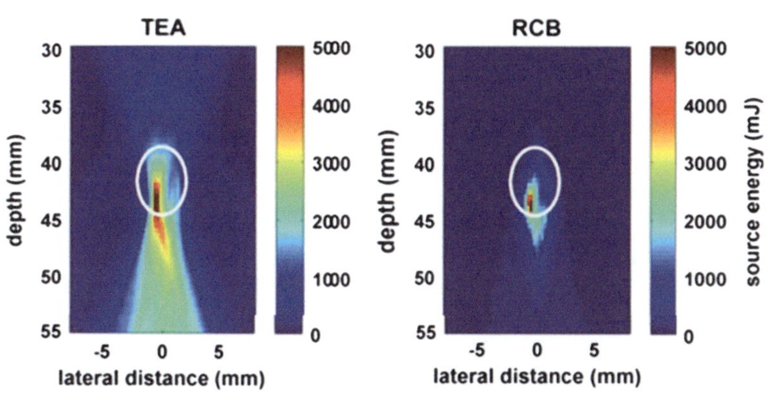

Fig. 7.27 Comparison of time exposure acoustic (TEA) and RCB. The *white ellipse* indicates the focus of HIFU. Reproduced with permission from Coviello et al. (2013). © 2013 Proceedings of Meetings on Acoustics

7.7 Cavitation Imaging Based on Super-Resolution Reconstruction

Usually, cavitation imaging is performed under condition where multiple cavitation bubbles are present. As such, some investigations are aimed at estimating the spatial and/or size distribution of cavitation bubbles. Cavitation bubbles can have a very large size distribution, such that bubbles can be less than a millimeter or even several micrometers in diameter. Although a single cavitation bubble cannot be distinguished from another due to the resolution limitation of current ultrasound systems and techniques, single bubbles can be detected in tissues because cavitation bubbles have much higher nonlinearity than tissues. For example, it is possible to identify a single bubble when the spatial distribution of the cavitation bubbles is sparse enough. Therefore, alongside improving the performance of ultrasound hardware and beamforming techniques, we can also investigate post-processing methods to improve the resolution of cavitation images.

This might be especially important for ultrafast cavitation imaging, which produce cavitation images with a lower SNR and poorer resolution than conventional B-mode imaging with focused transmission, because ultrafast cavitation imaging involves no focusing of the plane wave field. Therefore, improving the spatial resolution of the plane wave images for precise ultrafast cavitation imaging is crucial. A direct way to improve the resolution performance of ultrafast plane wave imaging is to use several tilted transmitting plane waves, such that the backscattered signals are coherently summed to produce a fully dynamic image (in transmit and receive); this concept has already been introduced in Sect. 7.2.2.1 (Tanter and Fink 2014). Here, we introduce another method, called the super-resolution reconstruction (SRR) method, which can also improve the resolution of ultrafast cavitation images.

7.7.1 *Principles of Super-Resolution Reconstruction*

Besides transmission-based methods, approaches using signal post-processing are another option for overcoming spatial resolution limitations. The term "super-resolution reconstruction" can be used to describe the process of obtaining a high-resolution (HR) image or a sequence of HR images from a set of low-resolution (LR) observations. The basic premise for increasing the spatial resolution in SRR techniques is the availability of multiple LR images captured from the same scene, which represent different "looks" at the same scene. As the frame rate of ultrafast plane wave imaging is always on the order of thousand frames per second, the adjacent frames can be approximately regarded as images captured from the same scene. That is, LR images are subsampled (aliased) as well as shifted with subpixel precision. If the LR images are shifted by integer units, each image contains the same information; thus, there is no new information that can be used to reconstruct an HR image. Whereas if the LR images have different subpixel shifts from each other and aliasing is present, each image cannot be obtained from the others (Patanavijit 2009). In this case, the new information contained in each LR image can be exploited to form an HR image.

Generally, SRR techniques can be divided into frequency-domain algorithms and spatial-domain algorithms (Borman and Stevenson 1998). Frequency-domain methods are based on three fundamental principles: (1) the shifting property of the Fourier transform, (2) the aliasing relationship between the continuous Fourier transform and the discrete Fourier transform (DFT), and (3) the original scene is bandlimited. Many of the earlier frequency-domain works were developed using DFT (Tsai and Huang 1984; Kim et al. 1990; Kim and Su 1993), but more recently, SSR methods based on discrete cosine transform (Rhee and Kang 1999) and wavelet transform (Chan et al. 2003; Lertrattanapanich and Bose 2002) have also been proposed. SRR via the frequency-domain approach has significant advantages because it is theoretically simple and has low computational complexity and it is highly amenable to parallel implementation because the frequency-domain equations can be decoupled. Conversely, the disadvantages of frequency-domain methods include the limitation to global translational motion and space-invariant degradation models (Borman and Stevenson 1998). Also, SRR is generally a severely ill-posed problem and regularization is often required to achieve acceptable image quality.

Frequency-domain methods have limited ability to include a priori knowledge that can be used to constrain the solution space to affect regularization. Therefore, spatial-domain methods are superior to the frequency-domain super-resolution formulation (Park et al. 2003). In the spatial domain, typical reconstruction models include non-uniform interpolation (Ur and Gross 1992), iterative back-projection (Irani and Peleg 1991), projection onto convex sets (POCS) (Stark and Oskoui 1989; Tekalp et al. 1992), maximum likelihood (ML) (Tom and Katsaggelos 1994), maximum a posteriori probability (MAP) (Schultz and Stevenson 1996), and so on. Most spatial-domain methods are motion-based, and the high-resolution images are reconstructed by the inverse process of that shown in Fig. 7.28 (Park et al. 2003).

Figure 7.28 shows the relationship between the desired HR image X and the *k*th observed LR image Y_K. The *k*th observed LR image of the same scene results from warping, blurring, and subsampling operators performed on the designed HR image

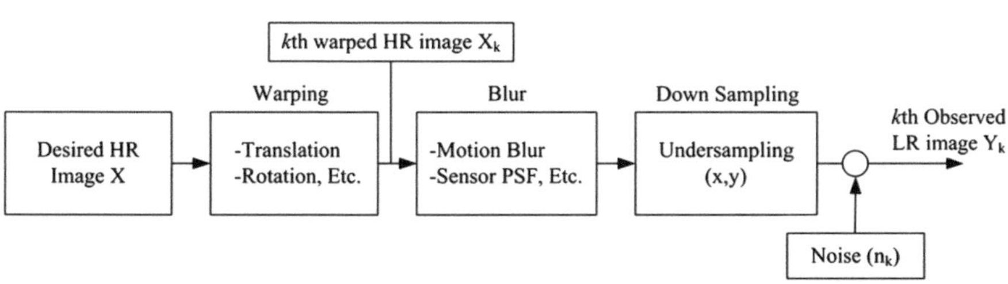

Fig. 7.28 Image formation process

X. Therefore, most of the HR image reconstruction methods proposed in the literature consist of three stages: registration, interpolation, and restoration. The registration stage involves the estimation of motion information, and it is extensively studied in various fields of image processing. In the registration stage, the relative displacement between kth observed LR image and the reference LR image is estimated with fractional pixel accuracy. Since the displacements between LR images are always subpixel, the registered HR image does not always match up to a uniformly spaced HR grid. Thus, the interpolation stage involves non-uniform interpolation to obtain a uniformly spaced HR image from a non-uniformly spaced composite of LR images. Finally, image restoration is applied to the upsampled image to remove blurring and noise. Within the class of spatial-domain SRR methods, two major techniques stand out as the most promising: the Bayesian MAP approach and the set theoretical POCS methods. To learn more about the spatial-domain SRR methods, the relevant reference can be consulted (Park et al. 2003); in the next section, we mainly focus on the POCS method.

7.7.2 POCS Reconstruction Approach

The POCS approach was first proposed by Stark and Oskoui (1989) to incorporate the prior knowledge of the solution into the reconstruction process (Stark and Oskoui 1989). With the estimation of registration parameters, this algorithm simultaneously solves the restoration and interpolation estimation problem. The implementation procedure for the POCS algorithm is shown in Fig. 7.29.

Let us assume that each image is measured under some geometric warping, blurring, and rational downsampling of non-stationary objects, in addition to degradation by some additive noise uncorrelated with the measurement. The image formation process is then expressed by:

$$Y_k(m_1, m_2) = \sum_{(n_1, n_2)} \widetilde{X}(n_1, n_2) \cdot W_k(m_1, m_2; n_1, n_2) + \eta_G(m_1, m_2) \tag{7.18}$$

where $Y_k(m_1, m_2)$ is the kth observed LR image; $\widetilde{X}(n_1, n_2)$ is the continuous real HR image; $W_k(m_1, m_2; n_1, n_2)$ is a linear blur mapping of the HR image to the LR image; and $\eta_G(m_1, m_2)$ is an additive white Gaussian noise. Assuming that the motion information is accurate, a data consistency constraint set based on the observation model is presented for each pixel within the LR images $Y_k(m_1, m_2)$ as follows:

$$C_{m_1, m_2} = \left\{ X(n_1, n_2) : \left| r^{(X)}(m_1, m_2) \right| \leq \delta(m_1, m_2) \right\} \tag{7.19}$$

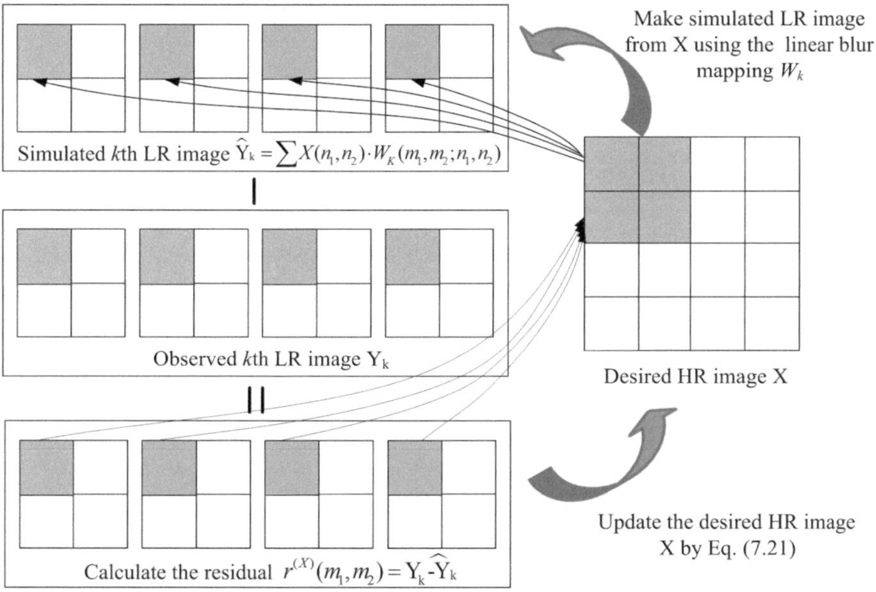

Fig. 7.29 Implementation process for POCS

where

$$r^{(X)}(m_1, m_2) = Y_k(m_1, m_2) - \sum_{(n_1,n_2)} X(n_1, n_2) \cdot W_K(m_1, m_2; n_1, n_2) \tag{7.20}$$

The residual $r^{(X)}(m_1, m_2)$ is bounded in magnitude by $\delta(m_1, m_2)$. Since the bound $\delta(m_1, m_2)$ is determined from the noise statistics, the ideal solution is a member of the set within a certain statistical confidence. The projection of the desired HR image $\mathrm{X}(n_1, n_2)$ onto C_{m_1,m_2} can be defined as follows:

$$X^{n+1}(n_1, n_2) = X^n(n_1, n_2) + \begin{cases} \frac{\left(r^{(X)}(m_1,m_2)-\delta(m_1,m_2)\right)\cdot W_k(m_1,m_2;n_1,n_2)}{\sum_{n_1,n_2} W_k^2(m_1,m_2,n_1,n_2)}, & r^{(X)}(m_1, m_2) > \delta(m_1, m_2) \\ 0, & \left|r^{(X)}(m_1, m_2)\right| \le \delta(m_1, m_2) \\ \frac{\left(r^{(X)}(m_1,m_2)+\delta(m_1,m_2)\right)\cdot W_k(m_1,m_2;n_1,n_2)}{\sum_{n_1,n_2} W_k^2(m_1,m_2,n_1,n_2)}, & r^{(X)}(m_1, m_2) < -\delta(m_1, m_2) \end{cases} \tag{7.21}$$

where

$$W_k(m_1, m_2; n_1, n_2) = \frac{\exp\left(\left((n_1 - X_0)^2 + (n_2 - Y_0)^2\right)/2\right)}{\sum_{n_1=X_0\pm2}\sum_{n_2=Y_0\pm2}\exp\left(\left((n_1 - X_0)^2 + (n_2 - Y_0)^2\right)/2\right)} \tag{7.22}$$

The POCS algorithm for SRR includes three primary steps:

1. Among the observed LR image sequences, the middle frame is chosen as the reference image (RI) and the resolution is doubled by cubic spline interpolation to create a desired HR image.
2. The relative movement between the RI and the *k*th observed LR image is estimated by the optical flow method. Then, by using the blur mapping shown in Eq. (7.22), the residual $r^{(X)}(m_1, m_2)$ between the *k*th observed LR image and *k*th simulated LR image can be estimated.
3. The desired HR image $X(n_1, n_2)$ is then iteratively revised according to Eq. (7.21) until the residual meets the required value. Then, the process returns to Step 2 to continue looping until the last LR image has been processed.

(X_0, Y_0) is the mapping point of (m_1, m_2) in the desired HR image $X(n_1, n_2)$. The superscript n represents the number of iterations. If the absolute value of the residual $r^{(X)}(m_1, m_2)$ is greater than $\delta(m_1, m_2)$, the desired HR image $X(n_1, n_2)$ needs to be revised as Eq. (7.21). Additional constraints, such as the gray scale, also can be used to improve the performance of the results. The super-resolution images shown in this section were reconstructed from three adjacent LR image sequences.

Typical SRR results of cavitation images taken in pig muscle in vitro are presented in Fig. 7.30. The electrical power of HIFU was 60 W with an exposure time of 5 ms; cavitation bubbles were imaged immediately after the HIFU transducer was completely off. Figure 7.30a is a low-resolution image acquired at 1.5 ms after the HIFU transducer was turned off by PWMVCF-based UACI. The SRR image is displayed in Fig. 7.30b. The focus areas of HIFU in Fig. 7.30a, b are outlined by white ellipses, which correspond to the cavitation areas in the tissue. The cavitation bubbles are not obvious because of the hyperechogenicity of the surrounding tissue. In order to suppress the signals of the surrounding tissue, the image acquired before HIFU exposure was subtracted from Fig. 7.30a, b. The corresponding results are shown in Fig. 7.30c, d. From this figure, we can see that the differential SRR image displays the cavitation bubbles more clearly, such that not only was the resolution of the cavitation areas improved, but also the signals of surrounding tissue were also suppressed. The CTRs of the cavitation area in Fig. 7.30c, d are 10.65 and 20.4 dB, respectively. The resolution of Fig. 7.30c, d was also evaluated by axial and lateral autocorrelation, of which the normalized −6 dB width was used to quantify the resolution (Abeyratne et al. 1995). The results are shown in Fig. 7.31. Compared to the PWMVCF-based UACI, the axial and lateral resolution gains of the SRR image were improved by at least 3.34× and 1.83×, respectively.

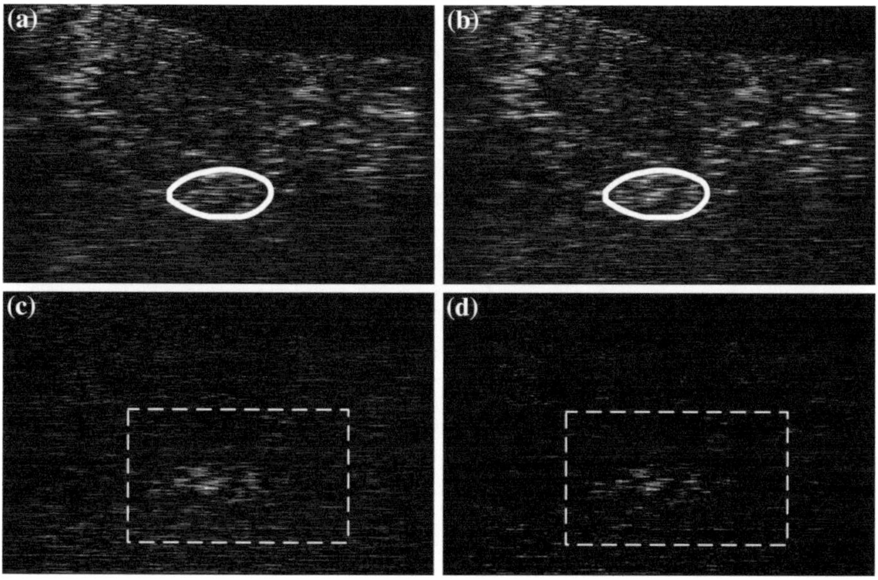

Fig. 7.30 Typical SRR results of cavitation images taken in a pig muscle in vitro. **a** Cavitation image acquired by PWMVCF-based UACI. **b** Corresponding SRR image of (**a**). Panels **c**, **d** are the differential cavitation images

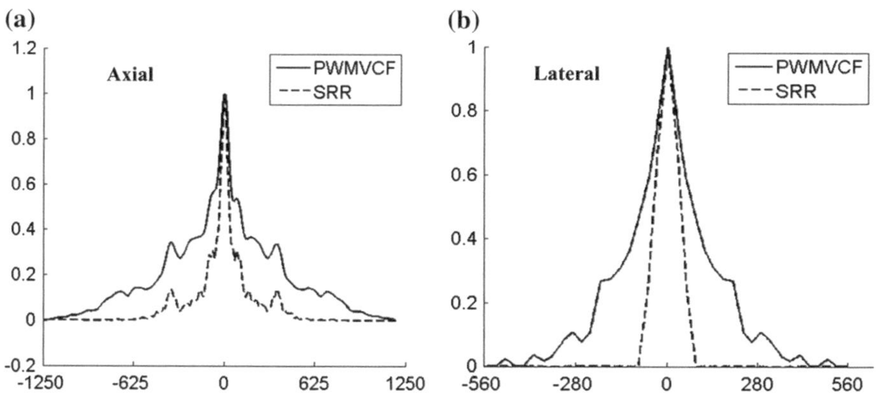

Fig. 7.31 Using autocorrelation to compare image resolution between PWMVCF-based UACI and SRR images in the **a** axial and **b** lateral directions

Next, differential cavitation images within the white dashed rectangular box in Fig. 7.30, were captured for different HIFU exposures, and these are displayed in Fig. 7.32. The electrical powers were 60, 80, and 100 W, and the pulse duration was 5 ms. The upper row of Fig. 7.32 shows the PWMVCF-based UACI results, while the bottom row shows the corresponding SRR images. Clearly, the area of the cavitation bubbles enlarged with higher HIFU excitation powers. After calculating

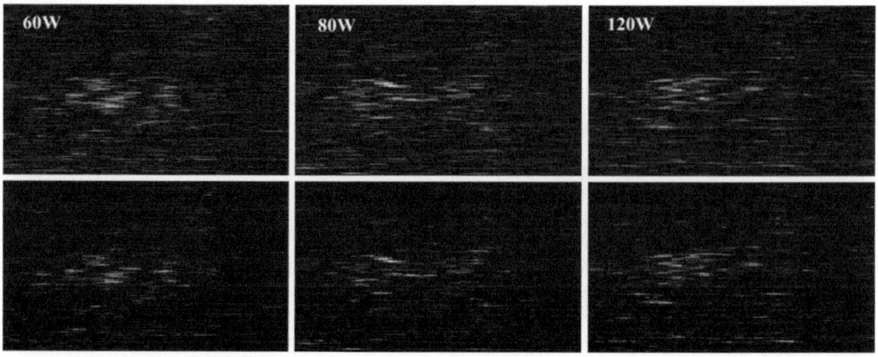

Fig. 7.32 Cavitation bubble images with different HIFU exposures, taken in pig muscle in vitro. *Upper row* PWMVCF-based UACI with powers of 60, 80, and 120 W. *Bottom row* corresponding to SRR cavitation images

Table 7.1 CTR of cavitation images under different HIFU exposures

Electrical power (W)	CTR of PWMVCF-based UACI (dB)	CTR of SRR (dB)
60	10.65	20.40
80	9.06	17.56
120	9.56	17.40

the CTRs for these images, we found that the SRR cavitation images were much better than those from PWMVCF-based UACI. The CTR values are given in Table 7.1.

7.8 3D Cavitation Imaging

7.8.1 Brief Description

2D cavitation imaging only provides information about cavitation activity in one spatial plane. However, cavitation bubbles are produced in 3D space. Thus, 3D cavitation imaging can offer more complete information of cavitation bubble distribution and is desired for monitoring cavitation activity. However, 3D cavitation imaging usually requires a 2D array transducer, which is expensive. Also, the aperture size and layout geometry of the 2D array (two key factors) determine the spatial resolution of the target. Consequently, to obtain satisfactory resolution in 3D images, the 2D array should have a large aperture, which would increase the cost because of the increase of the number of amplifiers, filters, A/D, etc. However, while acquiring the appropriate ultrasound instrument is costly, cavitation activity usually only occurs in a small area, especially for applications using focused ultrasound; thus, 3D cavitation imaging does not require nearly as many elements as does normal 2D cavitation imaging (usually 128 elements on one side). Several tens of elements on one side and in total only several hundred scan lines are enough for most 3D cavitation imaging situations. Here, we introduce some basic principles and methods about how to obtain 3D cavitation images and how to reduce the cost of 3D cavitation imaging.

First, we introduce the principle of 3D ultrafast ACI. As in 2D UACI, plane wave transmission can also be used for ultrafast real-time cavitation imaging, and the adaptive beamforming technique introduced in Sect. 7.2.2.2 can also be used to obtain 3D UACI with high SNR and CTR.

Making use of current diagnostic ultrasound systems to realize 3D cavitation imaging is reasonable, and this does not involve too much of a cost burden. Ultrasound diagnostic systems with 128 channels are available, and RF data from 128 channels can simultaneously be obtained from such systems. However, these 128-channel original data are not enough to form an acceptable 3D image. Thus, in this section, we introduce a compressed sensing and sparse processing method to form a 3D image by reconstructing more data (400 channels) from the original 128-channel data.

To reduce the number of elements from the number needed in the 2D array to that needed in 3D imaging, and to minimize the cost of the 3D imaging system, we introduce two newly designed types of sparse array transducers in this section. The corresponding beamforming algorithms based on these sparse arrays can potentially produce 3D cavitation images that are of acceptable quality. Furthermore, the beamforming algorithms introduced here can be used for both 3D PCI and also 3D ACI.

7.8.2 3D Ultrafast Cavitation Imaging Based on Adaptive Beamforming

3D ultrafast cavitation imaging can be performed based on a 2D array transducer. It combines 2D plane wave synthetic aperture imaging with minimum variance (MV) adaptive beamforming and coherence factor (CF) weighting to achieve wide imaging zone, high spatial–temporal resolution, and a high SNR.

Time delays should be calculated first as the propagation time from the transmitting element to the focus point and back to the receiving element, in the same way described for 2D ACI in Sect. 2.6. In a 3D coordinate system as shown in Fig. 7.33, assuming an active aperture of $M \times M$ elements determined by the given point $\vec{r}_p = (x, y, z)$, the delay of the receiving signal of the element in $(x_m, y_n, 0)$ is given by the following:

$$\tau_{mn}\left(\vec{r}_p\right) = \left(z + \sqrt{z^2 + (x - x_m)^2 + (y - y_n)^2}\right)/c \tag{7.23}$$

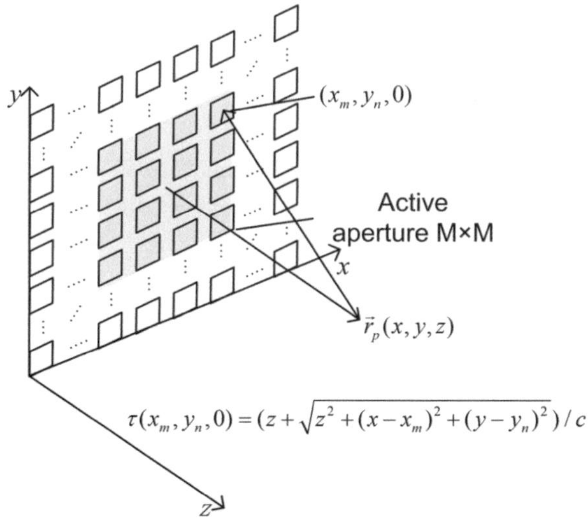

Fig. 7.33 Three-dimensional coordinate system for calculation of time delays

for $m = 1, \ldots, M, n = 1, \ldots, M$, and where c is the speed of sound.

The delayed signals of $M \times M$ elements are then given by the following:

$$\mathbf{RF}(t) = \begin{bmatrix} \mathrm{ch}_{11}(t + \tau_{11}(\vec{r}_p)) & \ldots & \mathrm{ch}_{1M}(t + \tau_{1M}(\vec{r}_p)) \\ \mathrm{ch}_{21}(t + \tau_{21}(\vec{r}_p)) & \ldots & \mathrm{ch}_{2M}(t + \tau_{2M}(\vec{r}_p)) \\ \vdots & \ddots & \vdots \\ \mathrm{ch}_{M1}(t + \tau_{M1}(\vec{r}_p)) & \ldots & \mathrm{ch}_{MM}(t + \tau_{MM}(\vec{r}_p)) \end{bmatrix} \tag{7.24}$$

where $\mathrm{ch}_{mn}(t)$ is the received channel-domain raw RF signal on the element in (x_m, y_n, 0).

The beamformed output of point $\vec{r}_p = (x, y, z)$ can be synthesized by weighting $\mathbf{RF}(t)$ with a fixed weighting and then summing it, which is considered to be delay-and-sum (DAS) beamforming.

As mentioned in Sect. 2.6, a widely used adaptive beamforming algorithm, MV beamforming (Capon 1969) can be used to compute the adaptive weightings by minimizing the variance of the beamformed output while maintaining a distortionless response of the focal point. 3D MV beamforming is decomposed into two steps: The first step is along the x-axis, and the second step is along the y-axis; the theory of MV beamforming for each step is the same as described in Sect. 2.6. The output of MV beamforming along the x-axis is given as follows:

$$s_m(t) = \frac{1}{M - L + 1} \sum_{l=1}^{M-L+1} \mathbf{w}_{\mathrm{opt\_X}}(t)^H \mathbf{RF}_{m\_l}(t) \tag{7.25}$$

for $m = 1, \ldots, M$, where L is the subarray length for spatial smoothing (Synnevag et al. 2007); $\mathbf{w}_{\text{opt\_X}}$ is the optimum MV weighting vectors along the x-axis; and $\mathbf{RF}_{m\_l}(t) = [\text{ch}_{m,l}(t + \tau_{m,l}(\vec{r}_p)) \ldots \text{ch}_{m,l+L-1}(t + \tau_{m,l+L-1}(\vec{r}_p))]^T$.

Then, the MV beamforming is performed along the y-axis, and the output is as follows:

$$S_{\text{MV}}(t) = \frac{1}{M - L + 1} \sum_{k=1}^{M-L+1} \mathbf{w}_{\text{opt\_Y}}(t)^H \mathbf{S}_k(t) \tag{7.26}$$

where $\mathbf{w}_{\text{opt\_Y}}$ is the optimum MV weighting vectors along the y-axis; and $\mathbf{S}_k(t) = [s_k \ldots s_{k+L-1}]^T$.

Furthermore, CF weighting (Asl and Mahloojifar 2009), another adaptive sidelobe reduction method, is applied to increase the robustness and the imaging quality of beamforming:

$$\text{CF}(t) = \frac{\left|\sum_{m=1}^{M} \sum_{n=1}^{M} \mathbf{RF}(m, n, t)\right|^2}{M^2 \sum_{m=1}^{M} \sum_{n=1}^{M} |\mathbf{RF}(m, n, t)|^2} \tag{7.27}$$

The final output is given by:

$$S_{\text{MVCF}}(t) = S_{\text{MV}}(t) \cdot \text{CF}(t) \tag{7.28}$$

By 3D MV beamforming and CF weighting, we can acquire the output of RF data for each point in 3D space. Then, the 3D cavitation image can be reconstructed using a 3D reconstruction algorithm.

7.8.3 Compressed Sensing and Sparse Modeling Method

In 3D cavitation imaging, a 2D array is required. However, to achieve the same resolution, 3D cavitation imaging requires significantly more than that required for a linear array; in fact, such resolution in 3D imaging would require the square of the number required for a 2D linear array, which would be incredibly costly. Moreover, 3D cavitation imaging produces a large amount of data, which might be a limitation for real-time imaging or even simply for data storage. However, a recent technique developed in 2006, called compressive sampling or compressed sensing (CS), is a novel theory aiming at reducing the amount of data collected during acquisition (Candès et al. 2006).

Compressed sensing utilizes the sparsity of the signal in signal acquisition, coding, and decoding (Candès and Wakin 2008). This theory indicates that as long as the signal is sparse or sparse in a transform domain, we can use a sampling matrix not related to the transform basis to acquire a series of projections from the signal. And then by solving an optimization problem, we can reconstruct the original signal with high probability from the small number of projections (Ender 2010). Therefore, using the CS theory allows us to reduce the number of receiving elements.

In this section, we describe how the CS method was used in 3D cavitation imaging to reduce the element number required in the 2D array, specifically for constructing a 128-element sparse array which can be conveniently integrated into current 128-channel diagnostic ultrasound systems.

7.8.3.1 Compressed Sensing Theory (Donoho 2006)

A complex signal $\mathbf{x} = [x(1)\, x(2) \ldots x(N)] \in C^N$ can be described as a linear combination of orthogonal bases:

$$\mathbf{x} = \sum_{i=1}^{N} \psi_i \alpha_i = \boldsymbol{\Psi\alpha} \tag{7.29}$$

where $\boldsymbol{\Psi} = \{\psi_i\}_{i=1}^{N}$ is the orthogonal basis and $\boldsymbol{\alpha} = [\alpha_1\, \alpha_2 \ldots \alpha_N]^T$ is the corresponding coefficient.

Signal $\mathbf{x}$ should be sparse or compressive in basis $\boldsymbol{\Psi}$, if $\boldsymbol{\alpha}$ has only $K(K \ll N)$ nonzero entries, which is equivalent to $\|\boldsymbol{\alpha}\|_0 = K$.

With the sparse signal $\mathbf{x}$ and a low-dimensional measurement matrix $\boldsymbol{\Phi}$, we can obtain an observation vector $\mathbf{y}$:

$$\mathbf{y} = \boldsymbol{\Phi}\mathbf{x} = \boldsymbol{\Phi\Psi\alpha} = \boldsymbol{\Theta\alpha} \tag{7.30}$$

where $\mathbf{y}$ is a vector of length $M(M \ll N))$ and $\boldsymbol{\Theta} = \boldsymbol{\Phi\Psi}$ is the sensing matrix of size $M \times N$.

Theoretical proof shows that when $\boldsymbol{\Theta}$ satisfies the restricted isometry property (RIP) condition (Candès and Wakin 2008), the most direct method to reconstruct the signal is the optimization problem:

$$\min\|\boldsymbol{\alpha}\|_0 \quad \text{s.t.} \quad \mathbf{y} = \boldsymbol{\Theta\alpha} \tag{7.31}$$

However, this is an non-deterministic polynomial (NP)-hard problem and it cannot be solved efficiently to obtain estimates of the sparse coefficient. However, it is already known that this problem can be relaxed into norm $p(0<p<1)$:

$$\min\|\boldsymbol{\alpha}\|_p \quad \text{s.t.} \quad \mathbf{y} = \boldsymbol{\Theta\alpha} \tag{7.32}$$

Then, the estimate of the sparse coefficient can be obtained by solving the formula in Eq. (7.32). Furthermore, with the additive white Gaussian noise, the signal can be expressed as follows:

$$\mathbf{y} = \boldsymbol{\Theta\alpha} + \mathbf{n} \tag{7.33}$$

Accordingly, the optimization model with noise reads as follows:

$$\min\|\boldsymbol{\alpha}\|_1 \quad \text{s.t.} \quad \|\mathbf{y} - \boldsymbol{\Theta\alpha}\| < \varepsilon \tag{7.34}$$

where ε is the noise level. By using the LASSO algorithm, the homotopy algorithm, etc., the problem can be effectively solved.

In order to guarantee the convergence of the reconstruction algorithm and the accuracy of K coefficients recovered from M measurements, the selection of the measurement matrix—which is usually chosen by a submatrix extracted from Gaussian random matrices or orthogonal matrices by line—must be simultaneously constrained by the sparse dictionary and the reconstruction matrix. Then, the reconstruction matrix, including the random measurement matrix $\boldsymbol{\Phi}$ and arbitrary fixed sparse matrix, can be proven to satisfy the RIP condition with high probability.

7.8.3.2 Compressed Sensing for 3D Cavitation Imaging

According to Fig. 7.34, assume that a planar array, which is composed of $M(m \times m)$ array elements, is located in the plane XOY. As a reference, the first element is located at the origin, and each element's coordinates are arranged as (x_i, y_j) for $i, j = 1, 2, \ldots, m$. Suppose that there are K far-field narrowband signals, and the pitching and azimuthal angle of the kth signal is (θ_k, ϕ_k).

If the array output signal is expressed as a vector, which is $\mathbf{x}(t) = [x_{1,1}(t)\ x_{1,2}(t) \ldots x_{m,m}(t)]^T = [x_1(t)\ x_2(t) \ldots x_M(t)]^T$, and $\alpha(\theta, \phi) = \sin\theta\cos\phi$ and $\beta(\theta, \phi) = \sin\theta\sin\phi$, for the noisy case, we have the following:

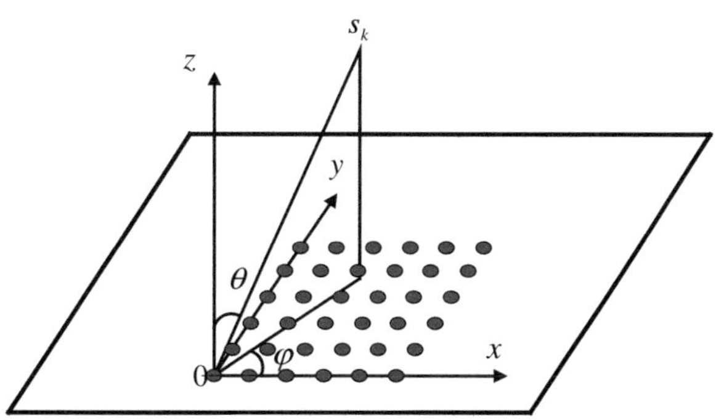

Fig. 7.34 A planar array receiving signals

$$x_l(t) = x_{(i-1)\times m+j}(t) = \sum_{k=1}^{K} s_k(t)\exp\{-j2\pi[x_i\alpha(\theta_k,\phi_k) + y_j\beta(\theta_k,\phi_k)]/\lambda\} + n_l(t) = h_l s_l(t) + n_l(t) \quad (l = 1, 2, \ldots, M) \tag{7.35}$$

This can be written as follows:

$$\mathbf{x}(t) = \mathbf{H}\mathbf{s}(t) + \mathbf{n} \tag{7.36}$$

where $\mathbf{s}(t)$ is the complex envelope for the signal, $\mathbf{H}$ is the guiding matrix (basis for CS) for the array, and $\mathbf{n}$ is the additive white Gaussian noise, which is uncorrelated with the signal received by the array. Here, the noise received by each element is supposed to be uncorrelated with each other and has the same power.

Hence, $\mathbf{s}(t)$ is thus the projection coefficient vector of the received signal $\mathbf{x}(t)$ in the transformation matrix $\mathbf{H}$. Generally, θ_k and ϕ_k would be one of $\theta_1\ \theta_2 \ldots \theta_M$ and $\phi_1\ \phi_2 \ldots \phi_M$, respectively.

Because $\mathbf{x}(t)$ is sparse, according to CS theory, only a few transducers are required for sampling to rebuild $\mathbf{x}(t)$. As a result, this would reduce the number of receiver channels needed and, subsequently, the cost of 3D cavitation imaging.

Assuming that the signal sampling of each element is stored into the matrix in turn as $\mathbf{x}(t) = [x_1(t)\ x_2(t) \ldots x_M(t)]$ and $\mathbf{x}(t)$ is extracted randomly by line via the extracting matrix $\mathbf{T}_{\text{ex}}$, the final extracting result can be expressed as follows:

$$\mathbf{r}(t) = \mathbf{T}_{\text{ex}}\mathbf{x}(t) \tag{7.37}$$

By the theory of CS, the Gaussian random matrix can be chosen as the measurement matrix:

$$\mathbf{y}(t) = \mathbf{\Gamma}\mathbf{r}(t) = \mathbf{\Gamma}\mathbf{T}_{\text{ex}}\mathbf{x}(t) = \mathbf{\Gamma}\mathbf{T}_{\text{ex}}\mathbf{H}\mathbf{s}(t) = \mathbf{\Theta}\mathbf{s}(t) \tag{7.38}$$

where $\mathbf{\Gamma}$ is the Gaussian random matrix, $\mathbf{\Phi} = \mathbf{\Gamma}\mathbf{T}_{\text{ex}}$, $\mathbf{\Theta} = \mathbf{\Phi}\mathbf{H}$, $\mathbf{x}(t)$ is the sparse signal in the sparse matrix $\mathbf{H}$, and $\mathbf{\Phi}$ is the submatrix randomly extracted from the Gaussian random matrix by column.

Thus, the sampling of an ultrasonic signal by random arrangement of the array elements can be regarded as a random projection of the sparse signal $\mathbf{x}(t)$ in the measurement matrix $\mathbf{\Phi}$. Then, the sensing matrix $\mathbf{\Theta}$ is the product of the submatrix randomly selected from the Gaussian random matrix by column, and the steering vector, which satisfies the RIP conditions for ensuring the effectiveness and robustness of the CS reconstruction algorithm.

For the actual signal, supposing that the total number of sampling points received by the array is N_s and the ith sampling point is y_i, whose orientation information can be expressed as s_i, we can conclude the formula as follows, in which the signal has noise vector n_i, subject to complex normal distribution as $n_i \sim \text{CN}(0, \sigma^2 I)$:

$$\mathbf{y} = \mathbf{\Theta s} + \mathbf{n} \tag{7.39}$$

where the received signal matrix is $\mathbf{y} = [y_1\ y_2 \ldots y_{Ns}]$ and the corresponding orientation matrix is $\mathbf{s} = [s_1\ s_2 \ldots s_{Ns}]$. In practice, the cavitation targets are sparse in space, which means $\|\mathbf{s}\|_1 = K(K \ll M)$ and $\mathbf{s}$ has only K nonzero entries.

Direct optimization of the underdetermined equations actually, like those shown above, can allow us to estimate the orientation domain information. Specifically, algorithms such as OMP and FOCUSS can be used for this estimation (Cotter et al. 2005).

Suppose that the number of elements for the planar array used in this system is 400 (20 × 20). An example of the extracted results of a CS array element distribution (with 128 elements) is shown in Fig. 7.35. This figure represents a random extraction from the original total of 400 (20 × 20) elements. According to the position of the CS array elements, the corresponding extraction matrix $\mathbf{T}_{\text{ex}}$ can be constructed for subsequent calculation.

Figure 7.36a, b shows the beamforming simulation for estimating a single bubble's position using either the entire planar array or the CS array (the construction of the CS array is shown in Fig. 7.35), respectively. The bubble's pitching angle is 45°, whereas the azimuthal angle is 90°. The system sampling rate was 40 MHz, and 10 dB of white Gaussian noise was added. For a single bubble, in this simulation, the MUSIC algorithm should be effective for estimating the bubble's position, given that this represents a simple situation in which the number of bubbles can be confirmed. The result shown in Fig. 7.36a was obtained by using the entire 400 (20 × 20) elements, while the result in Fig. 7.36b used only the 128 elements extracted randomly from the original 400 elements. Comparing these

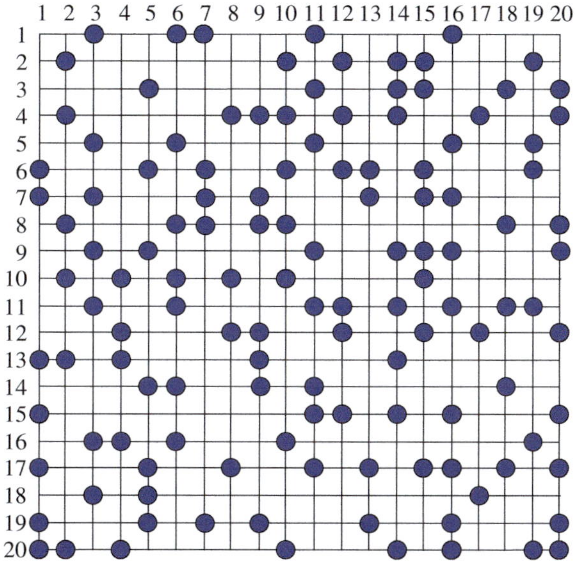

Fig. 7.35 An example of randomly extracted results of a CS array element distribution

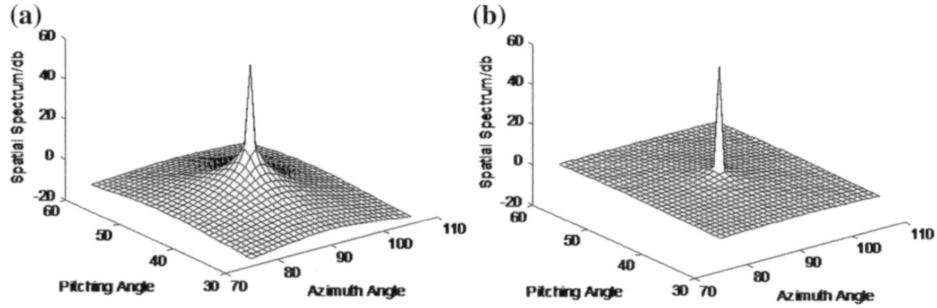

Fig. 7.36 Estimation of a bubble's position by **a** the entire planar array (20 × 20 elements) and **b** the CS array (128 elements)

figures, we see that the CS method does indeed clearly represent the position of a single bubble, which is almost identical to the bubble location found using the entire array.

7.8.4 Sparse Array Design

For 3D cavitation imaging, using the traditional rectangular 2D array would come at considerable cost because it would require many more elements and channels. To reduce costs, some sparse arrays have been developed, and when used with proper beamforming algorithms, they can produce images with acceptable quality.

7.8.4.1 Passive Cavitation Imaging with Thin-Film Sparse Boundary Arrays

Coviello et al. (2012) designed a novel 2D array that overcomes the shortcomings associated with traditional ultrasound array, such as the bandlimited characteristic of piezoceramics, to perform passive cavitation imaging with a smaller number of elements, thus reducing the cost of the array. This novel design takes advantage of the coarray technique (Hoctor and Kassam 1990; Kozick and Kassam 1991) to synthesize a fully filled 2D aperture. This technique, which would provide better performance as more elements are added to the array, reduces the complexity and cost of the array, but requires additional signal processing.

As mentioned in Sect. 2.9, the source power can be expressed as an integral of source strength on T:

$$\Psi(\mathbf{x}) = \frac{1}{4\pi\rho_0 cT} \int\limits_{t_0}^{t_0+T} (q(x,t))^2 \mathrm{d}t \tag{7.40}$$

where ρ_0 is the density of the medium, c is the speed of propagation, and $q(x, t)$ is the source strength field.

Previous studies have shown that apodization processing or weighting techniques, such as beamforming, can shape the received signal and reduce the effect of noise and unwanted waves (Szabo 2004). Therefore, combining this with the theory of PCI, two different weights $w_{1,j}$ and $w_{2,j}$ were applied to each array element to obtain two different calculations of the source strength (Hoctor and Kassam 1990):

$$q_k(x,t) = \beta \sum_{j=1}^{4(N-1)} w_{k,j} d_j(\mathbf{x}) s_j\big(t + d_j(\mathbf{x})/c\big) \tag{7.41}$$

where $k = 1, 2$ and $\beta = (4\pi)/(4(N - 1)\alpha)$. Then, the source power estimation can be rewritten as follows:

$$\Psi_W(\mathbf{x}) = \frac{1}{4\pi\rho_0 cT} \int_{t_0}^{t_{0+T}} q_1(\mathbf{x},t) q_2(\mathbf{x},t)^* \mathrm{d}t \tag{7.42}$$

where $[\cdot]^*$ denotes the complex conjugate.

Unlike normal transducers, which are made from traditional piezoceramic materials, the probes are designed with a relatively small bandwidth to optimize active imaging at a particular frequency. This means that a number of broadband components from the acoustic emissions are to be lost. To tackle this situation, Coviello et al. (2012) designed a 2D array in which thin-film materials (rather than sparse reconstruction) are used to recover information on the out-of-band frequencies for PCI. The thin-film materials, such as polyvinylidene fluoride (PVDF) and PVDF copolymers, have been used for a long time in a variety of ultrasound sensors. And, thin films have high sensitivity for receiving and are of low cost (Coviello et al. 2012). For these reasons, we describe a 2D array that is designed with a PVDF array made from multilayer printed circuit board (PCB).

Moreover, this 2D array is characterized by its geometric sparsity: The positions of its elements are only on the boundary of the rectangular aperture. The missing information associated with the interior of the array should be filled in using linear imaging techniques.

Specifically, supposing we have a rectangular boundary array with 5 elements on each side, as shown in Fig. 7.37a, meaning that there are a total of 16 elements in this array ($4(N - 1)$). Based on the point-spread function (PSF), the random arrays can produce a large aperture with few array elements, and this is usually characterized by a main lobe with a fairly uniform sidelobe level. The key point is that with additional data and signal processing, boundary array has the capacity to afford the same image as filled arrays do, since both the boundary and filled arrays share the same coarray. Therefore, each element on the boundary should have a different

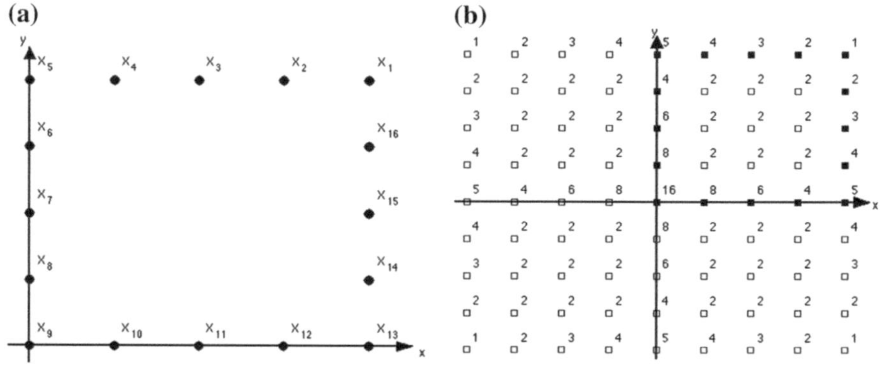

Fig. 7.37 **a** An $N = 5$ rectangular boundary array and **b** its difference coarray with multiplicity of each coarray point. Reproduced with permission from Coviello et al. (2012). © 2012 IEEE

coarray with multiplicity of each coarray point. For passive cavitation imaging, the vector of the coarray can be defined as follows:

$$C_{\mathrm{d}} = \{y|y = x_i - x_j,\ \forall x_i, x_j \in A\} \tag{7.43}$$

Then, according to C_{d}, the difference coarrays for the $N = 5$ boundary array can be depicted, as in Fig. 7.37b. Furthermore, this figure suggests that the difference coarray is the support of the inverse Fourier transform of the PSF and also the morphological autocorrelation of the array (Hoctor and Kassam 1990). With the weights applied to the receiving elements, the inverse Fourier transform of the resulting PSF can be as follows (Kozick and Kassam 1991):

$$\gamma(y) = \sum_{\{i,j:y=x_i-x_j\}} w_{1,i} w_{2,j}^*, \quad y \in C_{\mathrm{d}} \tag{7.44}$$

This is called the coarray weighting function, and $w_{1,i}$ and $w_{2,j}$ denote the array weights.

For a 2D weighting function, $\gamma(y)$ should be chosen over C_{d}, which is represented by a matrix $\mathbf{C}$ with the size $(2N - 1) \times (2N - 1)$. However, the weighting synthesis algorithm aims to decompose the weighting matrix $\mathbf{C}$ into a series of weight vectors, each with the size $4(N - 1) \times 1$, to make an intermediate image. To achieve this, we must construct a map from the weighting matrix $\mathbf{C}$ to matrix $\mathbf{C}_{\mathbf{w}}$, having a size $4(N - 1) \times 4(N - 1)$, with element–pair weights. Specifically, this is performed by taking the value of $\mathbf{C}$ at every normalized point and then mapping it to the row and column of $\mathbf{C}$. Then, $\mathbf{C}_{\mathbf{w}}$ can be decomposed by singular value decomposition (SVD):

$$\mathbf{C_w} = \mathbf{USV^H} = \sum_{l=1}^{4(N-1)} u_l \sigma_l v_l^H \tag{7.45}$$

where $[\cdot]^H$ denotes the conjugate transpose. Then, with the weighting vector $w_1(l) = u_l \sigma_l^{1/2}$ and $w_2(l) = u_l \sigma_l^{1/2}$, $l = 1, \ldots, 4(N-1)$, we will acquire

$$\psi_l(x) = \frac{1}{4\pi \rho_0 c T} \int_{t_0}^{t_{0+T}} q_1(x, t; l) q_2(x, t; l)^* \mathrm{d}t \tag{7.46}$$

At last, the output image is generated by summing all the subimages:

$$\Psi_{\text{out}}(x) = \sum_l \Psi_l(x) \tag{7.47}$$

To verify the algorithm and test the performance of the array in passive cavitation imaging, a homogenous 3 % aqueous agar gel prepared with deionized water was used as a tissue-mimicking material (TMM) in which the cavitation-nucleating particles (mixed with talc) were loaded (Coviello et al. 2012). The setup for the experiment is illustrated in Fig. 7.38.

A 500-kHz HIFU transducer aligned axially with an NDT transducer, which was 12.7 mm in diameter, operated at $f_0 = 5\,MHz$ and had a focal distance of 75 mm, was used to initiate cavitation in the TMM. The NDT transducer was the passive detector, used to confirm the presence of cavitation. Here, each resulting cavitation image was produced only during one 10-cycle pulse. Then, the signals from the bubbles were received by the sparse array (a total of 32 elements, with 9 elements on each side) and then sent through the custom filters, preamplified, and finally sampled.

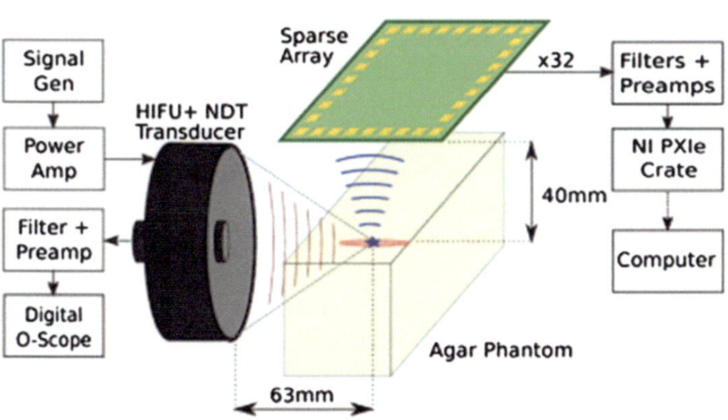

Fig. 7.38 Scheme for cavitation imaging in a phantom. Reproduced with permission from Coviello et al. (2012). © 2012 IEEE

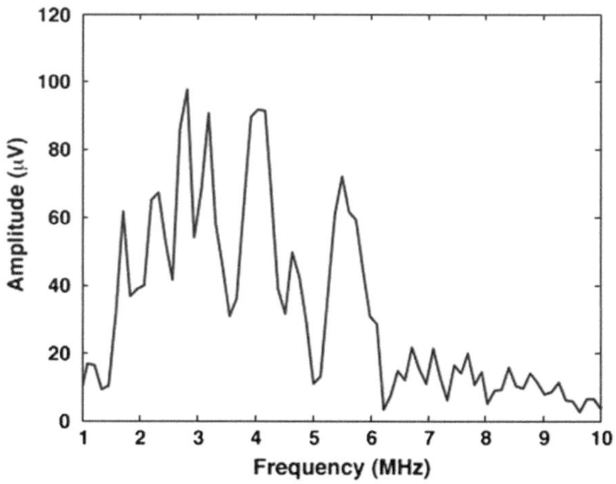

Fig. 7.39 Frequency spectrum of the signal received by an element in the sparse array. Reproduced with permission from Coviello et al. (2012). © 2012 IEEE

The receiving signal, shown in Fig. 7.39, had a bandwidth of approximately 3.5 MHz, and the system resolution was expected to be about 0.3 mm × 1.3 mm (transverse direction multiplied by the axial direction). The data were processed by both conventional time exposure acoustic (TEA) and the sparse aperture weighted algorithm to assess passive cavitation imaging.

Even though this array cannot provide sufficient information to accurately confirm the size of the region where inertial cavitation takes place, it does tackle the problem by predicting the size of the region using knowledge of the cavitation threshold for the TMMs(Faragher 2011). The results of cavitation imaging are shown in Fig. 7.40, in which we can see a continuous increase of the peak negative focal pressure (P_{PNFP}) from 2.23 to 3.35 MPa. Note here that the axial dimension starts from the HIFU transducer, and the HIFU in this experiment was traveling from top to bottom. Regions inside the black ellipses represent the predicted areas where inertial cavitation acted.

7.8.4.2 Hemispherical Random Sparse Array

O'Reilly et al. (2012) presented a design to fabricate a spherical sparse random receiver array integrated with an existing therapeutic array with a set of 1372 elements within a 30-cm-diameter hemisphere. They then evaluated the passive beamforming capabilities of the fabricated array when operating at frequencies relevant to transcranial therapy (O'Reilly et al. 2012). The transmit elements of this array were cylindrical, with an internal diameter of 7 mm, an external diameter of 10 mm, a height of 6 mm, and $f_0 = 306$ kHz (Song and Hynynen 2010). The receive elements, made out of a wideband polyvinylidene fluoride (PVDF) material with an active element diameter of 2.48 mm, were integrated with the existing transmit elements, and they were fixed using reusable adhesive. Figure 7.41 shows the array construction.

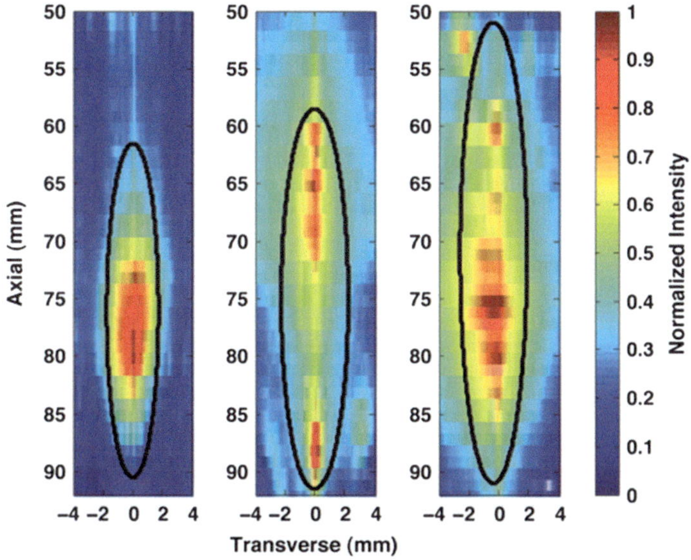

Fig. 7.40 The sparse aperture weighted algorithm for imaging the cavitation region in the tissue phantom at different P_{PNFP} values of 2.23, 2.62, and 3.35 MPa, respectively. Reproduced with permission from Coviello et al. (2012). © 2012 IEEE

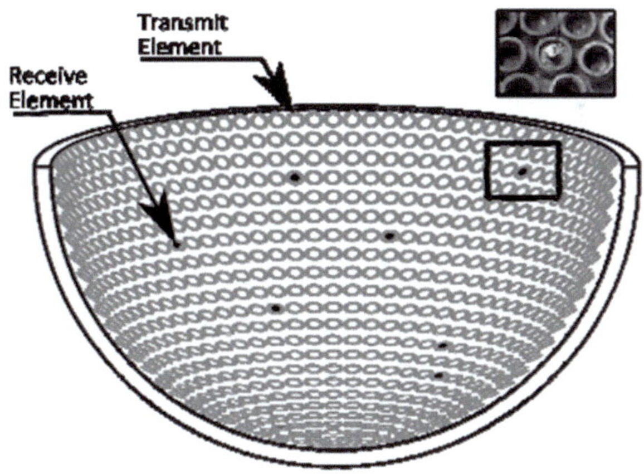

Fig. 7.41 The combined receiving and transmitting array. Reproduced with permission from O'Reilly et al. (2012). © 2012 American Institute of Physics

To evaluate the array performance, sparse random receiver array configurations with 32, 64, 128, 256, and 512 elements were simulated. Signal from a single point source was received, and the TEA algorithm was used for passive beamforming. The study showed that the grating lobes were effectively decreased when the number of elements increased, but there was little benefit of using a fully populated

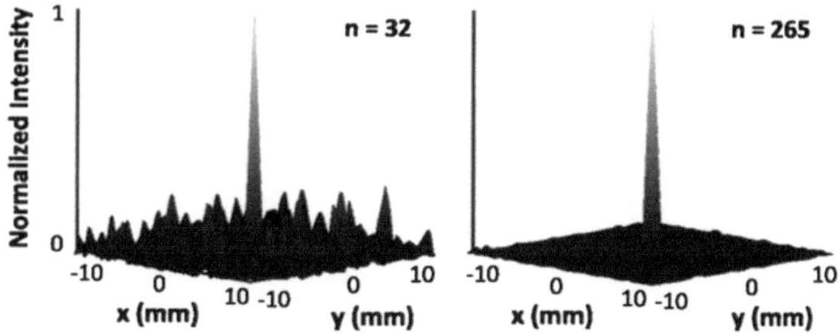

Fig. 7.42 Intensity map for a simulated 918-kHz point source with (*left*) a 32-element configuration and (*right*) a 256-element configuration. Reproduced with permission from O'Reilly et al. (2012). © 2012 American Institute of Physics

array over a 256-element array. Figure 7.42 shows simulated intensity maps for a point source with a 32-element and a 256-element configuration.

The simulation results for the 32-element array showed that this configuration was able to resolve point sources, emitting at an f_0 of the therapy array (306 kHz) and the second (612 kHz) and third harmonics (918 kHz), at a lateral resolution of 4.2, 2.1, and 1.4 mm, respectively, as shown in Fig. 7.43. The results also suggest that the TEA beamforming algorithm was robust for small numbers of elements when random noise with different amplitudes was introduced.

Considering large array in the transcranial ultrasound applications to treat brain diseases, O'Reilly used this hemispherical random sparse array to obtain transcranial measurement through a formalin-fixed human skullcap. Figure 7.44 shows passive signals from a 765-kHz, 1.5-mm narrowband source emitted through the skull and beamformed, in cases with and without phase correction techniques. Without the correction techniques, the intensity images were highly distorted, but those where hydrophone-based corrections were used were not distorted.

The two kinds of sparse arrays introduced in this section show that 2D arrays with fewer elements can be used to achieve acceptable resolution and can

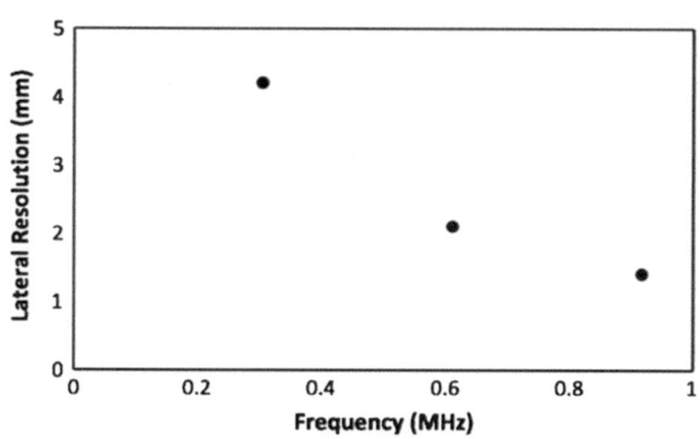

Fig. 7.43 Lateral resolution of the 32-element receiver array as a function of source frequency. Reproduced with permission from O'Reilly et al. (2012). © 2012 American Institute of Physics

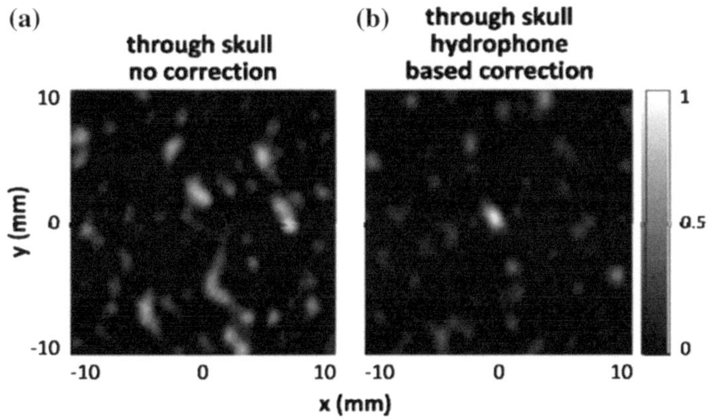

Fig. 7.44 Intensity images of a 765-kHz narrowband source emitting through a human skullcap (*left*) without phase correction and (*right*) with hydrophone-based phase correction. Reproduced with permission from O'Reilly et al. (2012). © 2012 American Institute of Physics

consequently reduce the cost of an array. Compared with compressed sensing (CS), methods based on the sparse array often do not utilize the sparsity of the signal, so some information needed for reconstruction or imaging is likely lost, especially with large-scale arrays.

7.9 Magnetic Resonance Cavitation Imaging

Magnetic resonance imaging (MRI) is a very important modality for monitoring HIFU therapy. In the early 1990s, Hynynen et al. (1993) studied the feasibility of using MRI to detect tissue necrosis induced by a focused ultrasound beam. MRI methods for monitoring temperature changes in the target region have also been investigated (Bohris et al. 2001; Cline et al. 1994). Because of its good performance in temperature mapping and superior contrast between necrotic and normal tissues, MR-guided focused ultrasound surgery has been applied in some clinical fields (Chapman and ter Haar 2007; Furusawa et al. 2006).

Temperature maps can be obtained by using T1-weighted fast spin echo (FSE) pulse sequences. Furthermore, because the contrast between the lesion and the tissue is excellent with either T1-weighted or T2-weighted FSE, lesions has been monitored by both techniques. However, cavitation activity is better detected by T2-weighted FSE, while T1-weighted FSE produces a poor contrast between air and necrotic tissues (Damianou et al. 2004; Damianou 2004). Damianou et al. (2004) found that there is a significant difference in signal intensity between air spaces (cavitation bubble) and necrosis tissues, and water-filled air spaces appear brighter than thermal lesions using T2-weighted FSE.

Figure 7.45 shows various MR images in a pig kidney with T2-weighted FSE using different ablation exposures. In Fig. 7.45a, the lesion was created based on

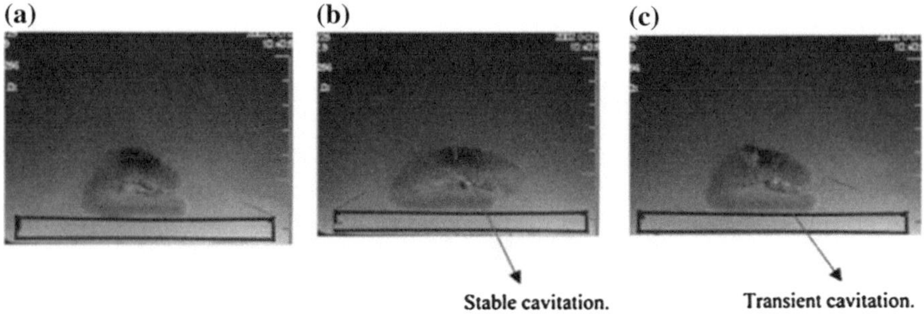

Fig. 7.45 MRI of HIFU ablation using T2-weighted FSE demonstrating the detection of various tissue destruction modes: **a** thermal, **b** stable cavitation, and **c** transient cavitation. Reproduced with permission from Damianou (2004). © 2004 World Federation for Ultrasound in Medicine & Biology

pure thermal mechanisms, and the treated region appears darker than normal tissue. The lesion in Fig. 7.45b was presumably created with stable cavitation. The cavities that appear bright are shown in the image. These bright cavities likely arose from stable cavitation because there was no unexpected tissue destruction induced by transient cavitation. Figure 7.45c shows a large lesion volume, which includes one lesion affected by transient cavitation. That lesion contains violent tissue disruption that was visible via MRI and in gross examination (Damianou et al. 2004).

As another example of MR imaging used for detecting cavitation, Fig. 7.46 shows the MR images of a large lesion in the kidney in the plane perpendicular to the beam using T2-weighted FSE. In Fig. 7.46a, the lesion was created by using a 3 × 3 grid of lesions, and the spacing between the lesions was 5 mm. The intensity used was 1000 W/cm^2 for 5 s. The panel shows that the necrosis coverage is

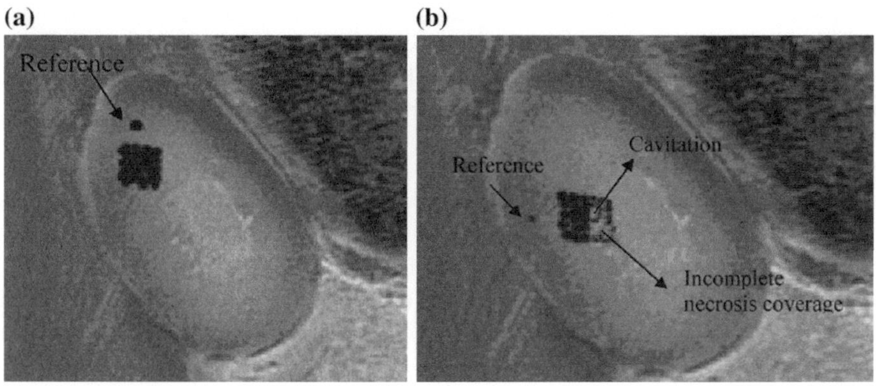

Fig. 7.46 MR images of a large lesion using T2-weighted FSE. Images were obtained in the plane perpendicular to the transducer beam. The spatial average intensity was **a** 1000 W/cm^2 for 5 s (pure thermal lesions); **b** 1500 W/cm^2 for 5 s (occurrence of cavitation). Reproduced with permission from Damianou (2004). © 2004 World Federation for Ultrasound in Medicine & Biology

continuous, leaving no untreated spaces, which implies that none of the lesions were created through cavitation at this intensity. Figure 7.46b shows the large lesion created using a 5 × 6 grid of lesions, and in this setup, the intensity was 1500 W/cm^2 for 5 s, and the spacing between the lesions was 1.5 mm. This panel shows that the coverage of necrosis is incomplete, leaving some spaces untreated. However, some of these small areas that look untreated were, in fact, lesions created by cavitation. This was verified by monitoring the acoustic emissions and temperature, and through gross examination. This figure suggests that it is difficult to discriminate between a cavitation lesion and untreated tissue when using MRI, because they both appear bright in MR images that use T2-weighted FSE (Damianou 2004).

We can see that although cavitation lesions have different properties than purely thermal lesions, cavitation activity was not able to be verified from MR images alone. In addition, MRI has poor temporal resolution (a few seconds per frame), which is insufficient for monitoring cavitation activity.

7.10 Summary

Being able to use imaging techniques to monitor what happens during ultrasound is very important for the development and eventual adoption of therapeutic ultrasound. As such, because cavitation is integral to most therapeutic ultrasound applications, cavitation imaging cannot be ignored. Ultrasound-based imaging methods have great advantages for cavitation detection and imaging, especially for opaque tissues, although they have limited capabilities for temperature monitoring.

In HIFU ablation treatment, which has been used clinically, cavitation is a very important mechanism besides thermal mechanisms. Cavitation bubbles induced by negative pressures of acoustic irradiation and/or boiling will be generated when the energy is high enough, and these bubbles can distort the shape of the thermal lesion. Thus, monitoring the dynamic changes of cavitation activity is necessary and valuable for understanding the relationship between the distorted lesion and cavitation activity. In pulsed HIFU treatments, including histotripsy (tissue erosion), sonoporation, and newer applications such as drug delivery and gene transfection, the cavitation mechanism dominates. Cavitation imaging could provide useful information for understanding the process of cavitation activity and for evaluating the effect of cavitation on these applications in various situations. Furthermore, cavitation imaging could provide the feedback information needed for controlling cavitation processes.

ACI and PCI are two kinds of important methods for cavitation imaging. ACI can provide superior spatial resolution, but the images can be only acquired when HIFU is off. Conversely, PCI can provide real-time monitoring, but it has poor axial resolution. Thus, ACI and PCI are complementary.

Conventional B-mode imaging is the simplest method of ACI. However, it has very low CTR and cannot separate cavitation information from thermal lesion information. Cavitation is difficult to identify when the cavitation signal is not very strong, and some tissues with hyperechogenicity exist. Therefore, improving the CTR of cavitation images in tissue is the most important task for ACI. The differential imaging method can be used for this purpose, and it has the potential to be realized on a low-cost ultrasound platform due to its relatively simple algorithm. Integral differential imaging can improve the CTR by eliminating the signals from hyperechogenic tissue outside the HIFU-irradiated area, whereby the data frames acquired before HIFU are subtracted from those acquired after HIFU irradiation. This technique is effective when tissues are still or move only a little. However, movement artifacts are often produced in the integral differential images due to the radiation force induced by HIFU. Also, integral differential imaging cannot eliminate the information from thermal lesions. To solve this problem, transient differential imaging could be used to exclude thermal lesion information by subtracting the two consecutive data frames acquired between HIFU exposures; this can also significantly reduce movement artifacts. AD, SAD, and SSD are three algorithms that can be used in differential imaging, whereby SSD offers cavitation images with the highest CTR, but at the cost of losing some small signals.

In order to acquire pure cavitation information and a high CTR, the differences between the signals from cavitation bubbles and those from tissues should be stimulated and/or detected, so they are as large as possible. Pulse-inversion-based second-harmonic and subharmonic imaging methods make use of the fact that cavitation bubbles have a much higher nonlinearity than tissues, and this can significantly improve the CTR of cavitation images. The CBWT cavitation imaging method introduced in this chapter can also produce high-CTR cavitation images, and this method utilizes the bubble oscillation model to distinguish cavitation bubbles from tissues. It can also be combined with the pulse-inversion technique to further improve CTR. Recently, the bubble Doppler technique was proposed for ACI, and the sensitivity of this technique depends on the dissolution, nonlinearity, movement, and/or collapse of cavitation bubbles for three different pulse sequences. The three modes of bubble Doppler can be considered complementary because they provide different information about the bubble distribution.

PCI was developed based on the PCD technique, but by using the multielement array instead of the single-element transducer that is used in PCD. A significant advantage of PCI is that it can provide real-time cavitation information of the treated area at the beginning of boiling or when dramatic inertial cavitation occurs. Thus, by monitoring the HIFU focus location with PCI, irreversible damage or overtreatment may be avoided. By comparing the results of different PCI methods, we can see that method processing in time domain can produce better resolution, while method processing in the Fourier domain, which was derived by numerically solving the Rayleigh–Sommerfeld integral under the Fresnel approximation, affords more information related to the frequency of interest. To improve the resolution of PCI, the deconvolution technique and the adaptive beamforming technique can be utilized, but the deconvolution technique is restricted to the assumption of temporal sparsity.

Ultrafast and 3D cavitation imaging techniques are the focuses of further investigation into cavitation imaging because cavitation activity is a rapid change process and occurs in 3D spaces. In ultrafast cavitation imaging, the most important problem is how to get high-CTR cavitation images with a high frame rate. The coherent compounding-based and adaptive beamforming-based UACI methods are both feasible for this purpose. In 3D cavitation imaging, a very important problem to be considered is that acceptable quality 3D cavitation images need to be produced, but at a reasonable cost. The compressed sensing technique may have potential for use in a sparse 2D array design that has 128 elements. This method can potentially connect easily to an available 128-channel diagnostic ultrasound system to realize 3D cavitation imaging. Two kinds of 2D sparse arrays were introduced in this chapter, and they can potentially offer acceptable performance with a small number of elements and consequently reduce the cost of the array, but some information for reconstruction or imaging would probably be lost, especially with a large-scale array. However, the investigation of ultrafast and 3D cavitation imaging techniques is still in their initial stages, so more work must be done in order for ultrafast and 3D cavitation imaging to be realized in the clinic.

References

Abeyratne UR, Petropulu AP, Reid JM. Higher-order spectra based deconvolution of ultrasound images. IEEE Trans Ultrason Ferroelectr Freq Control. 1995;42:1064–75.

Allen JS, May DJ, Ferrara KW. Dynamics of therapeutic ultrasound contrast agents. Ultrasound Med Biol. 2002;28:805–16.

Alter J, Sennoga CA, Lopes D, Eckersley RJ, Wells DJ. Microbubble stability is a major determinant of the efficiency of ultrasound and microbubble mediated in vivo gene transfer. Ultrasound Med Biol. 2009;35:976–84.

Asl BM, Mahloojifar A. Minimum variance beamforming combined with adaptive coherence weighting applied to medical ultrasound imaging. IEEE Trans Ultrason Ferroelectr Freq Control. 2009;56:1923–31.

Bailey MR, Couret LN, Sapozhnikov OA, Khokhlova VA, ter Haar G, Vaezy S, Shi X, Martin R, Crum LA. Use of overpressure to assess the role of bubbles in focused ultrasound lesion shape in vitro. Ultrasound Med Biol. 2001;27:695–708.

Becker S, Bobin J, Candès EJ. NESTA: a fast and accurate first-order method for sparse recovery. SIAM J Imaging Sci. 2011;4:1–39.

Bohris C, Jenne JW, Rastert R, Simiantonakis I, Brix G, Spoo J, Hlavac M, Nemeth R, Huber PE, Debus J. MR monitoring of focused ultrasound surgery in a breast tissue model in vivo. Magn Reson Imaging. 2001;19:167–75.

Borman S, Stevenson RL. Super-resolution from image sequences-a review. In: Proceedings of IEEE midwest symposium on circuits and systems, Notre Dame, France; 1998. pp. 374–378.

Candès EJ, Wakin MB. An introduction to compressive sampling. IEEE Signal Process Mag. 2008;25:21–30.

Candès EJ, Romberg J, Tao T. Robust uncertainty principles: exact signal reconstruction from highly incomplete frequency information. IEEE Trans Inf Theory. 2006;52:489–509.

Capon J. High-resolution frequency-wavenumber spectrum analysis. Proc IEEE. 1969;57:1408–18.

Chan RH, Chan TF, Shen L, Shen Z. Wavelet algorithms for high-resolution image reconstruction. SIAM J Sci Comput. 2003;24:1408–32.

Chapman A, ter Haar G. Thermal ablation of uterine fibroids using MR-guided focused ultrasound —a truly non-invasive treatment modality. Eur J Radiol. 2007;17:2505–11.

Chatterjee D, Sarkar K. A Newtonian rheological model for the interface of microbubble contrast agents. Ultrasound Med Biol. 2003;29:1749–57.

Chen SS, Donoho DL, Saunders MA. Atomic decomposition by basis pursuit. SIAM J Sci Comput. 1998;20:33–61.

Choi JJ, Carlisle RC, Coviello C, Seymour L, Coussios CC. Non-invasive and real-time passive acoustic mapping of ultrasound-mediated drug delivery. Phys Med Biol. 2014;59:4861–77.

Church CC. The effects of an elastic solid surface layer on the radial pulsations of gas bubbles. J Acoust Soc Am. 1995;97:1510–21.

Church CC, Yang X. A theoretical study of gas bubble dynamics in tissue. In: Innovations in nonlinear acoustics: ISNA17-17th international symposium on nonlinear acoustics including the international Sonic Boom Forum. vol 1. AIP Publishing; 2006. pp. 217–224.

Cline HE, Hynynen K, Hardy CJ, Watkins RD, Schenck JF, Jolesz FA. MR temperature mapping of focused ultrasound surgery. Magn Reson Med. 1994;31:628–36.

Cotter SF, Rao BD, Engan K, Kreutz-Delgado K. Sparse solutions to linear inverse problems with multiple measurement vectors. IEEE Trans Signal Process. 2005;53:2477–88.

Coussios CC, Holland CK, Jakubowska L, Huang S-L, MacDonald RC, Nagaraj A, McPherson DD. In vitro characterization of liposomes and Optison® by acoustic scattering at 3.5 MHz. Ultrasound Med Biol. 2004;30:181–90.

Coussios CC, Farny CH, Ter Haar G, Roy RA. Role of acoustic cavitation in the delivery and monitoring of cancer treatment by high-intensity focused ultrasound (HIFU). Int J Hyperth. 2007;23:105–20.

Coviello C, Kozick RJ, Choi JJ, Gyöngy M, Collin J, Jensen C, Smith PP, Coussios CC. Passive acoustic mapping using optimal beamforming for real-time monitoring of ultrasound therapy. In: Proceedings of meetings on acoustics 2013. Vol 19; 2013. p. 075024.

Coviello CM, Faragher SR, Coussios CC. Robust Capon beamforming for passive cavitation mapping during high-intensity focused ultrasound therapy. J Acoust Soc Am. 2010;128:2280.

Coviello CM, Song RJ, Hurrell A, Smith PP, Coussios C. Thin-film sparse boundary array design for passive acoustic mapping during ultrasound therapy. IEEE Trans Ultrason Ferroelectr Freq Control. 2012;59:2322–30.

Damianou C. MRI monitoring of the effect of tissue interfaces in the penetration of high intensity focused ultrasound in kidney in vivo. Ultrasound Med Biol. 2004;30:1209–15.

Damianou C, Pavlou M, Velev O, Kyriakou K, Trimikliniotis M. High intensity focused ultrasound ablation of kidney guided by MRI. Ultrasound Med Biol. 2004;30:397–404.

Daubechies I. Ten lectures on wavelets. Philadelphia: SIAM; 1992.

de Jong N, Hoff L. Ultrasound scattering properties of Albunex microspheres. Ultrasonics. 1993;31:175–81.

de Jong N, Hoff L, Skotland T, Bom N. Absorption and scatter of encapsulated gas filled microspheres: theoretical considerations and some measurements. Ultrasonics. 1992;30:95–103.

Donoho DL. Compressed sensing. IEEE Trans Inf Theory. 2006;52:1289–306.

Ender JH. On compressive sensing applied to radar. Sig Process. 2010;90:1402–14.

Faragher S. Cavitational methods for characterising and testing clinical high intensity focused ultrasound systems. Oxford: Oxford University; 2011.

Farny CH. Identifying and monitoring the roles of cavitation in heating from high-intensity focused ultrasound. Boston: Boston University; 2007.

Farny CH, Holt RG, Roy RA. Temporal and spatial detection of HIFU-induced inertial and hot-vapor cavitation with a diagnostic ultrasound system. Ultrasound Med Biol. 2009;35:603–15.

Furusawa H, Namba K, Thomsen S, Akiyama F, Bendet A, Tanaka C, Yasuda Y, Nakahara H. Magnetic resonance guided focused ultrasound surgery of breast cancer: reliability and effectiveness. J Am Coll Surg. 2006;203:54–63.

Gateau J, Aubry JF, Chauvet D, Boch AL, Fink M, Tanter M. In vivo bubble nucleation probability in sheep brain tissue. Phys Med Biol. 2011a;56:7001–15.

Gateau J, Aubry JF, Pernot M, Fink M, Tanter M. Combined passive detection and ultrafast active imaging of cavitation events induced by short pulses of high-intensity ultrasound. IEEE Trans Ultrason Ferroelectr Freq Control. 2011b;58:517–32.

Goldberg BB, Raichlen JS, Forsberg F. Ultrasound contrast agents: basic principles and clinical applications. London: Martin Dunitz; 2001.

Gyöngy M, Coviello CM. Passive cavitation mapping with temporal sparsity constraint. J Acoust Soc Am. 2011;130:3489–97.

Gyöngy M. Passive cavitation mapping for monitoring ultrasound therapy. Oxford: Oxford University; 2010.

Hoctor RT, Kassam SA. The unifying role of the coarray in aperture synthesis for coherent and incoherent imaging. Proc IEEE. 1990;78:735–52.

Hoff L. Acoustic characterization of contrast agents for medical ultrasound imaging. New York: Springer; 2001.

Hoff L, Sontum PC, Hovem JM. Oscillations of polymeric microbubbles: effect of the encapsulating shell. J Acoust Soc Am. 2000;107:2272–80.

Hu H, Xu S, Yuan Y, Liu R, Wang S, Wan MX. Spatial-temporal ultrasound imaging of residual cavitation bubbles around a fluid-tissue interface in histotripsy. J Acoust Soc Am. 2015;137.

Hynynen K, Darkazanli A, Unger E, Schenck J. MRI-guided noninvasive ultrasound surgery. Med Phys. 1993;20:107–15.

Irani M, Peleg S. Improving resolution by image registration. CVGIP Graph Models Image Process. 1991;3:231–9.

Jensen CR, Ritchie RW, Gyöngy M, Collin JR, Leslie T, Coussios CC. Spatiotemporal monitoring of high-intensity focused ultrasound therapy with passive acoustic mapping. Radiology. 2012;262:252–61.

Jensen CR, Cleveland RO, Coussios CC. Real-time temperature estimation and monitoring of HIFU ablation through a combined modeling and passive acoustic mapping approach. Phys Med Biol. 2013;58:5833–50.

Kennedy JE. High-intensity focused ultrasound in the treatment of solid tumours. Nat Rev Cancer. 2005;5:321–7.

Kim Y, Nelson P. Spatial resolution limits for the reconstruction of acoustic source strength by inverse methods. J Sound Vib. 2003;265:583–608.

Kim S, Su W. Recursive high-resolution reconstruction of blurred multiframe images. IEEE Trans Image Process. 1993;2:534–9.

Kim S, Bose N, Valenzuela H. Recursive reconstruction of high resolution image from noisy undersampled multiframes. IEEE Trans Acoust Speech Signal Process. 1990;38:1013–27.

Kozick RJ, Kassam SA. Linear imaging with sensor arrays on convex polygonal boundaries. IEEE Trans Syst Man Cybern. 1991;21:1155–66.

Krim H, Viberg M. Two decades of array signal processing research: the parametric approach. IEEE Signal Process Mag. 1996;13:67–94.

Leighton T. The acoustic bubble. London: Academic press; 1994.

Lertrattanapanich S, Bose NK. High resolution image formation from low resolution frames using Delaunay triangulation. IEEE Trans Image Process. 2002;11:1427–41.

Li T, Khokhlova TD, Sapozhnikov OA, O'Donnell M, Hwang JH. A new active cavitation mapping technique for pulsed HIFU applications-bubble doppler. IEEE Trans Ultrason Ferroelectr Freq Control. 2014;61:1698–708.

Liao AH, Shen CC, Li PC. Potential contrast improvement in ultrasound pulse inversion imaging using EMD and EEMD. IEEE Trans Ultrason Ferroelectr Freq Control. 2010;57:317–26.

Lu W, Sapozhnikov OA, Bailey MR, Kaczkowski PJ, Crum LA. Evidence for trapped surface bubbles as the cause for the twinkling artifact in ultrasound imaging. Ultrasound Med Biol. 2013;39:1026–38.

Mallat SG, Zhang Z. Matching pursuits with time-frequency dictionaries. IEEE Trans Signal Process. 1993;41:3397–415.

Marengo EA, Devaney AJ, Ziolkowski RW. Inverse source problem and minimum-energy sources. J Acoust Soc Am. 2000;17:34–45.

Maresca D, Emmer M, van Neer PL, Vos HJ, Versluis M, Muller M, de Jong N, van der Steen AF. Acoustic sizing of an ultrasound contrast agent. Ultrasound Med Biol. 2010;36:1713–21.

Montaldo G, Tanter M, Bercoff J, Benech N, Fink M. Coherent plane-wave compounding for very high frame rate ultrasonography and transient elastography. IEEE Trans Ultrason Ferroelectr Freq Control. 2009;56:489–506.

Morgan KE, Allen JS, Dayton PA, Chomas JE, Klibaov A, Ferrara KW. Experimental and theoretical evaluation of microbubble behavior: effect of transmitted phase and bubble size. IEEE Trans Ultrason Ferroelectr Freq Control. 2000;47:1494–509.

O'Reilly MA, Rahman S, Hynynen K. Cavitation monitoring and passive beamforming using a hemispherical random sparse array. In: 11th International symposium on therapeutic ultrasound. vol 1. New York: AIP Publishing; 2012. pp. 271–75.

Park SC, Park MK, Kang MG. Super-resolution image reconstruction: a technical overview. IEEE Signal Process Mag. 2003;20:21–36.

Pasovic M, Danilouchkine M, Faez T, van Neer PL, Cachard C, van der Steen AF, Basset O, de Jong N. Second harmonic inversion for ultrasound contrast harmonic imaging. Phys Med Biol. 2011;56:3163.

Patanavijit V. Super-resolution reconstruction and its future research direction. Assumption Univ J Technol. 2009;12:149–63.

Pennes HH. Analysis of tissue and arterial blood temperatures in the resting human forearm. J Appl Physiol. 1948;1:93–122.

Rabkin BA, Zderic V, Vaezy S. Hyperecho in ultrasound images of HIFU therapy: involvement of cavitation. Ultrasound Med Biol. 2005;31:947–56.

Rhee S, Kang MG. Discrete cosine transform based regularized high-resolution image reconstruction algorithm. Opt Eng. 1999;38:1348–56.

Salgaonkar VA, Datta S, Holland CK, Mast TD. Passive cavitation imaging with ultrasound arrays. J Acoust Soc Am. 2009a;126:3071–83.

Salgaonkar VA, Datta S, Holland CK, Mast TD. Passive imaging of cavitational acoustic emissions with ultrasound arrays. In: Proceeding of the AIP conference; 2009b. pp. 73–7.

Sandrin L, Catheline S, Tanter M, Hennequin X, Fink M. Time-resolved pulsed elastography with ultrafast ultrasonic imaging. Ultrason Imaging. 1999;21:259–72.

Schultz RR, Stevenson RL. Extraction of high-resolution frames from video sequences. IEEE Trans Image Processing. 1996;5:996–1011.

Seo J, Tran B, Hall T, Fowlkes J, O'Donnell M, Cain C. Evaluation of ultrasound tissue damage based on changes in image echogenicity. In: Proceedings of 2002 IEEE ultrasonics symposium, Munich, Germany; 2002. pp. 1431–4.

Shen CC, Li PC. Pulse-inversion-based fundamental imaging for contrast detection. IEEE Trans Ultrason Ferroelectr Freq Control. 2003;50:1124–33.

Sijl J, Gaud E, Frinking PJ, Arditi M, de Jong N, Lohse D, Versluis M. Acoustic characterization of single ultrasound contrast agent microbubbles. J Acoust Soc Am. 2008;124:4091–7.

Simpson DH, Chin CT, Burns PN. Pulse inversion Doppler: a new method for detecting nonlinear echoes from microbubble contrast agents. IEEE Trans Ultrason Ferroelectr Freq Control. 1999;46:372–82.

Song J, Hynynen K. Feasibility of using lateral mode coupling method for a large scale ultrasound phased array for noninvasive transcranial therapy. IEEE Trans Biomed Eng. 2010;57:124–33.

Stark H, Oskoui P. High-resolution image recovery from image-plane arrays using convex projections. J Opt Soc Am A: Opt Image Sci A. 1989;6:1715–26.

Synnevag JF, Austeng A, Holm S. Adaptive beamforming applied to medical ultrasound imaging. IEEE Trans Ultrason Ferroelectr Freq Control. 2007;54:1606–13.

Szabo TL. Diagnostic ultrasound imaging: inside out. London: Academic Press; 2004.

Tanter M, Fink M. Ultrafast imaging in biomedical ultrasound. IEEE Trans Ultrason Ferroelectr Freq Control. 2014;61:102–19.

Tekalp AM, Ozkan MK, Sezan MI. High-resolution image reconstruction from lower-resolution image sequences and space-varying image restoration. In: Proceedings of IEEE international conference on acousics speech and signal processing, San Francisco, USA; 1992. pp. 169–72.

ter Haar G, Coussios C. High intensity focused ultrasound: physical principles and devices. Int J Hyperth. 2007;23:89–104.

ter Haar G, Sinnett D, Rivens I. High intensity focused ultrasound-a surgical technique for the treatment of discrete liver tumours. Phys Med Biol. 1989;34:1743–50.

Tom BC, Katsaggelos AK. Reconstruction of a high-resolution image from multiple-degraded misregistered low-resolution images. In: Visual communications and image processing, Chicago, USA; 1994. pp. 971–81.

Tsai R, Huang TS. Multiframe image restoration and registration. Advances in computer vision and image processing. Greenwich: JAI Press Inc.; 1984.

Ur H, Gross D. Improved resolution from subpixel shifted pictures. CVGIP Graph Models Image Process. 1992;54:181–6.

Xu Z, Ludomirsky A, Eun LY, Hall TL, Tran BC, Fowlkes JB, Cain CA. Controlled ultrasound tissue erosion. IEEE Trans Ultrason Ferroelectr Freq Control. 2004;51:726–36.

Xu Z, Fowlkes JB, Ludomirsky A, Cain CA. Investigation of intensity thresholds for ultrasound tissue erosion. Ultrasound Med Biol. 2005a;31:1673–82.

Xu Z, Fowlkes JB, Rothman ED, Levin AM, Cain CA. Controlled ultrasound tissue erosion: The role of dynamic interaction between insonation and microbubble activity. J Acoust Soc Am. 2005b;117:424–35.

Xu Z, Hall TL, Fowlkes JB, Cain CA. Effects of acoustic parameters on bubble cloud dynamics in ultrasound tissue erosion (histotripsy). J Acoust Soc Am. 2007a;122:229–36.

Xu Z, Hall TL, Fowlkes JB, Cain CA. Optical and acoustic monitoring of bubble cloud dynamics at a tissue-fluid interface in ultrasound tissue erosion. J Acoust Soc Am. 2007b;121:2421–30.

Xu Z, Raghavan M, Hall TL, Mycek MA, Fowlkes JB, Cain CA. Evolution of bubble clouds induced by pulsed cavitational ultrasound therapy-histotripsy. IEEE Trans Ultrason Ferroelectr Freq Control. 2008;55:1122–32.

Yang X, Church CC. A model for the dynamics of gas bubbles in soft tissue. J Acoust Soc Am. 2005;118:3595–606.

Zhang S, Wan MX, Zhong H, Xu C, Liao Z, Liu H, Wang S. Dynamic changes of integrated backscatter, attenuation coefficient and bubble activities during high-intensity focused ultrasound (HIFU) treatment. Ultrasound Med Biol. 2009;35:1828–44.

Zhao X, Zhong H, Wan MX, Shen L. Ultrasound contrast imaging based on a novel algorithm combined pulse inversion with wavelet transform. Ultrasound Med Biol. 2011;37:1292–305.

Zhong H, Wan MX, Jiang Y, Wang S. Differential ultrasonic imaging for the characterization of lesions induced by high intensity focused ultrasound. Ultrasonics. 2006;44:e285–8.

Zhong H, Wan MX, Jiang Y, Wang S. Monitoring imaging of lesions induced by high intensity focused ultrasound based on differential ultrasonic attenuation and integrated backscatter estimation. Ultrasound Med Biol. 2007;33:82–94.

Zhong H, Ding G, She Y, Zong Y, Wan MX. The study of sub-harmonic cavitation imaging based on pulse inversion technique for the monitoring of high intensity focused ultrasound therapies. J Med Imaging Health Inf. 2014;4:410–5.

Chapter 8
Laser-Induced Cavitation and Photoacoustic Cavitation

Yi Feng, Dui Qin and Mingxi Wan

Keywords Photodisruption · Photoacoustic cavitation · Nanomaterials · Nucleation modeling · Bubble dynamics · Photoacoustic signal analysis · Theranostics

8.1 Introduction

As a physical phenomenon, cavitation can be induced by several factors, including heating, acoustic pressure, and high-energy light, for which a focused laser has been widely used.

Laser-induced cavitation (LIC) has been studied more thoroughly than other types of cavitation, illuminating various laser-induced cavitation phenomena, their process descriptions, and related mechanisms. Laser-induced cavitation has also been used for several purposes (Rink et al. 1992; Lubatschowski et al. 2000; Hellman et al. 2007; Dijkink and Ohl 2008), such as mixing, pumping, and examination, and it has also been used for biomedical diagnosis and therapy. However, LIC has an inherent disadvantage: It is limited by the depth of light transmission. Applying laser-induced cavitation to opaque objects is severely restricted.

Acoustic cavitation was discovered several years ago, as early as when ultrasound was used to measure distances and detect cracks (Achenbach et al. 1992).

Y. Feng · D. Qin · M. Wan (✉)
The Key Laboratory of Biomedical Information Engineering of Ministry of Education, Department of Biomedical Engineering, School of Life Science and Technology, Xi'an Jiaotong University, Xi'an 710049, People's Republic of China
e-mail: mxwan@mail.xjtu.edu.cn

Y. Feng
e-mail: fengyi@mail.xjtu.edu.cn

D. Qin
e-mail: duiqin2011@gmail.com

M. Wan et al. (eds.), *Cavitation in Biomedicine*,
DOI 10.1007/978-94-017-7255-6_8

In most cases, cavitation is regarded as a phenomenon that needs to be avoided because the microbubbles formed during cavitation can interfere with the desired process (Farny et al. 2009). In the early stages of studying cavitation for biomedical imaging and therapy, researchers had the same attitude. Later, however, they realized that microbubbles could be useful in biomedical theranostics, for example in molecular imaging (Klibanov 2005), synergistically with high-intensity focused ultrasound (HIFU) (Fry et al. 1995; Holt and Roy 2001), and to monitor imaging (Coussios et al. 2007). Acoustic cavitation occurs when a high negative pressure is applied to a cavitation nucleus and its surrounding material. And, because ultrasound transmission can penetrate much deeper into tissues than laser light can, acoustic cavitation is more applicable to biomedical situations. However, in contrast to laser-induced cavitation, it is almost impossible to predict the exact region where acoustic cavitation forms, making it difficult to utilize and control.

Moreover, regardless of whether cavitation is laser-induced or acoustic, the high flux of energy can cause unintended effects on the targeted niduses and its surrounding regions (Maatz et al. 2000; Chen et al. 2003). In some cases, especially biomedical therapy, these effects are unacceptable. Therefore, to prevent unwanted consequences, cavitation should be induced at lower energy thresholds. Importantly, this is where photoacoustic cavitation (PAC) becomes interested. PAC is the process where a focused laser and ultrasound are used synchronously to induce cavitation. A focused laser (lower than the threshold of LIC) produces the potential nucleus and provides the initial conditions for acoustic cavitation microbubble to form. In this situation, it is easier to form cavitation microbubble than when only ultrasound is applied, since the laser increases the temperature of "potential nucleus." At the same time, acoustic irradiation at intensity lower than the threshold of acoustic cavitation creates negative pressure within the region. Thus, the synchronous effects of applying a focused laser and ultrasound can induce PAC at lower intensities of both the laser and ultrasound.

Studies on PAC have explored its physical mechanisms, the key techniques for application, and how it is applied in scientific research and in the clinic. These topics will be introduced in this chapter.

8.2 Laser-Induced Cavitation

8.2.1 Brief Description

LIC, also called laser-induced optical breakdown (LIOB) or photodisruption, is caused by nonlinear energy absorption at laser pulse intensities above a threshold defined by the targeted material (Chiou et al. 2010). Because LIC mostly causes the formation of a bubble, several researchers use the term "laser-induced bubble" to describe the process of bubble formation, growth, collapse, and rebound. LIC has been utilized in mixing, pumping, and examination and in some technologies such as high-speed microfluidic switches, fluorescence-activated cell sorters, and a high-speed droplet-on-demand device.

Laser-induced bubble formation and collapse involve water surface tension, plasma formation, and repetitive evaporation–condensation within a short time (Chiou et al. 2010). As shown in Fig. 8.1, when a laser pulse is highly focused in a liquid medium, the laser energy breaks down the molecules in the liquid and generates ionized hot plasma (6000–15,000 K) near the focal point (Fig. 8.1, panel 1). On nanosecond time scales, the hot plasma recombines, instantly heats the liquid, and creates an explosive vapor bubble that expands at a speed greater than 100 m/s. Pressure inside the bubble can be as high as tens of megapascals. Triggering these bubbles takes nanoseconds or even femtoseconds, depending on the pulse duration.

During the process, two damaging mechanisms have been clarified as the source of cavitation damage. One is the shock waves emitted into the liquid during bubble nucleation (Fig. 8.1, panel 2) and the collapse of spherical bubbles. The other is the impingement of liquid jets penetrating through bubbles near the rigid boundaries. Moreover, laser-induced bubbles show several specific patterns, which will be introduced subsequently.

Using high-speed photography, Lee et al. (2011) described the normal process of laser-induced cavitation and the emission of a shock wave, shown in Fig. 8.2. It shows a spherical bubble generated after the plasma recombination by the laser breakdown in distilled water and the process of bubble collapse with very weak rebound. During the process, the focused laser pulse created plasma, having a high temperature and pressure, in the water. The liquid around the plasma was compressed, and then, a very short time later, it flowed outward. A compression wave or a primary shock wave propagated through the liquid at the time, as shown in

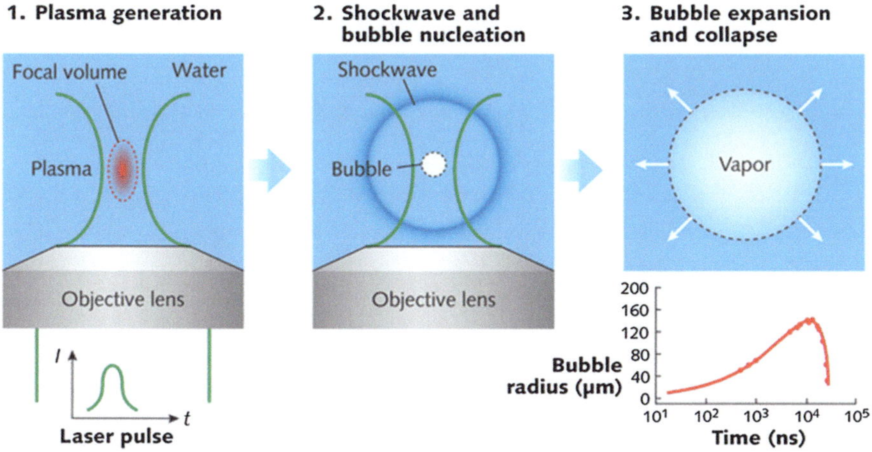

Fig. 8.1 Schematic diagram of laser-induced cavitation (LIC). Reproduced with permission from Chiou et al. (2010). © 2010 SPIE

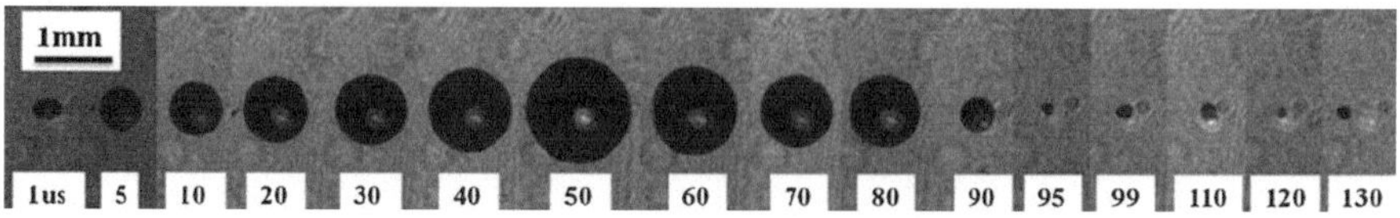

Fig. 8.2 An image sequence of laser-induced bubble dynamics with the first collapse at ~93 μs. Reproduced with permission from Lee et al. (2011). © 2011 Springer-Verlag

Fig. 8.3. Then, upon the breakdown of water, a single spherical bubble was generated and expanded during the isothermal expansion, as shown in Fig. 8.2.

The LIC process involves four stages (Lee et al. 2011). In the first stage, which occurs after the bubble forms, the bubble wall velocity continues to drive wall expansion because the pressure is higher inside the bubble. As the bubble radius increases, the bubble wall velocity decreases because the kinetic energy passed onto the bubble wall–liquid interface is distributed into the surrounding liquid. Thermal diffusion causes the temperature inside the bubble to cool down until it matches the temperature of the liquid outside the bubble wall. At this temperature, the pressure inside the bubble reaches its saturated vapor pressure.

In the second stage, as the bubble interior becomes filled mostly with water vapor and only contains a small quantity of permanent gas (air), it reaches its maximum radius. Then, the bubble wall motion reaches equilibrium after the pressure jump of ~100×. Then, the bubble starts to contract at the rate of vapor condensation. This strong compression of the bubble is accompanied with another rise in the vapor temperature and pressure.

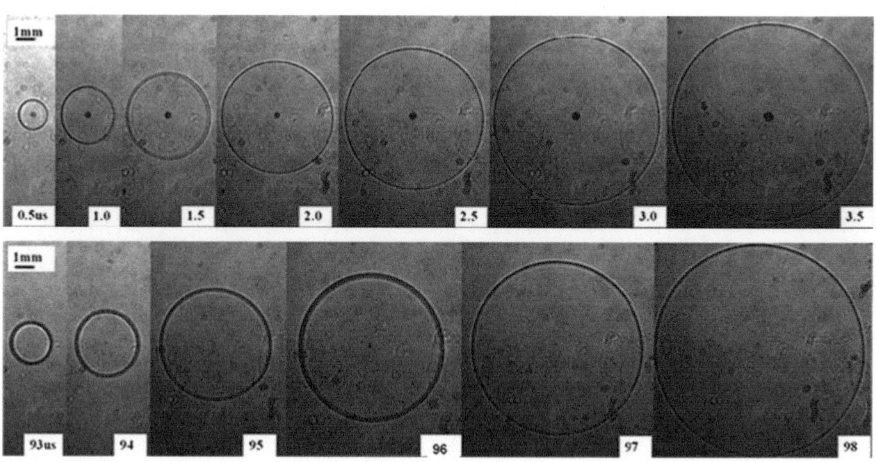

Fig. 8.3 A sequence of a shock wave induced by laser breakdown (0.5–3.5 μs) in water and a secondary shock wave following the collapse of a bubble (93–98 μs). Reproduced with permission from Lee et al. (2011). © 2011 Springer-Verlag

The third stage involves isothermal bubble contraction, which occurs as long as the vapor temperature and pressure are below their critical points ($T_v < T_{cr} \approx 647$ K and $p_v < p_{cr} \approx 22$ MPa). In this stage, the liquid and vapor velocities at the bubble surface are different from the velocity of the bubble wall due to phase transitions (evaporation and condensation). That is, the portion of water vapor inside the bubble wall is slowly converted into liquid compared to the bubble wall inward velocity. The liquid near the bubble wall is also heated due to this phase transition (as vapor turns into water) during the compression stage, which, in turn, forms a thin supercritical pressure and temperature region around the bubble wall. Finally, because the condensation rate of the bubble interior can no longer keep up with the rate that the bubble is decreasing in volume (on the order of the speed of sound), the condensation process stops naturally and then the liquid temperature and pressure near the bubble wall cool down to their original values.

However, in the fourth stage, the substantial inward force continuously compresses the non-condensable air under an adiabatic process, which, in turn, exceeds the critical points of temperature and pressure, causing the center of the bubble to reach maximum values of 10 GPa and 10^4 K, respectively, at collapse (Fujikawa and Akamatsu 1980; Akhatov et al. 2001). This pressure difference between the center of the bubble wall and the surrounding liquid generates the tremendous reverse outgoing force within a very short period. The outward force opposes the inward motion and causes bubble to implode, and it also generates the secondary shock wave in distilled water, as shown in Fig. 8.3.

8.2.2 *LIC Modeling*

8.2.2.1 Laser-Induced Breakdown in Liquids

It has been assumed that laser-induced cavitation is a consequence of the generation of plasma following dielectric breakdown of a medium, known as laser-induced breakdown (LIB). LIB in pure and transparent liquids (such as water) is a combination of photoionization, inverse bremsstrahlung absorption, and cascade (or avalanche) ionization. It creates a large amount of free electrons, and together with the positive ions, they are able to form plasma. The processes that take place during breakdown are depicted in Fig. 8.4 (Vogel et al. 2005). Water is assumed to be an amorphous semiconductor (Williams et al. 1976; Sacchi 1991).

During the whole breakdown process, energy loss exists because of certain loss mechanisms, such as inelastic collisions and diffusion of electrons out of the laser beam, which then result in an irradiance threshold for breakdown. However, it is difficult to determine the breakdown threshold due to the statistical nature of the breakdown and varying experimental conditions in different experiments. To determine the breakdown threshold, we can calculate the evolution of the free electron density under laser radiation. The generic form of such a rate equation is as follows (Kennedy 1995; Noack and Vogel 1999):

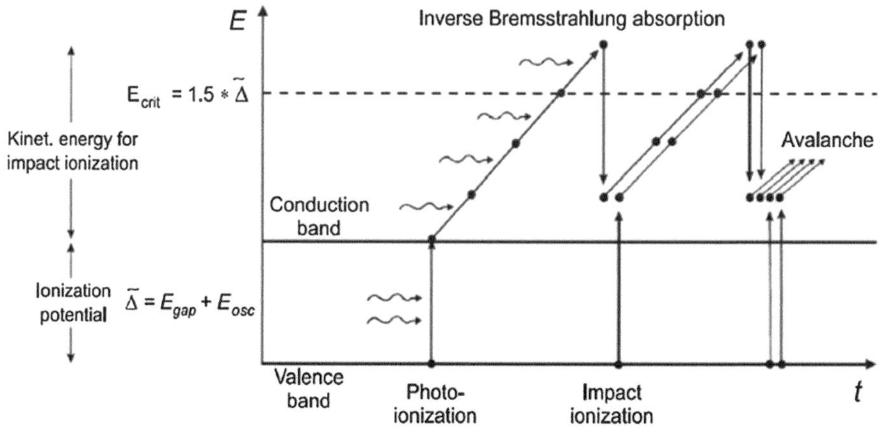

Fig. 8.4 Visualization of the breakdown and plasma formation in LIC. Reproduced with permission from Vogel et al. (2005). © 2005 Springer-Verlag

$$\frac{\mathrm{d}\rho_\mathrm{e}}{\mathrm{d}t} = \left(\frac{\mathrm{d}\rho_\mathrm{e}}{\mathrm{d}t}\right)_\mathrm{mp} + \eta_\mathrm{casc}\rho_\mathrm{e} - g\rho_\mathrm{e} - \eta_\mathrm{rec}\rho_\mathrm{e}^2, \tag{8.1}$$

where ρ_e is the free electron density and η_casc, g, and η_rec represent the cascade ionization (CI), diffusion, and recombination rates, respectively. The first two terms describe electron generation through multiphoton ionization (MPI) and CI, respectively. The last two terms denote the diffusion of electrons out of the focal volume and the recombination losses, respectively.

According to the rate equation model, the threshold irradiance I_th required to produce breakdown for a given wavelength λ and pulse duration τ_L was iteratively solved for different irradiance values until the maximum electron density during the laser pulse equaled the critical electron density for optical breakdown ρ_cr, as shown in Fig. 8.5 (Noack and Vogel 1999).

To match the experimental results for the breakdown thresholds, the critical free electron density in laser-induced plasmas was calculated for two assumed values of the critical electron density: $\rho_\mathrm{cr} = 10^{21}\,\mathrm{cm}^{-3}$ for femtosecond or low-picosecond laser pulses, and $\rho_\mathrm{cr} = 10^{20}\,\mathrm{cm}^{-3}$ for nanosecond laser pulses (Noack and Vogel 1999).

8.2.2.2 Shock Wave Emission at Breakdown

The energy of a free electron is transferred to molecules by collisions and electron–hole recombination, which leads to a rapid rise in temperature and pressure at the laser focus. Thus, when investigating high-speed photographs or shadowgraphs of

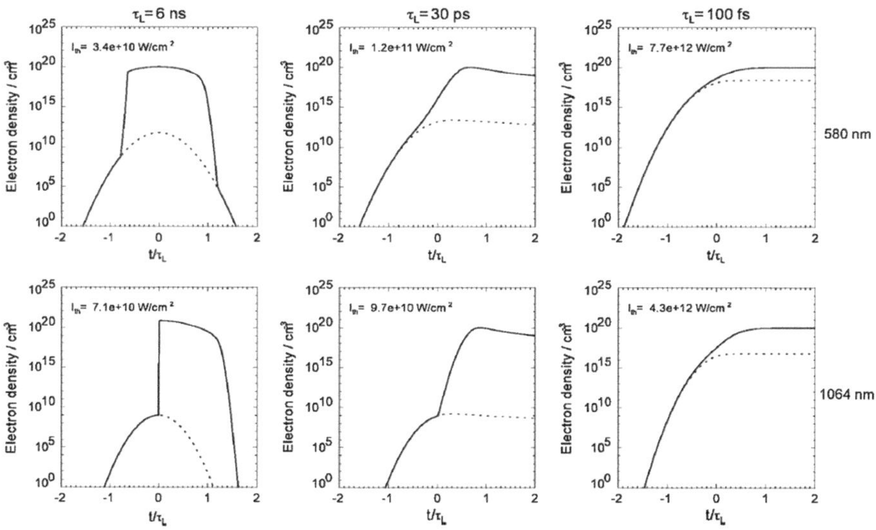

Fig. 8.5 Evolution of the free electron density at the breakdown threshold for different pulse durations (*left* to *right*) and different wavelengths (*top* to *bottom*). The calculations were performed for breakdown in pure water, $\rho_{cr} = 10^{20}$ cm^{-3}, and a spot size of 5 μm. In addition to the total free electron concentration (*solid curve*), the concentration due to multiphoton absorption alone (*dotted curve*) is plotted as a function of time. The time axis has been normalized to the laser pulse duration τ_L. Reproduced with permission from Noack and Vogel (1999). © 1999 IEEE

the LIB area, we can observe that the shock waves are produced by the pressure rise following the breakdown.

In a previous study, shock wave emission and cavitation bubble expansion were investigated by taking series of photographs where the delay time—the time between optical breakdown and the exposure of the photograph—was increasing (Docchio et al. 1986). The distances r traveled by the shock wave and the bubble wall were plotted as a function of the delay time t, and curves $r(t)$ were fitted through the measurement points from photographs using a curve-fitting program, as shown in Fig. 8.6a (Vogel et al. 1996). From the slope of the $r(t)$ curves, the shock wave velocity $u_s(t)$ and bubble wall velocity $u_b(t)$ were derived, as shown in Fig. 8.6b. The time axis gives the delay between laser pulse and the illumination pulse. A Nd:YAG laser pulse with a 1064-nm wavelength, 10-mJ pulse energy, 6-ns pulse duration, and a focal cone angle in water of 22° was used.

The shock wave's peak pressure p_s can be estimated by the following formula (Vogel et al. 1996):

$$p_s = c_1 \rho_l u_s \left(10^{(u_s - c_l)/c_2} - 1 \right) + p_\infty, \tag{8.2}$$

where c_l is the sound speed, ρ_l is the density of water, c_1 and c_2 are empirical constants, and p_∞ is the ambient pressure. Shock waves provide important

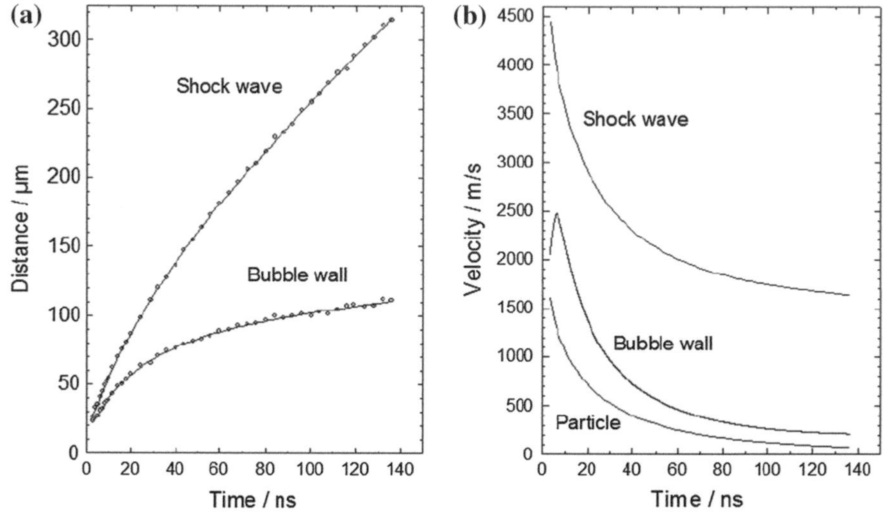

Fig. 8.6 Shock wave propagation and corresponding **a** bubble wall expansion and **b** velocity perpendicular to the optical axis after laser-induced breakdown in water. Reproduced with permission from Vogel et al. (1996). © 1996 Acoustical Society of America

mechanical effects during LIB, as they are critical for energy dissipation and for applications in biomedicine.

8.2.2.3 Dynamics of an LIC Bubble

During LIB, the high-density energy is used in evaporation of the focal volume, plasma radiation, and mechanical processes, such as shock wave emission and cavitation. However, when the shock wave has been emitted from the plasma, there is still a volume of high pressure and temperature that remains out of equilibrium with the surroundings. This volume expands more slowly than the shock wave and drives the early stage of bubble expansion until a new equilibrium reached. As for water, disregarding chemical reactions inside the plasma, the expanding volume will soon be filled with water vapor of pressure p_{v}. When the energy of the plasma is used up, the final volume is assumed to be a spherical bubble of maximum radius $R_{\max}$, as often observed in experiments. The interior vapor pressure p_{v} (2.33 kPa in water at 20 °C) is much smaller than the outer pressure p_{∞} (normally $\sim$100 kPa), and the bubble is compressed back to a small volume. Then, as energy is lost, the compressed bubble rebounds to a smaller maximum radius.

A model has been introduced to describe the collapse and rebound of a laser-induced cavitation bubble in water (Akhatov et al. 2001, 2002). It starts from the instant when the bubble has reached its maximum radius $R_{\max}$. A model of a pure vapor bubble is proposed first, and then, it is extended for the case where a small amount of non-condensable gas is inside the bubble.

1. Vapor bubble model

First, a model of a pure vapor bubble is proposed. In the absence of viscosity, diffusion, and non-mechanical energy loss (due to radiation), the spherically symmetric equations for the conservation of mass, momentum, and energy, which describe the vapor dynamics inside the bubble, are as follows:

$$\frac{\partial \rho_{\mathrm{v}}}{\partial t}+\frac{1}{r^2}\frac{\partial}{\partial r}\left(\rho_{\mathrm{v}}u_{\mathrm{v}}r^2\right)=0, \tag{8.3}$$

$$\frac{\partial}{\partial t}(\rho_{\mathrm{v}}u_{\mathrm{v}})+\frac{1}{r^2}\frac{\partial}{\partial r}\left(\rho_{\mathrm{v}}u_{\mathrm{v}}^2r^2\right)+\frac{\partial p_{\mathrm{v}}}{\partial r}=0, \tag{8.4}$$

$$\frac{\partial e_{\mathrm{v}}}{\partial t}+\frac{1}{r^2}\frac{\partial}{\partial r}\left(u_{\mathrm{v}}r^2(e_{\mathrm{v}}+p_{\mathrm{v}})-\lambda_{\mathrm{v}}r^2\frac{\partial T_{\mathrm{v}}}{\partial r}\right)=0, \tag{8.5}$$

here, $\rho_{\mathrm{v}}, u_{\mathrm{v}}, p_{\mathrm{v}}, T_{\mathrm{v}}$, and $e_{\mathrm{v}}=\rho_{\mathrm{v}}\varepsilon_{\mathrm{v}}+\rho_{\mathrm{v}}u_{\mathrm{v}}^2/2$ are the density, velocity, pressure, temperature, and the total energy of vapor per unit volume, respectively; ε_{v} is the internal energy of vapor per unit mass, and λ_{v} is the heat conductivity coefficient of vapor.

Liquid and vapor velocities at the bubble surface and the velocity of the bubble wall $\dot{R}$ differ due to phase transitions (evaporation and condensation) according to the following formula:

$$u_{\mathrm{v}}|_{r=R}=\dot{R}-\frac{j_{\mathrm{v}}}{\rho_{\mathrm{v}}},\quad u_{\mathrm{l}}|_{r=R}=\dot{R}-\frac{j_{\mathrm{v}}}{\rho_{\mathrm{l}}}, \tag{8.6}$$

where $u_{\mathrm{v}}|_{r=R}$ and $u_{\mathrm{l}}|_{r=R}$ are the liquid and vapor velocities at the bubble surface, respectively; j_{v} is the evaporation rate from a unit surface of the bubble wall. According to the Hertz–Knudsen–Langmuir formula, j_{v} is as follows (Schrage 1953; Fujikawa and Akamatsu 1980):

$$j_{\mathrm{v}}=\frac{\alpha}{\sqrt{2\pi B_{\mathrm{v}}}}\left(\frac{p_{\mathrm{sat}}(T_{\mathrm{l}}|_{r=R})}{\sqrt{T_{\mathrm{l}}|_{r=R}}}-\frac{\Gamma_{\mathrm{v}}p_{\mathrm{v}}(T_{\mathrm{v}}|_{r=R})}{\sqrt{T_{\mathrm{v}}|_{r=R}}}\right), \tag{8.7}$$

where α is the accommodation coefficient that shows which portion of vapor molecules striking the liquid surface is absorbed by this surface. The p_{sat} term is the saturation vapor pressure at temperature T_{l}. The correction factor Γ_{v} is obtained according to the modified Hertz–Knudsen–Langmuir formula (Fujikawa and Akamatsu 1980).

The pressure jump at $r = R$ is described by the standard boundary condition:

$$p_{\mathrm{v}}|_{r=R}=p_{\mathrm{l}}|_{r=R}+\frac{2\sigma}{R}+\frac{4\mu_{\mathrm{l}}}{R}\left(\dot{R}-\frac{j_{\mathrm{v}}}{\rho_{\mathrm{l}}}\right). \tag{8.8}$$

Here, σ and μ_l are the surface tension and dynamic viscosity of water, respectively.

In the low-Mach region ($u_v|_{r=R}/c_v \leq M_{cr} = 0.3$), where c_v is the speed of sound in the vapor, a homobaric scheme for the vapor inside the bubble and an incompressibility condition for the surrounding liquid are used to decrease the computational efforts. In this case, the vapor parameters satisfy the ideal gas equation of state:

$$p_v = \rho_v B_v T_v, \quad \varepsilon_v = B_v T_v/(\gamma - 1), \tag{8.9}$$

where B_v is the gas constant and γ is its adiabatic index.

So, the bubble radius $R(t)$ can be defined by a generalized Rayleigh–Plesset equation taking into account acoustic radiation of the oscillating bubble and the effect of mass transfer through the bubble wall:

$$\begin{aligned} &\left(1 - \frac{u_l|_{r=R}}{c_l}\right) R \frac{du_l|_{r=R}}{dt} + \frac{3}{2}\left(1 - \frac{u_l|_{r=R}}{3c_l}\right) u_l^2|_{r=R} + 2\left(1 - \frac{u_l|_{r=R}}{4c_l}\right) \frac{u_l|_{r=R} j_v}{\rho_{l0}} \\ &= \left(1 + \frac{\dot{R}}{c_l}\right) \frac{p_l|_{r=R} - p_{l0}}{\rho_{l0}} + \frac{R}{\rho_{l0} c_l} \frac{d}{dt} p_l|_{r=R}. \end{aligned} \tag{8.10}$$

Here, ρ_{l0} and p_{l0} are the initial liquid density and pressure, respectively, and c_l is the speed of sound in water.

In the high-Mach region, the vapor inside the bubble is modeled as a hard-core van der Waals gas:

$$p_v = \frac{\rho_v B_v T_v}{1 - b_1 \rho_v} - b_2 \rho_v^2, \quad \varepsilon_v = \frac{B_v T}{\gamma - 1} - b_2 \rho_v, \tag{8.11}$$

where b_1 and b_2 are the vapor parameters.

The basic results for the numerical solution describing the dynamics of a laser-induced bubble with $R_{max} = 1$ mm are shown in Fig. 8.7 (Akhatov et al. 2001). The system of partial differential equation (8.1)–(8.3) is solved here only in the high-Mach region to give the solution between two dots; all other times, we use the simplified homobaric model to describe vapor dynamics, using the incompressible liquid assumption for water. The results in Fig. 8.7 have been obtained for an accommodation coefficient $\alpha = 0.075$. Based on these results, the calculated minimum bubble radius was ~14 μm, and at collapse, the mass decreased down to 6 % of the initial value. Numerical simulations of vapor bubble motion showed that the evaporation and condensation on the bubble surface play an essential role in this process.

2. Gas–vapor bubble model

A real laser-induced bubble, however, also contains a small amount of non-condensable gas. So, a modified mathematical model based on the vapor bubble model is proposed to describe the gas–vapor bubble. In such a case, the

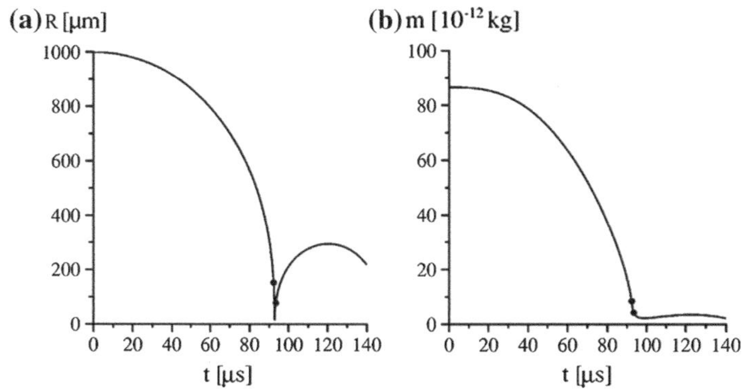

Fig. 8.7 The **a** vapor radius and **b** mass as functions of time. Dots show moments in time when the high-Mach period starts (*first dot*) and ends (*second dot*). Results are presented for $R_0 = 1\,\mathrm{mm}$, $\alpha = 0.075$. Reproduced with permission from Akhatov et al. (2001). © 2001 American Institute of Physics

governing equations for the gas–vapor mixture dynamics are Eqs. (8.3) and (8.4) and the energy equation is in the following form:

$$\frac{\partial e}{\partial t} + \frac{1}{r^2}\frac{\partial}{\partial r}\left(ur^2(e+p)\right) = \frac{1}{r^2}\frac{\partial}{\partial r}\left(\lambda r^2\frac{\partial T}{\partial r}\right) + \frac{1}{r^2}\frac{\partial}{\partial r}\left(\rho r^2 D\frac{\partial k}{\partial r}\left(i_v - i_g\right)\right), \quad (8.12)$$

where k and D are the vapor mass concentration and the coefficient of binary diffusion, respectively, and $i = \varepsilon + p/\rho$ is the enthalpy.

The vapor mass concentration k is calculated by the diffusion equation:

$$\frac{\partial k}{\partial t} + u\frac{\partial k}{\partial r} = \frac{1}{\rho r^2}\frac{\partial}{\partial r}\left(\rho r^2 D\frac{\partial k}{\partial r}\right). \quad (8.13)$$

The coefficient of binary diffusion of vapor in the air is chosen to be a function of temperature according to the following formula (Akhatov et al. 2002):

$$D = 2.16 \times 10^{-5}\,\mathrm{m}^2/\mathrm{s} \times \left(\frac{T}{273}\right)^{1.8}. \quad (8.14)$$

Numerical investigations based on the two-component gas-phase model were carried out with an initial vapor concentration of 99 % (1 % of non-condensable gas). The other parameters were chosen to be the same as in the previous calculations for the pure vapor bubble. By comparing these results with those obtained for the vapor bubble without gas, Akhatov et al. (2001) concluded that the small addition of a non-condensable gas strongly influences the bubble dynamics. To explain this phenomenon, the spatial distribution of the vapor concentration during the compression stage of the bubble was considered, as shown in Fig. 8.8. There is

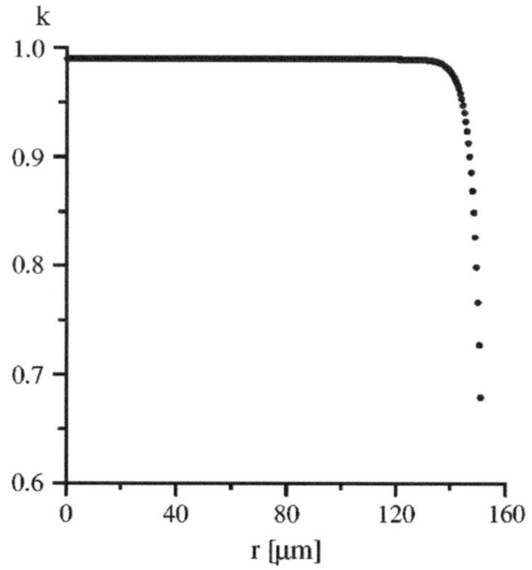

Fig. 8.8 The typical vapor concentration distribution during the bubble compression stage ($t = 529.58$ ns before collapse). Results are presented for $R_0 = 1$ mm, $\alpha = 0.075$. Reproduced with permission from Akhatov et al. (2001). © 2001 American Institute of Physics

a strongly uniform distribution of vapor concentration inside the bubble: $k = 0.99$ near the center and $k = 0.65$ at the boundary. The partial vapor pressure near the bubble surface decreases, essentially decreasing the condensation processes on the bubble surface, which can be explained by a gas shell hypothesis.

8.2.2.4 Shock Wave Emission at Bubble Collapse

Shock wave emission is a typical phenomenon that occurs during cavitation bubble collapse. For a laser-induced cavitation bubble, similar shock wave emission can be observed photographically using extremely high framing rates. The mechanism of the shock wave emission at bubble collapse is similar to that at breakdown, which is also caused by high temperature and pressure. The difference is in how these high temperatures and pressures are generated. At breakdown, the shock wave conditions are caused by LIB and plasma formation, but those at bubble collapse result from strong bubble compression in a very short time. The shock wave produced by the collapse of a laser-induced bubble can be measured using a fiber optic hydrophone. From these measurements, it has been found that the shock wave amplitudes after plasma formation and those after bubble collapse are similar.

8.2.3 LIC in Water

8.2.3.1 Classic Experimental Setup

Most studies of laser-induced cavitation are performed in water. The typical experimental system includes one or more laser generators, focusing optics system, one or two super-high-speed cameras, enough light sources, and a timing system. A typical system is shown in Fig. 8.9. The image sequences from a high-speed camera show the dynamics of cavitation nucleus formation and bubble growth over time.

8.2.3.2 Specific Patterns of an LIC Bubble Observed by High-Speed Photography

Along with spherical bubbles generated in LIC, several patterns of LIC bubbles have also been investigated for different laser irradiation patterns (Zwaan et al. 2007; Dijkink et al. 2008; Sasaki et al. 2009). These patterns include bubbles near a wall, bubbles inside limited space, and interactions between bubbles and non-spherical bubbles.

1. Near the wall

When laser-induced cavitation occurs near a wall, Dijkink et al. (2008) observed different bubble dynamics for different standoff distances, as shown in Fig. 8.10.

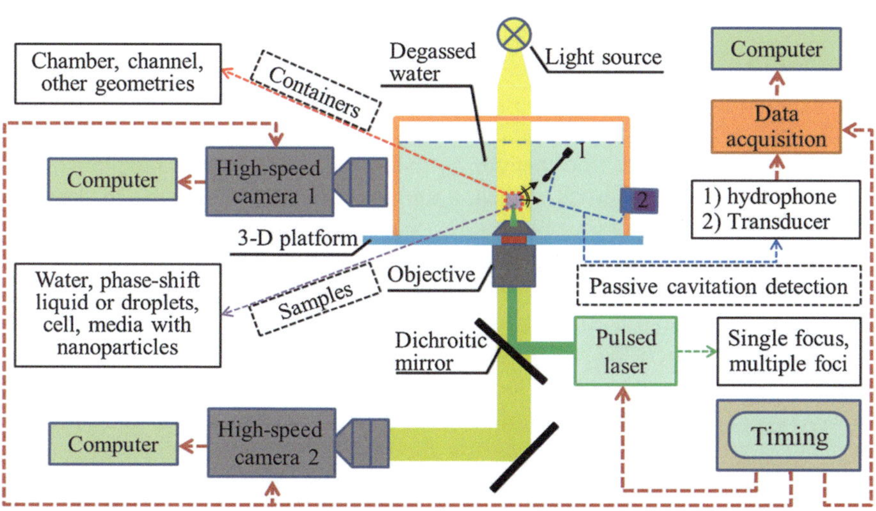

Fig. 8.9 A classic experimental setup for laser-induced cavitation

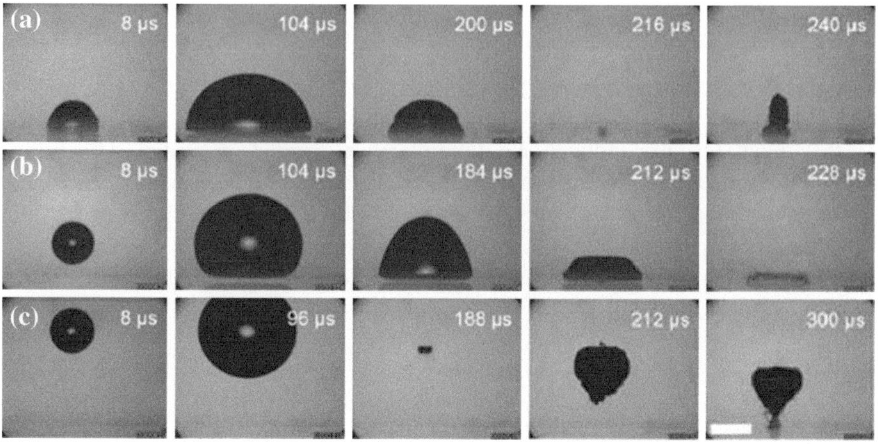

Fig. 8.10 Three examples of the bubble dynamics for different standoff distances. Reproduced with permission from Dijkink et al. (2008). © 2008 Institute of Physics and Engineering in Medicine

Here, the distance of the bubble nucleation point from the boundary is the shortest in row (a), longer in row (b), and the longest in row (c).

Figure 8.10a depicts a bubble very close to the boundary (standoff distance $\gamma_0 = 0.1$). The bubble expands in a hemispherical shape and then shrinks. At about 200 μs after its creation, the bubble collapses. Then, the bubble rebounds into a cigarlike shape directed upward. When the standoff distance increases, for example to $\gamma_0 = 0.9$ in Fig. 8.10b, the upper side of the bubble expands into a spherical shape. However, the lower side facing the boundary becomes flat. During shrinkage, the bubble obtains an approximately triangular shape in projection and finally forms a vortex ring because of the jet flow generation. Here, the jet flow is generated starting from the top more curved part of the bubble and is accelerated toward the boundary. Then, the impact and spreading of the jet flow generates a vortex ring, which stabilizes the re-expanding bubble into a torus. When the standoff distance is large, for example $\gamma_0 = 2.0$ in Fig. 8.10c, the bubble expands and collapses almost spherically; then, a thin and fast jet flow is created, which is visualized by the shape of the re-expanding bubble on top of the jet flow. It causes a cusp shape at the bubble's lower side, and the jet impacts on the surface at about 300 μs. During the process, the bubble moves toward the boundary, such that the bubble's second collapse could take place on the boundary.

Figure 8.11a depicts the stages of bubble expansion and shrinkage at the channel boundary (Zwaan et al. 2007). During shrinkage, the bubble develops into a triangular shape. Then, the top part of the bubble interface flips toward the boundary and thereby creates a jet, which impacts the boundary with a velocity of roughly 25 m/s. When the radial flow from the shrinking bubble interacts with the outward-spreading jet flow, the liquid is accelerated away from the wall, as shown

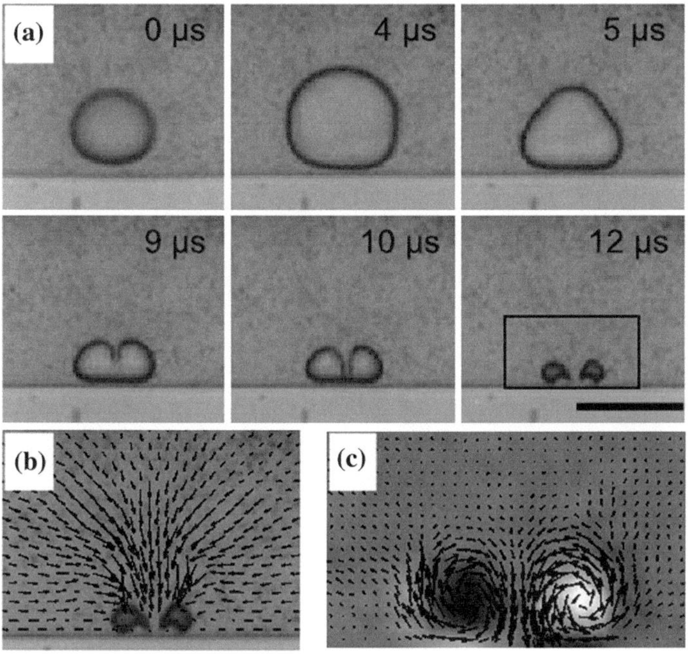

Fig. 8.11 Planar bubble dynamics close to a channel wall. **a** The formation of a liquid jet toward the wall. **b** Velocity field of the flow after the bubble has split up due to the impacting jet flow. **c** Vorticity plot after the bubble has collapsed. Reproduced with permission from Zwaan et al. (2007). © 2007 The American Physical Society

in Fig. 8.11b. In the later stage, the two tiny remains of the bubbles are entrained in two counter-rotating vortex rings, as shown in Fig. 8.11c.

2. Limited in different shapes

When the space for laser-induced cavitation is strictly limited, for example, in a small (compared to the maximum radius of the laser-induced bubble) channel with a rectangular-shaped cross section, Zwaan et al. (2007) observed that expanding bubbles become ellipsoidal, and then, two jets develop in opposite directions and flow along the channel (150 μm wide in their study), as shown in Fig. 8.12a. These jets impact and reflect toward the channel walls, causing vorticity to be transported to the walls at a later stage. The jetting is found from interface areas where liquid is most easily "accessible." For example, a bubble oscillating inside a triangular structure quickly develops high-speed jets originating from the feeding capillaries at the corners (Fig. 8.12b). Jets develop similarly in a square geometry (Fig. 8.12c), where four jets originate from the corners and two of them are connected to feeding capillaries. This geometry leads to a bubble shape resembling a four-leaved clover. At a later stage, the complex bubble breaks up because the triangular and the square chambers reflect the inflowing jets.

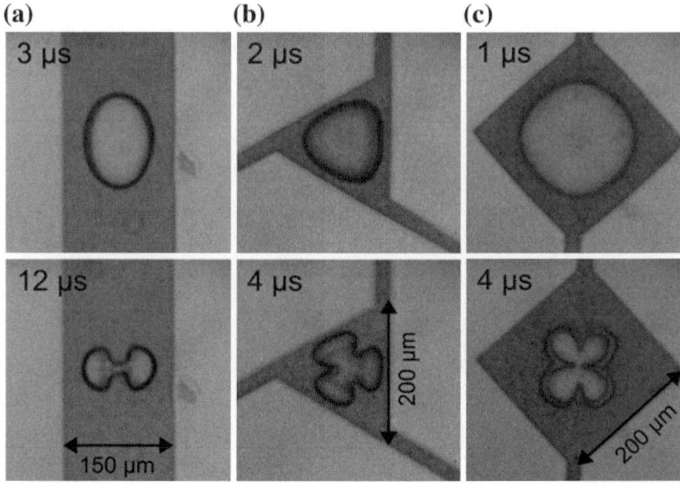

Fig. 8.12 LIC bubble in three limited geometries. Reproduced with permission from Zwaan et al. (2007). © 2007 The American Physical Society

3. Interactions between two bubbles

When two cavitation bubbles are generated with pulsed focus laser at a distance, the bubble patterns, including their growth and collapse, as well as their interactions with the surrounding interface, are closely related to the distance between bubbles. One study showed that within their 14-μs lifetime, bubbles could expand to a maximum radius of $R_{\max} = 38\,\mu\text{m}$. The same study also showed the patterns for bubbles having a non-dimensional standoff distance $\gamma_0 = \text{d}/2R_{\max}$ of 0.65–2, which are shown in Fig. 8.13a–c (Quinto-Su and Ohl 2009).

For bubbles with weak ($\gamma_0 = 2$) and intermediate ($\gamma_0 = 1.3$) interactions, occurring when the bubbles were created at a distance of 93 and 140 μm, respectively, from each other, the maximum radii of two bubbles are ~35 μm. The bubble centers (*) have moved several μm away from each other compared to laser foci (o) during the time of bubble generation. During shrinkage and collapse, bubbles move toward each other, where the total change in distance between the bubbles was 75 % higher for when $\gamma_0 = 1.3$ than for when $\gamma_0 = 2$. In the bottom frames of Fig. 8.13a, b, the outer edges of the bubbles appear blurred due to the formation of fast counter-propagating jets. Except the case, the bubbles remain cylindrical when $\gamma_0 = 2$, and the proximal bubble walls show slight deformation when $\gamma_0 = 1.3$.

Bubbles show a strong interaction ($\gamma_0 = 0.6$) when they were generated 46 μm apart. At first, the bubbles repelled each other, and then, the distal bubble walls remained cylindrical, while the proximal walls flattened and formed a straight channel between the bubbles. The thickness of this channel decreased over time, but no coalescence occurred in this initial separation. In addition, they produced two counter-propagating jets, visible in the bottom frame of Fig. 8.13c.

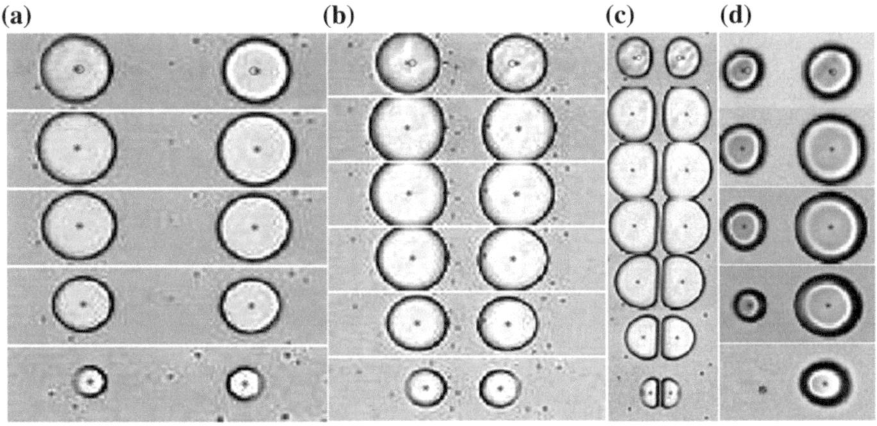

Fig. 8.13 Four patterns of interaction between two laser-induced bubbles. **a** $\gamma = 0$. **b** $\gamma = 1.3$. **c** $\gamma = 0.6$. **d** unequal sizes. Reproduced with permission from Quinto-Su and Ohl (2009). © 2009 Cambridge University Press.

When two bubbles of different sizes were formed (Fig. 8.13d), the larger bubble moved considerably less than the smaller one, until the smaller bubble collapsed. The smaller bubble collapsed first, and then, jets were generated toward the larger bubble, as we can see by the shape and leftward-pointing direction of the large bubble in the bottom frame.

4. Non-spherical bubble

Multiple laser foci can create multiple distinct bubbles. Depending on the laser energy attributed to each focus and their distances, these bubbles may coalesce, allowing us to create non-spherical bubbles through bubble–bubble coalescence in the LIC experiments. One such study produced several patterns of non-spherical bubbles, as shown in Figs. 8.14, 8.15, 8.16, and 8.17 (Lim et al. 2010). The numerical simulations for the non-spherical bubbles are also shown in each figure (each panel (b) for Figs. 8.14, 8.15, 8.16, and 8.17).

Focusing the light on a line leads to an essentially elliptical bubble, as shown in Fig. 8.14. The bubble was fragmented at first and then coalesced quickly into one bubble. The horizontal length of the bubble was larger than the vertical one. Then, the bubble became almost spherical, after which the vertical dimension grew larger than the horizontal one until the bubble collapsed. This process is typical for all elliptic bubbles. From the results of the numerical simulations, shown in Fig. 8.14b, along with the consistent bubble patterns, we can see more details of a ring jet that formed along the horizontal direction during bubble collapse. In the last frame of the simulation, we can see that the curvature was relatively high; however, the physics were still dominated by inertial effects because of the high velocities that occurred at that instant. Thus, the effects of surface tension were still small.

When two linear focusing regions are connected at an angle, more complex bubble dynamics ensue for the V-shaped bubble, shown in Fig. 8.15 (Lim et al. 2010).

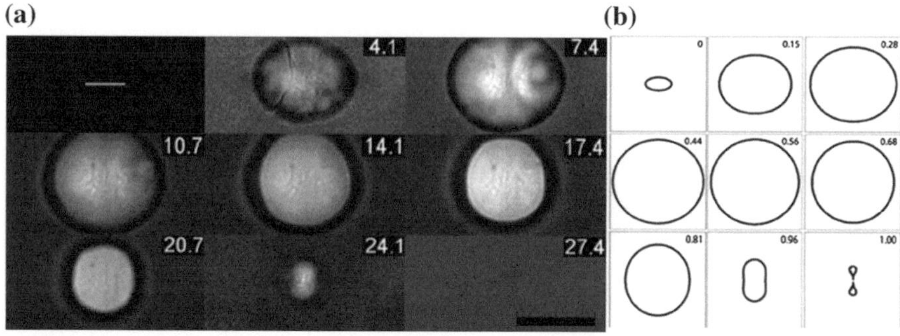

Fig. 8.14 An elliptical bubble with eccentricity inversion. Reproduced with permission from Lim et al. (2010). © 2010 The American Physical Society

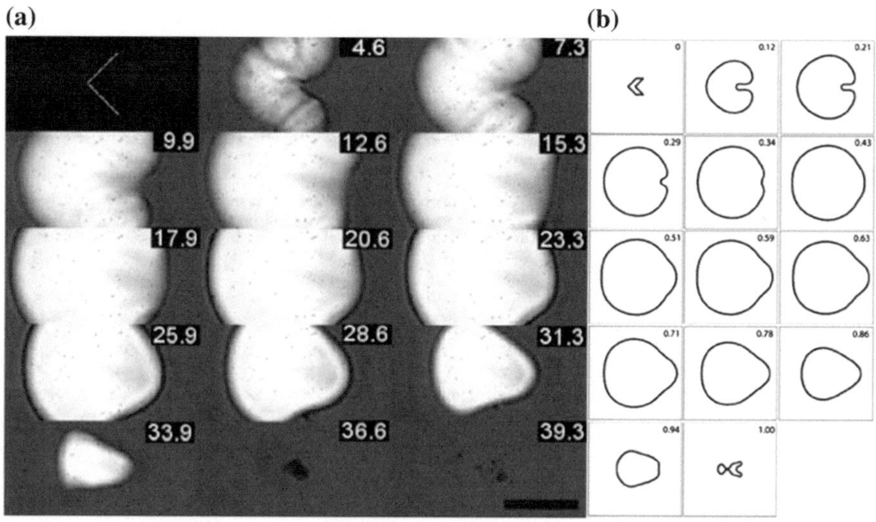

Fig. 8.15 A V-shaped bubble. Reproduced with permission from Lim et al. (2010). © 2010 The American Physical Society

In the experiment, the bubble expanded to its maximum size with some cleavages visible. Then, the bubble transformed into a triangular shape pointing in a direction opposite to the initial orientation of the bubble. Until bubble collapse, it remained triangular. The simulation results (Fig. 8.15b) show that the collapse process is rather complex. It includes the formation of multiple jets, one from the right pointed tip, and two counter-propagating jets from below and above.

A toroidal bubble is created by shaping the beam into a circle, as shown in Fig. 8.16a (Lim et al. 2010). A non-smooth bubble surface was observed first, indicating that the torus consisted of distinct and rapidly expanding bubbles. After the bubbles coalesced, the toroidal bubble continued to expand until it reached its maximum size, and then, it started to collapse. The simulation results (Fig. 8.16b)

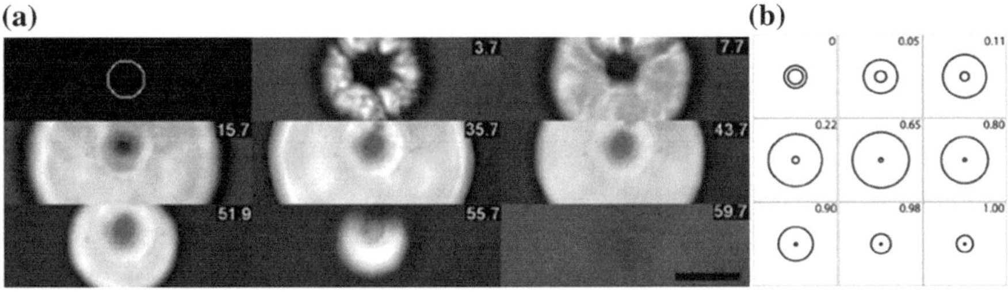

Fig. 8.16 Toroidal bubble. Reproduced with permission from Lim et al. (2010). © 2010 The American Physical Society

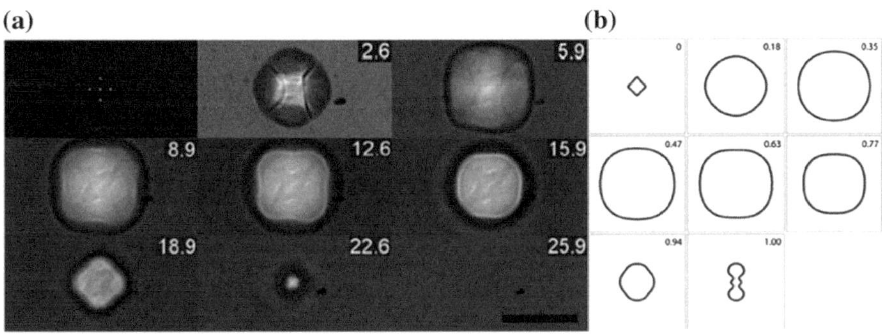

Fig. 8.17 Square bubble generated with multiple laser foci. Reproduced with permission from Lim et al. (2010). © 2010 The American Physical Society

show that the toroidal bubble developed an annular jet toward its center. This can be explained by the center of mass moving toward the torus center during shrinkage.

The dynamics of a square bubble are shown in Fig. 8.17, which was formed from five nucleated bubbles by (Lim et al. 2010). The square shape was stable and expanded quickly, and then, it collapsed and its shape rotated by 45°. During the collapse, the bubble became visibly rounded, then it developed a diamond shape shortly after, and then jets quickly emerged from all four corners. The last frame in the simulation does not match well with the experiment, as it is depicting the already re-expanding bubbles. The reason for this discrepancy may be that the simulated bubbles have higher gas pressure, resulting in less compression. Therefore, the simulations show that bubble re-expands before it develops the four jets which are weakly visible in the experiment.

8.2.4 LIC in a Phase-Shift Liquid

A high-flux femtosecond laser is always needed for LIC in water. This process depends on a high-quality instrument, but the high flux of energy can cause unintended effects on the targeted nidus and its surrounding regions. This consequence is unacceptable in some cases, especially in biomedical therapy. Thus, some experiments have been performed in phase-shift liquids, which have lower phase transition temperatures. Such liquids include perfluoropentane (PFP, C_5F_{12}, 29 °C) and perfluorohexane (PFH, C_6F_{14}, 62 °C). These liquids have been widely studied because of their potential applications in biomedical ultrasound diagnosis and therapy (Dayton et al. 2006; Rapoport et al. 2009; Schad and Hynynen 2010). Most importantly, the introduction of these liquids makes it possible to induce cavitation by the laser pulse in nanoseconds with low energy, promoting new "theranostic" approaches.

Compared to LIC in water, in phase-shift liquids, a focused laser with lower intensity can induce cavitation bubbles with longer lifetimes. Three patterns of LIC always emerge in PFP, and two patterns occur in PFH, as shown in Fig. 8.18. In Fig. 8.18a, we can see that for one pattern in PFP (Pattern I), a single bubble was generated, and it shrank relatively slowly until it collapsed gently. In another pattern (Pattern II), the bubble first oscillated slightly without shrinking, and then, it expanded until the volume reached a maximum value. In the last pattern (Pattern III) in PFP, multiple LIC bubbles were produced and then shrank at different survival times, whereby bubble survival time was directly proportional to its maximum radius. In a phase-shift liquid with a higher boiling temperature, such as PFH, Patterns I and III are observed, but not Pattern II, as shown in Fig. 8.18b.

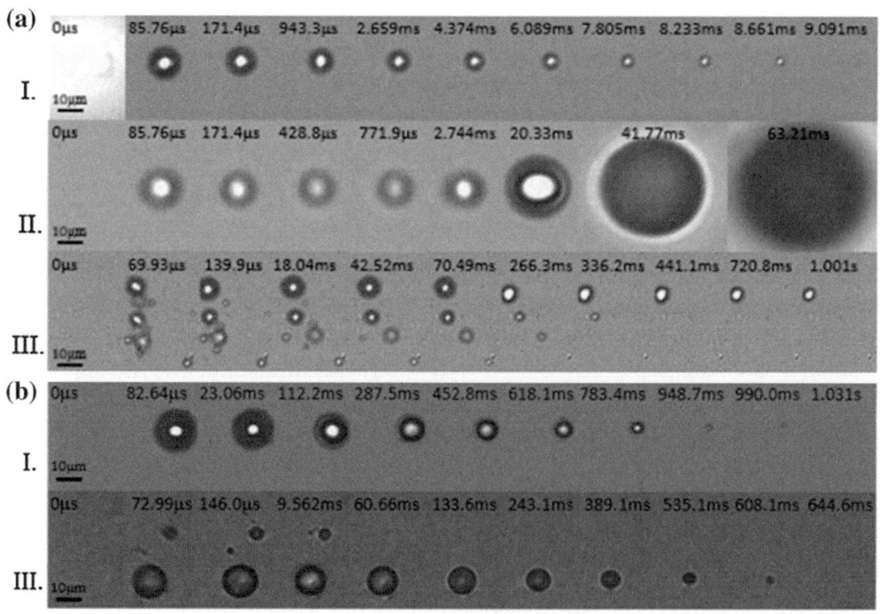

Fig. 8.18 Different patterns of high-speed photographs of LIC in **a** perfluoropentane (PFP) and **b** perfluorohexane (PFH)

During LIC in PFP, the laser produces a super-high electric field at the focus, which causes liquid breakdown, and then, the liquid swells to form a bubble. Comparing the three patterns observed in PFP, temperature may be the reason for the differences. In relatively low temperatures, the LIC bubble would shrink, but not in relatively high temperatures. This may be because temperature is closed related to the vapor pressure inside an LIC bubble. At relatively high temperatures, the vapor pressure is higher than the threshold for bubble shrinking/condensation, so the bubble keeps growing. However, in normal cases, where the temperature is higher than 24 °C, multiple LIC bubbles are produced (Pattern III). This might be because near the phase-shift temperature, the liquid can break down more easily, producing more potential cavitation cavities. The shock waves from every LIC bubble interact and produce multiple bubbles removed from the laser focus.

In another experiment on LIC in PFP, original RF data were recorded by a polyvinylidene difluoride (PVDF) needle hydrophone (Precision Acoustic Inc, UK), shown as the passive acoustic signals in Fig. 8.19a. In this panel, the first major peak indicates the shock wave emitted from the rapid expansion of the plasma. Its location on the time axis reveals when the shock wave was formed. The second major peak has a similar pattern but is slightly smaller in amplitude. In every experiment, the time interval between these two major peaks in PFP is always ~2 μs, indicating that the second peak may be caused by reflection when the traveling acoustic signal encounters a certain fixed interface, for example an objective lens. The third peak may be the acoustic signal emitted from the first bubble collapse, and the fourth peak is the echo of the third peak. The spectrum analysis of obtained acoustic signals is displayed in Fig. 8.19b. Here, we see that the frequency spectrum was mainly distributed between 0 and 2 MHz.

A time–frequency analysis of cavitation signals gives a more intuitive understanding of the changes that occur on the cavitation spectrum over time. For the analysis, a short-time FFT was applied in PFP and PFH, as shown in Fig. 8.20a, b, respectively. The significant frequency components from 0 to 2 MHz are observed at the time corresponding to the peaks in the time domain.

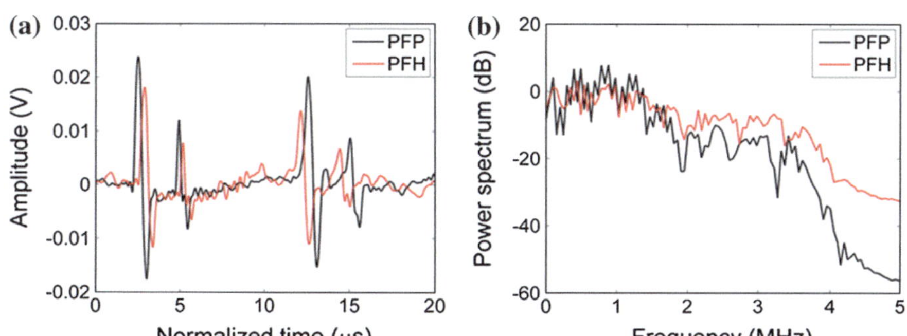

Fig. 8.19 The **a** original passive acoustic signals and **b** spectrum analysis of LIC in PFP

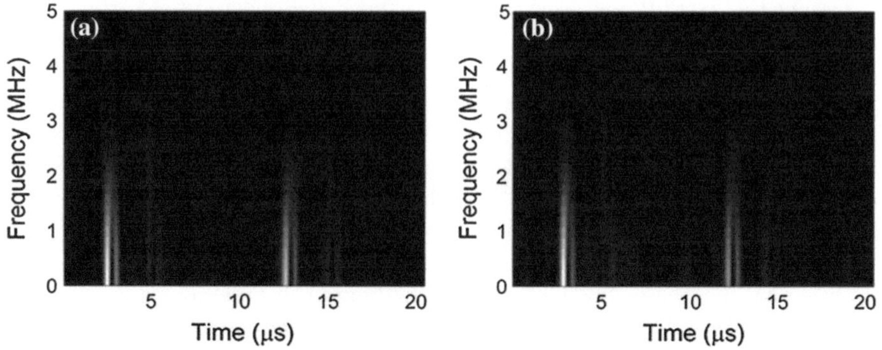

Fig. 8.20 Time–frequency diagram of cavitation signals in **a** PFP and **b** PFH

As previously mentioned, temperature is closely related to the patterns of laser-induced cavitation. Furthermore, it is also associated with the time that cavitation forms. Figure 8.21 shows the time that cavitation formed in PFP and PFH as a function of temperature. As temperature increased in these perfluorinated compounds (PFCs), the time for formation of laser-induced cavitation decreased, revealing that relatively higher temperatures are helpful to produce laser-induced cavitation in phase-shift liquids.

At a fixed temperature (20 °C), the threshold of LIC in PFC is shown as the peak-to-peak value of acoustic emissions detected with a passive transducer as a function of laser energy. As shown in Fig. 8.22, with increasing laser energy from 36.7 to 43.3 μJ, no laser-induced cavitation in PFP was observed, and no passive cavitation detection (PCD) signal was acquired. At laser energies of 43.3 μJ and higher, cavitation bubbles were captured by a super-high-speed camera and the PCD signal was acquired. In this experiment, the PCD signal revealed that the

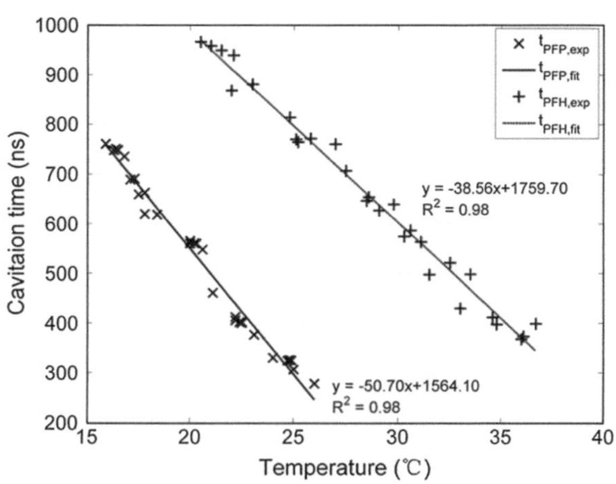

Fig. 8.21 The time for cavitation to form in PFP and PFH as a function of temperature

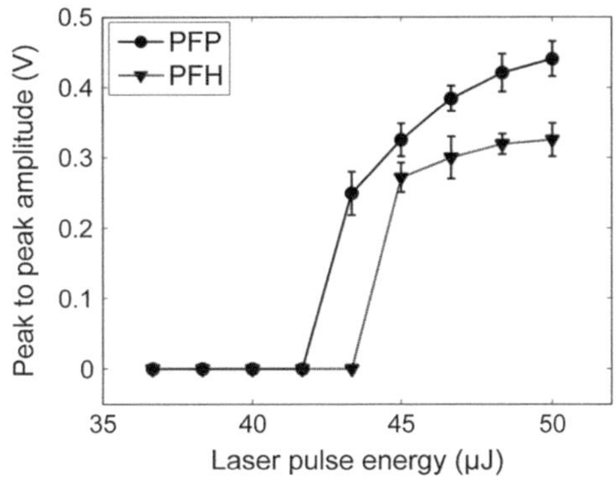

Fig. 8.22 The peak-to-peak value of acoustic emissions detected with the passive transducer as a function of laser power (temperature, 20 °C)

threshold for LIC in PFP was ~43.3 μJ. Correspondingly, the threshold of LIC in PFH was ~45 μJ.

8.2.5 *LIC of Droplets*

Phase-shift liquids are always microemulsified as nanodroplets, which have many applications in theranostics. When a focused laser acts on the PFP droplets, three patterns of cavitation are observed, as shown in Fig. 8.23. In the first pattern (a), the laser induced two small bubbles in a big droplet. The two bubbles mixed together and then dissipated gradually. In Fig. 8.23b, one big bubble was formed in the droplet and then dissipated gradually. Finally, multiple bubbles can form when a laser is focused on a small PFH droplet. As shown in Fig. 8.23c, multiple bubbles each expanded to their maximum radius and then dissipated with different lifetimes.

The PCD signal and frequency spectrum obtained during LIC in PFP droplets are shown in Fig. 8.24a. The time-domain signal has four peaks, and this signal is similar to the acoustic signal emitted during LIC in pure PFP liquid. The frequency domain shows that the frequency is mainly distributed between 0 and 4 MHz. The corresponding time–frequency diagram is shown in Fig. 8.24b. This clearly shows that the PCD signal has frequency components of 0–3 MHz at 2.91 and 12.92 μs, as well as 0–3.5 MHz at 4.75 and 14.75 μs.

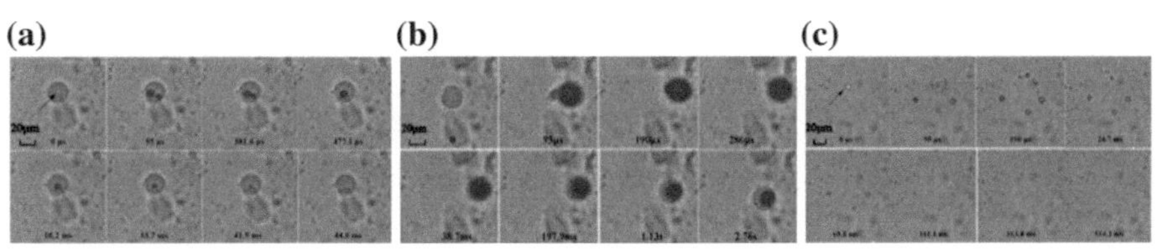

Fig. 8.23 Three patterns of LIC bubbles formed by a PFP droplet

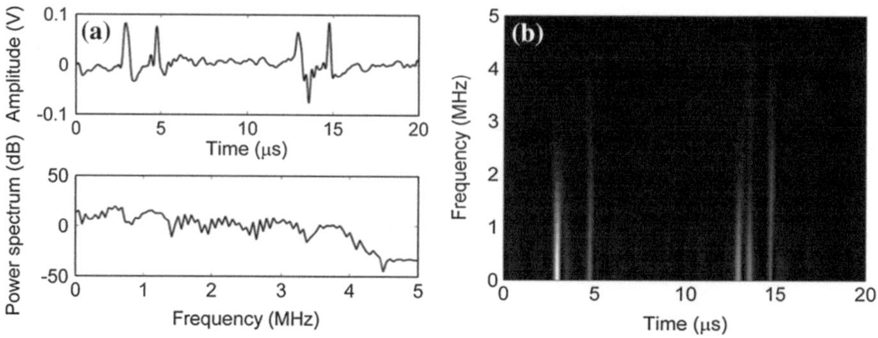

Fig. 8.24 **a** Normalized passive acoustic signals and the corresponding frequency spectrum of LIC bubbles made of PFP droplets. **b** Time–frequency diagram of LIC bubbles made of PFP droplets (pulse laser energy, 45 μJ; temperature, 24.7 °C)

8.2.6 LIC in a Cell

Laser-induced cavitation (breakdown or photodisruption) can be used to generate microbubble(s) within an individual cell. This process is highly precise and controllable, and it is spatially confined by tightly focusing the laser beam (Zohdy et al. 2006). The process is similar to LIC in water, because it can excite ions to higher energy levels, create a self-absorbing plasma that rapidly expands, vaporize the medium, etc. The optical breakdown can create precise damage within the volume of a linearly transparent medium; its application in a single cell has high potential for use in precise surgical procedures in bulk biological tissues.

One setup for LIC in a single cell is introduced by Zohdy et al. (2006), and two patterns of LIC in cells were observed for different laser fluence, as shown in Fig. 8.25. The arrow points to the location of photodisruption. With a laser fluence

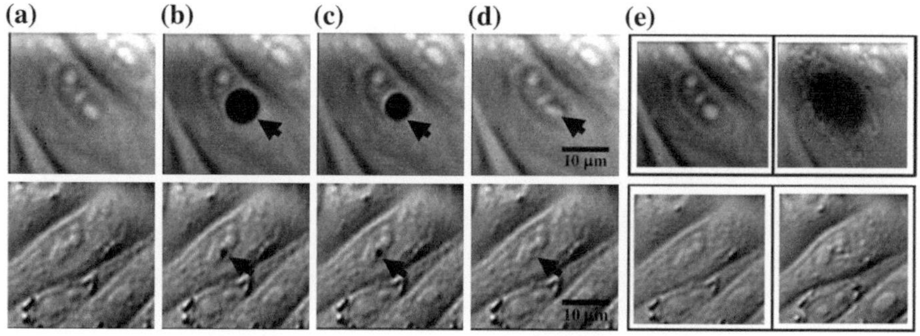

Fig. 8.25 Optical detection of destructive photodisruption in a single cell. *Upper row* 5 J/cm^2, 2.6 s; *Lower row* 9 J/cm^2, 2.6 ms. **a** Before laser exposure. **b–d** During and after laser exposure, respectively. **e** Comparison of a cell before (*left panel*) and after (*right panel*) laser exposure by live/dead staining. Reproduced with permission from Zohdy et al. (2006). © 2006 IEEE

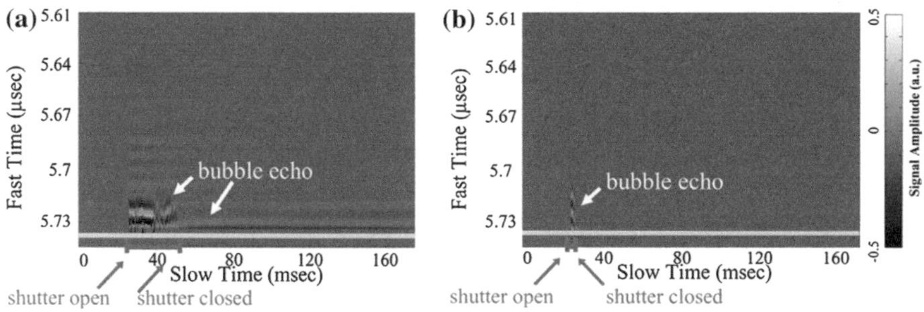

Fig. 8.26 Acoustic wave field plot of a laser-induced **a** destructive and **b** nondestructive bubble in the cell. Reproduced with permission from Zohdy et al. (2006). © 2006 IEEE

of 5 J/cm^2 and a total laser exposure time of 2.6 s, destructive photodisruption in a single cell was observed, as the target cell was stained darkly, indicating cell death. However, the neighboring cell remained unstained, indicating it was still viable. At a higher laser intensity (9 J/cm^2) and a shorter exposure time (2.6 ms), nondestructive photodisruption in a single cell was induced. Some morphology changes were induced by the laser fluence, but the target cell remained unstained, meaning that it was viable after photodisruption.

Photodisruption in different parts of a cell, for example in the cell nucleus or cytosol, were also investigated in this study (Zohdy et al. 2006). It revealed that photodisruption in the cell nucleus, with a laser fluence of 9 J/cm^2 and a total laser exposure time of 263 ms, caused the target cell to die. However, applying the same laser fluence in the cell cytosol only caused some morphology changes in the target cell, but it remained viable after irradiation.

The acoustic wave field plot of LIC in a cell differs when the applied LIC is destructive and nondestructive, as shown in Fig. 8.26 (Zohdy et al. 2006). In the case where 100 laser pulses were applied at 46 J/cm^2, a destructive bubble emitted a strong bubble echo during the laser exposure, and a smaller residual echo was apparent after the laser exposure ended. However, for a nondestructive bubble that was formed by 10 laser pulses at 5 J/cm^2, a fainter and much short-lived bubble echo was detected. This proves that even smaller short-lived bubbles (with lifetimes of <10 ms) created with nondestructive laser parameters can be detected by high-frequency acoustic detection. Thus, a potentially interesting research area might be monitoring laser microsurgery in cell via acoustic methods.

8.2.7 *Plasmonic Nanoparticle-Generated Photothermal Bubbles*

Plasmonic nanoparticles (NPs) strongly absorb and scatter specific wavelengths of light due to their surface plasmon resonance properties. They can be easily produced and can be made stable by adding surface ions or molecules. Furthermore, they are relatively biocompatible. Thus, NPs have been widely used as theranostic probes in biomedical applications. As Fig. 8.27 shows, when NPs are exposed to a laser, the absorption of visible light by a plasmonic NP results in the instantaneous heating of the particle. This heating has a profound effect both on the particle itself and on the surrounding medium (Hashimoto et al. 2012). Effects to the particle involve electron dynamics, coherent acoustic lattice vibrations, melting and evaporation from the particle surface, and explosive fragmentation. Effects to the surrounding medium include cavitation and stress-wave generation that occur as heat transfers from the particle to the surroundings. Thus, the introduction of plasmonic NPs to a laser irradiation system can improve the optical absorption of the laser and also help generate transient vapor bubbles or photothermal bubbles (PTBs) around the plasmonic NPs. Comparatively in water, NPs considerably decrease the cavitation threshold of LIC.

8.2.7.1 Formation and Detection of Photothermal Bubbles Around Plasmonic NPs

The evaporation of the medium around a nanoparticle involves several processes. First, thermalization of the NP (in about 1 ps) is induced by the laser pulse and makes the temperature of the NP reach or even exceed its melting point. This thermal energy can be delivered from the NP to the surrounding medium, causing it to evaporate. During the process, a thin vapor layer around the NP is formed and creates the PTB nucleus. Then, once there is sufficient initial potential energy deposited to allow the PTB to overcome opposing forces of surface tension and viscosity, the PTB develops from the nucleus. After the bubble emerges, it undergoes expansion to its maximum diameter and then collapses (Lapotko 2009). In the case where NPs are functionalized with an antibody to target certain regions of the body, these functionalized NPs normally form a cluster around a target molecule; when they are excited with a short laser pulse, then it acts as a heat source, and an intracellular photothermal vapor bubble can be generated (Lapotko 2009).

PTBs around a single plasmonic NP or multiple NPs can be observed by optical scattering (Lapotko 2009), as shown in Fig. 8.28. In one study, to evaluate the PTB duration, responses were obtained simultaneously (Fig. 8.29) with the pumped laser pulse, and then, the authors measured the widths of the dip-shaped signals (typical of PTBs) at the level of 0.5× their maximums. A single NP exposed to no pumping pulse corresponded to undetectable optical scattering (Fig. 8.28a). Conversely, a single NP exposed to one pumped laser pulse resulted in the appearance of bright diffraction-limited spots in the images. In well-separated single NPs, either fusion or overlapping of the PTBs was observed, which may mean that isolated bubbles can be generated and detected. Compared to single NPs, clusters of NPs had much stronger scattered brightness on account of their increased size, as shown in Fig. 8.28c. A single pumping laser pulse applied to clustered NPs produced bright

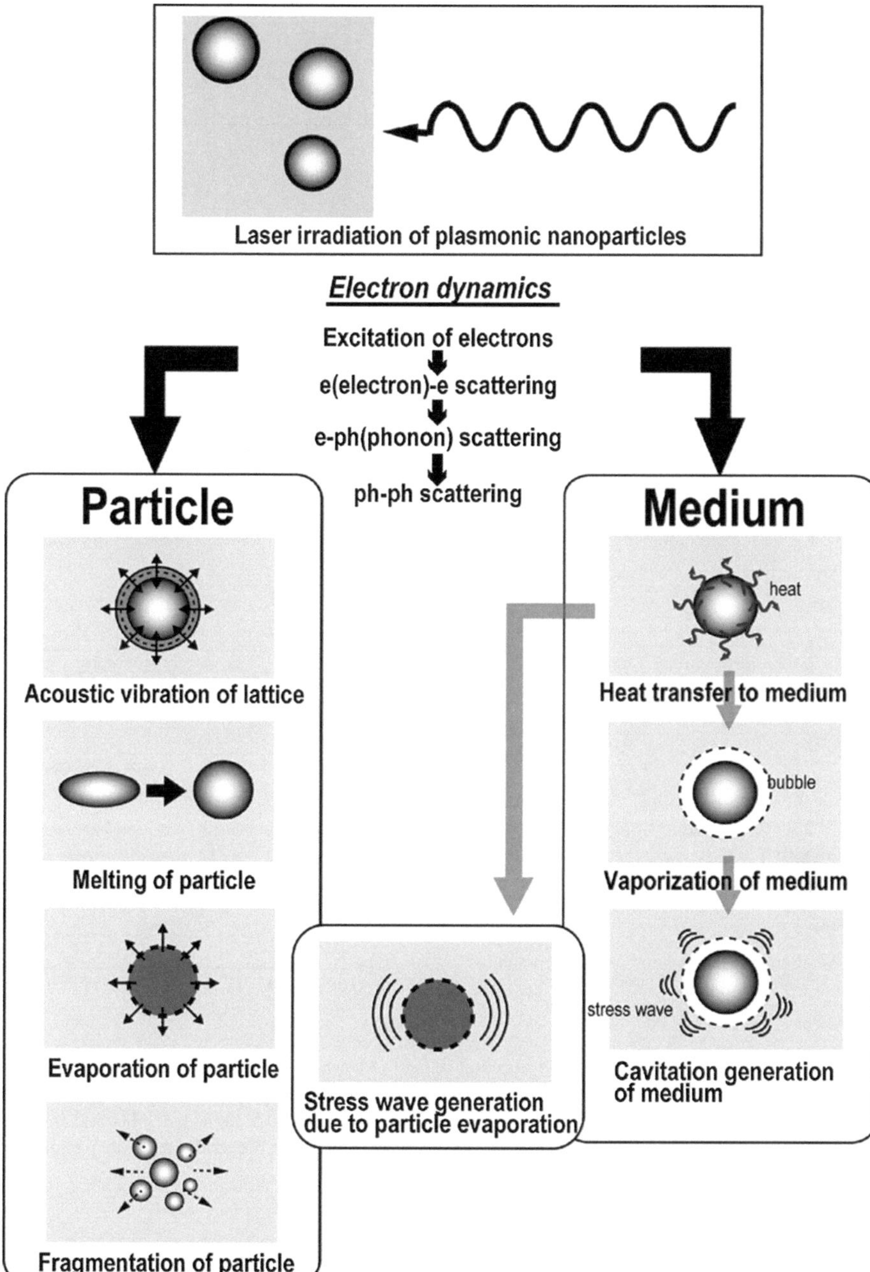

Fig. 8.27 Interaction of pulsed lasers with plasmonic nanoparticles and related phenomena. Reproduced with permission from Hashimoto et al. (2012). © 2012 Elsevier B.V.

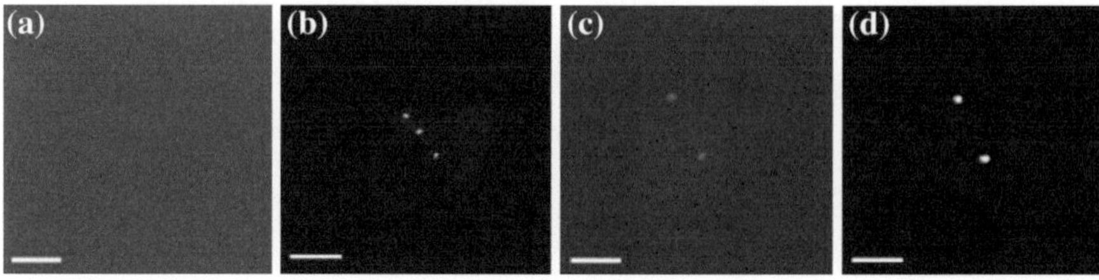

Fig. 8.28 Time-resolved optical scattering images obtained for 30-nm gold NPs: **a** single NPs (no pulse); **b** single NPs exposed to a pumped laser pulse (532 nm, 0.5 ns, 0.9 J/cm^2); **c** clustered NPs (no pulse); **d** clustered NPs exposed to a pumped laser pulse (532 nm, 0.5 ns, 0.12 J/cm^2). Reproduced with permission from Lapotko (2009). © 2009 Optical Society of America

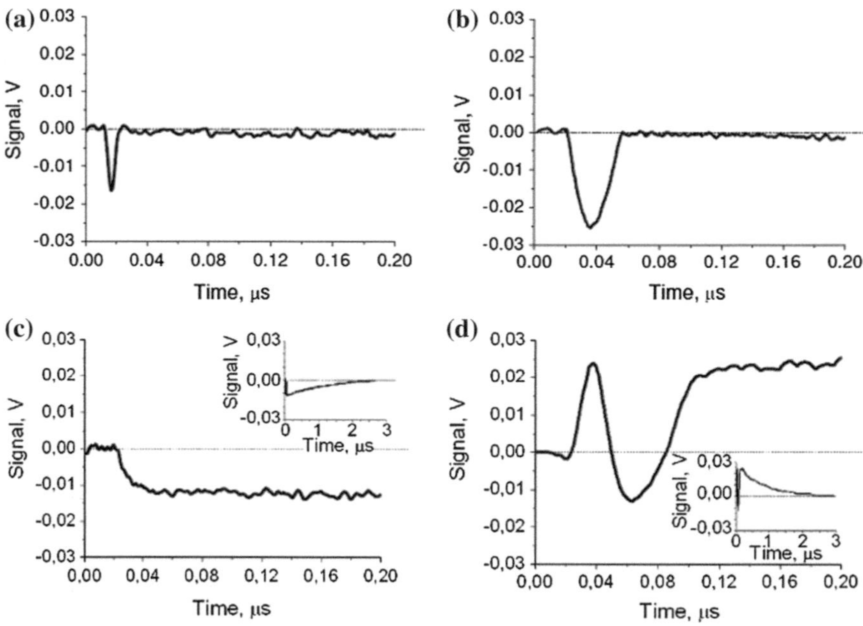

Fig. 8.29 Time responses for a single pumped pulse (532 nm, 0.5 ns): **a** PTB around 30-nm gold NP (0.9 J/cm^2); **b** PTB around a cluster of 30-nm gold NPs (0.9 J/cm^2); **c** thermal signal in the sample of 30-nm gold NPs at a fluence below the PTB threshold (0.5 J/cm^2); **d** PTB and thermal signal in the sample with hemoglobin as a homogenous absorber. Reproduced with permission from Lapotko (2009). © 2009 Optical Society of America

spots in the scattering image (Fig. 8.28d) that spatially coincided with the NP clusters. Compared to the single NP-generated PTBs, cluster-generated PTBs have longer lifetime and lower threshold fluence. This implies an increase in the maximum diameter of cluster-generated PTBs.

8.2.7.2 Theranostic Applications

In microtissue surgery, NP-generated PTBs have been widely used, e.g., to mechanically disrupt and remove plaque tissue by creating a channel in the plaque (Lukianova-Hleb et al. 2009). Furthermore, NP-generated PTBs have greater potential in theranostic therapies for targeted cells. For example, because of their well-known and outstanding photothermal and optical scattering properties, plasmonic NPs can improve the signal-to-noise ratio in tissue during photothermal imaging diagnosis, thereby enhancing contrast and the ability to delineate tissue types (Boyer et al. 2002; Huang et al. 2006; Cognet et al. 2003). The generation and imaging of PTBs from NPs in living cells and tissues can also significantly increase scattering amplification at the cell level, and small PTBs can perform highly sensitive, noninvasive imaging of target cells. Moreover, bigger PTBs can also produce selective mechanical, nonthermal damage to individual target cells because they can rapidly disrupt cellular membranes. The theranostic applications of PTBs are summarized in Fig. 8.30.

Based on the characteristics of NPs and their generated PTBs, a theranostic protocol was proposed by Lukianova-Hleb et al. (2010). In the first step, a buffer suspension of gold NP is injected and the conjugates selectively cauterize the area around cancer cells. In the diagnosis stage, PTBs around NP clusters are selectively created by applying a short pumped pulse to the targeted cancer cells. The small PTBs act as optical scattering probes for the detection of cancer cells. In the therapy stage, selective mechanical destruction of the cancer cells is achieved when bigger PTBs are generated with the second pumped laser pulse. As a result of this two-pulse process, the cancer cells are selectively eliminated without damage occurring to the surrounding normal cells. This protocol was developed for gold NPs and their generated PTBs, as shown in Fig. 8.31. The results of diagnosis are shown here, depicting the optical scattering of the pulsed probe beam (690 nm) by

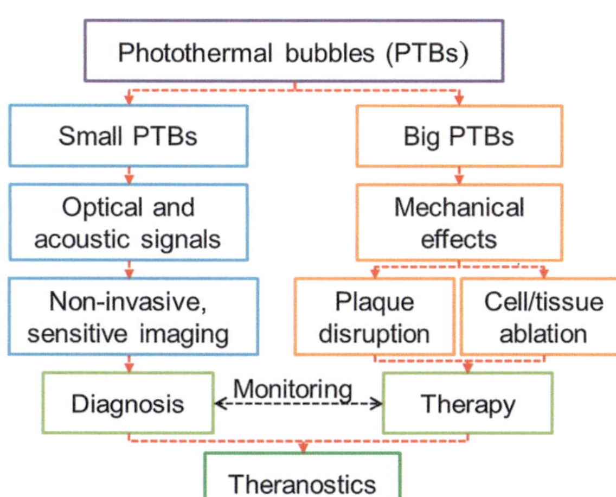

Fig. 8.30 A schematic diagram of PTBs in theranostic applications

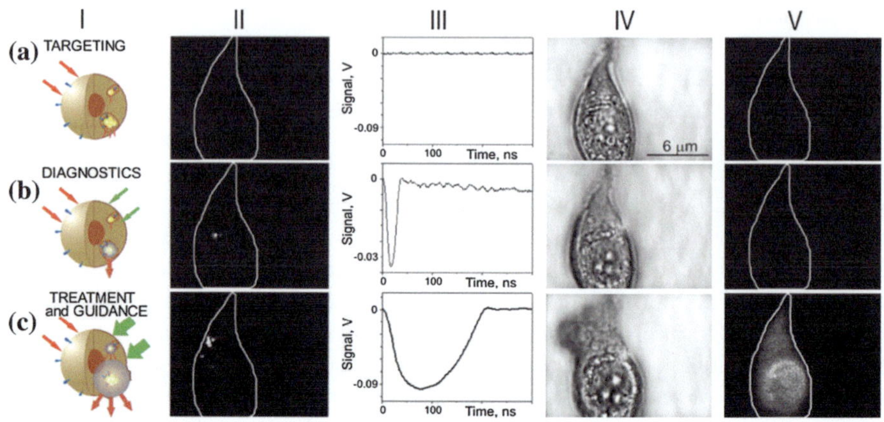

Fig. 8.31 **a** Targeting a cell with gold NPs and the optical generation and detection of intracellular PTBs. **b** PTBs induced by a lower laser fluence noninvasively amplify optical scattering. **c** PTBs mechanically damage the cell with increasing laser fluence. *I* Stages of the PNB theranostic action; *II* optical pulsed scattering images; *III* optical time response of the PNB that shows its lifetime; *IV* bright-field and *V* fluorescent (ethidium bromide specific) images of the cell. Reproduced with permission from Lukianova-Hleb et al. (2010). © 2010 IOP Publishing Ltd.

the gold NPs and by the PNBs in the cells (Fig. 8.31-II). The scattering by gold NPs that accumulated in targeted cells was quite low, and the corresponding image amplitudes are close to the scattering image amplitudes associated with cellular organelles. However, the sensitivity of PTB diagnosis was found to be much higher than that for NP diagnosis. Also, the lifetime of the PNB was measured by the simultaneously obtained duration of a PNB-specific time response (Fig. 8.31-III). In Fig. 8.31b, the fluence level was only able to generate a PTB around the biggest clusters of NPs, as it was below the PTB generation threshold for the smaller NP clusters and single NPs. Thus, only one PTB was detected. This indicates that PNB generation is highly specific. At an increased fluence (Fig. 8.31c-II), PNB was much brighter and its scattering amplitude was much longer. The damage to the individual cells after exposure to the laser pulse was monitored by bright-field (Fig. 8.31-IV) and fluorescent (Fig. 8.31-V) imaging, where uptake of ethidium bromide (which stains cells that have a disrupted membrane) and blebbing (associated with cytoskeleton damage) were monitored. The fluorescent images show that cell death occurred (due to the penetration of the dye inside the cell), as did blebbing. This indicates that the cellular membrane and possibly the cytoskeleton were disrupted. Therefore, inducing intracellular PTBs by varying the laser fluence from levels that result in noninvasive imaging to cell disruption can be utilized in biomedical theranostics.

8.3 Photoacoustic Cavitation

8.3.1 Definition

Cavitation can be induced either by depositing energy or by lowering the liquid pressure. Laser-induced cavitation works through energy deposition, while acoustic cavitation lowers the liquid pressure. The formation of cavities or bubbles in liquids or nanodroplets by synchronously using a focused laser and a pre-established ultrasound is called PAC. The combination of laser and ultrasound allows for bubble formation at a lower laser irradiance and peak negative acoustic pressure than when using either modality alone. Thus, it is an effective method to avoid the unsafe high-energy exposures and to control the occurrence of cavitation at intended locations in theranostic applications.

Figure 8.32 illustrates the principles of PAC, including the formation and dynamics of a PAC bubble in a pure liquid. A laser pulse focused in water creates a strong optical field to ionize water molecules and generates plasma in the confocal volume of the laser and ultrasound. The water in the confocal volume is moved to a metastable region because both the liquid temperature T_l is elevated due to energy absorption by the plasma and the liquid pressure p_l is reduced due to the tensile effects of ultrasound and thermoelastic waves. Then, a nucleus is generated in the metastable water due to spontaneous fluctuations, accompanied by shock wave generation. The nucleus will grow explosively to a macroscopic cavitation bubble under both the vapor pressure p_v inside the bubble and the negative liquid pressure p_l outside the bubble.

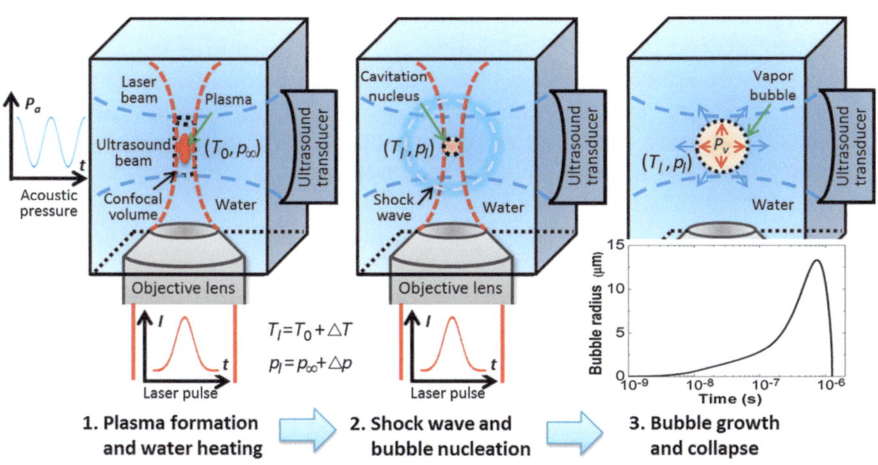

Fig. 8.32 Schematic diagram of photoacoustic cavitation (PAC)

8.3.2 PAC Modeling

8.3.2.1 Laser-Induced Temperature Rise and Acoustic Pressure

When sufficiently high-power laser pulses are focused into a transparent medium, breakdown can be induced (as stated in Sect. 8.2.2.1). However, the laser cannot cause the breakdown or formation of a bubble if the laser irradiance is below the breakdown threshold. The medium is partially ionized through multiphoton and CI, and then, a low-density plasma is formed. The free electron density of the low-density plasma can be calculated by introducing low laser irradiance into Eq. (8.1). The laser pulse represents a Gaussian time variation (Noack and Vogel 1999) and is focused into water by an objective lens with a numerical aperture (NA). The confocal volume is considered to approximate a sphere with a volume equal to the volume of a cylinder with a diameter $d = 2.24\lambda/2NA \cdot n_{\mathrm{l}}$ and length $l = \pi d^2/4\lambda$, where n_{l} denotes the refractive index of water.

The increase in the water temperature is assumed to be caused by absorption of the laser energy because the ultrasonic absorption coefficient of water is very small (Hall 1948). The deposition of laser pulse energy into the optically absorbing plasma is mediated by the generation and subsequent acceleration of free electrons. The temperature rise can be determined by calculating the volumetric energy density gained by the low-density plasma during the laser pulse. The corresponding temperature rise ΔT in the interaction volume after a single laser pulse is as follows:

$$\Delta T = \frac{E_{\mathrm{p}}}{C_{\mathrm{l}}}, \tag{8.15}$$

where C_{l} is the heat capacity of the water.

For femtosecond pulses, recombination and collisional losses during the laser pulse play only a minor role because the pulse duration is considerably shorter than the electron cooling and recombination times [a few picoseconds and tens of picoseconds, respectively (Noack and Vogel 1999)]. Thus, the low plasma energy density E_{p} at the end of the laser pulse is as follows:

$$E_{\mathrm{p}} = \frac{3}{2}\rho_{\mathrm{cr}}\Delta E, \tag{8.16}$$

For nanosecond laser pulses, Eq. (8.16) cannot be used to calculate the volumetric energy density because the electrons have enough time to transfer significant amounts of energy to the heavy plasma particles through collisional losses and recombination. A more general estimate for the energy density is therefore obtained by integrating the terms for collisional and recombination losses as follows:

$$E_{\mathrm{p}} \approx \frac{3}{2}\Delta E \int_{-\infty}^{\infty} \left(\frac{m_{\mathrm{e}}}{m_{\mathrm{l}}} \frac{\omega^2 \tau}{\omega^2 \tau^2 + 1} \rho(t) + \eta_{\mathrm{rec}} \rho(t)^2 \right) \mathrm{d}t, \tag{8.17}$$

where ΔE is the band gap of water (6.5 eV), m_{e} and m_{l} are the electron mass and water molecule mass, respectively, ω is the laser frequency, and τ is the time between electron–heavy particle collisions. The first term represents the collisional energy deposition, while the second denotes the energy deposition by recombination.

The acoustic pressure p_{a} is as follows:

$$p_{\mathrm{a}} = P_{\mathrm{A}} \cos(2\pi f t + \varphi), \tag{8.18}$$

where P_{A} is the amplitude, f is the frequency of the ultrasound, and φ is the relative acoustic phase when the laser is introduced. Thus, the liquid pressure p_{l} in the confocal volume is determined by the following:

$$p_{\mathrm{l}} = p_{\infty} + p_{\mathrm{a}} + p_{\mathrm{h}}, \tag{8.19}$$

where $p_{\infty}, p_{\mathrm{a}}$, and p_{h} denote the atmospheric pressure, the applied acoustic pressure, and the thermoelastic pressure caused by the water temperature rise, respectively.

For this simulation performed in pure water with $\lambda = 532\,\mathrm{nm}$, $\tau_{\mathrm{L}} = 3\,\mathrm{ns}$, and $\mathrm{NA} = 1.3$, the temperature rise ΔT was plotted, and it is presented in Fig. 8.33. The temperature rise increased with time during one laser pulse, and its maximum increased with increased laser irradiance.

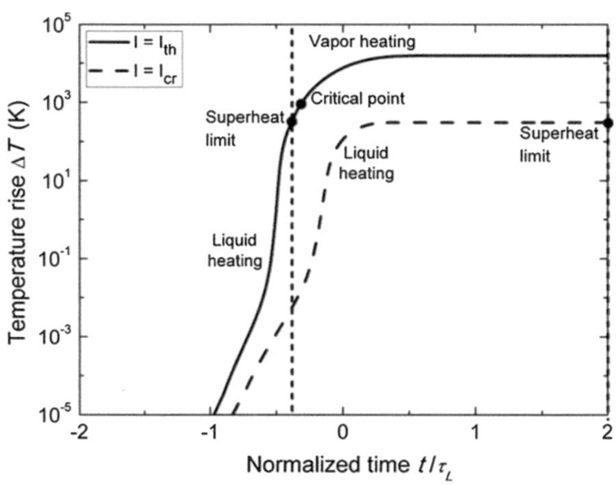

Fig. 8.33 Evolution of temperature rise for two laser irradiances. I_{th} is the laser-induced breakdown threshold, and the critical electron density ρ_{cr} is 10^{20} cm^{-3}. I_{cr} is a critical laser irradiance that elevates T_{l} from T_0 to T_{sh}, where T_{sh} is the superheat limit of water at atmospheric pressure

When the laser irradiance was sufficiently high (Fig. 8.33, solid line), the total energy used in these processes can be approximately given by the following (Byun and Kwak 2004):

$$E_L = \frac{4}{3}\pi R_f^3 \rho_l \left[C_l(T_{sh} - T_\infty) + h_{fg} + C_v(T - T_{sh})\right], \tag{8.20}$$

where E_L is the laser energy absorbed by the plasma, C_l and C_v are the heat capacities of water and its vapor, respectively, and h_{fg} is the latent heat.

When the laser irradiance equals the critical laser irradiance I_{cr} (Fig. 8.33, dashed line), the last two terms in Eq. (8.20) vanish, which denote the energy required for water evaporation and vapor temperature rise, respectively. In PAC, the laser irradiance threshold is reduced because of the combination of ultrasound (Ju et al. 2013; Gerold et al. 2011). Therefore, the applied laser irradiance is smaller than the I_{cr}, and the absorbed laser energy is only used for liquid water heating. Meanwhile, the plasma has low density and its expansion is small, which can result in a relatively accurate estimate of the laser-induced temperature rise.

8.3.2.2 PAC Nucleation

In the model, a modified classical nucleation theory (CNT) has been proposed to describe the homogeneous nucleation of PAC in pure degassed water. It is based on the Kashchiev's scaling function to avoid the approximation of surface tension in CNT (Talanquer 1997; Kusaka 2003), that is,

$$\phi(\xi) = \frac{W^*}{W^*_{CNT}} = \left(\frac{\sigma_R}{\sigma_\infty}\right)^3, \tag{8.21}$$

where $\phi(\xi) = (1 - \xi)(1 + 1/2\xi)^2$, and W^* denotes the energy barrier of critical nucleus formation. σ_R and σ_∞ are the surface tension of the nucleus and the flat interface used in CNT, respectively. The normalized degree of metastability ξ at the same temperature is as follows (Kashchiev 2003):

$$\xi = \frac{p_l(T) - p_v(T)}{p_{spin}(T) - p_v(T)}, \tag{8.22}$$

where p_{spin} is the spinodal pressure of the metastable phase. Thus, the CNT is modified by

$$J(T_l, p_l) = \sqrt{\frac{2\sigma_R(T_l, p_l)\rho_l^2}{\pi m_l^3}} e^{-\frac{W^*(T_l, p_l)}{k_B T_l}}, \tag{8.23}$$

$$W^*(T_\mathrm{l}, p_\mathrm{l}) = \frac{4\pi}{3} r^*(T_\mathrm{l}, p_\mathrm{l})^2 \sigma_\mathrm{R}(T_\mathrm{l}, p_\mathrm{l}), \tag{8.24}$$

$$r^*(T_\mathrm{l}, p_\mathrm{l}) = \frac{2\sigma_\mathrm{R}(T_\mathrm{l}, p_\mathrm{l})}{p_\mathrm{v} - p_\mathrm{l}}, \tag{8.25}$$

where r^* and k_B are the radius of the critical nucleus and the Boltzmann constant, respectively. Furthermore, a correction between vapor pressure p_v and saturation pressure p_sat is introduced as follows (Delale et al. 2003):

$$p_\mathrm{v} = \eta(T_\mathrm{l}) \cdot p_\mathrm{sat}(T_\mathrm{l}), \tag{8.26}$$

where $\eta(T_\mathrm{l}) = \exp -\left(\frac{p_\mathrm{sat}(T_\mathrm{l}) - p_\mathrm{l}}{\rho_\mathrm{l} R_\mathrm{g} T_\mathrm{l}}\right)$ and R_g is the specific gas constant.

The cavitation threshold is obtained by defining a probability of 50 % for cavitation bubble formation

$$\int_0^{\tau_L} J(T_\mathrm{l}, p_\mathrm{l}) dt \cdot 4\pi R_\mathrm{f}^3 / 3 = 0.5. \tag{8.27}$$

For the simulation, we assume that the acoustic pressure is constant during laser irradiation because the laser pulse duration is much shorter than the ultrasound period (Wu et al. 2011). Moreover, the pressure p_h in Eq. (8.19) is ignored because the stress confinement time is shorter than τ_L.

A smaller critical radius corresponds to a larger nucleation rate, which means nucleation is more likely to occur. In Fig. 8.34, both were calculated by considering

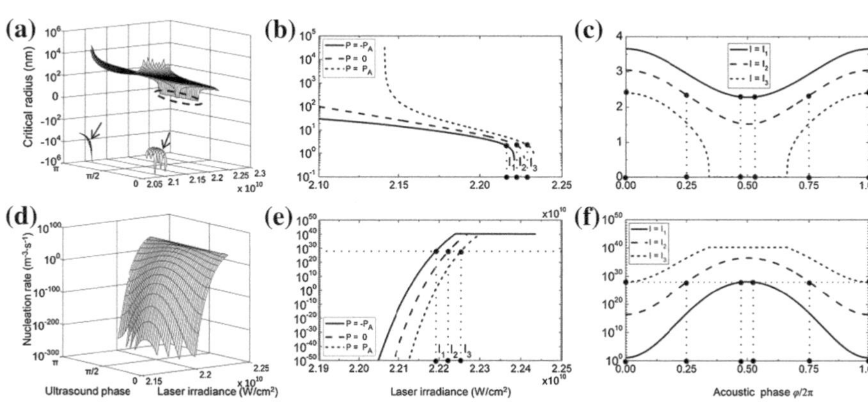

Fig. 8.34 The **a** critical radius and **d** nucleation rate calculated by modified CNT at different laser irradiances and acoustic pressures. Figures **b** and **c** are the projections of (**a**) at three specific pressures and laser irradiances, respectively. Similarly, figures **e** and **f** are the projections of (**d**). The points are the critical points where nucleation occurs

the pre-established ultrasound and the temperature rise. The results reveal that the nucleation of PAC is facilitated by increased laser irradiance or decreased acoustic pressure.

When water is stable, no nucleus is formed and the critical radius is negative because the vapor pressure p_v is smaller than the liquid pressure p_l, as displayed by the arrows in Fig. 8.34a. Meanwhile, when the superheated liquid is unstable, the free energy barrier W^* disappears and the critical radius becomes zero, as displayed by the dashed ellipse.

8.3.2.3 PAC Bubble Dynamics

A spherical pure vapor bubble is assumed to be generated in the center of the confocal region, and its dynamics could be described by the Rayleigh–Plesset equation (Plesset and Prosperetti 1977):

$$\rho\left[R\ddot{R}+\frac{3}{2}\dot{R}^2\right]=p_b-p_l-\frac{2\sigma}{R}-4\mu_l\frac{\dot{R}}{R} \tag{8.28}$$

where R refers to the radius of the bubble and p_b is the pressure in the bubble. Here, p_b and σ depend strongly on temperature, and their initial conditions are determined by laser irradiance.

In PAC, the bubble dynamics are divided into two stages as shown in Fig. 8.35. In the isobaric dynamics process, the pressure in the bubble nearly equals the vapor pressure $p_b(t)=p_v(T_l)$. In the adiabatic dynamics process, the polytropic relation

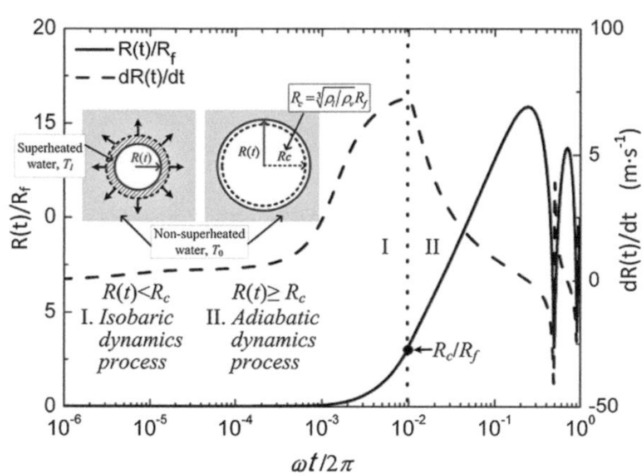

Fig. 8.35 Diagram of PAC bubble dynamics. The normalized bubble radius $R(t)/R_f$ (*solid line*) and the velocity of the bubble radius $dR(t)/dt$ (*dashed line*) are divided into two stages (isobaric dynamics process and adiabatic dynamics process) by the critical value R_c, which is calculated by $R_c=\sqrt[3]{\rho_l/\rho_v}R_f$

$p_b(t) = p_v \cdot (R_c/R(t))^{3\gamma}$ is employed, where R_c is the critical value separating two stages and γ is the adiabatic index. R_c is calculated by the equation $R_c = \sqrt[3]{\rho_l/\rho_v} R_f$, where ρ_v is the vapor density and R_f is the radius of the confocal volume. The surface tensions in these two stages are $\sigma(T_l)$ and $\sigma(T_0)$, respectively, where T_0 is the room temperature.

The bubble dynamics without and with the pre-established ultrasound are shown in Fig. 8.36a, b, respectively. When no acoustic pressure forces are on the bubble, it first expands and then collapses with subsequent rebounds, as shown in Fig. 8.36a. The bubble radius is slightly larger at a higher irradiance. The bubble dynamics process is in accordance with the results described by (Akhatov et al. 2001).

Figure 8.36b shows the bubble dynamics when the laser is introduced at four characteristic points of the ultrasound phase (φ = 0, π/2, π, and 3π/2) with corresponding threshold irradiances. In the isobaric dynamics process (upper), the bubble radius is smallest in case 1 (solid line) and largest in case 3 (dash-dotted line). This means that the bubble generated at negative pressure grows faster than one generated at positive pressure, although the laser irradiance is smaller at negative pressure. Moreover, the bubble radius in case 2 is slightly larger than that in case 4 (inset), revealing that the subsequent negative acoustic pressure leads to easier bubble growth compared with the subsequent positive acoustic pressure. In the whole dynamics process, the bubble grows initially and then collapses with subsequent rebounds when the initial acoustic pressure is positive (cases 1 and 4). The amplitude of the rebound will increase significantly when the negative acoustic pressure is forced. If the bubble grows to a sufficiently large radius, no subsequent

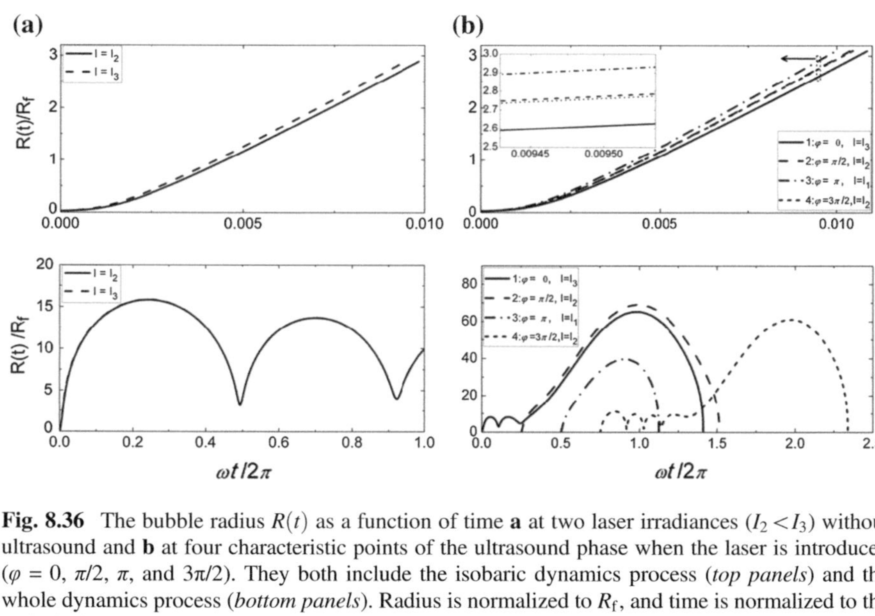

Fig. 8.36 The bubble radius $R(t)$ as a function of time **a** at two laser irradiances ($I_2 < I_3$) without ultrasound and **b** at four characteristic points of the ultrasound phase when the laser is introduced (φ = 0, π/2, π, and 3π/2). They both include the isobaric dynamics process (*top panels*) and the whole dynamics process (*bottom panels*). Radius is normalized to R_f, and time is normalized to the period of ultrasound

rebound will occur, probably because the contraction cannot be stopped. It is also shown that the maximum value of the radius in case 3 is significantly smaller than that in the other three cases, which may be because case 3 involved the shortest duration of exposure to negative acoustic pressure.

Here, we can see that the effects of both laser and acoustic pressure on the first isobaric dynamics process are small. A reasonable explanation might be that the isobaric dynamics process is too short for the whole dynamics process ($\omega t/2\pi \approx 0.01 \ll 1$). Comparing Fig. 8.36a, b, we can see that positive acoustic pressure limited the growth as well as the rebound, while negative acoustic pressure facilitated these processes. Consequently, it might be reasonable to conclude that the effect of a laser on bubble dynamics is insignificant, while the effect of ultrasound is significant.

8.3.3 Experimental System

Figure 8.37 shows a schematic representation of a classic experimental PAC system. The ultrasound is generated by a focused ultrasound transducer. The laser is focused by the objective on microscope, and the high-speed photography is confocal to the laser. In the chamber or microtubes where PAC happens, it is very important to permit irradiation of the ultrasound focus with the laser pulse, which is

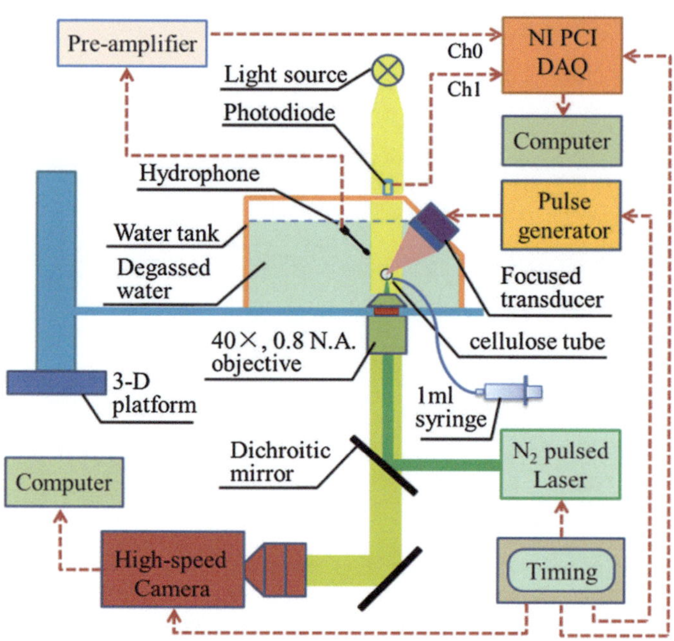

Fig. 8.37 A classic experimental system of PAC

confocal to the high-speed photography system. If a transducer is utilized in the PCD system, it should also be confocal to the ultrasound and the optical focus. A three-dimensional microtranslation stage is equipped to focus the three parts of the system. Moreover, to ensure synchronization of the laser pulse, the flash illumination, and operation of the high-speed camera, each component is electronically triggered with appropriately delayed TTL pulses from a timing system. The entire system is constructed on a self-leveling and vibration-damping optical table.

8.3.4 PAC in Water

To make the onset of nucleation predictable, Gerold et al. (2011) introduced an instrument in which acoustic cavitation can be forced to occur at a specific location in degassed water using a laser to nucleate a microcavity in a pre-established ultrasound field. The high-speed photographic observations of the LIC and PAC study prove that cavitation can be nucleated below the optical cavitation threshold under ambient pressure conditions in a pre-established ultrasound field and that PAC forms a bubble with a relatively smaller maximum radius. Figure 8.38 shows a PAC in an ultrasound field with different acoustic pressures. As the mechanical index (MI) increased, the cavitation clouds showed a larger number of distinct cavities interacting, coalescing, and reforming, but also exhibiting collective behavior as a single entity. The morphology changes of the cavitation clouds potentially came about because of the primary radiation forces.

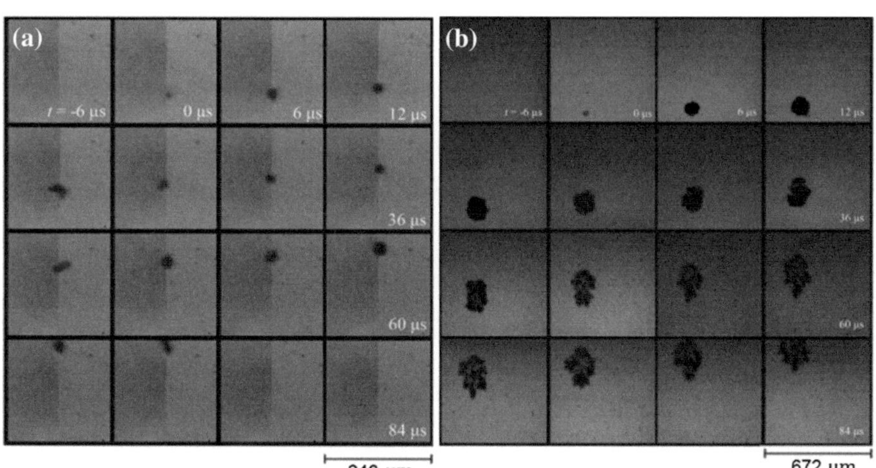

Fig. 8.38 High-speed photographic images of PAC with a 0.95-mJ laser pulse in pre-established ultrasound field in **a** MI = 1.1 and **b** MI = 3.4. Reproduced with permission from Gerold et al. (2011). © 2011 American Institute of Physics

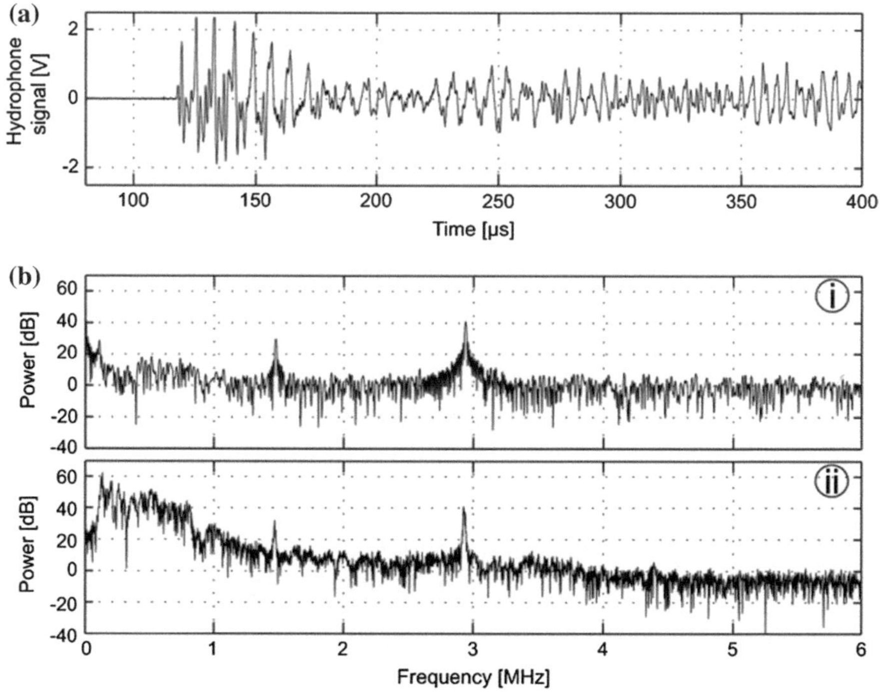

Fig. 8.39 Acoustic signals and spectrum analysis. **a** The acoustic signal notch-filtered at 1.47 MHz. **b** The spectra of (*i*) no cavitation and (*ii*) PAC. Reproduced with permission from Gerold et al. (2011). © 2011 American Institute of Physics

In this study, the acoustic signal was recorded using a PVDF needle hydrophone, and its spectrum analysis is shown in Fig. 8.39. The sudden increase in signal corresponds to high-speed observations of the onset of cavitation activity. The signal decayed as the cluster translated away from the hydrophone tip position, under the primary radiation force. In the spectrum of the pre-established ultrasound field (Fig. 8.39b-i), the presence of a second harmonic peak and the absence of broadband emissions suggest stable cavitation outside the field of view before the laser pulse was generated. Comparing this spectrum with that of PAC (Fig. 8.39b-ii), which also exhibits a harmonic, we see a strong increase in acoustic emissions from 100 kHz to 1 MHz. This suggests that the PAC in Fig. 8.39 is inertial.

8.3.5 *PAC in a Phase-Shift Liquid*

In a phase-shift liquid, PAC can be induced by a nanosecond laser pulse at low intensity with a pre-established ultrasound. PAC nucleation is synchronously formed by the laser and ultrasound. However, the cavitation bubble mainly interacts with ultrasound when multiple ultrasound pulses are applied. In a pre-established ultrasound field, a PAC bubble moves along the ultrasonic propagation direction at the frequency of RPF, and then, it moves back after it grows to a critical size. The bubble movement follows an S shape, which is the result of the balancing between the ultrasound irradiation force and the buoyant force that corresponds to the bubble diameter.

Single ultrasound pulse has less effect on the dynamics of PAC bubbles. As shown in Fig. 8.40a, the ultrasound is forced in the first two frames, and thus, the PAC bubble in the following frames shrinks without the ultrasound field. Two patterns of PAC bubbles were acquired (Fig. 8.40), namely the generation of one bubble or the generation of multiple bubbles, both followed by shrinking. Compared to LIC bubbles in PFP liquid, no growing bubble (Pattern II, Fig. 8.40) was observed for PAC in PFP using a single ultrasound pulse.

The PCD signal of PAC with a single ultrasound pulse and its corresponding frequency-domain diagram are shown in Fig. 8.41a. Compared to LIC in a phase-shift liquid, there seem to be no significant differences in the time domain; however, only one ultrasound pulse was able to change the frequency domain,

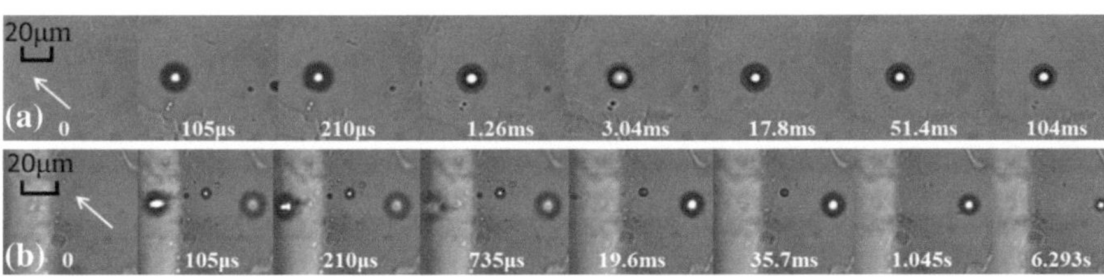

Fig. 8.40 PAC in a PFP liquid with a single ultrasound pulse (5 MHz, acoustic pressure of 0.89 MPa, a pulse laser energy of 45 μJ, and a temperature of 20.8 °C)

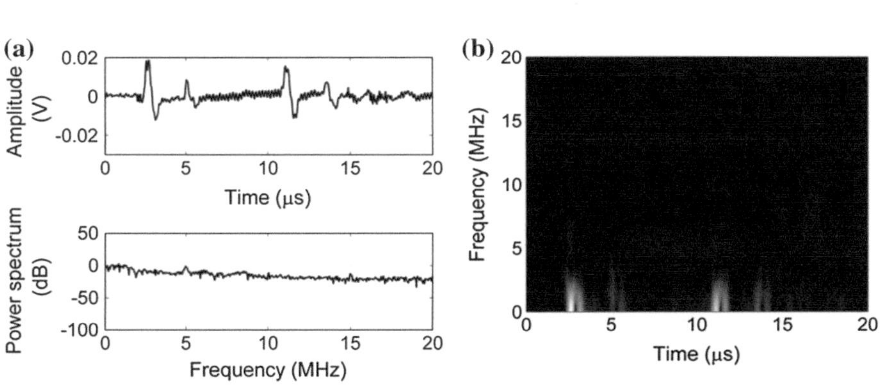

Fig. 8.41 Filtered PCD signals and time–frequency diagram of PAC in PFP liquid

producing small peaks at 5, 10, and 15 MHz. The same can be seen in the time–frequency diagram, presented in Fig. 8.41b. This reveals that during the PAC nucleation, the ultrasonic field used to form the PAC bubble has a significant effect on the characteristics of bubble dynamics.

8.3.6 *PAC of Nanodroplets*

8.3.6.1 Threshold of Acoustic Droplet Vaporization (ADV) and PAC

Acoustic droplet vaporization (ADV) is the process by which superheated liquid droplets are phase-transitioned into gas bubbles by means of ultrasound. These gas bubbles can be used for many applications, such as occlusion therapy, drug delivery, and aberration correction, among others. Compared to ADV, PAC introduces laser energy (lower than the threshold of LIC) to form a nucleus for ADV, which can reduce the ADV threshold significantly. Figure 8.42 shows the threshold of ADV for droplets of different diameters with a single ultrasound pulse (5 MHz, 10 cycles). When the acoustic pressure was lower than 0.82 MPa, no ADV happened, even though the maximum diameter of the droplet reached ~16 μm. With increasing acoustic pressure, the diameter of the ADV droplets decreased. This plot reveals the close relationship between ADV and the original diameter of the droplet.

For droplets with a diameter of ~6 μm, Fig. 8.43 shows the peak-to-peak value of acoustic emission as a function of laser pulse energy in a pre-established ultrasound field at different acoustic pressures. Here, we can see that the laser energy threshold increases with the decrease in acoustic pressure. Compared to ADV, the introduction of the laser decreased the threshold of the acoustic pressure. Correspondingly, we can see that the laser energy threshold for PAC is significantly decreased due to the acoustic pressure, even when the ultrasound was as low as 0.09 MPa.

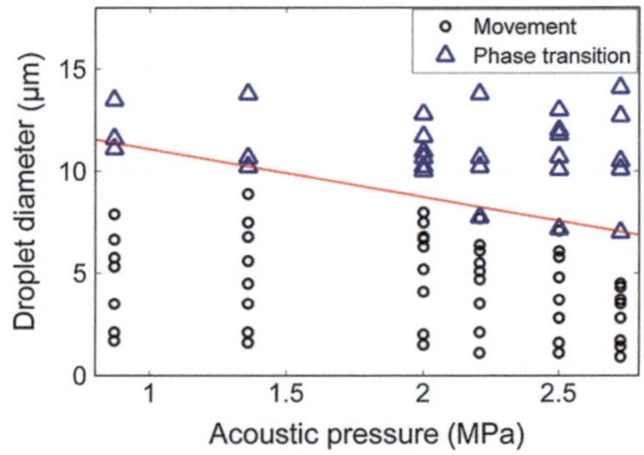

Fig. 8.42 Threshold of ADV for droplets of different diameters ($p < 0.05$)

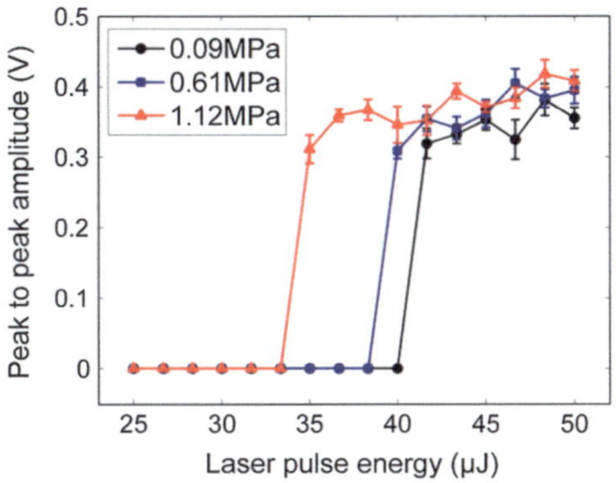

Fig. 8.43 The peak-to-peak value of acoustic emissions (detected with a passive transducer) as a function of laser pulse energy in different pre-established ultrasound fields (PFP droplet, 21.5 °C)

8.3.6.2 PAC of a PFP Droplet

Figure 8.44 shows the dynamics of a PAC bubble made of a PFP droplet under a single ultrasound pulse. The acoustic pressure was 0.61 MPa, lower than the ADV threshold of a droplet with a diameter of 18 μm (a relatively big droplet). The figure shows that PAC formed a small bubble inside the droplet, and then, the small bubble grew upon vaporization and reached the border of the droplet, finally growing to a large PAC bubble with a diameter of ~60 μm.

The differential PCD signal of PAC for the PFP droplet and its frequency spectrum are shown in Fig. 8.45a. There are two peaks in the time domain, representing the primary signal emitted during PAC and the reflection of the primary

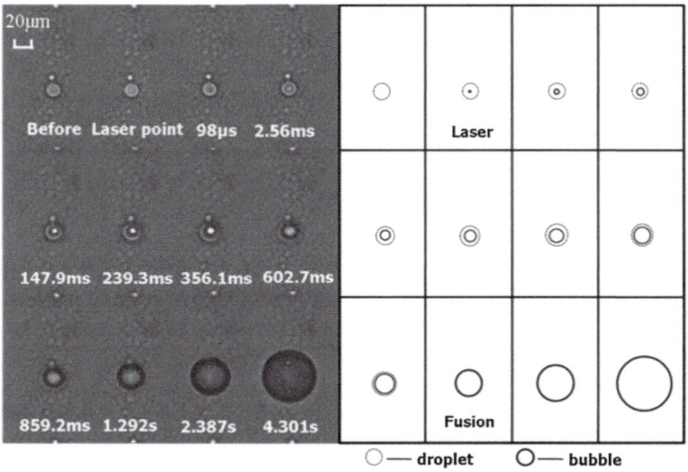

Fig. 8.44 PAC for a PFP droplet irradiated with a single ultrasound pulse

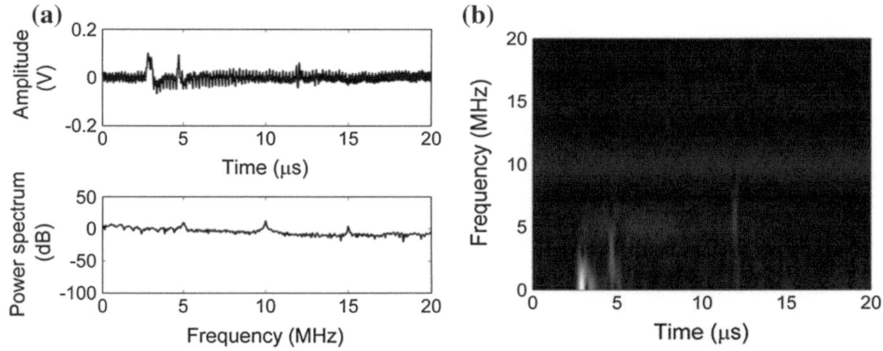

Fig. 8.45 PCD signals and spectrum analysis of PAC of a PFP droplet: **a** PCD signal of PAC (*top*) and spectrum of PAC (*bottom*); **b** time–frequency diagram of PAC

signal from the objective lens, respectively. Broadband noise at a low frequency is shown in the spectrum analysis, and small peaks appear at 5, 10, and 15 MHz, revealing that PAC bubble dynamics are inertial and highly affected by a single ultrasound pulse. The same conclusion can be made after observing the time–frequency diagram of the PCD signal shown in Fig. 8.45b.

It is currently hypothesized that ADV occurs by two main mechanisms. The first mechanism is vaporization within the core of the droplet, which happens when the transmitted ultrasonic field interacts with the dispersed medium. The second mechanism is vaporization due to inertial cavitation in or near the droplet. During inertial cavitation, a secondary shock wave is emitted, which causes vaporization of the droplet. In PAC of a droplet, it seems the former mechanism dominates.

8.3.7 PAC with Nanoparticles

8.3.7.1 Phenomenon and Detection of PAC with Nanoparticles

Because nanoparticles absorb significant amounts of optical radiation, this radiation can be efficiently converted into heat, leading to localized heating of the medium immediately surrounding the nanoparticle. This phenomenon can be used in PAC because nanoparticles can reduce the threshold of cavitation, given their special optical and thermal properties. In addition, the nanoparticles can be used as pre-existing nuclei (McLaughlan et al. 2010).

To clarify the effects of ultrasound on PAC of nanoparticles, Farny et al. (2005) measured the HIFU cavitation threshold at four laser energy settings, whereby the HIFU pressure was increased by 0.09-MPa increments from 0.8 to 1.4 MPa. Figure 8.46 shows that cavitation was detected at peak negative pressures as low as 0.92 MPa when the laser pulse was synchronized with the HIFU signal, as

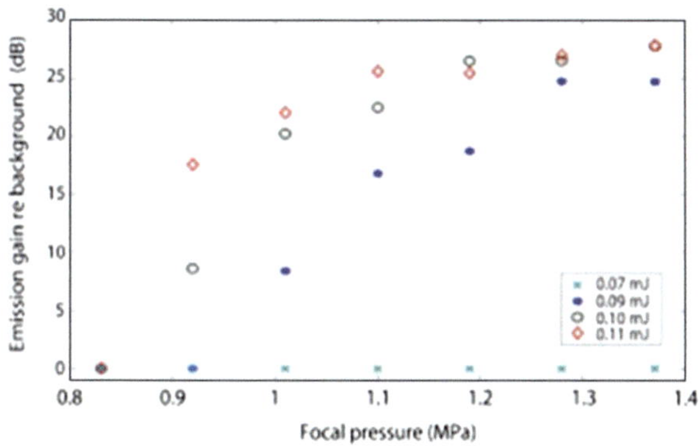

Fig. 8.46 The emission gain relative to the background as a function of HIFU peak negative pressure and laser energy. The laser was fired 0.15 ms after the fifth HIFU cycle arrived at the focus. Reproduced with permission from Farny et al. (2005). © 2005 Acoustical Society of America

compared to 4.50 MPa with no laser irradiation. Also, there was no detectable reduction in the cavitation threshold pressure when the laser was set to fire at 0.07 mJ/pulse; this suggests that the threshold optical energy for generating a vapor cavity sufficient for cavitation nucleation lies between 0.07 and 0.09 mJ/pulse for the laser wavelength and particle sizes employed in this study. Above this critical energy, increasing the laser energy served to lower the cavitation threshold and increase the amount of cavitation activity at a given pressure amplitude—at least up to ~0.12 mJ/pulse. The emission gain plotted in Fig. 8.46 shows that laser illumination of the gold nanoparticles proved effective in significantly lowering the HIFU cavitation threshold pressure. The improved efficacy of HIFU cavitation nucleation can be attributed to the vapor-filled nanobubbles produced from the laser-heated gold nanoparticles.

In another study, gold nanorods were used for PAC. Ju et al. (2013) examined the PAC events—the acoustic signals of which are shown in Fig. 8.47a—around mixed gold nanorods (5.7 × 108 nanoparticles/mL) at a peak negative pressure of 1.8 MPa and fluencies of 1.3 mJ/cm^2 (black), 4.1 mJ/cm^2 (red), and 13.0 mJ/cm^2 (blue). The corresponding peak-to-peak amplitude of the acoustic signals from these PAC events is shown in Fig. 8.47b. As the laser fluence increased, the amplitude of the acoustic emission also increased. This effect is most likely due to an increase in the number of nanoparticles within the interaction region that produce bubbles. In addition, the shape of the acoustic signal broadened, and at higher fluencies, acoustic arrivals associated with bubble-rebound events were observed. This is consistent with an increase in the volume of liquid within which the pressure–fluence combination can produce cavitation events. The ultrasound pressure decreased rapidly outside the focal region, essentially limiting the volume of liquid within which the critical conditions (fluence and pressure) were satisfied. Notably, even at the smallest laser fluence, the inertial collapse of bubbles produced an

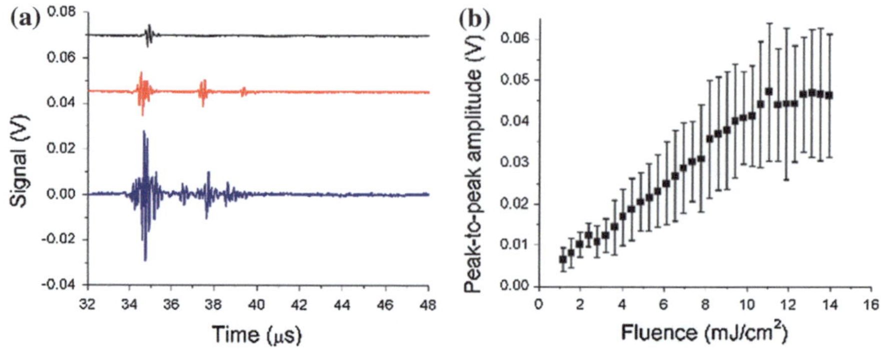

Fig. 8.47 PCD signals from PAC around mixed gold nanorods and the corresponding analysis. Reproduced with permission from Ju et al. (2013). © 2012 Optical Society of America

exceptionally strong photoacoustic signal. Moreover, it indicates that the inertial collapse amplitude is primarily dictated by the applied ultrasound field rather than the laser fluence and that the amplitude shows relatively large fluctuations about the mean, particularly at the larger peak negative pressure amplitude. Thus, inertial collapse monitoring may be useful for therapeutic imaging.

8.3.7.2 Description of the Physical Process

Several physical processes occur during optical irradiation, and they are important for understanding the function of nanoparticles in PAC. The absorption of light by the nanoparticle generates significant heat, a portion of which diffuses into the liquid. Due to surface tension, the diffused energy will superheat a thin film of liquid around the particle surface. Above a certain threshold, explosive vaporization occurs, forming a vapor cavity that will continue to expand due to evaporation of the outer liquid boundary. Once the temperature in the vapor layer decreases sufficiently, the bubble layer will collapse because of surface tension and condensation effects.

As previously mentioned, gold NPs are usually used as the "seeds" for optical cavitation because they have excellent optical absorption properties. To get a further understanding about the function of NPs and the physical process that occurs in PAC with NPs, some models have been introduced (Wu et al. 2011; Krasovitski et al. 2007) and are subsequently described.

1. Optical absorption and heating of the surrounding liquid

For simplicity, consider only a single spherical gold nanoparticle suspended in water. The nanoparticle is irradiated by a laser pulse, and the laser energy absorption across the entire nanoparticle is assumed to be uniform. Thus, the absorption power per unit time $\dot{E}$ can be calculated as follows (Wu et al. 2011):

$$\dot{E} = Q_{\text{abs}} \pi R_s^2 I_0, \tag{8.29}$$

where Q_{abs} is the absorption efficiency, R_s is the radius of the nanoparticle, and I_0 is the intensity of the incident laser beam.

Heat loss to the surrounding liquid will occur primarily through thermal conduction. Thus, the particle temperature T_s is calculated due to thermal conduction:

$$\frac{4}{3} \pi R_s^3 \rho_s C_{\text{ps}} \frac{\partial T_s}{\partial t} = q - 4\pi R_s^2 \left(-k_1 \frac{\partial T_1}{\partial r} |_{r=R_s} \right), \tag{8.30}$$

in which ρ_s and C_{ps} are the density and heat capacity of the particle, k_1 and T_1 are the thermal conductivity and the temperature of the surrounding liquid, respectively, and t and r are the time and space coordinates, respectively. The heat source term q is expressed as follows:

$$q = \begin{cases} \dot{E} & 0 < t < \tau_L \\ 0 & t > \tau_L \end{cases}, \tag{8.31}$$

where τ_L is the laser pulse duration.

According to the heat diffusion equation, the temporal and spatial temperature distribution of the surrounding liquid is expressed as below:

$$\frac{\partial T_1}{\partial t} = \frac{D_1}{r^2} \frac{\partial}{\partial r} \left(r^2 \frac{\partial T_1}{\partial r} \right), \quad 0 < t < t_0, R_s \leq r < \infty, \tag{8.32}$$

where D_1 is the thermal diffusivity in the water and t_0 is the moment when the evaporation begins.

2. Nucleation and bubble growth by synergy of laser and ultrasound

Under the assumption that the nanoparticle is uniformly heated, energy and mass conservation principles are used to derive the threshold conditions of vapor cavity nucleation. The evaporation process was first investigated by examining a thin layer of water on the nanoparticle surface. By conservation of mass, the thickness of the original liquid layer δ_1 must satisfy the following relationship (Wu et al. 2011):

$$\frac{4\pi}{3} \left((\delta_1 + R_s)^3 - R_s^3 \right) \rho_1 = \frac{4\pi}{3} \left((\delta_v + R_s)^3 - R_s^3 \right) \rho_v, \tag{8.33}$$

where δ_1 and δ_v are the thicknesses of the water layer and vapor layer, respectively. The δ_v is defined as follows:

$$\delta_v = \frac{k_B T_{\text{vp}}(p_1)}{\sqrt{2} \pi d^2 p_1}, \tag{8.34}$$

where T_{vp} is the temperature at which water evaporation initiates, d is the diameter of the water molecule, and p_l is the local pressure in the liquid, which is given by

$$p_l = p_\infty + \frac{2\sigma(T_{vp})}{R_s + \delta_v} - P_A \sin(2\pi ft + \varphi), \tag{8.35}$$

where P_A is the amplitude, f is the frequency of the ultrasound incident on the particle, and φ labels the relative acoustic phase at the instant when particle heating commences ($t = 0$). For a nanosecond-long pulsed laser application and a microsecond-long ultrasound exposure, the acoustic pressure can be regarded as constant, so that $p_a = P_A \sin\varphi$ during the laser radiation time.

The spatially averaged temperature $\overline{T}$ within the original liquid layer is given by

$$\overline{T} = \frac{\int_{R_s}^{R_s+\delta_l} T_l(r) 4\pi r^2 \mathrm{d}r}{\frac{4\pi}{3}\left((\delta_l + R_s)^3 - R_s^3\right)}, \tag{8.36}$$

Evaporation will start when $\overline{T}$ is equal to or above T_{vp}. A phase change will not occur if this condition is not satisfied during the period when laser energy is incident on the nanoparticle.

If the evaporation starts at time $t = t_1$, the heat will transfer from the heated nanoparticle to the surrounding liquid to support the evaporation and drive the bubble expansion. For simplicity, only the thermal conductivity is considered. The energy conservation equation is as follows:

$$\dot{m} h_{fg} = 4\pi R_s^2 \left(-k_v \frac{\partial T}{\partial r}\right), \tag{8.37}$$

where $\dot{m}$ is the mass flux of the phase change and h_{fg} is the latent heat. During evaporation, the pressure of the saturated vapor and liquid mixture is constant. However, the Laplace pressure decreases due to volume expansion. The unbalanced pressure across the liquid–vapor interface will force the liquid outward under the excess pressure p_e

$$p_e = \frac{2\sigma}{R_s} - \frac{2\sigma}{R}, \tag{8.38}$$

So, the velocity of liquid motion u due to the p_e can be calculated from

$$\frac{1}{2}\rho_l 4\pi R^3 u^2 = \int_{R_s}^{R} p_e 4\pi r^2 \mathrm{d}r. \tag{8.39}$$

Consequently, the interface velocity $\dot{R}$ is as follows:

$$\dot{R} = u + \frac{\dot{m}}{\rho_l 4\pi R^2}\frac{\rho_l}{\rho_v} \tag{8.40}$$

in which the second term on the right side of the equation denotes the velocity due to vaporization or condensation on the bubble's boundary.

3. Vapor bubble dynamics in an acoustic pressure field

When the vapor cavity fully grown, at time $t = t_2$, we can consider the dynamics of a vapor bubble in liquid at room temperature with the effect of the acoustic pressure. The dynamics of the bubble can be described by the Keller–Miksis equation.

$$\left(1 - \frac{\dot{R}}{c_l}\right)R\ddot{R} + \frac{3}{2}\left(1 - \frac{\dot{R}}{3c_l}\right)\dot{R}^2 = \frac{1}{\rho_l}\left(1 + \frac{\dot{R}}{c_l} + \frac{R}{c_l}\frac{\mathrm{d}}{\mathrm{d}t}\right)[p_B - (p_\infty + p_a)], \tag{8.41}$$

where

$$p = p_B + \frac{2\sigma}{R} + 4\mu_l\frac{\dot{R}}{R}, \tag{8.42}$$

where c_l is the sound speed in the liquid, p is the bubble internal pressure, p_B is the pressure on the liquid side of the interface, and μ_l is the viscosity of water.

The numerical results of the model introduced by Wu et al. (2011) are shown in Fig. 8.48. This figure shows the time-dependent radial motion (R–t) of the liquid–vapor interface of the vapor cavity, which is formed from a 40-nm particle.

In Fig. 8.48a, results are shown for a laser exposure of 3.4 mJ/cm^2 accompanied by acoustic sonication of with peak negative pressures 0.68, 0.80, 0.90, and

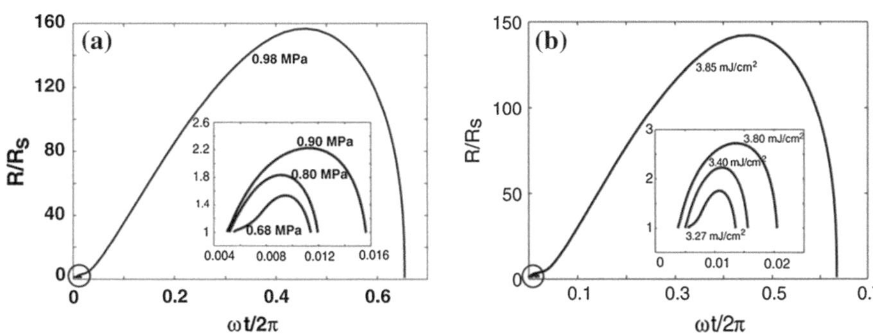

Fig. 8.48 The numerical results of the model. **a** Laser exposure is 3.4 mJ/cm^2, and the acoustic peak negative pressures are 0.68, 0.80, 0.90, and 0.98 MPa. The *inset* is a magnified version of the region in the *circle* in the main plot. **b** Peak acoustic negative pressure is 0.9 MPa. Laser exposures are 3.27, 3.40, 3.80, and 3.85 mJ/cm^2. Reproduced with permission from Wu et al. (2011). © 2011 Acoustical Society of America

0.98 MPa. When the sound pressures are below 0.68 MPa, there is no macroscopic phase change on the particle surface, indicating that a cavity is not formed. For pressures between 0.68 and 0.98 MPa, a vapor cavity forms and expands to ~2–3 times the original particle size, depending on the pressure amplitude. With increasing pressure, cavity nucleation occurs earlier, R_{max} becomes larger, and the life span of the cavity gradually increases. When the pressure exceeds 0.98 MPa, the vapor cavity undergoes a characteristic "explosive" expansion, termed inertial cavitation.

Therefore, we can obtain two thresholds from this numerical simulation: (1) The threshold for the formation of a vapor cavity, or the nucleation threshold, is 0.68 MPa for a laser fluence of 3.4 mJ/cm^2; and (2) the threshold for inertial vapor cavitation is 0.98 MPa for a laser fluence of 3.4 mJ/cm^2. Similarly, from Fig. 8.48b, we can conclude that the nucleation threshold for laser exposure is 3.27 mJ/cm^2 and the inertial threshold is 3.85 mJ/cm^2 for a fixed acoustic peak pressure of 0.9 MPa.

The total local pressure on the particle surface determines the temperature to which the liquid must be elevated in order to nucleate. Figure 8.49 illustrates the effects of acoustic pressure by plotting the maximum radius as a function of acoustic phase for a transient vapor cavity (Wu et al. 2011). The phase φ starts from $\pi/2$, which implies heating at the peak positive pressure.

During compression, R_{max}/R_s is approximately unity, meaning that no significant bubble growth is occurring. When the acoustic pressure is negative, the conditions are favorable for creation and expansion of a vapor cavity, so R_{max} gradually increases. There is a sharp transition where the vapor cavity has a lengthy exposure to the negative stress, so R_{max} reaches a maximum under these conditions. As φ

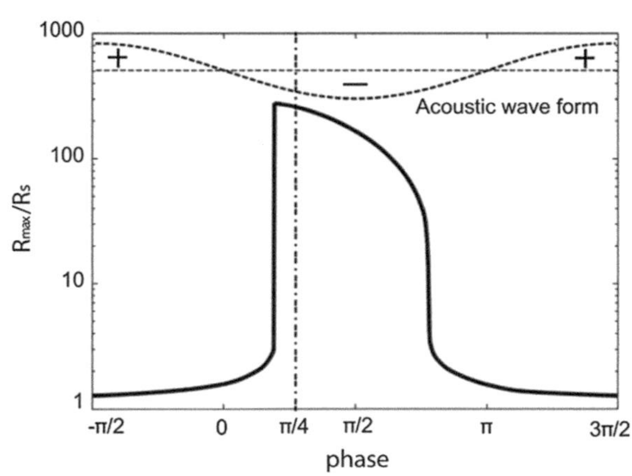

Fig. 8.49 The normalized maximum radius of the vapor cavity as a function of φ. $R_s = 40$ nm. The laser exposure is 0.12 mJ/cm^2, and the peak negative acoustic pressure is 1.4 MPa. Reproduced with permission from Wu et al. (2011). © 2011 Acoustical Society of America

continues to increase, R_{max} decreases and returns to a very small expansion ratio after $\varphi = 0.8\pi$. It is evident that a threshold for φ exists, whereby the relative timing can both induce vaporization and also bring about the greatest potential for expansion.

8.3.7.3 PAC with Nanoparticles Used in Imaging of an Inertial Collapse Signal

Compared to the photoacoustic imaging of NPs, using spontaneous confocal ultrasound irradiation to generate PAC with NPs can significantly increase imaging quality by using the inertial collapse signals, as shown in Fig. 8.50 (McLaughlan et al. 2010). Images (a) and (b) show that with laser fluence alone, the photoacoustic signals from the NPs are low. However, as images (c) and (d) show, the inertial collapse emissions from the same samples under the same laser fluencies but with the application of ultrasound (peak negative pressure of 1.5 MPa) result in a dramatic improvement in the ability to detect NPs in the phantom. Thus, a new imaging protocol should be developed based on the inertial collapse emissions, which, under the experimental conditions, were only observed in the presence of both ultrasound and laser fields.

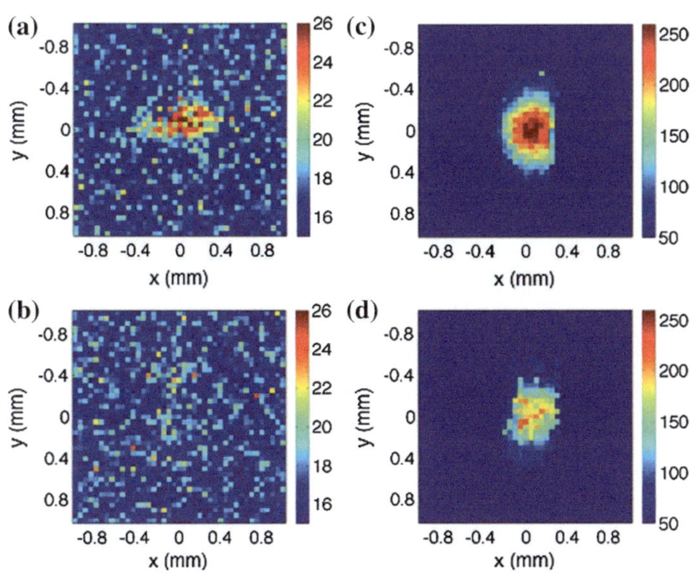

Fig. 8.50 Images of nanoparticle-doped spheres generated from the photoacoustic response to a laser fluence of **a** 22.6 mJ/cm^2 and **b** 11.4 mJ/cm^2. Images created using the inertial collapse signals with an applied 1.5-MPa ultrasound field are shown for a laser fluence of **c** 22.6 mJ/cm^2 and **d** 11.4 mJ/cm^2. Reproduced with permission from McLaughlan et al. (2010). © 2010 Optical Society of America

8.4 Summary

Compared with acoustic cavitation, LIC and PAC have different physical mechanisms and biomedical applications. LIC is a well-established approach that has been used in cell surgery. It also allows us to study single-bubble dynamics in a range of fluidic environments, with the assistance of high-speed cameras. As stated in Sects. 8.2.2–8.2.6, the representative experimental results of LIC in several kinds of liquids, in PFC droplets, and in cells have been studied experimentally via high-speed optical images and additional acoustic detections. The relevant theoretical explanations were also presented in these sections. Furthermore, LIC can potentially be used to perform precise surgical procedures in bulk tissues. Section 8.2.7 introduced PTBs, generated by the interactions between a laser and NPs, and then discussed their applications in theranostics.

To avoid unsafe high-energy exposure in LIC, the emerging cavitation technology of PAC has been recently used, and it was introduced in Sect. 8.3. The combination of laser and a pre-established ultrasound field allows for bubble formation at lower laser fluencies and peak negative ultrasound pressures that can be achieved using either modality alone. In addition, the dynamics of PAC bubbles are different than the dynamics associated with each individual modality. Thus, Sect. 8.3.2 introduces an alternative theoretical description for PAC in water, where this description includes prenucleation heating induced by laser, nucleation, and the resulting bubble dynamics closely related to ultrasound. Then, Sects. 8.3.3–8.3.6 present, via detecting optical and acoustic signals, the threshold of PAC, and the characteristics of PAC bubble dynamics in a variety of mediums, such as water, PFC, and PFC droplets. Finally, PAC with NPs is discussed, which is aimed at improving the efficiency of laser absorption. This discussion consisted of three parts: describing the experimental phenomenon and detection of PAC with nanoparticles, presenting the theoretical description of this physical process, and then reviewing its applications in imaging.

LIC and PAC can induce many physical, chemical, and biological effects. They both have potential applications in theranostics, especially in PAC imaging and nanosurgery. In this developing research field, several scientific and technological questions have not been illuminated yet, but the application of these techniques in theranostics remains promising.

References

Achenbach JD, Komsky IN, Lee Y, Angel YC. Self-calibrating ultrasonic technique for crack depth measurement. J Nondestr Eval. 1992;11(2):103–8.

Akhatov I, Lindau O, Topolnikov A, Mettin R, Vakhitova N, Lauterborn W. Collapse and rebound of a laser-induced cavitation bubble. Phys Fluids. 2001;13(10):2805.

Akhatov I, Vakhitova N, Topolnikov A, Zakirov K, Wolfrum B, Kurz T, Lindau O, Mettin R, Lauterborn W. Dynamics of laser-induced cavitation bubbles. Exp Thermal Fluid Sci. 2002;26 (6):731–7.

Boyer D, Tamarat P, Maali A, Lounis B, Orrit M. Photothermal imaging of nanometer-sized metal particles among scatterers. Science. 2002;297(5584):1160–3.

Byun KT, Kwak HY. A model of laser-induced cavitation. Jpn J Appl Phys. 2004;43(2R):621.

Chen WS, Lafon C, Matula TJ, Vaezy S, Crum LA. Mechanisms of lesion formation in high intensity focused ultrasound therapy. Acoust Res Lett Online. 2003;4(2):41–6.

Chiou P-Y, Wu T-H, Park S, Chen Y Pulse laser driven ultrafast micro and nanofluidics system. In: SPIE NanoScience + Engineering, 2010. International Society for Optics and Photonics, pp. 77590Z-77590Z–77598.

Cognet L, Tardin C, Boyer D, Choquet D, Tamarat P, Lounis B. Single metallic nanoparticle imaging for protein detection in cells. Proc Natl Acad Sci. 2003;100(20):11350–5.

Coussios C, Farny C, Ter Haar G, Roy R. Role of acoustic cavitation in the delivery and monitoring of cancer treatment by high-intensity focused ultrasound (HIFU). Int J Hyperth. 2007;23(2):105–20.

Dayton PA, Zhao S, Bloch SH, Schumann P, Penrose K, Matsunaga TO, Zutshi R, Doinikov A, Ferrara KW. Application of ultrasound to selectively localize nanodroplets for targeted imaging and therapy. Mol Imaging. 2006;5(3):160.

Delale CF, Hruby J, Marsik F. Homogeneous bubble nucleation in liquids: the classical theory revisited. J Chem Phys. 2003;118(2):792–806.

Dijkink R, SeL Gac, Nijhuis E, Avd Berg, Ia Vermes, Ae Poot, Ohl C-D. Controlled cavitation–cell interaction: trans-membrane transport and viability studies. Phys Med Biol. 2008;53 (2):375.

Dijkink R, Ohl CD. Laser-induced cavitation based micropump. Lab Chip. 2008;8(10):1676–81.

Docchio F, Sacchi C, Marshall J. Experimental investigation of optical breakdown thresholds in ocular media under single pulse irradiation with different pulse durations. Lasers Ophthalmol. 1986;1:83–93.

Farny CH, Holt RG, Roy RA. Temporal and spatial detection of HIFU-induced inertial and hot-vapor cavitation with a diagnostic ultrasound system. Ultrasound Med Biol. 2009;35 (4):603–15.

Farny CH, Wu T, Holt RG, Murray TW, Roy RA. Nucleating cavitation from laser-illuminated nano-particles. Acoust Res Lett Online. 2005;6(3):138–43.

Fry FJ, Sanghvi NT, Foster RS, Bihrle R, Hennige C. Ultrasound and microbubbles: their generation, detection and potential utilization in tissue and organ therapy—experimental. Ultrasound Med Biol. 1995;21(9):1227–37.

Fujikawa S, Akamatsu T. Effects of the non-equilibrium condensation of vapour on the pressure wave produced by the collapse of a bubble in a liquid. J Fluid Mech. 1980;97(03):481–512.

Gerold B, Kotopoulis S, McDougall C, McGloin D, Postema M, Prentice P. Laser-nucleated acoustic cavitation in focused ultrasound. Rev Sci Instrum. 2011;82(4):044902.

Hall L. The origin of ultrasonic absorption in water. Phys Rev. 1948;73(7):775–81.

Hashimoto S, Werner D, Uwada T. Studies on the interaction of pulsed lasers with plasmonic gold nanoparticles toward light manipulation, heat management, and nanofabrication. J Photochem Photobiol C. 2012;13(1):28–54.

Hellman AN, Rau KR, Yoon HH, Bae S, Palmer JF, Phillips KS, Allbritton NL, Venugopalan V. Laser-induced mixing in microfluidic channels. Anal Chem. 2007;79(12):4484–92.

Holt RG, Roy RA. Measurements of bubble-enhanced heating from focused, MHz-frequency ultrasound in a tissue-mimicking material. Ultrasound Med Biol. 2001;27(10):1399–412.

Huang X, El-Sayed IH, Qian W, El-Sayed MA. Cancer cell imaging and photothermal therapy in the near-infrared region by using gold nanorods. J Am Chem Soc. 2006;128(6):2115–20.

Ju H, Roy RA, Murray TW. Gold nanoparticle targeted photoacoustic cavitation for potential deep tissue imaging and therapy. Biomed Optics Express. 2013;4(1):66–76.

Kashchiev D. Thermodynamically consistent description of the work to form a nucleus of any size. J Chem Phys. 2003;118(4):1837–51.

Kennedy PK. A first-order model for computation of laser-induced breakdown thresholds in ocular and aqueous media. I. Theory. IEEE J Quantum Electron. 1995;31(12):2241–9.

Klibanov AL. Ligand-carrying gas-filled microbubbles: ultrasound contrast agents for targeted molecular imaging. Bioconjug Chem. 2005;16(1):9–17.

Krasovitski B, Kislev H, Kimmel E. Modeling photothermal and acoustical induced microbubble generation and growth. Ultrasonics. 2007;47(1):90–101.

Kusaka I. A scaling function of nucleation barrier based on the diffuse interface theory. J Chem Phys. 2003;119(3):1808–12.

Lapotko D. Optical excitation and detection of vapor bubbles around plasmonic nanoparticles. Opt Express. 2009;17(4):2538–56.

Lee H, Gojani AB, T-h Han, Yoh JJ. Dynamics of laser-induced bubble collapse visualized by time-resolved optical shadowgraph. J Visual. 2011;14(4):331–7.

Lim KY, Quinto-Su PA, Klaseboer E, Khoo BC, Venugopalan V, Ohl CD. Nonspherical laser-induced cavitation bubbles. Phys Rev E. 2010;81(1):016308.

Lubatschowski H, Maatz G, Heisterkamp A, Hetzel U, Drommer W, Welling H, Ertmer W. Application of ultrashort laser pulses for intrastromal refractive surgery. Graefe's Arch Clin Exp Ophthalmol. 2000;238(1):33–9.

Lukianova-Hleb E, Hanna E, Hafner J, Lapotko D. Tunable plasmonic nanobubbles for cell theranostics. Nanotechnology. 2010;21(8):085102.

Lukianova-Hleb EY, Mrochek AG, Lapotko DO. Method for disruption and re-canalization of atherosclerotic plaques in coronary vessels with photothermal bubbles generated around gold nanoparticles. Lasers Surg Med. 2009;41(3):240–7.

Maatz G, Heisterkamp A, Lubatschowski H, Barcikowski S, Fallnich C, Welling H, Ertmer W. Chemical and physical side effects at application of ultrashort laser pulses for intrastromal refractive surgery. J Opt A Pure Appl Opt. 2000;2(1):59.

McLaughlan JR, Roy RA, Ju H, Murray TW. Ultrasonic enhancement of photoacoustic emissions by nanoparticle-targeted cavitation. Opt Lett. 2010;35(13):2127–9.

Noack J, Vogel A. Laser-induced plasma formation in water at nanosecond to femtosecond time scales: calculation of thresholds, absorption coefficients, and energy density. IEEE J Quantum Electron. 1999;35(8):1156–67.

Plesset MS, Prosperetti A. Bubble dynamics and cavitation. Annu Rev Fluid Mech. 1977;9 (1):145–85.

Quinto-Su PA, Ohl CD. Interaction between two laser-induced cavitation bubbles in a quasi-two-dimensional geometry. J Fluid Mech. 2009;633:425–35.

Rapoport NY, Kennedy AM, Shea JE, Scaife CL, Nam K-H. Controlled and targeted tumor chemotherapy by ultrasound-activated nanoemulsions/microbubbles. J Control Release. 2009;138(3):268–76.

Rink K, Delacretaz G, Salathe R. Fragmentation process induced by nanosecond laser pulses. Appl Phys Lett. 1992;61(22):2644–6.

Sacchi C. Laser-induced electric breakdown in water. J Opt Soc Am B. 1991;8(2):337–45.

Sasaki K, Nakano T, Soliman W, Takada N. Effect of pressurization on the dynamics of a cavitation bubble induced by liquid-phase laser ablation. Appl Phy Express. 2009;2(4):046501.

Schad KC, Hynynen K. In vitro characterization of perfluorocarbon droplets for focused ultrasound therapy. Phys Med Biol. 2010;55(17):4933.

Schrage RW. A theoretical study of interphase mass transfer. Columbia: Columbia University Press; 1953.

Talanquer V. A new phenomenological approach to gas–liquid nucleation based on the scaling properties of the critical nucleus. J Chem Phy. 1997;106(23).

Vogel A, Busch S, Parlitz U. Shock wave emission and cavitation bubble generation by picosecond and nanosecond optical breakdown in water. J Acoust Soc Am. 1996;100(1):148–65.

Vogel A, Noack J, Hüttman G, Paltauf G. Mechanisms of femtosecond laser nanosurgery of cells and tissues. Appl Phys B. 2005;81(8):1015–47.

Williams F, Varma S, Hillenius S. Liquid water as a lone-pair amorphous semiconductor. J Chem Phys. 1976;64(4):1549–54.

Wu T, Farny CH, Roy RA, Holt R. Modeling cavitation nucleation from laser-illuminated nanoparticles subjected to acoustic stress. J Acoust Soc Am. 2011;130(5):3252–63.

Zohdy MJ, Tse C, Ye JY, O'Donnell M. Optical and acoustic detection of laser-generated microbubbles in single cells. IEEE Trans Ultrasonics Ferroelectr Freq Control. 2006;53 (1):117–25.

Zwaan E, Le Gac S, Tsuji K, Ohl CD. Controlled cavitation in microfluidic systems. Phys Rev Lett. 2007;98(25):254501.

Chapter 9
Cavitation Mechanobiology and Applications

Yi Feng, Lu Zhao, Gail ter Haar and Mingxi Wan

Keywords Cavitation mechanobiology · Effect of cavitation on tissue · Effect of cavitation on cell · Molecular signaling pathway · Sonodynamic therapy · Effect of cavitation on molecule

9.1 Introduction

9.1.1 Bioeffects of Ultrasound and Cavitation

Ultrasound wave can produce a variety of biological effects based on its pressure parameters (Fig. 9.1) through thermal or nonthermal mechanisms in vitro and in vivo (Dalecki 2004). In fact, the thermal and nonthermal effects are not separable in most physical therapies.

When ultrasound waves are absorbed by tissue, the ultrasound energy is converted to heat, which may significantly increase the temperature of the tissue. This gives rise to a thermal mechanism (Fig. 9.2). For high-intensity focused ultrasound (HIFU) therapy, thermal ablation and erosion are the main mechanisms of action. However, in cancer treatment, thermal effects can also increase cancer-specific

Y. Feng · L. Zhao · M. Wan (✉)
The Key Laboratory of Biomedical Information Engineering of Ministry of Education, Department of Biomedical Engineering, School of Life Science and Technology, Xi'an Jiaotong University, 710049 Xi'an, People's Republic of China
e-mail: mxwan@mail.xjtu.edu.cn

Y. Feng
e-mail: fengyi@mail.xjtu.edu.cn

L. Zhao
e-mail: lulupersist@gmail.com

G. ter Haar
Division of Radiotherapy and Imaging, The Institute of Cancer Research and the Royal Marsden NHS Foundation Trust, 123 Old Brompton Road, London SW7 3RP, UK
e-mail: Gail.terHaar@icr.ac.uk

M. Wan et al. (eds.), *Cavitation in Biomedicine*,
DOI 10.1007/978-94-017-7255-6_9

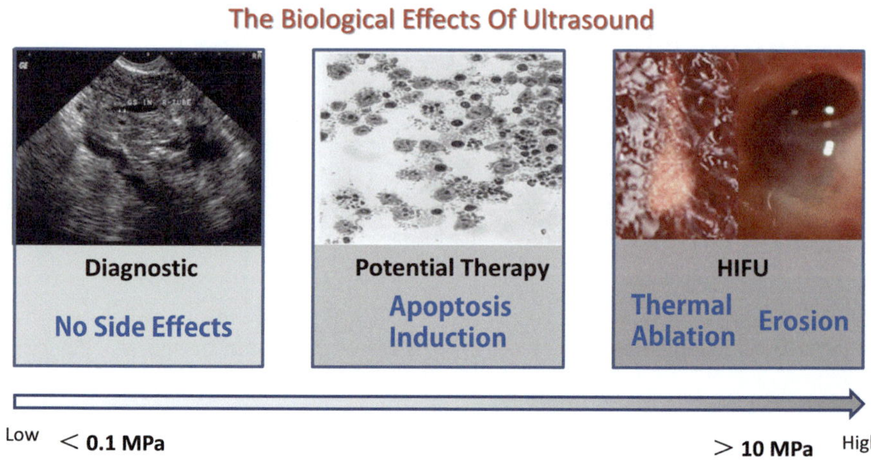

Fig. 9.1 Relationship between the biological effects of ultrasound and the ultrasound pressure. Low-power ultrasound is used in medical diagnosis, without significant clinical side effects. High-power (intensity)-focused ultrasound (*HIFU*) can thermally ablate tissue (e.g., in breast, liver, or prostate cancer) or be used to perform histotripsy by eroding tissue. Between these extremes, a moderate ultrasound dose might offer a therapeutic window for cancer therapy by inducing specific molecular signaling, including apoptosis and autophagy

immunity by inducing the release of heat shock proteins (HSPs), which are known to enhance the T-cell-mediated immune response and cytotoxic T-cell activity.

In contrast to the thermal effects of ultrasound, nonthermal effects are regarded as occurring when ultrasound produces a temperature increase less than 1 °C above normal physiological temperatures (Stratmeyer et al. 2008). These include mechanical and cavitation effects, which are closely related to each other.

The mechanical effects under consideration of ultrasound are always associated with ultrasonically mediated forces that act on cells or tissues, such as radiation forces, microstreaming, and shear stresses (Dalecki 2004). Acoustic cavitation involves more than just these forces: It describes the formation and activity of a gas bubble in an acoustic field (Leighton 2012). When exposed to an ultrasonic field, a bubble in liquid will experience cycles of negative and positive pressures and oscillate around an equilibrium radius. This response produces several interesting physical phenomena, such as shock waves, high local shear stress associated microstreaming, microjets, and implosions (Brujan 2004) (Fig. 9.2). The shear forces normally result from either stable cavitation (oscillations of preexisting gas bodies in response to varying acoustic pressure) or inertial cavitation (which occurs when acoustic pressures are high enough to fracture the liquid or grow bubble nuclei, such that bubbles grow and implode violently after reaching a critical size) (Miller 2007). Regardless of how a shearing force is produced—by either stable or inertial cavitation—it has mechanical effects on cells, which can induce cell membrane

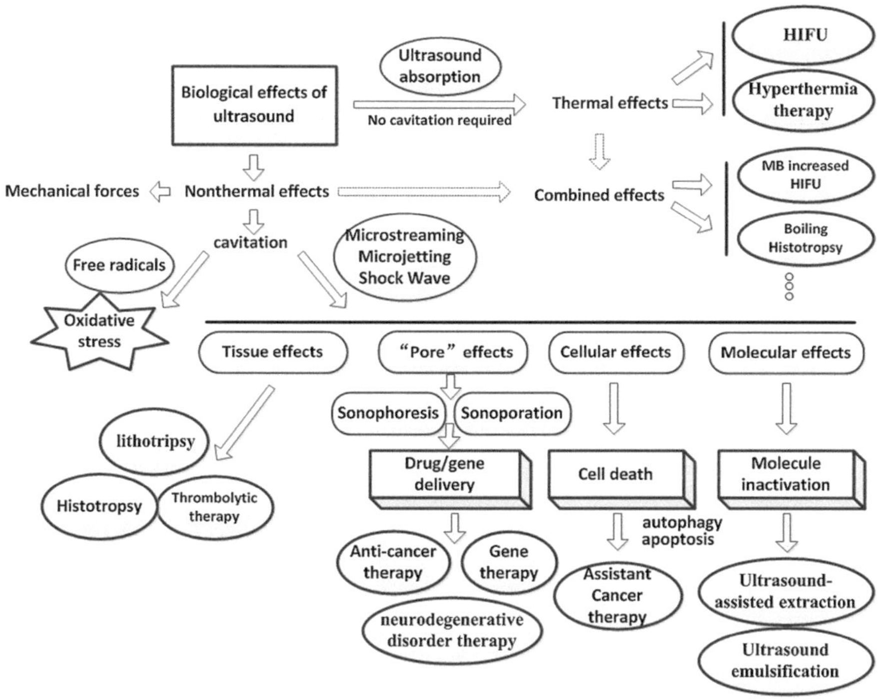

Fig. 9.2 Schematic describing the biological effects of ultrasound. *MB* microbubble

permeability changes or causing cell membranes to rupture. There is thus a close relationship between mechanical and cavitation effects.

Of the nonthermal effects arising from ultrasound, acoustic cavitation is perhaps the most widely studied and is associated with a variety of biological effects, including sonoporation, membrane permeability changes, and free radicals. (Miller et al. 1996) (Fig. 9.2). In this chapter, we will discuss the biological effects related to acoustic cavitation from the cellular to the tissue level.

9.1.2 Cavitation Mechanobiology

Mechanobiology is an interdisciplinary study that is concerned with the biological responses of cells to mechanical loads, and the mechanotransduction mechanisms by which these loads are transduced into a cascade of molecular and cellular events (Ingber 1998). Mechanical loads induce changes in the structure, composition, and function of living tissues. These changes are mediated at a cellular level and may cause physiological or pathological alternations in cell function (Wang and Thampatty 2006).

Human beings experience a wide variety of mechanical forces, with tissues growing and remodeling in response to changes in mechanical forces. The fundamental reason for this tissue response is that mechanical forces regulate some cell functions, such as gene induction, protein synthesis, cell growth, differentiation, and death. Moreover, several major cellular constituents are involved in mechanotransduction mechanisms, including the cytoskeleton, integrins, G proteins, receptor tyrosine kinases (RTKs), mitogen-activated protein kinases (MAPKs), and stretch-activated ion channels (Wang and Thampatty 2006). For example, mechanotransduction plays a major role in the extracellular matrix (ECM)–integrin–cytoskeleton pathway, as shown in Fig. 9.3. This is one of the most thoroughly studied signaling pathways (Juliano and Haskill 1993; Maniotis et al. 1997). Cells attach to the ECM via integrins, which are transmembrane proteins. The mechanical forces acting on the cell then introduce conformational changes to intracellular G proteins, which help modulate the mechanotransduction pathway (Hansen et al. 1994). Once the conformation of G proteins changes, a signaling cascade associated with cell growth is initiated. Shear stress in particular has been shown to activate G proteins (Gudi et al. 1996): It triggers a cascade of downstream signaling events,

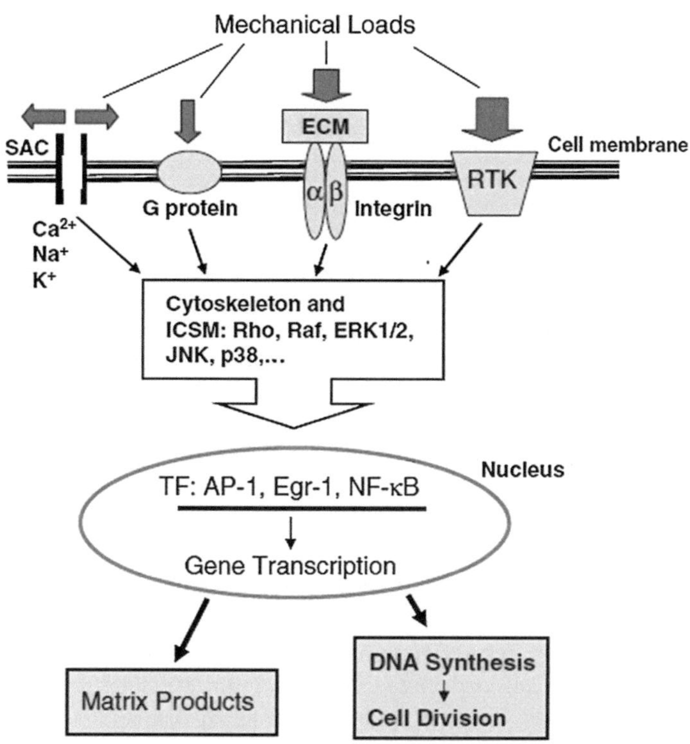

Fig. 9.3 A conceptual diagram of cellular mechanotransduction mechanisms. Reproduced with permission from Wang et al. © Springer

generating second messengers such as RTKs, a group of transmembrane proteins that play a major role in integrin-mediated signaling pathways. Overall, this structural connection involves transmitting mechanical signals from the ECM to the cell. Although it is well known that the cellular response to mechanical forces is closely associated with physiological and pathological changes to the organism, it remains unknown how the cell senses mechanical forces and converts these "mechanical signals" into biological responses.

Acoustic cavitation would introduce mechanical forces and initiate associated chemical reactions, on nearby molecules or cells. Although the forces associated with acoustic cavitation are different (e.g., in the time domain) from the mechanical loads that are often referred to in traditional mechanobiology, they can still be considered to be mechanical forces and can trigger similar cellular signaling cascades to the mechanotransduction mechanisms studied in mechanobiology research. Therefore, we call this phenomenon and its related mechanisms cavitation mechanobiology. However, the mechanical forces associated with cavitation are complicated and not easily controlled, thus producing more interesting cellular effects in vivo than traditional mechanical forces. In this chapter, we introduce cavitation mechanobiology and focus on its effects on tissue, cells, and molecules, and we discuss its potential molecular mechanisms and applications.

9.1.3 Physical Mechanisms

The vibration and collapse of acoustic cavitation bubbles in water near a surface can induce significant surface modifications in terms of mechanical damage. The vibrating/collapsing bubble and its resulting surface damage are related to a variety of physical phenomena. Stable oscillations may create significant liquid flow around microbubbles. This is called microstreaming (Rooney 1972) and results in shear stresses on cell membranes. As microbubbles grow rapidly and collapse, they may generate a local shock wave in the fluid (Lentacker et al. 2009). Finally, microjets can occur when a collapsing microbubble is located close to a surface, such as a cell membrane (Ohl et al. 2006) (Fig. 9.4). In addition to the shock waves and high local shear stresses associated with microstreaming and microjets, the asymmetric collapse of bubbles could also cause erosion and abrasion.

Microstreaming produces a bulk non-periodic motion of fluid; it is a nonlinear phenomenon that is directly related to the second-order terms of the sound field (Nyborg 2005). Cavitation microstreaming was initially proposed by Elder, who examined the effects around a bubble on a wall, at various oscillation amplitudes and fluid viscosities (Elder 1959). Subsequent studies examined a streaming field near a free gas bubble resting on a boundary (Miller and Nyborg 1978), and later researchers found that oscillations of free bubbles were sufficient to rupture lipid membranes (Marmottant and Hilgenfeldt 2003). These studies suggested that the main cause of this rupture was microstreaming generated by free bubble oscillation.

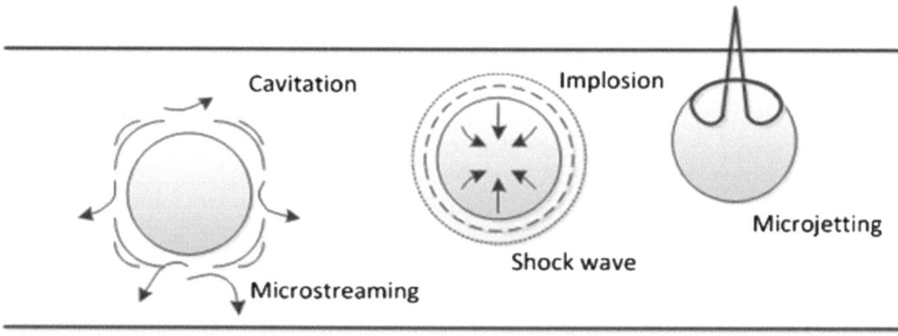

Fig. 9.4 Schematic representation of microstreaming, shock wave formation and microjets associated with cavitation bubble dynamics

Cavitation microstreaming plays a vital role in the therapeutic action of microbubbles driven by ultrasound, such as in sonoporation and sonothrombolysis. In reparable sonoporation, for example, cavitation microstreaming caused by ultrasound applied to a cell suspension can cause the cell membrane to temporarily "open" to small and large molecules and then "close" again when the ultrasound is removed (Wu et al. 2002). It has been shown that sonoporation could be caused by a spatial peak acoustic pressure amplitude as low as 0.1–0.2 MPa (1 or 2 MHz continues or pulsed ultrasound) in the presence of microbubbles (Optison® or Albunex®) in a cell suspension (Bao et al. 1997; Ward et al. 1999, 2000). Sonoporation was also studied in an in vitro experiment by Tran et al. (2007) who demonstrated that during sonoporation, the membrane of a marked cell (mammary breast cancer cell line MDA-MB-231) became hyperpolarized (1 MHz, 0.15 MPa negative peak pressure ultrasound, SonoVue® microbubbles). They concluded that ultrasound-activated oscillations of the microbubbles modified the cell's electrophysiological activities by tapping into their "cellular messaging," thus enhancing the cell's permeability for macroparticle uptake. Here, cellular messaging may have arisen from the moderate shear stresses generated by nearby oscillating bubbles, which is related to microstreaming around a microbubble.

The impact of a liquid microjet on a boundary can also explain many of the features associated with cavitation effects. When a bubble is located near a rigid boundary, a liquid microjet develops during the final stage of bubble collapse. The jet flow concentrates much of the bubble's energy at a small site far away from the initial location of the bubble. The migration of the bubble during collapse and thus the jet flow are both directed toward the boundary and increase the damage range in this direction (Brujan 2004). However, Chen et al. (2011) reported something different. They observed liquid jets with distention of blood vessels in which they were present (outward motion from the surrounding tissue) and vessel invagination (inward motion toward the lumen). They also proved that liquid jets were directed away from the nearest vessel wall during invagination, which led to more to the vessel damage (Chen et al. 2012).

In clinical situations, cavitation microjets are used for a number of medical procedures, such as ultrasound phacoemulsification (Steffen et al. 1994), dental tartar removal (Rink et al. 1995), and lithotripsy of renal stones (Philipp et al. 1993). However, cavitation microjets may also be associated with several damaging effects. Because the microjets generate a relatively high number of bubbles, interactions between a nearby acoustic transient signals emitted by collapsing bubbles and the jetting bubble may cause the jet velocity to accelerate (Philipp et al. 1993).

9.1.4 Chemical Mechanisms

Acoustic cavitation also produces chemical effects. It may create oxidizing species, such as $^{\cdot}$OH radicals and $^{\cdot}$H atoms, from the sonochemical decomposition of water (Suslick 1988). As a well-known example, short-lived oxygen-free radicals are formed when water breaks down inside a bubble (Mason et al. 2011). Upon bubble collapse, these radicals can be destroyed by reactions in the bubble or can migrate into the bulk solution and produce peroxide.

$$\text{H-O-H} \rightarrow \text{H}^{\cdot} + \text{OH}^{\cdot}$$

During the water decomposition process, the initial step is the production of hydrogen and hydroxyl radicals. Other species, such as hydrogen peroxide, singlet oxygen, and superoxide ions, are formed later, depending on the specific conditions.

Subsequently, the oxygen-free radicals may participate in a variety of cellular events, such as DNA degradation, inactivation of enzymes, lipid peroxidation, and cell apoptosis (Riesz and Kondo 1992). These free radicals are highly unstable and reactive because they have an odd number of electrons in the outermost orbit of the atoms, which can, when they come in contact with a protein, alter the charge distribution on the protein surface and thus cause it to lose activity (Ríos-Arrabal et al. 2013). Furthermore, in instances where ultrasound exposes microbubbles to form hydrogen peroxide, it is suggested that the hydrogen peroxides would cause these cell membranes to become transiently permeable (Juffermans et al. 2006).

Oxygen-free radicals, more generally known as reactive oxygen species (ROS), can be formed by acoustic cavitation. It is well recognized that ROS play a dual role as both deleterious and beneficial species in biological systems (Valko et al. 2006) (Fig. 9.5). This two-faced character of ROS is substantiated by a growing body of evidence that shows that in cells, ROS act as secondary messengers in intracellular signaling cascades; however, ROS can also cause damage to cellular structures, such as lipids and membranes, proteins, and nucleic acids. The cumulative production of ROS through either endogenous or exogenous insults is termed oxidative stress and commonly occurs in many types of cancer cells that have altered redox regulation of cellular signaling pathways, inducing a cellular redox imbalance. It

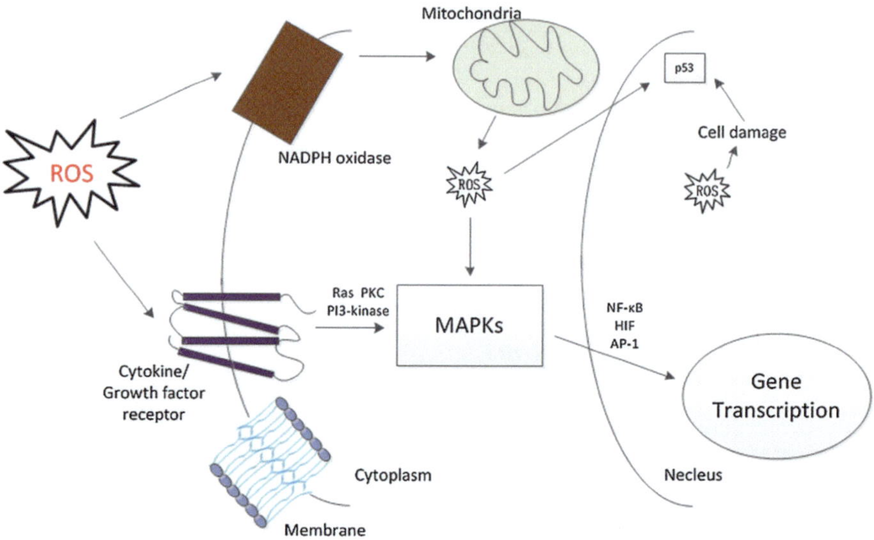

Fig. 9.5 The two-faced nature of ROS-induced cell signaling

has been found to be present in various cancer cells compared with normal cells. Thus, this redox imbalance may be related to oncogenic simulation.

ROS also act in all three stages of carcinogenesis: initiation, promotion, and progression (Klaunig and Kamendulis 2004) (Fig. 9.6). Initiation involves a non-lethal DNA mutation that produces a diseased cell followed by at least one round of DNA synthesis to fix the damage (Loft and Poulsen 1996). During this process, ROS can cause oxidative damage to the DNA. The next stage, promotion, is characterized by the clonal expansion of diseased cells by the induction of cell proliferation and the inhibition of cell apoptosis. This process results in the formation of an identifiable focal lesion. ROS production during this stage of carcinogenesis is the main mechanism for ROS-related tumor promotion (Dreher and Junod 1996). Finally, during progression, the final stage of carcinogenesis, ROS production may lead to genetic instability and disruption of chromosome integrity.

Although the aforementioned effects have not been specifically associated with the ROS formed by cavitation, ROS that arise from cavitation may potentially be damaging to cells via these effects which we include in the term cavitation mechanobiology.

9.1.5 Basic Applications

Ultrasound can induce many biological effects, which are closely associated with several important applications in biomedicine via acoustic cavitation. It is worth

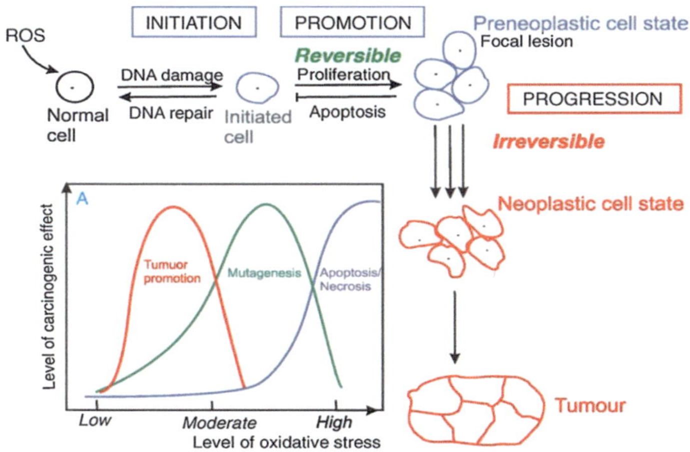

Fig. 9.6 The three stages of carcinogenesis and the level of carcinogenetic effect versus level of free radicals at each stage of carcinogenesis. Reproduced with permission from Valko et al. © Elsevier

noting that cavitation-related biological effects occur when bubbles reside near surfaces and as such have been modeled as bubbles surrounded by an infinite, homogenous liquid body (Kimmel 2006). These biological effects and their applications can be divided into four main classes:

I. Direct effects of mechanical forces on tissues

The mechanical forces induced during cavitation can directly interact with tissues. For example, they can interact with a thrombus in thrombolytic therapy, with diseased tissue or organs in histotripsy, with vessels in techniques designed to open the blood–brain barrier (BBB), with stones in lithotripsy, etc. (see Sect. 9.2).

In sonothrombolysis, ultrasound can aid the dissolution of blood clots when used in combination with microbubbles and fibrinolytic drugs (ter Haar 2007). In this situation, cavitation microstreaming may enhance the thrombolytic effect because inertial cavitation causes mechanical disruptions due to bubble rebound, microstreaming, and microjetting. This disruption may open pores and channels in the clot allowing drugs to more readily access the fibrin. In lithotripsy, there are several exiting theories for stone fragmentation, such as cavitation, tear and shear forces, spallation, quasi-static squeezing, and dynamic squeezing (Bustamante et al. 2013). In the theory of cavitation, the interaction of a high-amplitude acoustic shock wave with a bubble can result in the formation of a high-speed liquid jet, which concentrates part of the shock wave energy onto a small area. When the shock wave is concentrated on a stone, the stone's surface becomes deformed and eventually erodes.

II. “Pore” effects

Cavitation has another important effect in which “pores” are formed in nearby interfaces. In biomedicine, the pores can be formed in cell membrane. This is called sonoporation and is widely studied for gene or drug delivery therapy. Such pores are also implicated in sonophoresis applications on the skin to increase the penetration of drugs and for targeted anticancer therapy in tissues.

In sonophoresis, acoustic cavitation has been shown to enhance transdermal transport of various pharmacologically active drugs, including macromolecules (Mitragotri and Kost 2004). When acoustic cavitation occurs in the vicinity of living cells, the mechanical interactions between ultrasound and microbubble gas cavities exert a force that is substantial enough to create pores in the cell membranes. This sonoporation (see Sect. 9.3.1) is generally accepted as a transient phenomenon in which the cells’ transmembrane permeability is temporarily increased before the membrane pores reseal (Bao et al. 1997). Sonoporation has been considered as a potential strategy for facilitating cellular uptake of external molecules such as drugs, genes, and proteins (Mitragotri 2005; Newman and Bettinger 2007; Pua and Zhong 2009; Nomikou and McHale 2010). For applications in gene and drug therapy, acoustic cavitation causes bubbles to collapse, which may release energy that temporarily increases the internalization of exogenous genes or drugs into living cells. Because ultrasound can be focused almost anywhere in the body, this technique offers the additional possibility of noninvasive targeting of therapies to specific tissue or location.

III. Cavitation mechanobiological effects on cells

While cavitation produces effects on whole tissues, it also produces physical and chemical changes in cells. In response to these changes, cells can lose viability and undergo structural changes (see Sect. 9.3.2). Moreover, if subcellular organelles (mitochondria, endoplasmic reticulum, etc.) are damaged (see Sect. 9.3.4), this can trigger the cell to undergo one or more several death pathways, of which apoptosis and autophagy (see Sect. 9.3.3) are the most investigated. These two programmed cell death mechanisms offer the possibility of depressing cancer growth via triggering cellular signaling at different stages of cancer developments. The molecular mechanism of cavitation mechanobiological effects on cells, which might offer a therapeutic window for cancer therapy, has also been investigated (see Sect. 9.3.5).

IV. Cavitation on small molecules and macromolecules

Acoustic cavitation has also been used to extract small molecular substances and interact with macromolecules, including proteins and DNA (see Sect. 9.4). The effect on the structure and function of these molecules is another example of cavitation mechanobiology.

9.2 Mechanical Effects of Cavitation on Tissues

In cavitation mechanobiology, two main problems must be investigated. First, we must determine the physical mechanisms of the mechanical forces arising from cavitation, including microstreaming, microjets, shear stresses, and shock waves. With shock waves, for example, much work has already been carried out using both high-speed photomicrography and simulations to learn how they form and propagate, and to study the direction of the microjet (detailed information can be found in Chap. 5). The second issue, however, involves their biological effects. We need to understand how these mechanical forces act on tissue interfaces and affect the underlying tissue biology, especially when they are used for therapeutic applications. In this section, we introduce the potential therapeutic applications for cavitation and explain their mechanical effects on tissues.

9.2.1 Thrombolysis and Thrombolytic Therapy

Ultrasound has been shown to enhance thrombolysis when microbubbles are present, and this is typically carried out in conjunction with intravenous administration of thrombolytic drugs (Alexandrov et al. 2004; Molina et al. 2006; Unger et al. 2004). The first report that showed that ultrasound (26.5 kHz) on its own could recanalize thrombosed arteries was published in 1976 (Trubestein et al. 1976). Several later studies have confirmed this, showing that low-frequency (25–37 kHz) ultrasound and microbubbles alone can enhance thrombolysis (Birnbaum et al. 1998; Nishioka et al. 1997; Porter and Xie 2001). Further, Tachibana and Tachibana (1995) demonstrated that they could achieve enhanced fibrinolysis in vitro using ultrasound and perfluorocarbon-exposed sonicated dextrose albumin microbubbles.

Despite a number of successful studies using ultrasound on its own or with the addition of microbubbles (Rosenschein et al. 1997, 1999; Yock and Fitzgerald 1997), thrombolytic enhancement is also achieved when ultrasound exposures are combined with fibrinolytic drugs, such as streptokinase, urokinase, or tissue plasminogen activator (t-PA) (Daffertshofer and Hennerici 2003; Luo et al. 1996, 1998; Akiyama et al. 1998; Behrens et al. 2001). However, the mechanisms for this enhancement are poorly understood. It has been proposed that cavitation-induced microstreaming may break apart the clot and facilitate the permeation of the drug into the clot and that the mechanical action of the ultrasound affects the fibrin mesh, allowing better transit of the drug. Microstreaming occurring around oscillating in the vicinity of the cell membrane will also alter the cellular environment. When inertial cavitation occurs, mechanical disruption due to bubble rebound, involution, and microjetting may open pores and channels in the clot, thus allowing fibrinolytic drugs to directly access the fibrin (ter Haar 2007). Schäfer et al. (2005) examined the effect of varying ultrasonic operating parameters on thrombolytic enhancement.

They found that in the absence of drugs or microbubbles, thrombolytic efficiency increased with increasing exposure duration, intensity, and pulse length, but it was inversely dependent on frequency, such that lower frequencies were more effective than the higher frequencies of 2–5 MHz. Low ultrasound frequencies also allow increased penetration through the skull, which may be advantageous in stroke applications. An example of this was reported by Riggs et al. (1997), who found that ultrasound (2 W/cm^2) at 1 MHz actually produced prothrombotic effects.

With regard to ultrasound intensity, Nilsson et al. (1995) found that intensities of 0.5–1 W/cm^2 (0.5–2.3 MHz, 1 kHz pulse repetition frequency (PRF), and 10-μs pulse duration for 15–60 min) produced clot lysis, but an intensity of 4 W/cm^2 resulted in less clot lysis than when fibrinolytic drugs were applied alone. The most promising clinical results have resulted from the use of transcranial Doppler (ultrasound) with the drug t-PA (ter Haar 2007). Normally, t-PA is given intravenously to stroke patients, but recovery may be incomplete, due in part to inadequate thrombolysis. In one study for treatment of brain ischemia caused by thrombi, the combined lysis approach using transcranial ultrasound and systemic t-PA applied 2-MHz transcranial Doppler ultrasound with t-PA was applied 3 h after the onset of symptoms. Complete recanalization was achieved in 49 % of patients, compared to 30 % in patients receiving t-PA alone. At 3 months, patients who had been treated with ultrasound and t-PA had a more favorable outcome than the control group (Alexandrov et al. 2004).

9.2.2 Cavitation-Mediated Macromolecule Delivery

Ultrasound-mediated delivery of macromolecules has shown promise for facilitating the intracellular transport of genes, drugs, and other therapeutic agents, which are otherwise unable to permeate the cells (Tachibana 2004; Dalecki 2004). Traditional drug delivery methods, such as injections and oral medications, are often unsuitable for highly active therapeutic agents such as chemotherapeutic drugs and genes. However, on account of local drug administration that can minimize toxicity to normal tissues (Wang and Yuan 2006), ultrasound is now being used extensively for drug and gene delivery (Ng and Liu 2002). This is possible primarily because of ultrasound-induced acoustic cavitation, particularly inertial cavitation.

Cavitation can increase capillary permeability and delivery of materials to the interstitial tissue. It has been used to facilitate the delivery of small molecules, as well as macromolecules, including drugs and DNA, into tissue (Paliwal and Mitragotri 2006). Since ultrasound can be focused in many places in the body using an extracorporeal device, it offers the additional benefit of targeting therapies to specific tissues or locations in the body noninvasively (Schlicher et al. 2006). While ultrasound shows promise for delivering drugs to whole organs, it may also be suitable for cellular delivery. One of the greatest barriers to the delivery of genes, drugs, and other macromolecules into cells and tissue, the cell membrane can be reversibly disrupted by acoustic cavitation, to facilitate intracellular uptake of exogenous molecules (Bao et al. 1997; van Wamel et al. 2003). This process, known as "sonoporation," in which sound energy is used to enhance the permeability of plasma membranes through the creation of pores, facilitates intracellular transport of drugs, genes, and other therapeutic agents that cannot otherwise get into cells (Fechheimer et al. 1986). The ability of ultrasound to disrupt these barriers transiently may be an important tool, especially for delivering macromolecules into the cytosol.

Whereas non-inertial cavitation can be used to alter vascular permeability to give increased extravasation and hence improve drug delivery to whole tissues, inertial cavitation can be used to alter the permeability of individual cells. Although sonoporation is reversible and therefore nondestructive, the activity of cavitation (especially inertial cavitation) is non-uniform and difficult to control. Ultimately, some of the exposed cells will experience irreversible damage that leads to cell death (Frenkel et al. 1999).

Cavitation may also impart a ballistic effect in blood vessels to drive delivered drugs or genes from the vasculature into or through the vessel wall, allowing for extravasation of the pharmaceutical agent (Price et al. 1998). During this process, an ultrasound contrast agent with an attached or incorporated bioactive substance is introduced into the vasculature and thus distributed throughout the capillaries. Ultrasound can then destroy the microbubbles in the target region, releasing the transported substance into the surrounding tissue.

Microbubbles are uniquely suited for local drug/gene delivery because they can be forced to oscillate within a region with instantaneous dimensions as small as hundreds of microns, or as large as centimeters (Ferrara et al. 2007). If these bubbles can be made to accumulate within a target volume, ultrasound can destroy them locally and release the therapeutic agents. Several modes have been investigated for increasing local drug delivery (Fig. 9.7). For example, drugs can be incorporated into the microbubbles and then released locally by rupturing the microbubbles selectively in the desired vessels (Chen et al. 2006) (Fig. 9.7b). Targeted microbubbles can also be combined with ultrasound exposure to promote selective drug delivery locally in areas showing surface marker expression (Tartis et al. 2006) (Fig. 9.7c). In general, this approach is most likely to be effective for highly active macromolecules that do not require large concentrations for therapeutic efficacy, such as cytokines, growth factors, genes, and chemotherapeutic agents. Finally, another option is that microbubbles can be used alongside drugs to aid in delivery (Price et al. 1998) (Fig. 9.7a). This mode has been used in an attempt to facilitate the transport of drugs across the BBB (Hynynen et al. 2006).

9.2.3 Gene Therapy

The goal of gene therapy applications is to transfer genetic material to the intended target, where the transgenes will either alter the phenotypes or produce local

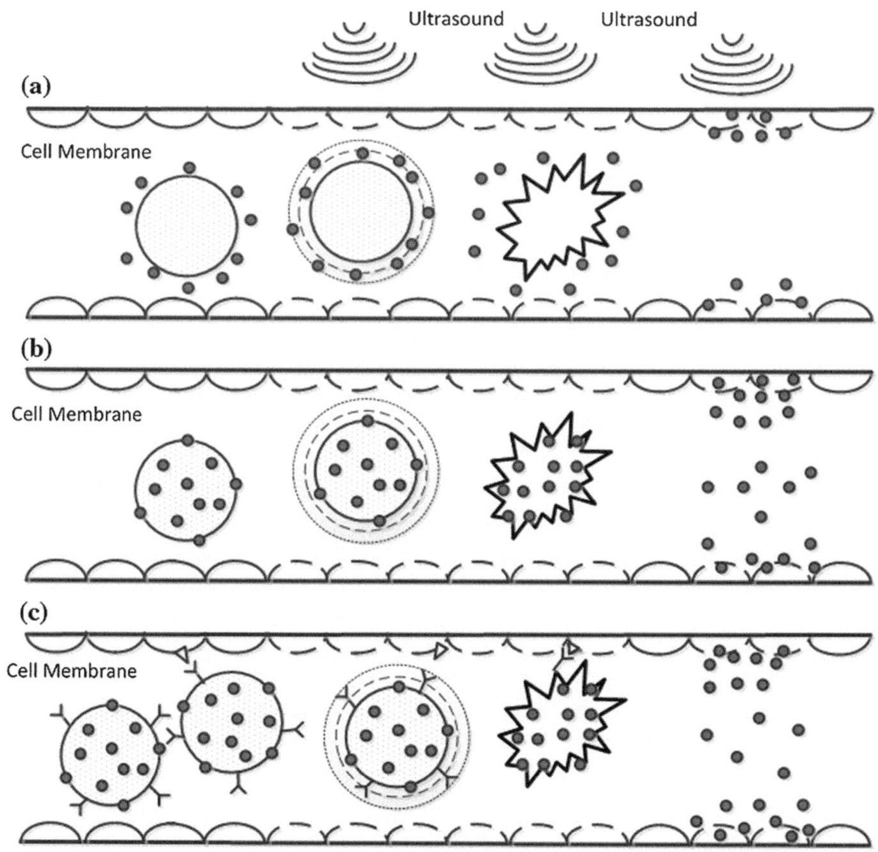

Fig. 9.7 Schematic representation of local drug delivery using ultrasound microbubble destruction. **a** Microbubbles are freely circulating in small vessels along with drug/gene. Once sufficient ultrasound energy is applied to the targeted area, and the microbubbles rupture the endothelial lining. **b** Drug-loaded microbubbles are freely circulating throughout the vasculature. An ultrasound wave is applied and ruptures the microbubbles, thereby liberating the drug payload. Because ultrasound is only applied in the region of interest, the drug is preferentially delivered locally. **c** Drug-loaded microbubbles bearing surface ligands targeted to specific endothelial receptors are freely circulating. The ligand preferentially binds the microbubble in the target region, increasing local agent accumulation. An ultrasound wave is then applied to liberate the drug payload

therapeutic agents, depending on the treatment (Hallow et al. 2006). There is considerable interest in facilitating the transfer of genes into diseased tissues and organs. Meanwhile, ultrasound-enhanced gene therapy is a rapidly evolving field. Cavitation plays an important role in the ultrasound-enhanced delivery of gene-based therapies.

The ideal system would enhance gene expression in the target while having no effect in nontarget tissues. Significantly, better transfection is achieved in the presence of cavitation. Enhanced gene transfer is found either when the exposed bubbles are in the vicinity of the genetic material, or when the genes are encapsulated within, or bound to, the bubbles. All of these strategies have been investigated experimentally in vitro (Frenkel et al. 2002; Lawrie et al. 2000; Miller et al. 2003) and in vivo (Shohet et al. 2000; Tachibana and Tachibana 1999; Manome et al. 2000; Taniyama et al. 2002a, b).

The strategy for enhancing gene transfection is to administer DNA systemically, followed by ultrasound exposure of a target tissue or organ. Such a mode has been proposed for treating cardiovascular disease, where transthoracic ultrasound focused on the heart could assist in the delivery of genes to increase vascularization of the myocardium in order to ameliorate congestive heart failure (Unger et al. 2001). Increased gene delivery can also be achieved where DNA is introduced into a large blood vessel and then followed by ultrasound exposure. One study showed that in an isolated blood vessel, combining DNA with microbubbles enhanced local gene expression (Hashiya et al. 2004).

Non-specific regional delivery of adenoviral and plasmid receptor genes has also been achieved using ultrasound-directed methods. These vectors can easily be attached to microbubble shells, thus facilitating coinjection. Several studies have developed techniques for incorporating DNA into lipid microbubble shells with local transfection after intravenous injection and insonation. Lawrie et al. (2000) showed enhanced transgene expression of approximately 300× after naked DNA transfection when microbubbles were present (956-kHz ultrasound, 2.0 mechanical index, 6 % duty cycle with or without 10 % microbubbles for 60 s), compared to a 10× enhancement in their absence. Tachibana and Tachibana (1999) also showed that this method enhanced transfection efficiency of plasmid DNA into blood vessels without any apparent toxicity. Moreover, Frenkel et al. (2002) showed that plasmid DNA attached to albumin microbubbles (20 kHz, 8 s) gave 2× more enhancement of gene expression and transfection than when the DNA and the albumin microbubbles were not bound together. Shohet et al. (2000) injected perfluoropropane microbubbles into rats. The bubbles were bound recombinant adenovirus and contained the complementary DNA encoding the β-galactosidase gene. In rat hearts, after exposure to 1.3-MHz pulsed ultrasound (transmit 1.3 MHz/received 2.6 MHz in second harmonic mode, mechanical index 1.5, and a depth of 4 cm) for 4 days, a 10× enhancement of β-galactosidase was found. In addition, microbubbles carrying plasmids encoding VEGF have been used to induce angiogenesis in the myocardium of rats after ultrasound exposure (20 kHz, 16 s) (Korpanty et al. 2005a). Co-injection of plasmid DNA and microbubbles into the renal artery produced local gene expression in the kidney when combined with transient vascular compression and ultrasound (1-MHz ultrasound at 5 % power output, for a total of 60 s in 30 s intervals) (Koike et al. 2005). In another example, coinjection of plasmid DNA and microbubbles into the cerebrospinal fluid, coupled with ultrasound (1 MHz, 20 % duty cycle, and 0.5 W/cm^2), produced DNA transfer into the rat central nervous system (Shimamura et al. 2005).

Moreover, to overcome the low efficiency of traditional microspheres caused by its negatively charged characteristics, Tiukinhoy et al. (2000) developed a positively charged liposome with ultrasonographically detectable echogenic properties. Using an intravascular ultrasound system, they were able to deliver and detect luciferase gene expression in HUVEC cells following ICAM-1 targeted, ultrasound-directed gene transfection.

9.2.4 Anticancer Therapy

Cancer is a leading cause of death in most countries. For most advanced cancers, chemotherapy remains the most common and effective form of treatment. However, while the inherently toxic nature of chemotherapy drugs is essential for cancer cell killing, this is also why patients experience detrimental side effects. In addition, chemotherapy agents are effective for killing tumor cells in monolayers that are grown in culture, and they are unable to reach all tumor cells in vivo (Minchinton and Tannock 2006). Many factors have been identified in the microenvironment of tumors that are responsible for non-uniform and inadequate levels of anticancer agents being delivered.

For years, lipid-soluble anticancer agents have been incorporated into delivery vehicles so as to avoid systemic toxicity. Several non-viral approaches have been proposed for drug and gene delivery, such as electroporation, chemical methods, liposomal delivery, and gene gun-mediated transfer. These techniques show potential for drug and gene delivery, but they do not offer delivery that is as specific or efficient as desired. Ultrasound-mediated techniques may fill this gap, however. In this approach, ultrasound exposure is used to induce acoustic cavitation, which could cause sonoporation which transiently increases cell membrane permeability and facilitate mechanical delivery of macromolecules, such as drugs and genes, to cells (Larina et al. 2005). In vivo applications of this method are highly attractive because ultrasound exposure can be focused on many parts of the body.

In ultrasound, acoustic cavitation is responsible for releasing drugs from their carriers. Preclinical studies have shown promising developments for systemically administered drug carriers, where extracorporeal ultrasound exposures are used to create reversible changes in the carriers and consequently release the chemotherapeutic agents that they incorporate. The types of carriers used include microbubbles, polymeric micelles (Husseini et al. 2007), and liposomes (Allen 1997). The obvious benefit of using this type of drug/device combination is that it increases local drug delivery in a targeted region for improved efficacy, while concomitantly minimizing systemic exposure to the drug and the subsequent side effects associated with its use. As described above, it is now possible to incorporate hydrophobic agents into the outer lipid layer of the microbubbles (Shortencarier et al. 2004) or to attach hydrophilic molecules to the bubble shell (Eder et al. 2005). Alternatively, it is possible to immerse a hydrophobic drug into an oil layer of acoustically active liposomes (AALs) (Tartis et al. 2006). Toxicity studies indicate that a 10× reduction in toxicity was achieved when AAL-encapsulated paclitaxel was systemically administered to mice (compared to unencapsulated paclitaxel).

To reach the cancer cells in a tumor, the therapeutic molecules must make their way into the blood vessels of the tumors and across the vessel wall into the interstitium, and finally migrate through the interstitium. Compared to normal tissues, blood vessels in tumors are leaky, possessing large gaps between endothelial cells (McDonald and Choyke 2003). Ultrasound combined with microbubbles could significantly affect the integrity of the endothelial monolayer by destabilizing its tight junctions. In addition, angiogenesis is an essential process for the growth and spread of cancer (Folkman 1986). Thus, endothelial cells and receptors associated with angiogenesis are readily accessible to intravascular agents with specific ligands. Microbubble surfaces have been conjugated to several key ligands for angiogenesis in order to target the microbubbles to sites of angiogenesis. For example, monoclonal antibodies and RGD peptides (Leong-Poi et al. 2003) targeted to endoglin (CD105) (Duff et al. 2003) and integrin $\alpha_v\beta_3$ (Nisato et al. 2003), respectively, and incorporated with the avidin–biotin system (Korpanty et al. 2005b) have been conjugated to the surface of microbubbles to target them to where angiogenesis is occurring. These microbubbles used in conjunction with ultrasound could be introduced as anticancer therapies in the future.

9.2.5 BBB and Neurodegenerative Disorder Therapy

Neurodegenerative diseases are difficult to treat, partly because of the BBB, a composite of highly specialized endothelial and perivascular structures that limit transport of molecules into the brain. More recent developments for opening the BBB have shown that the stably oscillating cavitation bubbles generate mechanical stress in the adjacent blood vessel walls while causing little or no damage to brain tissue. The bubbles increase vascular permeability both through structural and physiological processes (Hynynen 2007). As a consequence, the BBB is reversibly disrupted, allowing transit of therapeutic agents. This might be particularly beneficial for treating Alzheimer's disease, a neurodegenerative disorder typified by the accumulation of misfolded proteins—beta-amyloid and hyperphosphorylated tau—which aggregate and form amyloid plaques and tau tangles (Ingelsson and Hyman 2002). Many therapeutic agents for reducing amyloid plaques have limited efficacy in vivo because of poor transport across the BBB. However, Raymond et al. (2008) found that low-intensity ultrasound administered with microbubbles can transiently disrupt the BBB, and they further showed that ultrasound-mediated transient BBB disruption allowed noninvasive and localized delivery of both therapeutic and molecular imaging agents directly to amyloid plaques.

9.2.6 Ultrasound Tissue Erosion and Histotripsy

There are two general mechanisms by which focused ultrasound damages biological tissues and promotes cell death (Parsons 2006). One is thermal ablation, which occurs when tissues in the focal volume absorb acoustic energy. Thermal ablation can either be used in hyperthermia, by producing a low and sustained temperature rise, or in HIFU therapy, where rapid heating can be used to induce nearly instantaneous tissue coagulation. The other mechanism by which ultrasound damages tissue is through mechanical ablation, which can arise from non-cavitational forces (e.g., radiation forces or acoustic microstreaming) or direct cavitational activity, leading to severe physical disruption of tissue microstructure (e.g., cytorrhexis).

In mechanical ablation, tissue erosion is achieved in a localized and controlled manner by very short, relatively intense ultrasound pulses with optimum PRF. The mechanism of this action is closely related to cavitation (Xu et al. 2004). Normally, cavitation refers to the inception and the subsequent behavior of gas or vapor cavities (microbubbles) that are either preexisting or arise due to the large tensile forces produced by ultrasound in a tissue volume. These microbubbles are in the order of several microns to tens of microns in diameter at typical therapeutic ultrasound frequencies (0.5–5 MHz). They can be produced endogenously in tissues or introduced exogenously by using ultrasound contrast agents. Then, during pulsed ultrasound, the microbubbles undergo inertial cavitation and collapse. These microbubble collapses are high-energy events that can generate a wide range of potentially destructive effects. Thus, cavitation, especially inertial cavitation, is considered the primary mechanism for tissue erosion, and the efficiency of tissue erosion depends on the dynamics of the microbubbles.

High-intensity pulsed ultrasound can produce significant mechanical tissue fractionation with sharp boundaries, a process called histotripsy. At a tissue–fluid interface, histotripsy produces clearly demarcated tissue erosion depending on the pulse parameters (Xu et al. 2007). The pulse parameters determine the dynamic patterns of the microbubble cloud formed during histotripsy. For example, Xu et al. (2007) proved that the collapse cycle of the bubble cloud—that is the time for the bubble cloud to grow and then collapse—increases with increases in certain factors, such as pulse duration, peak rarefactional pressure, and gas concentration. Another factor, the PRF, also affects the collapse cycle such that it increases with the PRF only when the next pulse arrives after the first bubble cloud has collapsed. If the next pulse arrives before the first bubble cloud has fully collapsed, for example when the PRF is too high, the previous cloud does not collapse effectively. Another issue affecting the collapse cycle of the bubble cloud is the tissue interface: The collapse cycle is longer at a tissue–water interface than in free water. Maxwell et al. (2011) had discussed the origin of these bubble clouds. They found that bubble clouds are initiated from the single cavitation bubbles that are formed during the initial cycles of the ultrasound pulse and that they grow along the acoustic axis in a direction opposite to that of propagation. They therefore hypothesized that the bubble clouds form as a result of the large negative pressure generated by the backscattering of the shock waves from a single bubble.

Histotripsy has been used to perform noninvasive controlled tissue erosion for a variety of conditions. It has been used to treat hypoplastic left heart syndrome by perforating the neonatal atrial septum (Xu et al. 2004), in rabbit kidneys (Roberts et al. 2006) and in prostates (Roberts 2014), and has also been used for thrombolysis (Maxwell et al. 2009, 2011). Recently though, some new experimental therapies for noninvasive focused ultrasound therapies have been put forward that are based on histotripsy. For example, in boiling histotripsy, shocked millisecond-length pulses are applied to achieve mechanical disintegration of a targeted tissue. Here, the localized delivery of high-amplitude shocks causes rapid heating, resulting in boiling of the tissue (Maxwell et al. 2014). Undoubtedly, histotripsy has become a potential therapeutic approach with extremely high promise. This is discussed in more detail in Chap. 5.

9.3 Cavitation and Cells

Exposure to ultrasound has long been known to produce a variety of biological effects in cells. These effects have been extensively described for enhancing the delivery of local energy to cells and for inducing cell lysis, membrane disruptions, and lipid peroxidation (Frenkel 2008). To induce the cellular responses, microbubbles are often added to enhance the cavitation activity and to reduce the intensity threshold at which cavitation will occur. Cavitation bubbles may undergo sustained growth and oscillations over several acoustic cycles (non-inertial cavitation) or violent growth and collapse in less than a cycle (inertial cavitation). In a biomedical context, both inertial cavitation and non-inertial cavitation generate local fluid microstreaming and high-speed fluid microjets that can exert significant mechanical forces on nearby cells, leading to disruption of the cell membrane (sonoporation), changing the cell viability, or even causing the cell death.

9.3.1 Sonoporation

We know that ultrasound can transiently enhance the extracellular membrane permeability. This mechanobiological process is called sonoporation (Park et al. 2011) and is probably caused by acoustic cavitation and associated microstreaming and concurrently generated shear stresses. Despite its applications in drug/gene delivery and therapy, sonoporation can also have significant effects on cell viability and cell death. While the temporary opening of cell membrane (which later reseals to its original state) is called reparable sonoporation (Ward et al. 2000), the term lethal sonoporation refers to the case when a cell suffers lethal damage with its membrane becoming permanently open. Wu et al. (2002) found that the

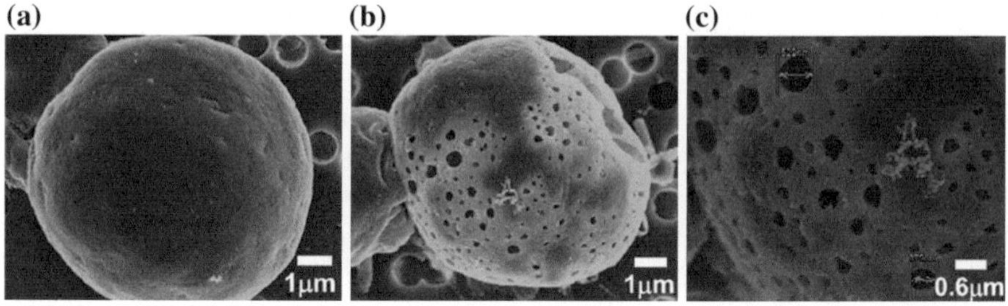

Fig. 9.8 Scanning electron microscopy images of HL-60 leukemia cells. **a** Sham exposure. **b** Three-second ultrasound exposure (1 MHz center frequency, 10 % duty cycle, 1 kHz pulse repetition frequency, 0.5 MPa peak negative pressure, 1 % microbubbles). **c** The enlarged (**b**), the largest holes have a diameter about 0.5 μm. Reproduced with permission from Zhong et al. © Elsevier

mechanisms required to produce reparable and lethal sonoporation are similar, but they differ in the severity of the damage incurred by the cell membranes. When the damage is minor, cells can repair the membrane. However, when the damage is beyond a cell's self-repair capacity, they suffer lethal sonoporation and often undergo cell lysis.

Sonoporation of the cell membrane is often examined using scanning electron microscopy (SEM). Figure 9.8 shows three scanning electron micrographs of representative HL-60 leukemia cells that were exposed to ultrasound. It can be seen that for the sham exposure, the cell membrane has remained intact and is spherical (Fig. 9.8a). In contrast, when the cell was fixed immediately after a 3-s ultrasound exposure, the SEM photographs show that the cell has randomly positioned pores on its membrane (Fig. 9.8b, c).

9.3.2 Cell Viability

Ultrasound can alter cell viability. This is associated with the physical stress created by acoustic cavitation. Mechanical wounding of cells and their associated resealing capacity are important processes. For example, in one study, the effective wound radius was found to be as large as 150–300 nm for all test molecules immediately after exposure to acoustic cavitation (Zarnitsyn et al. 2008). In general, ultrasound has been shown to cause cellular damage, cell lysis (Rapley and Walker 1998), loss of viability (Riesz and Kondo 1992), loss of colony-forming ability (Kondo and Kano 1987, 1988) and growth retardation (Jernberg et al. 2001). The emissions generated by acoustic cavitation may cause the disruption of the cell machinery (Vivino et al. 1985). For example, because it is a self-assembled structure, the lipid bilayer of the cell membrane can easily be disordered. Such disordering can result in increased membrane permeability (Liebeskind et al. 1981). Cells exposed to ultrasound, specifically where there is acoustic cavitation, may experience transient breaches to their plasma membrane that allow transport of exogenous materials into cells, as previously mentioned, but acoustic cavitation can also lead to a number of other effects such as membrane blebbing, perikarya formation, nuclear ejection, and instant cell lysis, depending on the degree of wounding incurred (Schlicher et al. 2010) (Figs. 9.9 and 9.10). Furthermore, sonoporation may cause various cellular biological effects, such as an increase in free radical formations, alteration of the cell membrane potential, and modulation of intracellular calcium ion signals.

In addition to harming the plasma membrane, ultrasound can also affect migration by disrupting the cell adhesion. Cell adhesion, on which migration depends, is strongly influenced by the integrity of the cytoskeleton–focal contact complex (Ruoslahti 1991). Recent studies indicate that therapeutic ultrasound treatment could disrupt the cytoskeleton–focal contact complex, affect the integrin-mediated adhesion and migration, and cause the structural changes to the cytoskeleton (Belkin et al. 1990; Juliano and Haskill 1993). In addition, one study offered evidence that ultrasound-induced cavitation microstreaming can cause

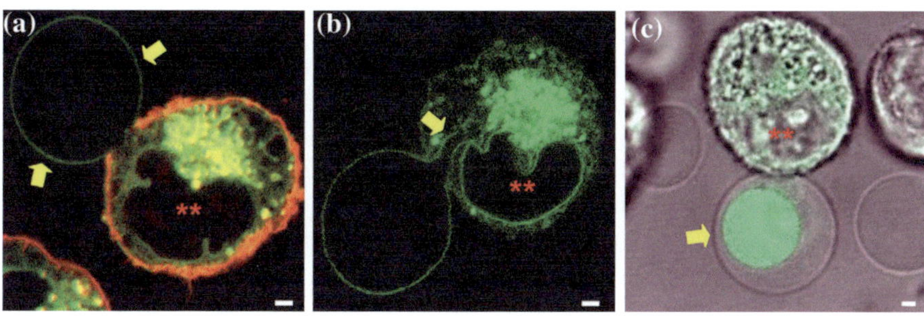

Fig. 9.9 Cellular membrane blebs created by cavitation wounding (*arrowed*) (24 kHz, 20 acoustic pulses, 0.1 s pulse length at 10 % duty cycle). Reproduced with permission from Schlicher et al. © Elsevier

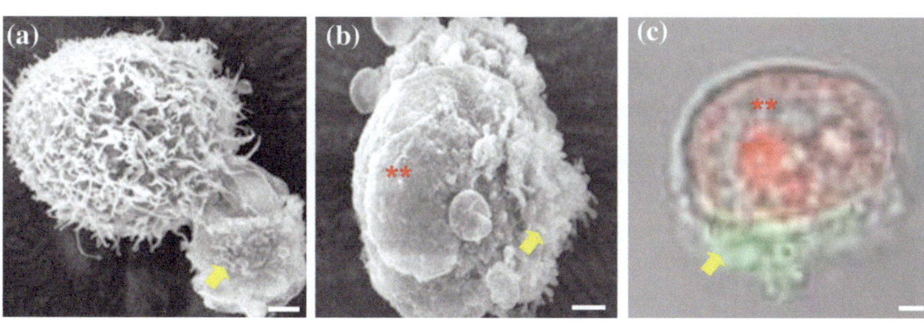

Fig. 9.10 Formation of cavitation wound-derived perikarya after plasma membrane damage (24 kHz, 20 acoustic pulses, 0.1 s pulse length at 10 % duty cycle). **a** Scanning electron microscopy (*SEM*), *arrow* indicates cytosol released from cell. **b** SEM, *double asterisk* indicates nucleus. **c** Bright field microscopy (*BFM*), *arrow* indicates plasma membrane (*PM*) remnants, *double asterisk* shows propodium iodide (*PI*)-stained nucleus. Reproduced with permission from Schlicher et al. © Elsevier

F-actin cytoskeleton rearrangement and calcium influx in human umbilical vein endothelial cells (Juffermans et al. 2009). Another study showed that ultrasound also inhibits the motion of smooth muscle cells (SMC) and reduces their ability to adhere to plastic or fibronectin-coated culture dishes (Alter et al. 1998). Interestingly, their ability to migrate decreases significantly a few minutes after ultrasound exposure. These results suggest that ultrasound could be used to decrease the tumor migration.

9.3.3 Cell Death

Therapeutic ultrasound can be used to modify the membrane structure and functional properties of cells and induce irreversible cell damage (Miller et al. 1996; Vivino et al. 1985). Acoustic cavitation resulting from acoustic pressure-induced collapse of microbubbles may be a critical cause for cell damage following ultrasonic exposure. Because acoustic cavitation involves physical wounding of the cell's structure, it may disrupt the homeostasis of living cells, thereby decreasing their viability and leading to the programmed cell death.

There are three main kinds of cell death: apoptosis, autophagy, and necrosis (Edinger and Thompson 2004) (Fig. 9.11). Apoptosis (type I programmed cell death) is the best characterized. This is a mechanism of cellular self-destruction, having uniquely defined morphological and molecular characteristics (Cohen 1993). Apoptosis ensures tissue homeostasis during development, host defense, and aging and occurs in response to a large variety of stimuli including γ-irradiation and

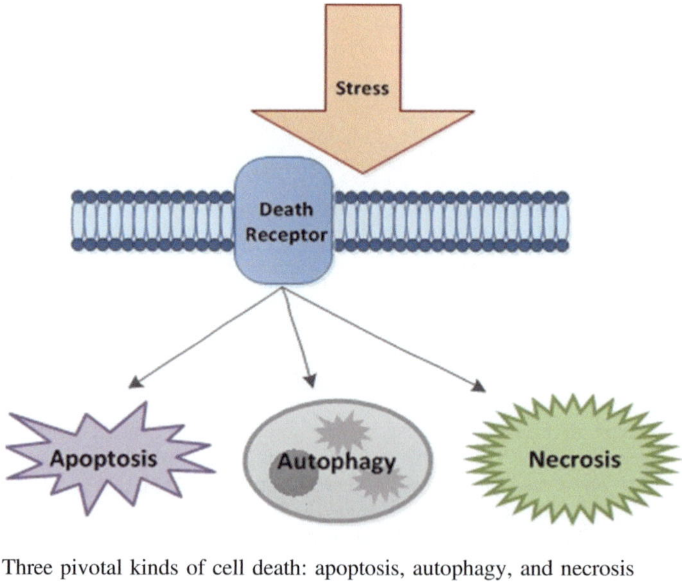

Fig. 9.11 Three pivotal kinds of cell death: apoptosis, autophagy, and necrosis

ultrasound exposure (Kulms and Schwarz 2000; Belyaev et al. 2009; Carloni et al. 2001). The morphological criteria that characterize apoptosis include cell shrinkage, membrane blebbing, chromatin condensation, nuclear fragmentation, and apoptotic body formation (Honda et al. 2004). In contrast, autophagy (type II programmed cell death) is the selective degradation of intracellular targets and sometimes leads to cell death. Autophagy, a process where cells "eat" themselves, is a major housekeeping mechanism that facilitates recycling of cellular materials and exploits them for energy production during emergency conditions and starvation. Unlike apoptosis, autophagic cell death is morphologically defined as a type of cell death that occurs in the absence of chromatin condensation but accompanied by massive autophagic vacuolization of the cytoplasm. In addition, necrosis is a form of cell injury that results in the premature death of cells by autolysis. It can be caused by infection and toxins.

9.3.3.1 Apoptosis

The induction of cell apoptosis is a goal of some therapeutic cancer applications, and so identifying how ultrasound affects apoptosis induction is growing in importance, especially now that using ultrasound for cancer therapy is being extensively studied. Normally low-energy ultrasonic treatments, for example, with the parameters Lagneaux et al. (2002) used (1.8 MHz, 7 mW/ml, 5.5/3 ms cycle), can cause a sequence of events recognized as hallmarks of apoptosis: a drop in mitochondrial potential, loss of phosphatidylserine asymmetry, morphological variations, DNA fragmentation, and finally, loss of the plasma membrane (Fig. 9.12). Moreover, low-energy ultrasound can modulate the Bcl-2/Bax ratio and

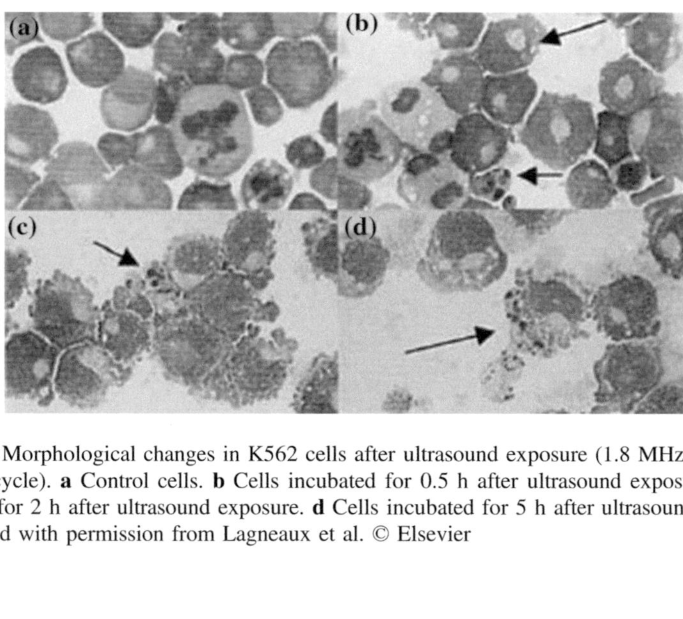

Fig. 9.12 Morphological changes in K562 cells after ultrasound exposure (1.8 MHz, 7 mW/ml, 5.5/3 ms cycle). **a** Control cells. **b** Cells incubated for 0.5 h after ultrasound exposure. **c** Cells incubated for 2 h after ultrasound exposure. **d** Cells incubated for 5 h after ultrasound exposure. Reproduced with permission from Lagneaux et al. © Elsevier

activate caspase-3, and this results in proteolytic degradation of the caspase substrate. Furthermore, after ultrasound exposure, the intracellular glutathione level is significantly decreased, suggesting that oxidative stress may play an important role in ultrasound-induced apoptosis. Ultrasound has also been shown to activate the mitochondria-mediated endogenous apoptosis pathway (Honda et al. 2002).

In vitro studies (750 kHz, 103.7 and 54.6 W/cm^2) have shown that therapeutic ultrasound-induced apoptosis is possible in cultured cells by the process of cavitation (Ashush et al. 2000). For example, inertial cavitation can cause cell lysis by producing shear stresses that disrupt the cell membranes. These changes in the cell membrane are some of the earliest signs of cellular apoptosis. Miller et al. reported that the chemical and biological effects of inertial cavitation can be enhanced by using ultrasound contrast agents, which provide nuclei for inertial cavitation (Miller and Thomas 1995). While, after inertial cavitation, cell membranes are damaged, acoustic cavitation (1 MHz, 0–6.5 W/cm^2) also promotes early apoptosis and secondary necrosis (Honda et al. 2002).

For example, Zhong et al. (2011) found that cells undergoing sonoporation may commit apoptosis and cell-cycle arrest through cytoplasmic signaling (Fig. 9.13). They found that the proliferation of sonoporated cells is suppressed temporarily due to the disrupted expression of checkpoint proteins (different forms of Cdk and cyclin) that are crucial for cells to progress through the cell cycle. Tian et al. (2005) found that low-intensity ultrasound exposure (1 MHz, 2.0–17.0 W/cm^2 for 10 min) up-regulates cellular tumor antigen protein 53 (p53) and Bid expression but decreases Bcl-2 levels. These ultrasound-induced key proteins are involved in two

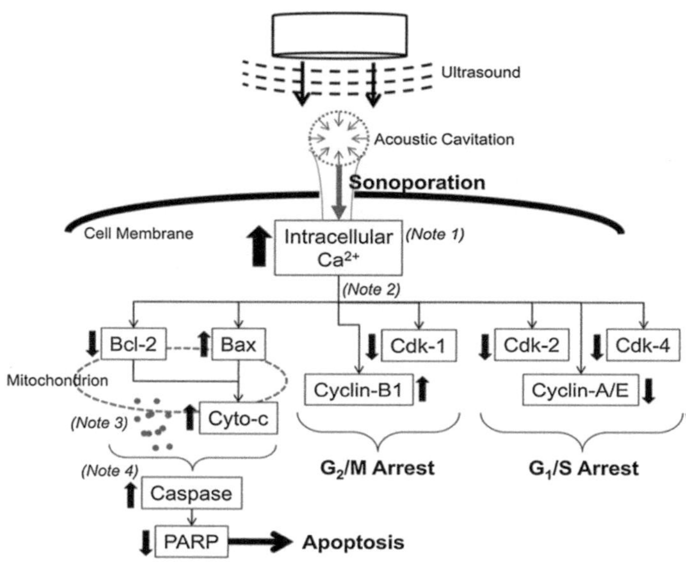

Fig. 9.13 Possible model for cytoplasmic transduction of sonoporation-induced cell apoptosis and cell-cycle arrest [Yu et al.]. Reproduced with permission from Zhong et al. © Elsevier

potential molecular signaling pathways associated with mitochondria and intracellular oxidative stresses, respectively. During apoptosis initiation, p53 helps activate a series of proapoptotic proteins, such as Bid, and induces mitochondrial membrane permeabilization, which promotes the release of apoptotic proteins such as cytochrome C to the cytoplasm. In addition, HO1, HSP70, HSP40, and OSR1 are up-regulated after low-intensity ultrasound exposure. Among these proteins, heme oxygenase 1 (HO1) is predominantly induced under various oxidative stress conditions, and its up-regulation may serve as an adaptive factor to protect cells from oxidative stress and apoptosis (Abraham et al. 2003). Recently, Zhong et al. (2013) found that ultrasound cavitation may upset ER homeostasis via sonoporation, and it may ultimately result in the initiation of mitochondria-mediated apoptosis. They observed that the cells exhibited a progressive loss of functional ER mass after exposure to ultrasound cavitation. In addition, they found the expression changes of ER stress relevant proteins, such as up-regulation of ER-resident enzymes responsible for catalyzing protein folding, activation of trans-ER-membrane stress sensors, and increased expression of ER-induced regulatory proteins that mediate proapoptotic signals to the mitochondria.

9.3.3.2 Autophagy

Autophagy is an evolutionarily degradative mechanism that is essential for growth regulation and maintenance of homeostasis in all eukaryotic cells (Li et al. 2009; Ito et al. 2007). It is a process whereby the cytosol and organelles become sequestered into double-membrane vesicles called autophagic vacuoles (AVOs) or autophagosomes. Autophagosomes ultimately fuse with lysosomes, thereby generating single-membrane autophagolysosomes and degrading their contents (Klionsky and Emr 2000; Levine and Klionsky 2004). Thus, the main role of autophagy is to promote cell survival. In cancer treatment, autophagy is a programmed cell death pathway that describes a cellular response to various cancer therapies. However, whether autophagy occurs in response to therapies in order to promote death or to promote survival is still uncertain and likely depends on the specific stimuli under given conditions.

In 2006, Kessel et al. reported that both apoptosis and autophagy can occur after photodynamic therapy (PDT), where autophagy was associated with cell survival at low levels of PDT and became a cell death pathway at higher PDT doses (Kessel et al. 2006; Kessel and Oleinick 2009). These results indicate that autophagy can become a cell death pathway if apoptosis is inhibited or when cells attempt to recycle damaged constituents beyond their capability for recovery. Recently, It has been found that autophagy participates in sonodynamic therapy (SDT)-induced cell death (Wang et al. 2010) (Fig. 9.14). In addition, Wang et al. (2011) observed autophagosomes using transmission electron microscope (TEM) in CNE2 cells treated with low-intensity ultrasound (1.7 MHz, 1.35 W/cm^2) (Fig. 9.15). However, these results have not yet been observed for autophagy induced by acoustic cavitation.

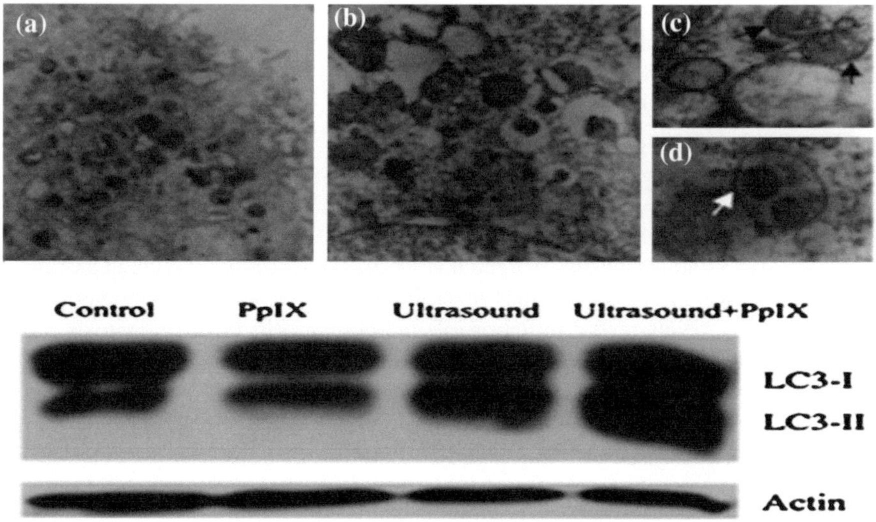

Fig. 9.14 Detection of SDT-induced autophagy in S180 cells. **a** Representative electron micrographs of autophagic cells at 1 h post-treatment. The control cells have very few autophagic vacuoles. **b** Western blot analysis of LC3-I conversion to LC3-II at 1 h after exposure. Reproduced with permission from Wang et al. © Elsevier

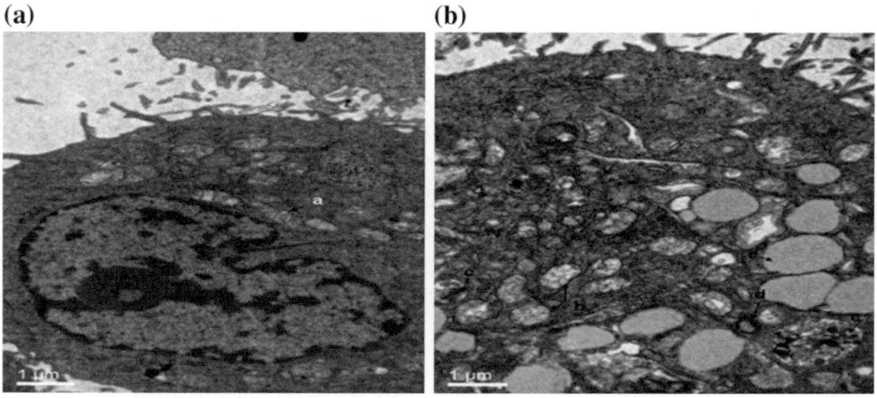

Fig. 9.15 Investigation of ultrastructural morphology and autophagy of CNE2 cells after ultrasound treatment. **a** Normal mitochondria. **b** Swollen mitochondria. Reproduced with permission from Wang et al. © Spandidos Publications

9.3.4 Subcellular Effects

Several studies have reported that ultrasound has subcellular structural effects on organelles, including mitochondria, the endoplasmic reticulum, attached ribosomes, lysosomes, and nuclei (Stephens et al. 1980; ter Haar et al. 1979). Many investigators have demonstrated that some subcellular effects are always related to ultrasound cavitation, such as cellular degranulation, changes in the function of the cell membrane, increased levels of intracellular calcium, and stimulation of fibroblast activity resulting in increased protein synthesis, angiogenesis, and also stronger collagen (Lirani-Galvao et al. 2006; De Oliveira et al. 2008). Furthermore, because treating cells with low-intensity pulsed ultrasound can stimulate the intracellular cytoskeleton (Alter et al. 1998), ultrasonic vibrations might offer a convenient way of disrupting the cytoplasm (Waddington and Perry 1962).

Electron microscopy has shown that ultrasound exposure can also produce ultrastructural alterations at the subcellular level. These ultrastructural changes will be discussed in the following sections.

9.3.4.1 Mitochondria

Ultrastructural investigations have shown that mitochondria are altered after ultrasound exposure. For example, intact cells show mitochondrial damage due to ultrasound cavitation. Mitochondria appear to be the most sensitive organelle, exhibiting swelling, decrease of matrix density, electron-dense condensations in the matrix, loss of cristae, and the formation of lamellar membranes, all of which eventually disrupt the outer membrane (Stephens et al. 1980) (Fig. 9.16). Interestingly, the size and extent of the mitochondrial electron-dense condensations can be useful indicators of the extent of damage to the cell. A decrease in the electron-dense condensations in mitochondria may be ascribed to a dilution of matrix proteins by an influx of water, and the appearance of mitochondria swelling supports this theory. This mechanism may also explain the mitochondrial membrane disruption and secondary autolytic disruption induced by ultrasound cavitation. In a study by Dmitrieva (1964), ultrasound produced irreversible disruption of mitochondria in Brown-Pearce tumor cells, in which mitochondria were initially severely swollen and had few remaining cristae.

9.3.4.2 Endoplasmic Reticulum

Electron microscopy has shown that ultrasound exposure can disrupt cellular endoplasmic reticulum into irregularly rounded vesicles (Selman and Jurand 1964). Compared to mitochondria, the endoplasmic reticulum is less sensitive, but it also has a graded response to increasing ultrasound exposure, including dilation of the cisternae, loss of ribosomes from the surface, and vesiculation. A typical cell exposed to ultrasound at ambient pressure has dilated rough endoplasmic reticulum and a damaged plasma membrane (Stephens et al. 1980) (Fig. 9.17). The endoplasmic reticulum is often associated with the ability of cells to synthesize proteins for export, and thus, this capacity may be diminished by ultrasound exposure, and Oliveira et al. (2009) reported that low-intensity pulsed ultrasound actually enhanced reticulum activity and increased protein synthesis.

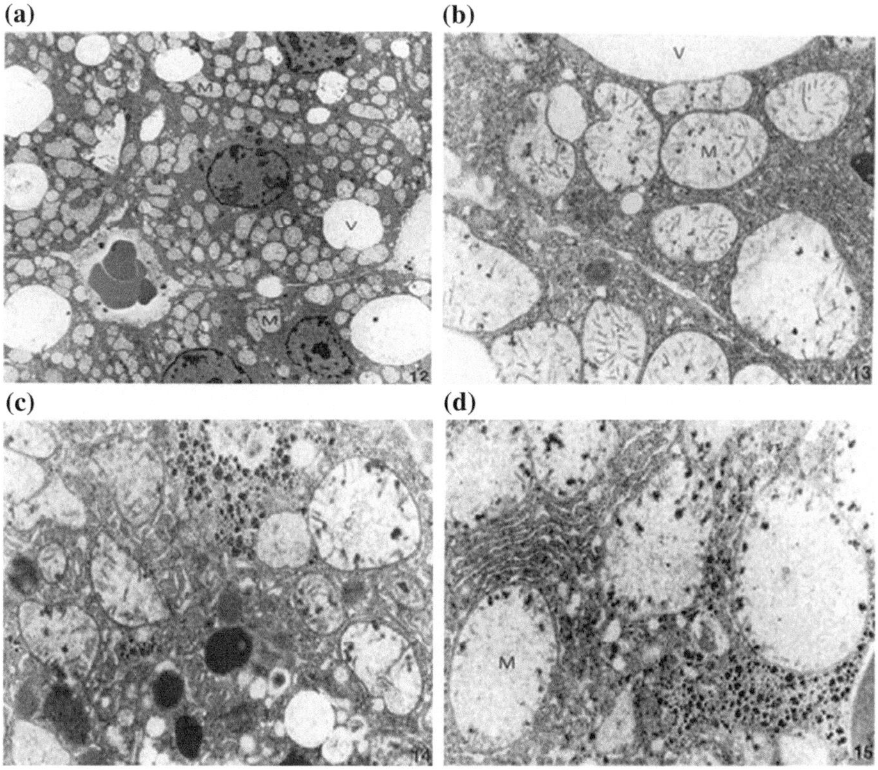

Fig. 9.16 Mitochondria (*M*) are more swollen and more electron-lucid with abnormal condensations. **a** Some mitochondrial cristae have disappeared. **b** Increased injury to the mitochondria with the formation of lamellar membranes and abnormal condensations. **c** The mitochondria have almost no remaining cristae, with the condensation granules located around the mitochondrial periphery. **d** The outer membranes are frequently ruptured. Reproduced with permission from Stephens et al. © Elsevier

9.3.4.3 Lysosomes

Previous studies have indicated that lysosomes appear to retain their structural integrity better than other cytoplasmic organelles under ultrasound exposure (Stephens et al. 1980). However, Taylor and Pond (1972) reported that the lysosome membrane is the primary site of ultrasonic damage, and Dyson et al. (1974) also reported that ultrasound exposure damaged the lysosomal membranes of chick embryos. These discrepancies might result from variations in the field intensity at the positions of fragile lysosomes. The damaged lysosomes are likely to be autolyzed, leaving intracellular vacuoles of low electron density, and the remaining lysosomes may appear to be undamaged.

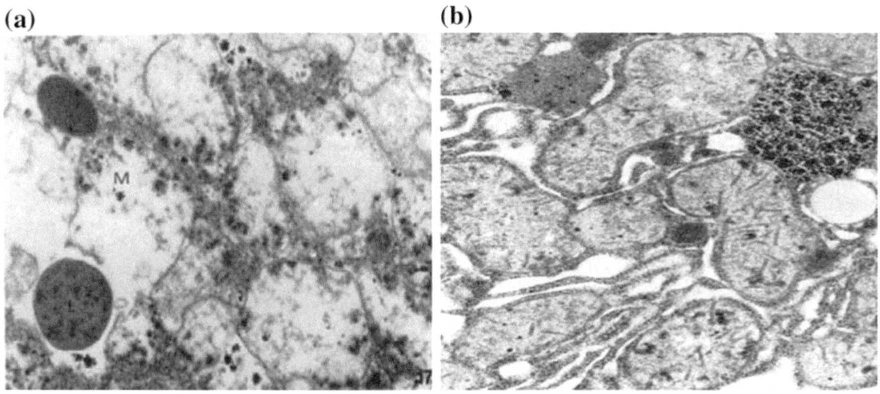

Fig. 9.17 The endoplasmic reticulum after ultrasound exposure. **a** Dilation of the cisternae of the endoplasmic reticulum. **b** The vesiculation of the endoplasmic reticulum. Reproduced with permission from Stephens et al. © Elsevier

9.3.4.4 Nuclei

Apoptotic cells can exhibit up to a 16-fold increase in ultrasound backscatter intensity in comparison with viable cells (Taggart et al. 2007). The most striking morphologic features of apoptosis are the condensation and fragmentation of the nucleus (ter Haar et al. 1979) (Fig. 9.18) and blebbing of the cell membrane. It has previously been shown that ultrasound cavitation is effective in causing the structural changes that occur in cells during apoptosis, which is characterized by early changes in the nuclear membrane and by chromatin condensation, followed by DNA fragmentation.

9.3.5 Molecular Mechanisms

No matter which kind of phenomenon is induced by ultrasound cavitation in cells, they all end up triggering intracellular signaling processes. The related signaling processes determine how cells respond to ultrasound cavitation, and this always occurs through up- or down-regulation of the expression of key genes or proteins.

9.3.5.1 Genetic Mechanisms

When ultrasound cavitation acts on cells, the intracellular and extracellular environments are changed by the physical and chemical events, producing a corresponding change in the cell's genetic profile. On the one hand, such genetic responses help cells to resist the damage caused by ultrasound and help them

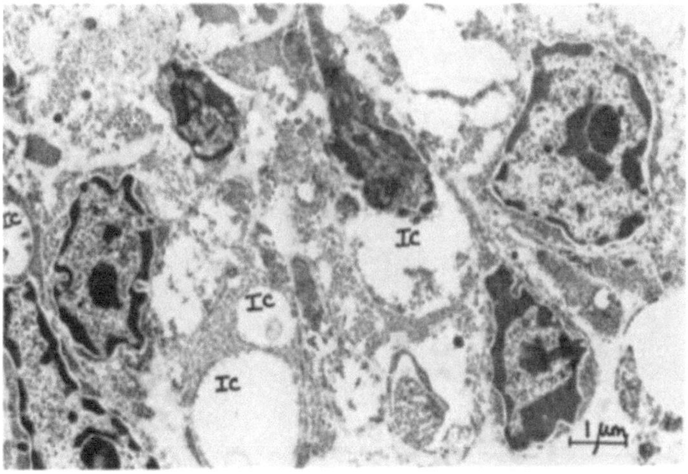

Fig. 9.18 Extensive intracellular cavities in myometrial smooth muscle cells treated with ultrasound exposure. The nuclei are more irregular in shape than usual, the chromatin is condensed, and the perinuclear cisternae are dilated. Reproduced with permission from ter Haar et al. © Elsevier

survive. On the other hand, however, they may initiate a cellular death pathway such as apoptosis or autophagy to limit side effects, for example inflammation, as far as possible. Thus, it is important to determine which key genes respond to ultrasound cavitation and to analyze their functions so we can start to develop the genetic network involved in ultrasonic cavitation. This would shed light on the molecular mechanisms of cavitation in cells and further help us apply the technology to cancer therapies.

To this end, one study identified the key genes that respond to low-intensity focused ultrasound in human lymphoma U973 cells (Tabuchi et al. 2008). Among the approximately 16,600 genes analyzed, some gene expression was up-regulated, including the expression of Bcl-2-associated athanogene 3 (*BAG3*), DnaJ (Hsp40) homolog, subfamily B, member 1 (*DNAJB1*), heat shock 70 kDa protein 1B (HSPA1B), and heat shock 70 kDa protein 6 (*HSPA6*). Meanwhile, some gene expression was down-regulated in the cells 3 h after the ultrasound treatment (0.3 W/cm^2). Expression was down-regulated for isopentenyl diphosphate delta isomerase (*IDI1*) and 3-hydroxy-3-methylglutaryl–coenzyme A synthase 1 (*HMGCS1*). This was the first report about how genes respond to ultrasound exposure, specifically to cavitation.

In addition to exploring the changes in gene expression, the genetic networks involved in the human lymphoma U937 cells' response to low-intensity focused ultrasound have also been explored. Several key genes, including 193 that were down-regulated and 201 that were up-regulated >1.5-fold, were identified. The genetic network is shown in Fig. 9.19. For the down-regulated genes, the significant genetic network D was found to be associated with cellular growth and

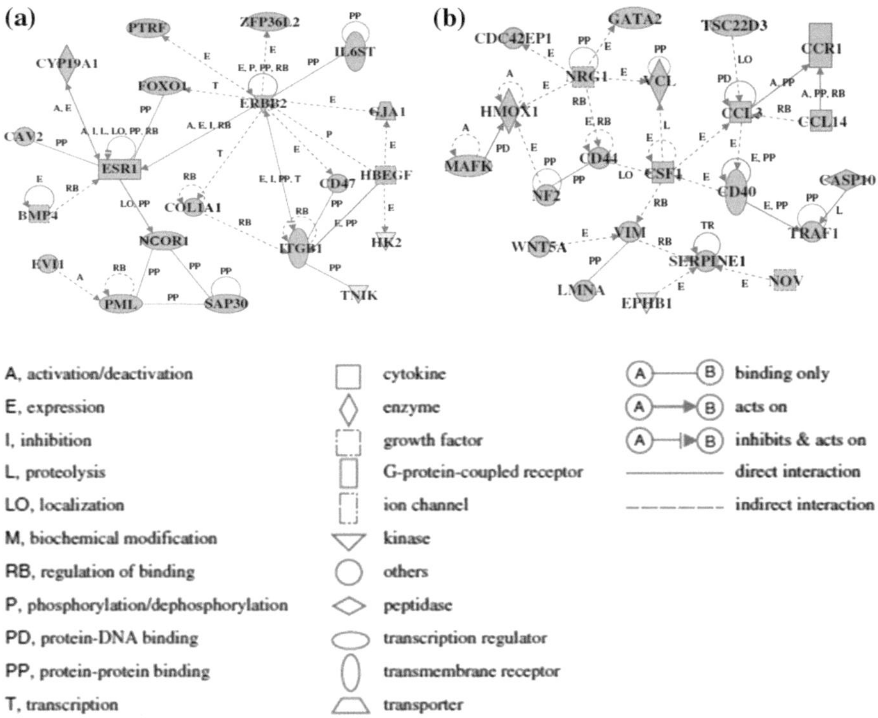

Fig. 9.19 *Notes* Genetic network D for down-regulated genes and genetic network U for up-regulated genes. **a** Genetic network D. **b** Genetic network U. Reproduced with permission from Tabuchi et al. © Elsevier

proliferation, gene expression, and cellular development. For the up-regulated genes, the significant genetic network U was found to be associated with cellular movement, cell morphology, and cell death.

In a different study, cDNA microarrays were used to analyze the gene response to ultrasound exposure and ultrasound cavitation in three closely related human Tk6 lymphoblast cell lines with different p53 status (Abdollahi et al. 2004). The study found that a complex signaling cascade exists with up-regulation of the proapoptotic genes, such as p53, p21, and Thy1 (CD90), and down-regulation of cellular survival components, such as Bcl-2 and SOD.

All the genetic research carried out thus far indicates that genes are up- or down-regulated in response to ultrasound cavitation and they interact with each other to form the genetic network that leads to a cell death pathway. However, there is the possibility that the up- or down-regulated genes are not translated to proteins, but it is proteins, not genes, that directly participate in the cell signaling cascades. For example, during the apoptotic cascades that occur in response to ultrasound exposure, groups of enzymes and several proteins are involved. These proteins may function as integrators of damage signals and may be the final decision points as to whether a cell lives or dies. Thus, their up- or down-regulation would also modulate the apoptotic cascades. This means that the complex pathways of cell death are not solely dictated by encoded genomic information, but they are also determined by execution molecules and proteins. Therefore, to understand the molecular mechanisms of ultrasound cavitation-induced apoptosis in depth, we must also examine the proteome.

9.3.5.2 Proteomics

The proteome is the expressed protein complement of a cell, matrix, organelle, tissue, or organ. It includes all isoforms and posttranslational variants, and it changes with time. The goal of comparative proteomics is to analyze how the proteome changes in different biological states (e.g., control vs. treatment or healthy versus diseased). Thanks to some pioneering proteomic studies, a few apoptosis-modulating proteins have been identified and have been applied as clinical apoptotic drug therapies to treat cancer. For example, cellular tumor antigen protein 53 (p53) has been used as a biochemical evaluation index for apoptotic cancer therapeutics modulated by anticancer drugs. However, for ultrasound-induced apoptotic cancer therapeutics, proteomic analyses have seldom been reported.

In one of the only proteomic studies by Feng et al. (2010), cells were examined by flow cytometry to determine the percentage of apoptosis after low-intensity focused ultrasound exposure. In the apoptotic group, proteins were extracted and separated by two-dimensional electrophoresis. Differently expressed proteins were separated if they showed up- or down-regulation greater than 1.3-fold in individual experiments or 1.5-fold on average. All the key proteins were then identified by matrix-assisted laser desorption/ionization time-of-flight mass spectrometry (MALDI-TOF-MS).

In Prof. Wan's laboratory, we have irradiated different cell lines with low-intensity focused ultrasound. In Walker 256 cells, $\sim$420 new protein spots were found in the gel after sonication (Tian et al. 2005). Among them, Hsp60 and Bcl-2-like protein 13 were found to be involved in cell apoptosis. This shows that ultrasound-induced apoptosis of the cells was probably achieved through a pathway that promotes the activation of caspase-3.

Conversely, in SMMC-7721 hepatocarcinoma cells, several proteins, such as cellular tumor antigen protein 53, BH3-interacting domain death agonist, apoptosis regulator Bcl-2, and heme oxygenase 1, were found to respond to ultrasound exposure on a molecular level, suggesting that in these cells, mitochondrial dysfunction and oxidative stresses were involved in ultrasound-induced apoptosis (Fig. 9.20). In addition, low-intensity ultrasound-induced apoptosis may be caused by mitofilin-regulated crista remodeling, which might cause permeabilization of the mitochondrial membrane and thus pore formation, leading to apoptosis (Fig. 9.21).

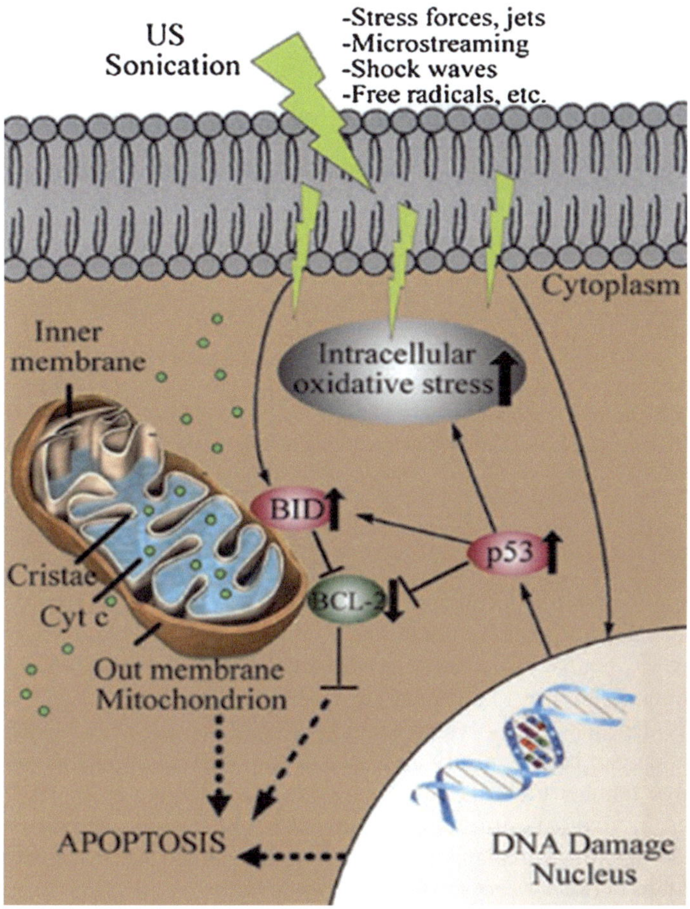

Fig. 9.20 Schematic of potential low-intensity ultrasound-induced molecular signaling cascades on activation of the apoptotic machinery in SMMC-7721 cells. Reproduced with permission from Feng et al. © American Institute of Ultrasound in Medicine

Studying the proteomics of cells exposed to low-intensity ultrasound (where ultrasonic cavitation plays an important role) provides abundant information about the proteins involved in the complex signaling cascades for cell death, e.g., apoptosis and autophagy. Knowing the proteomics will help us design an assisted therapeutic schedule that makes use of cavitation-induced cell death.

9.3.6 Sonodynamic Therapy

One accepted form of clinical cancer therapy, called PDT, utilizes a laser light source to induce chemical reactions in cells that produce free radicals. PDT uses

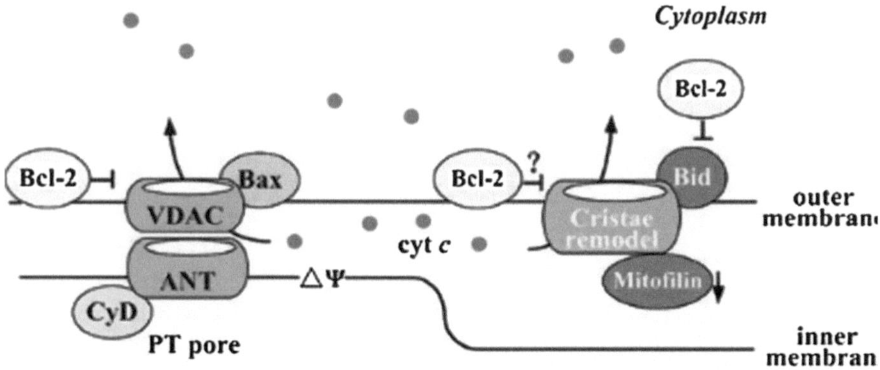

Fig. 9.21 Schematic of potential signaling pathways for mitochondrial membrane permeabilization (*MMP*) pore formation in mitochondria during low-intensity ultrasound-initiated apoptosis. ANT indicates adenine nucleotide translocase; *CyD* cyclophilin D; and *PT* permeability transition. Reproduced with permission from Feng et al. © American Institute of Ultrasound in Medicine

these free radicals to therapeutically destroy biological tissues. Similarly, SDT also induces free radical formation, but it does so using ultrasound exposure instead of a light source (Rosenthal et al. 2004). In SDT, therapeutic outcomes are achieved via the nonthermal effects of ultrasound, especially cavitation.

In SDT, sonosensitizers, which include porphyrins and their derivatives, are chemicals that can be chemically activated by exposure to ultrasound and thereby cause a large number of active ions to be produced (Kuroki et al. 2007). In both in vitro and in vivo experiments, SDT has been confirmed to be effective in suppressing the growth of malignant cells/tissues, and although it is alone, the ultrasound and the sonosensitizers showed very low cytotoxicity (Rosenthal et al. 2004). In vitro investigations have involved SDT in erythrocytes, sarcoma 180, L1210, and HL-60; however, the in vivo animal experiments suggested that this therapy was effective in treating sarcoma 180 and colon carcinoma 26.

The sonosensitizers, e.g., hematoporphyrin, pheophorbide A, photofrin, photofrin II, ATX-70, and ATX-S10, have been widely used with SDT. Other compounds, such as merocyanine 540 and some nonsteroidal anti-inflammatory drugs, e.g., tenoxicam and piroxicam, have also been introduced in SDT. When used for SDT, the structure of the sonosensitizer has an impact on the treatment's efficacy. Investigations into the mechanism of SDT have revealed that active oxygen, especially the singlet oxygen, mediates SDT induced by porphyrins and nonsteroid chemicals (Rosenthal et al. 2004). In this case, SDT efficiency can be blocked by histidine, the scavenger for singlet oxygen. Investigators found that free radical scavengers and antioxidants, such as mannitol, vitamin C, vitamin E, and superoxide dismutase (SOD), can reduce cavitation-induced tissue damage. These chemicals might be able to protect normal tissues from being destroyed.

SDT has the potential to act as a targeted therapy. For example, insonation can be administered at the time at which the chemical concentration in the cancerous tissue has reached a maximum compared to the concentration in the plasma. Conversely, ultrasound could be precisely focused on the target volume, which would limit the generation of active radicals to a defined area and ensure that only preselected tissues are damaged. However, the distribution of different sonosensitizers in tissue depends on the specific agent, and thus, achieving a therapeutic concentration of the sonosensitizer in the target malignant tissue might be difficult, as a concentration gradient of the agent between the target and adjacent normal tissue. Furthermore, having a higher concentration of sonosensitizers in the blood plasma than in the tissue might cause hemolysis.

To solve these problems, the use of endoscopic ultrasound and/or intracatheter ultrasound might be necessary, such that the ultrasonic energy can be delivered directly to the malignant lesion. Also, investigation of a series of sonosensitizers with specific pharmacokinetics characteristics would allow physicians to select a specific agent for any given tissue or organ. Another way to make SDT more effective is to introduce antibodies. An antibody can be coupled with the sonosensitizer such that they link to a target cell membrane specifically and efficiently. As a result, target cells will be destroyed efficiently during insonation. Taken together, all these approaches hold great promise for SDT.

Coupling SDT with other therapeutic approaches or active substances has also shown the great potential in cancer therapy. For example, a combination of PDT and SDT has been reported by Jin et al. (2000) for treating murine skin squamous cell carcinoma. In that study, the median survival in animals receiving combination therapy (>120 days) was longer than that for mice receiving only PDT or SDT (77–95 days). Furthermore, pathological examination revealed that the combination of SDT and PDT induced more extensive tumor necrosis.

9.4 Effects of Cavitation on Macromolecules and Small Molecules

Ultrasound cavitation can also be used in several applications because it interacts with macromolecules (e.g., protein and gene therapies as well as active enzymes) and small molecules. Because high-intensity ultrasound, which is inevitably associated with cavitation, interacts with these molecules, it is important to determine whether cavitation impairs the structure and function of molecules or modifies them, as this might prevent the use of ultrasound for cancer therapy.

9.4.1 Effects of Cavitation on Small Molecules

Ultrasound has been used to extract several small molecules that are active substances, such as active components for drugs or oil, from parts of plants, including the branches, roots, and leaves. This technique, called ultrasound-assisted extraction (UAE), employs the mechanical vibrations produced by ultrasound. Ultrasound can rapidly extract such substances because it increases the permeability of the cell wall, spontaneously forms bubbles in the liquid by cavitation induction, and increases the mechanical stress, i.e., the internal friction, of the cells. Compared with the conventional extraction methods, such as aqueous and ethanol extraction with infusion, decoction, maceration, and percolation; hydrodistillation; steam distillation; and turbohydrodistillation, UAE significantly increases extraction efficiency. However, it is unclear whether high-intensity ultrasound, which involves cavitation, might denature or inactivate the extracted substance. In one study that addressed this issue, Périno-Issartier et al. (2013) showed that for the essential oil extracted from lavandin flowers, ultrasonic-assisted extraction was quantitatively and qualitatively similar to conventional extraction methods and that there were no differences between extraction methods in the sensory perception of the oil. In a different study, natural and high-purity genistein was extracted from soybean seeds using an ultrasound technique (Tian et al. 2004b). After characterization using molecular ion $[M+H]^+$, UV spectra, retention time, IR spectra, 1HNMR, 14CNMR spectra, and tandem mass spectrometric analysis, the authors showed that UAE produced 99.6 % (w/w) genistein that had not significantly declined in activity.

While examining the structure and function of extracted substances is important, it might be even more important to evaluate their activity in vivo. In one such study, Devgan et al. compared the activity of extracts of the anti-diabetic drug *Pterocarpus marsupium* Roxb. heartwood that were obtained by either ultrasound or conventional extraction methods. The authors analyzed how well the extracts reduced blood glucose levels in alloxan-induced diabetic rats (Devgan et al. 2013). Interestingly, the UAE method actually enhanced the anti-diabetic activity of *Pterocarpus marsupium*. This study offers a new approach for examining the effects of ultrasonic cavitation on the function of small molecules extracted by ultrasound.

Because ultrasound cavitation has been proven useful for extracting small molecules, some new potential applications have now come to light. For example, UAE has been used to remove imazamox and its metabolites from the leaves of wheat plants, which were identified and analyzed by LC-TOF/MS and LC-UV (Rojano-Delgado et al. 2013). In summary, ultrasound can undoubtedly improve extraction efficiencies of small molecules. Furthermore, from the studies available, ultrasound seems to have few effects on the structure and function of extracted small molecules. Thus, it will likely be developed into a common method for sample preparation to be used with various analysis techniques.

9.4.2 *Effects of Cavitation on Macromolecules*

The primary way macromolecules interact with cavitation is used in the ultrasound emulsification technique. The ultrasound emulsification technique can generate a homogenous dispersion of macromolecules and has high encapsulation efficiency, and so it has been widely used for macroencapsulating proteins and other macromolecules (Suslick and Grinstaff 1990; Yan et al. 1994). However, when ultrasound is applied in the frequency range of 1–27 MHz, cavitation can cause degradation of polysaccharides and proteins (Henglein 1956). Similarly, DNA molecules are degraded by ultrasound exposure at 0.8 MHz (Barteri et al. 1996). Consequently, macromolecules are inherently less stable and are easily denatured and disrupted during cavitation.

While encapsulating proteins in microspheres is useful for protein delivery, using the ultrasound emulsification technique can damage proteins. For example, Krishnamurthy et al. indicated that when it was irradiated by ultrasound (50 W, 20 kHz, 1 s pulses interspersed with 0.5 s between pulses), the relatively stable lysozyme was significantly denatured and inactivated (Krishnamurthy et al. 2000). Similarly, Tian et al. (2004a) irradiated the enzyme trypsin and revealed that its activity (examined by a spectrophotometric analysis) decreased with increasing ultrasound power from 100 to 500 W (20 kHz, burst mode, 0.5 duty factors) or when the exposure time was extended from 1 to 20 min. However, certain additives, such as Tween 80 and mannitol, protected trypsin from ultrasound-induced inactivation. In the end, the authors suggested that the loss of trypsin activity was related to two mechanisms: (1) Acoustic cavitation produces a large interfacial area between water and air, which disturbs the conditions around the trypsin molecules, such as their hydrogen bonds and hydrophobic interactions. (2) Acoustic cavitation also introduces free radicals, shear stresses, and shock waves, which can damage the molecular structure of trypsin.

While ultrasound disrupts many proteins, some studies have shown that there are proteins which can tolerate ultrasound exposure during emulsification. Barteri et al. (1996) showed that the reduced horse heart cytochrome c maintained its bioactivity after being exposed to a 38-kHz ultrasound (ultrasonic cleaning bath, 38 kHz, 150 W). Furthermore, Tian et al. (2003) found that human tumor necrosis factor alpha (TNF-α) maintained its biological activity and did not denature after ultrasound exposure (20 kHz frequency, burst mode, 0.5 duty factor, 100–500 W total electrical power, 0–20 min total treatment time) in an aqueous solution, which may be contributed by its long, flat shape and the formation of its aggregated form.

In summary, ultrasound cavitation can create extreme (although relatively short-lived) conditions for nearby molecules, as temperatures can reach 2000–5000 K and pressures can approach 1800 atm inside collapsing cavitation bubbles (Putterman 1995). Thus, ultrasound emulsification can damage macromolecules because of these extreme conditions. Furthermore, inertial cavitation could cause some significant physical effects, such as high local shear forces, jets, and shock waves. Under these extreme physical stresses, cavitation can disrupt hydrogen bonding and van der Waals interactions in proteins' polypeptide chains, leading to modifications in their secondary and tertiary structures and oxidation of SH-groups (Suslick and Grinstaff 1990; Barteri et al. 1996). Thus, as their secondary and tertiary structures change, proteins may lose bioactivity. However a protein's specific structure really determines whether these disruptions will affect bioactivity. For example, proteins with higher percentages of α-helix structures have higher molecular stiffness, meaning that for this type of protein, ultrasound exposure can easily change and damage the protein structure.

9.5 Summary

Ultrasound can interact with biological molecules, cells, or tissues to produce a variety of mechanobiological effects attributed to the activity of acoustic cavitation. Such effects include sonoporation, membrane permeability changes, free radical formation, and more. These effects become enhanced when microbubbles are present during ultrasound exposure. Regarding physical phenomena, acoustic cavitation produces shock waves, high local shear stresses, microstreaming, microjetting, and implosion, and most of these involve mechanical forces. Because mechanobiology describes the biological responses of cells to mechanical loads and the mechanotransduction mechanisms by which these loads are transduced into a cascade of cellular and molecular events, we describe the mechanical loads due to acoustic cavitation, as well as the biological responses of molecules, cells, and tissues to acoustic cavitation as "cavitation mechanobiology."

This chapter has provided an overview of the mechanobiological effects of cavitation, which is the key nonthermal mechanism by which ultrasonic microbubbles interact with tissue, cells, and molecules. The categories of the biological effects produced by ultrasound have been reviewed to provide insight into the range of the mechanobiological effects produced by ultrasonic cavitation in vitro and in vivo, and to highlight the potential applications for these mechanobiological effects, in terms of designing relevant therapeutic applications based on ultrasound cavitation for biological and medical purposes.

In Sect. 9.2, we described the interactions between acoustic cavitation and tissues, and we emphasized the widely studied therapeutic applications. In Sect. 9.3, we discussed the biological effects of cavitation on cells in detail, in terms of the effects of cavitation on cell structure, functions (including at the subcellular level), and molecules (at the DNA and protein levels). We also introduced SDT, which is closely related to cavitation. In Sect. 9.4, we examined the effects of cavitation on macromolecules and small molecules.

Cavitation mechanobiology is closely related to ultrasound's various therapeutic applications in biomedicine. Understanding the mechanism of cavitation can help us control cavitation, regulate its biological effects, and then lay the foundation for establishing associated therapeutic regimens in the clinic.

References

Abdollahi A, Domhan S, Jenne JW, Hallaj M, Dell AQua G, Mueckenthaler M, Richter A, Martin H, Debus J, Ansorge W. Apoptosis signals in lymphoblasts induced by focused ultrasound. J Fed Am Soc Exp Biol. 2004;18(12):1413–14.

Abraham N, Lafaro R, Wolin M, McClung J. Heme oxygenase-1 attenuates glucose-mediated cell growth arrest and apoptosis in human microvessel endothelial cells. Circ Res. 2003;108 (17):265.

Akiyama M, Ishibashi T, Yamada T, Furuhata H. Low-frequency ultrasound penetrates the cranium and enhances thrombolysis in vitro. Neurosurgery. 1998;43(4):828–32.

Alexandrov AV, Molina CA, Grotta JC, Garami Z, Ford SR, Alvarez-Sabin J, Montaner J, Saqqur M, Demchuk AM, Moyé LA. Ultrasound-enhanced systemic thrombolysis for acute ischemic stroke. N Engl J Med. 2004;351(21):2170–8.

Allen TM. Liposomes: opportunities in drug delivery. Drugs. 1997;54(Suppl 4):8–14.

Alter A, Rozenszajn LA, Miller HI, Rosenschein U. Ultrasound inhibits the adhesion and migration of smooth muscle cells in vitro. Ultrasound Med Biol. 1998;24(5):711–21.

Ashush H, Rozenszajn LA, Blass M, Barda-Saad M, Azimov D, Radnay J, Zipori D, Rosenschein U. Apoptosis induction of human myeloid leukemic cells by ultrasound exposure. Cancer Res. 2000;60(4):1014–20.

Bao S, Thrall BD, Miller DL. Transfection of a reporter plasmid into cultured cells by sonoporation in vitro. Ultrasound Med Biol. 1997;23(6):953–9.

Barteri M, Fioroni M, Gaudiano M. Oxidation of Fe (II) horse heart cytochrome c by ultrasound waves. Biochim Biophys Acta (BBA)—Biomembr. 1996;1296(1):35–40.

Behrens S, Spengos K, Daffertshofer M, Schroeck H, Dempfle CE, Hennerici M. Transcranial ultrasound-improved thrombolysis: diagnostic versus therapeutic ultrasound. Ultrasound Med Biol. 2001;27(12):1683–9.

Belkin VM, Belkin AM, Koteliansky VE. Human smooth muscle VLA-1 integrin: purification, substrate specificity, localization in aorta, and expression during development. J Cell Biol. 1990;111(5 Pt 1):2159–70.

Belyaev IY, Czene S, Harms-Ringdahl M. Changes in chromatin conformation during radiation-induced apoptosis in human lymphocytes. Radiat Res. 2009;156(4):355–64.

Birnbaum Y, Luo H, Nagai T, Fishbein MC, Peterson TM, Li S, Kricsfeld D, Porter TR, Siegel RJ. Noninvasive in vivo clot dissolution without a thrombolytic drug recanalization of thrombosed iliofemoral arteries by transcutaneous ultrasound combined with intravenous infusion of microbubbles. Circulation. 1998;97(2):130–4.

Brujan EA. The role of cavitation microjets in the therapeutic applications of ultrasound. Ultrasound Med Biol. 2004;30(3):381–7.

Bustamante J, Navarro P, Arana G, de Diego A, Madariaga JM. Ultrasound assisted dialysis of semi-permeable membrane devices for the simultaneous analysis of a wide number of persistent organic pollutants. Talanta. 2013;114:32–7.

Carloni M, Meschini R, Ovidi L, Palitti F. PHA-induced cell proliferation rescues human peripheral blood lymphocytes from X-ray-induced apoptosis. Mutagenesis. 2001;16(2):115–20.

Chen S, Ding J-H, Bekeredjian R, Yang B-Z, Shohet RV, Johnston SA, Hohmeier HE, Newgard CB, Grayburn PA. Efficient gene delivery to pancreatic islets with ultrasonic microbubble destruction technology. Proc Natl Acad Sci USA. 2006;103(22):8469–74.

Chen H, Brayman AA, Kreider W, Bailey MR, Matula TJ. Observations of translation and jetting of ultrasound-activated microbubbles in mesenteric microvessels. Ultrasound Med Biol. 2011;37(12):2139–48.

Chen H, Brayman AA, Evan AP, Matula TJ. Preliminary observations on the spatial correlation between short-burst microbubble oscillations and vascular bioeffects. Ultrasound Med Biol. 2012;38(12):2151–62.

Cohen JJ. Apoptosis. Immunol Today. 1993;14(3):126–30.

Daffertshofer M, Hennerici M. Ultrasound in the treatment of ischaemic stroke. Lancet Neurol. 2003;2(5):283–90.

Dalecki D. Mechanical bioeffects of ultrasound. Annu Rev Biomed Eng. 2004;6:229–48.

De Oliveira RF, Oliveira DA, Monteiro W, Zangaro RA, Magini M, Soares CP. Comparison between the effect of low-level laser therapy and low-intensity pulsed ultrasonic irradiation in vitro. Photomed Laser Surg. 2008;26(1):6–9.

Devgan M, Nanda A, Ansari SH. Comparative evaluation of the anti-diabetic activity of *Pterocarpus marsupium* Roxb. heartwood in alloxan induced diabetic rats using extracts obtained by optimized conventional and non conventional extraction methods. Pak J Pharm Sci. 2013;26(5):973–6.

Dmitrieva N. Electron microscopic study of changes in structure of the mitochondria in tumor cells after exposure to ultrasound. Biofizika. 1964;9:571–9.

Dreher D, Junod A. Role of oxygen free radicals in cancer development. Eur J Cancer. 1996;32 (1):30–8.

Duff SE, Li C, Garland JM, Kumar S. CD105 is important for angiogenesis: evidence and potential applications. J Fed Am Soc Exp Biol. 2003;17(9):984–92.

Dyson M, Pond J, Woodward B, Broadbent J. The production of blood cell stasis and endothelial damage in the blood vessels of chick embryos treated with ultrasound in a stationary wave field. Ultrasound Med Biol. 1974;1(2):133–48.

Eder IE, Haag P, Bartsch G, Klocker H. Gene therapy strategies in prostate cancer. Curr Gene Ther. 2005;5(1):1–10.

Edinger AL, Thompson CB. Death by design: apoptosis, necrosis and autophagy. Curr Opin Cell Biol. 2004;16(6):663–9.

Elder SA. Cavitation microstreaming. J Acoust Soc Am. 1959;31(1):54–64.

Fechheimer M, Denny C, Murphy R, Taylor D. Measurement of cytoplasmic pH in Dictyostelium discoideum by using a new method for introducing macromolecules into living cells. Eur J Cell Biol. 1986;40(2):242–7.

Feng Y, Tian Z, Wan MX. Bioeffects of low-intensity ultrasound in vitro: apoptosis, protein profile alteration, and potential molecular mechanism. J Ultrasound Med Off J Am Inst Ultrasound Med. 2010;29(6):963–74.

Ferrara K, Pollard R, Borden M. Ultrasound microbubble contrast agents: fundamentals and application to gene and drug delivery. Annu Rev Biomed Eng. 2007;9:415–47.

Folkman J. How is blood vessel growth regulated in normal and neoplastic tissue?—GHA clowes memorial award lecture. Cancer Res. 1986;46(2):467–73.

Frenkel V. Ultrasound mediated delivery of drugs and genes to solid tumors. Adv Drug Deliv Rev. 2008;60(10):1193–208.

Frenkel V, Kimmel E, Iger Y. Ultrasound-induced cavitation damage to external epithelia of fish skin. Ultrasound Med Biol. 1999;25(8):1295–303.

Frenkel PA, Chen S, Thai T, Shohet RV, Grayburn PA. DNA-loaded albumin microbubbles enhance ultrasound-mediated transfection in vitro. Ultrasound Med Biol. 2002;28(6):817–22.

Gudi SR, Clark CB, Frangos JA. Fluid flow rapidly activates G proteins in human endothelial cells involvement of G proteins in mechanochemical signal transduction. Circ Res. 1996;79(4):834–9.

Hallow DM, Mahajan AD, McCutchen TE, Prausnitz MR. Measurement and correlation of acoustic cavitation with cellular bioeffects. Ultrasound Med Biol. 2006;32(7):1111–22.

Hansen CA, Schroering AG, Carey DJ, Robishaw JD. Localization of a heterotrimeric G protein gamma subunit to focal adhesions and associated stress fibers. J Cell Biol. 1994;126(3):811–9.

Hashiya N, Aoki M, Tachibana K, Taniyama Y, Yamasaki K, Hiraoka K, Makino H, Yasufumi K, Ogihara T, Morishita R. Local delivery of E2F decoy oligodeoxynucleotides using ultrasound with microbubble agent (optison) inhibits intimal hyperplasia after balloon injury in rat carotid artery model. Biochem Biophys Res Commun. 2004;317(2):508–14.

Henglein A. The acceleration of chemical reactions of ultrasound in solutions of oxygen-rare gas mixtures. Naturwissenschaften (in German). 1956;43(12):277–8.

Honda H, Zhao Q-L, Kondo T. Effects of dissolved gases and an echo contrast agent on apoptosis induced by ultrasound and its mechanism <i> via </i> the mitochondria-caspase pathway. Ultrasound Med Biol. 2002;28(5):673–82.

Honda H, Kondo T, Zhao Q-L, Feril LB Jr, Kitagawa H. Role of intracellular calcium ions and reactive oxygen species in apoptosis induced by ultrasound. Ultrasound Med Biol. 2004;30 (5):683–92.

Husseini GA, de la Rosa MAD, Gabuji T, Zeng Y, Christensen DA, Pitt WG. Release of doxorubicin from unstabilized and stabilized micelles under the action of ultrasound. J Nanosci Nanotechnol. 2007;7(3):1028.

Hynynen K. Focused ultrasound for blood–brain disruption and delivery of therapeutic molecules into the brain. Expert Opin Drug Deliv. 2007;4(1):27–35.

Hynynen K, McDannold N, Clement G, Jolesz FA, Zadicario E, Killiany R, Moore T, Rosen D. Pre-clinical testing of a phased array ultrasound system for MRI-guided noninvasive surgery of the brain—a primate study. Eur J Radiol. 2006;59(2):149–56.

Ingber DE. Cellular basis of mechanotransduction. Biol Bull. 1998;194(3):323–7.

Ingelsson M, Hyman BT. Disordered proteins in dementia. Ann Med. 2002;34(4):259–71.

Ito S, Koshikawa N, Mochizuki S, Takenaga K. 3-Methyladenine suppresses cell migration and invasion of HT1080 fibrosarcoma cells through inhibiting phosphoinositide 3-kinases independently of autophagy inhibition. Int J Oncol. 2007;31(2):261–8.

Jernberg MER, Lewensohn H, Wiksell A, Brahme A. Cellular effects of high-intensity focused continuous wave ultrasound alone and in combination with X-rays. Int J Radiat Biol. 2001;77 (1):127–35.

Jin Z-H, Miyoshi N, Ishiguro K, Umemura S, Kawabata K, Yumita N, Sakata I, Takaoka K, Udagawa T, Nakajima S. Combination effect of photodynamic and sonodynamic therapy on experimental skin squamous cell carcinoma in C3H/HeN mice. J Dermatol. 2000;27 (5):294–306.

Juffermans LJ, Dijkmans PA, Musters RJ, Visser CA, Kamp O. Transient permeabilization of cell membranes by ultrasound-exposed microbubbles is related to formation of hydrogen peroxide. Am J Physiol Heart Circ Physiol. 2006;291(4):H1595–601.

Juffermans LJ, van Dijk A, Jongenelen CA, Drukarch B, Reijerkerk A, de Vries HE, Kamp O, Musters RJ. Ultrasound and microbubble-induced intra- and intercellular bioeffects in primary endothelial cells. Ultrasound Med Biol. 2009;35(11):1917–27.

Juliano RL, Haskill S. Signal transduction from the extracellular-matrix. J Cell Biol. 1993;120 (3):577–85.

Kessel D, Oleinick NL. Initiation of autophagy by photodynamic therapy. Methods Enzymol. 2009;453:1–16.

Kessel D, Vicente MGH, Reiners JJ. Initiation of apoptosis and autophagy by photodynamic therapy. Autophagy. 2006;38(5):482–8.

Kimmel E. Cavitation bioeffects. Crit Rev™ Biomed Eng. 2006;34(2):105–61.

Klaunig JE, Kamendulis LM. The role of oxidative stress in carcinogenesis. Annu Rev Pharmacol Toxicol. 2004;44:239–67.

Klionsky DJ, Emr SD. Cell biology—autophagy as a regulated pathway of cellular degradation. Science. 2000;290(5497):1717–21.

Koike H, Tomita N, Azuma H, Taniyama Y, Yamasaki K, Kunugiza Y, Tachibana K, Ogihara T, Morishita R. An efficient gene transfer method mediated by ultrasound and microbubbles into the kidney. J Gene Med. 2005;7(1):108–16.

Kondo T, Kano E. Enhancement of hyperthermic cell killing by non-thermal effect of ultrasound. Int J Radiat Biol Relat Stud Phys Chem Med. 1987;51(1):157–66.

Kondo T, Kano E. Effect of free radicals induced by ultrasonic cavitation on cell killing. Int J Radiat Biol. 1988;54(3):475–86.

Korpanty G, Chen S, Shohet R, Ding J-H, Yang B-Z, Frenkel P, Grayburn P. Targeting of VEGF-mediated angiogenesis to rat myocardium using ultrasonic destruction of microbubbles. Gene Ther. 2005a;12(17):1305–12.

Korpanty G, Grayburn PA, Shohet RV, Brekken RA. Targeting vascular endothelium with avidin microbubbles. Ultrasound Med Biol. 2005b;31(9):1279–83.

Krishnamurthy R, Lumpkin JA, Sridhar R. Inactivation of lysozyme by sonication under conditions relevant to microencapsulation. Int J Pharm. 2000;205(1):23–34.

Kulms D, Schwarz T. Molecular mechanisms of UV-induced apoptosis. Photodermatol Photoimmunol Photomed. 2000;16(5):195–201.

Kuroki M, Hachimine K, Abe H, Shibaguchi H, Kuroki M, Maekawa S-I, Yanagisawa J, Kinugasa T, Tanaka T, Yamashita Y. Sonodynamic therapy of cancer using novel sonosensitizers. Anticancer Res. 2007;27(6A):3673–7.

Lagneaux L, de Meulenaer EC, Delforge A, Dejeneffe M, Massy M, Moerman C, Hannecart B, Canivet Y, Lepeltier M-F, Bron D. Ultrasonic low-energy treatment: a novel approach to induce apoptosis in human leukemic cells. Exp Hematol. 2002;30(11):1293–301.

Larina IV, Evers BM, Esenaliev RO. Optimal drug and gene delivery in cancer cells by ultrasound-induced cavitation. Anticancer Res. 2005;25(1A):149–56.

Lawrie A, Brisken A, Francis S, Cumberland D, Crossman D, Newman C. Microbubble-enhanced ultrasound for vascular gene delivery. Gene Ther. 2000;7(23):2023–7.

Leighton T. The acoustic bubble. London: Academic press; 2012.

Lentacker I, De Smedt SC, Sanders NN. Drug loaded microbubble design for ultrasound triggered delivery. Soft Matter. 2009;5(11):2161–70.

Leong-Poi H, Christiansen J, Klibanov AL, Kaul S, Lindner JR. Noninvasive assessment of angiogenesis by ultrasound and microbubbles targeted to alpha(v)-integrins. Circulation. 2003;107(3):455–60.

Levine B, Klionsky DJ. Development by self-digestion: molecular mechanisms and biological functions of autophagy. Dev Cell. 2004;6(4):463–77.

Li J, Hou N, Faried A, Tsutsumi S, Takeuchi T, Kuwano H. Inhibition of autophagy by 3-MA enhances the effect of 5-FU-induced apoptosis in colon cancer cells. Ann Surg Oncol. 2009;16 (3):761–71.

Liebeskind D, Bases R, Koenigsberg M, Koss L, Raventos C. Morphological changes in the surface characteristics of cultured cells after exposure to diagnostic ultrasound. Radiology. 1981;138(2):419–23.

Lirani-Galvao AP, Jorgetti V, da Silva OL. Comparative study of how low-level laser therapy and low-intensity pulsed ultrasound affect bone repair in rats. Photomed Laser Surg. 2006;24 (6):735–40.

Loft S, Poulsen HE. Cancer risk and oxidative DNA damage in man. J Mol Med (Berlin, Ger). 1996;74(6):297–312.

Luo H, Nishioka T, Fishbein MC, Cercek B, Forrester JS, Kim C-J, Berglund H, Siegel RJ. Transcutaneous ultrasound augments lysis of arterial thrombi in vivo. Circulation. 1996;94(4):775–8.

Luo H, Birnbaum Y, Fishbein MC, Peterson TM, Nagai T, Nishioka T, Siegel RJ. Enhancement of thrombolysis in vivo without skin and soft tissue damage by transcutaneous ultrasound. Thromb Res. 1998;89(4):171–7.

Maniotis AJ, Chen CS, Ingber DE. Demonstration of mechanical connections between integrins, cytoskeletal filaments, and nucleoplasm that stabilize nuclear structure. Proc Natl Acad Sci USA. 1997;94(3):849–54.

Manome Y, Nakamura M, Ohno T, Furuhata H. Ultrasound facilitates transduction of naked plasmid DNA into colon carcinoma cells in vitro and in vivo. Hum Gene Ther. 2000;11 (11):1521–8.

Marmottant P, Hilgenfeldt S. Controlled vesicle deformation and lysis by single oscillating bubbles. Nature. 2003;423(6936):153–6.

Mason TJ, Cobley AJ, Graves JE, Morgan D. New evidence for the inverse dependence of mechanical and chemical effects on the frequency of ultrasound. Ultrason Sonochem. 2011;18 (1):226–30.

Maxwell AD, Cain CA, Duryea AP, Yuan L, Gurm HS, Xu Z. Noninvasive thrombolysis using pulsed ultrasound cavitation therapy—histotripsy. Ultrasound Med Biol. 2009;35(12):1982–94.

Maxwell AD, Wang TY, Cain CA, Fowlkes JB, Sapozhnikov OA, Bailey MR, Xu Z. Cavitation clouds created by shock scattering from bubbles during histotripsy. J Acoust Soc Am. 2011;130(4):1888–98.

Maxwell AD, Khokhlova TD, Schade GR, Wang Y-N, Kreider W, Yuldashev P, Simon JC, Sapozhnikov OA, Farr N, Partanen A. Boiling histotripsy: a noninvasive method for mechanical tissue disintegration. J Acoust Soc Am. 2014;136(4):2249.

McDonald DM, Choyke PL. Imaging of angiogenesis: from microscope to clinic. Nat Med. 2003;9(6):713–25.

Miller DL. Overview of experimental studies of biological effects of medical ultrasound caused by gas body activation and inertial cavitation. Prog Biophys Mol Biol. 2007;93(1–3):314–30.

Miller DL, Nyborg WL. Theory for bioeffects of ultrasound near gas-filled pores. J Acoust Soc Am. 1978;63:S14.

Miller DL, Thomas RM. Ultrasound contrast agents nucleate inertial cavitation in vitro. Ultrasound Med Biol. 1995;21(8):1059–65.

Miller MW, Miller DL, Brayman AA. A review of in vitro bioeffects of inertial ultrasonic cavitation from a mechanistic perspective. Ultrasound Med Biol. 1996;22(9):1131–54.

Miller DL, Dou C, Song J. DNA transfer and cell killing in epidermoid cells by diagnostic ultrasound activation of contrast agent gas bodies in vitro. Ultrasound Med Biol. 2003;29(4):601–7.

Minchinton AI, Tannock IF. Drug penetration in solid tumours. Nat Rev Cancer. 2006;6(8):583–92.

Mitragotri S. Healing sound: the use of ultrasound in drug delivery and other therapeutic applications. Nat Rev Drug Discov. 2005;4(3):255–60.

Mitragotri S, Kost J. Low-frequency sonophoresis: a review. Adv Drug Deliv Rev. 2004;56(5):589–601.

Molina CA, Ribo M, Rubiera M, Montaner J, Santamarina E, Delgado-Mederos R, Arenillas JF, Huertas R, Purroy F, Delgado P, Alvarez-Sabin J. Microbubble administration accelerates clot lysis during continuous 2-MHz ultrasound monitoring in stroke patients treated with intravenous tissue plasminogen activator. Stroke. 2006;37(2):425–9.

Newman CMH, Bettinger T. Gene therapy progress and prospects: ultrasound for gene transfer. Gene Ther. 2007;14(6):465–75.

Ng KY, Liu Y. Therapeutic ultrasound: its application in drug delivery. Med Res Rev. 2002;22(2):204–23.

Nilsson A-M, Odselius R, Roijer A, Olsson S. Pro-and antifibrinolytic effects of ultrasound on streptokinase-induced thrombolysis. Ultrasound Med Biol. 1995;21(6):833–40.

Nisato RE, Tille JC, Jonczyk A, Goodman SL, Pepper MS. alphav beta 3 and alphav beta 5 integrin antagonists inhibit angiogenesis in vitro. Angiogenesis. 2003;6(2):105–19.

Nishioka T, Luo H, Fishbein MC, Cercek B, Forrester JS, Kim C-J, Berglund H, Siegel RJ. Dissolution of thrombotic arterial occlusion by high intensity, low frequency ultrasound and dodecafluoropentane emulsion: an in vitro and in vivo study. J Am Coll Cardiol. 1997;30(2):561–8.

Nomikou N, McHale AP. Exploiting ultrasound-mediated effects in delivering targeted, site-specific cancer therapy. Cancer Lett. 2010;296(2):133–43.

Nyborg WL. Acoustic streaming near a boundary. J Acoust Soc Am. 2005;30(4):329–39.

Ohl C-D, Arora M, Ikink R, de Jong N, Versluis M, Delius M, Lohse D. Sonoporation from jetting cavitation bubbles. Biophys J. 2006;91(11):4285–95.

Oliveira DA, De Oliveira RF, Magini M, Zangaro RA, Soares CP. Assessment of cytoskeleton and endoplasmic reticulum of fibroblast cells subjected to low-level laser therapy and low-intensity pulsed ultrasound. Photomed Laser Surg. 2009;27(3):461–6.

Paliwal S, Mitragotri S. Ultrasound-induced cavitation: applications in drug and gene delivery. Expert Opin Drug Deliv. 2006;3(6):713–26.

Park J, Fan Z, Deng CX. Effects of shear stress cultivation on cell membrane disruption and intracellular calcium concentration in sonoporation of endothelial cells. J Biomech. 2011;44 (1):164–9.

Parsons, J.E. Histotripsy in focused ultrasound surgery: mechanical ablation of tissue using controlled acoustic cavitation;2006.

Périno-Issartier S, Giniesa C, Cravottob G, Chemata F. A comparison of essential oils obtained from lavandin via different extraction processes: ultrasound, microwave, turbohydrodistillation, steam and hydrodistillation. J Chromatogr A. 2013;1305:41–7.

Philipp A, Delius M, Scheffczyk C, Vogel A, Lauterborn W. Interaction of lithotripter-generated shock waves with air bubbles. J Acoust Soc Am. 1993;93(5):2496–509.

Porter TR, Xie F. Ultrasound, microbubbles, and thrombolysis. Prog Cardiovasc Dis. 2001;44 (2):101–10.

Price RJ, Skyba DM, Kaul S, Skalak TC. Delivery of colloidal particles and red blood cells to tissue through microvessel ruptures created by targeted microbubble destruction with ultrasound. Circulation. 1998;98(13):1264–7.

Pua EC, Zhong P. Ultrasound-mediated drug delivery range of ultrasonic interactions and applications in biological systems. IEEE Eng Med Biol Mag. 2009;28(1):64–75.

Putterman SJ. Sonoluminescence: sound into light. Sci Am. 1995;272(2):46–51.

Rapley R, Walker JM. Molecular biomethods hand book, vol. 1646. Berlin: Springer; 1998.

Raymond SB, Treat LH, Dewey JD, McDannold NJ, Hynynen K, Bacskai BJ. Ultrasound enhanced delivery of molecular imaging and therapeutic agents in Alzheimer's disease mouse models. PLoS ONE. 2008;3(5):e2175.

Riesz P, Kondo T. Free radical formation induced by ultrasound and its biological implications. Free Radic Biol Med. 1992;13(3):247–70.

Riggs P, Francis C, Bartos S, Penney D. Ultrasound enhancement of rabbit femoral artery thrombolysis. Cardiovasc Surg (Lond, Engl). 1997;5(2):201–207.

Rink K, Delacretaz G, Salathe RP. Fragmentation process of current laser lithotriptors. Lasers Surg Med. 1995;16(2):134–46.

Ríos-Arrabal S, Artacho-Cordón F, León J, Román-Marinetto E, del Mar Salinas-Asensio M, Calvente I, Núñez MI. Involvement of free radicals in breast cancer. Springerplus. 2013;2(1):1–12.

Roberts WW. Development and translation of histotripsy: current status and future directions. Curr Opin Urol. 2014;24(1):104–10.

Roberts WW, Hall TL, Ives K, Wolf JS Jr, Fowlkes JB, Cain CA. Pulsed cavitational ultrasound: a noninvasive technology for controlled tissue ablation (histotripsy) in the rabbit kidney. J Urol. 2006;175(2):734–8.

Rojano-Delgado AM, Priego-Capote F, Prado RD, Castro MDLD. Ultrasound-assisted extraction with LC–TOF/MS identification and LC–UV determination of imazamox and its metabolites in leaves of wheat plants. Phytochem Anal PCA. 2013;25(4):357–63.

Rooney JA. Shear as a mechanism for sonically induced biological effects. J Acoust Soc Am. 1972;52(6):1718–24.

Rosenschein U, Roth A, Rassin T, Basan S, Laniado S, Miller HI. Analysis of coronary ultrasound thrombolysis endpoints in acute myocardial infarction (ACUTE trial) results of the feasibility phase. Circulation. 1997;95(6):1411–6.

Rosenschein U, Gaul G, Erbel R, Amann F, Velasguez D, Stoerger H, Simon R, Gomez G, Troster J, Bartorelli A. Percutaneous transluminal therapy of occluded saphenous vein grafts can the challenge be met with ultrasound thrombolysis? Circulation. 1999;99(1):26–9.

Rosenthal I, Sostaric JZ, Riesz P. Sonodynamic therapy—a review of the synergistic effects of drugs and ultrasound. Ultrason Sonochem. 2004;11(6):349–63.

Ruoslahti E. Integrins. J Clin Invest. 1991;87(1):1–5.

Schäfer S, Kliner S, Klinghammer L, Kaarmann H, Lucic I, Nixdorff U, Rosenschein U, Daniel WG, Flachskampf FA. Influence of ultrasound operating parameters on ultrasound-induced thrombolysis in vitro. Ultrasound Med Biol. 2005;31(6):841–7.

Schlicher RK, Radhakrishna H, Tolentino TP, Apkarian RP, Zarnitsyn V, Prausnitz MR. Mechanism of intracellular delivery by acoustic cavitation. Ultrasound Med Biol. 2006;32(6):915–24.
Schlicher RK, Hutcheson JD, Radhakrishna H, Apkarian RP, Prausnitz MR. Changes in cell morphology due to plasma membrane wounding by acoustic cavitation. Ultrasound Med Biol. 2010;36(4):677–92.
Selman G, Jurand A. An electron microscope study of the endoplasmic reticulum in newt notochord cells after disturbance with ultrasonic treatment and subsequent regeneration. J Cell Biol. 1964;20(1):175–83.
Shimamura M, Sato N, Taniyama Y, Kurinami H, Tanaka H, Takami T, Ogihara T, Tohyama M, Kaneda Y, Morishita R. Gene transfer into adult rat spinal cord using naked plasmid DNA and ultrasound microbubbles. J Gene Med. 2005;7(11):1468–74.
Shohet RV, Chen S, Zhou Y-T, Wang Z, Meidell RS, Unger RH, Grayburn PA. Echocardiographic destruction of albumin microbubbles directs gene delivery to the myocardium. Circulation. 2000;101(22):2554–6.
Shortencarier MJ, Dayton PA, Bloch SH, Schumann PA, Matsunaga TO, Ferrara KW. A method for radiation-force localized drug delivery using gas-filled lipospheres. IEEE Trans Ultrason Ferroelectr Freq Control. 2004;51(7):822–31.
Steffen W, Fishbein MC, Luo H, Lee DY, Nita H, Cumberland DC, Tabak SW, Carbonne M, Maurer G, Siegel RJ. High intensity, low frequency catheter-delivered ultrasound dissolution of occlusive coronary artery thrombi: an in vitro and in vivo study. J Am Coll Cardiol. 1994;24(6):1571–9.
Stephens RJ, Hart CP, Torbit CA, Edmonds PD. Reproducible subcellular alterations in hepatocytes resulting from ultrasound. Ultrasound Med Biol. 1980;6(3):239–49.
Stratmeyer ME, Greenleaf JF, Dalecki D, Salvesen KA. Fetal ultrasound: mechanical effects. J Ultrasound Med. 2008;27(4):597–605 (quiz 606–599).
Suslick KS. Ultrasound: its chemical, physical, and biological effects. Science. 1988.
Suslick KS, Grinstaff MW. Protein microencapsulation of nonaqueous liquids. J Acoust Soc Am. 1990;112(21):7807–9.
Tabuchi Y, Takasaki I, Zhao Q-L, Wada S, Hori T, Feril LB Jr, Tachibana K, Nomura T, Kondo T. Genetic networks responsive to low-intensity pulsed ultrasound in human lymphoma U937 cells. Cancer Lett. 2008;270(2):286–94.
Tachibana K. Emerging technologies in therapeutic ultrasound: thermal ablation to gene delivery. Hum Cell. 2004;17(1):7–15.
Tachibana K, Tachibana S. Albumin microbubble echo-contrast material as an enhancer for ultrasound accelerated thrombolysis. Circulation. 1995;92(5):1148–50.
Tachibana K, Tachibana S. Application of ultrasound energy as a new drug delivery system. Nihon Yakurigaku Zasshi. 1999;38(5S):3014.
Taggart LR, Baddour RE, Giles A, Czarnota GJ, Kolios MC. Ultrasonic characterization of whole cells and isolated nuclei. Ultrasound Med Biol. 2007;33(3):389–401.
Taniyama Y, Tachibana K, Hiraoka K, Aoki M, Yamamoto S, Matsumoto K, Nakamura T, Ogihara T, Kaneda Y, Morishita R. Development of safe and efficient novel nonviral gene transfer using ultrasound: enhancement of transfection efficiency of naked plasmid DNA in skeletal muscle. Gene Ther. 2002a;9(6):372–80.
Taniyama Y, Tachibana K, Hiraoka K, Namba T, Yamasaki K, Hashiya N, Aoki M, Ogihara T, Yasufumi K, Morishita R. Local delivery of plasmid DNA into rat carotid artery using ultrasound. Circulation. 2002b;105(10):1233–9.
Tartis MS, McCallan J, Lum AF, LaBell R, Stieger SM, Matsunaga TO, Ferrara KW. Therapeutic effects of paclitaxel-containing ultrasound contrast agents. Ultrasound Med Biol. 2006;32(11):1771–80.
Taylor KJW, Pond JB. Primary sites of ultrasonic damage. Phys Med Biol. 1972;17(3):440.
ter Haar G. Therapeutic applications of ultrasound. Prog Biophys Mol Biol. 2007;93(1):111–29.

ter Haar G, Dyson M, Smith SP. Ultrastructural changes in the mouse uterus brought about by ultrasonic irradiation at therapeutic intensities in standing wave fields. Ultrasound Med Biol. 1979;5(2):167–79.

Tian ZM, Wan MX, Wang B, Wang SP, Wu XM, Ruan YS. Effects of ultrasound on the structure and function of tumor necrosis factor-alpha. Ultrasound Med Biol. 2003;29(9):1331–9.

Tian ZM, Wan MX, Wang SP, Kang JQ. Effects of ultrasound and additives on the function and structure of trypsin. Ultrason Sonochem. 2004a;11(6):399–404.

Tian ZM, Wan MX, Wang ZL, Wang B. The preparation of genistein and LC–MS/MS on-line analysis. Drug Dev Res. 2004b;61(1):6–12.

Tian ZM, Wan MX, Lu MZ, Wang XD, Wang L. The alteration of protein profile of Walker 256 carinosarcoma cells during the apoptotic process induced by ultrasound. Ultrasound Med Biol. 2005;31(1):121–8.

Tiukinhoy SD, Mahowald ME, Shively VP, Nagaraj A, Kane BJ, Klegerman ME, MacDonald RC, McPherson DD, Matsumura JS. Development of echogenic, plasmid-incorporated, tissue-targeted cationic liposomes that can be used for directed gene delivery. Invest Radiol. 2000;35(12):732–8.

Tran T, Roger S, Le Guennec J-Y, Tranquart F, Bouakaz A. Effect of ultrasound-activated microbubbles on the cell electrophysiological properties. Ultrasound Med Biol. 2007;33 (1):158–63.

Trubestein G, Engel C, Etzel F, Sobbe A, CremerH, Stumpff U. Thrombolysis by ultrasound. Clin Sci Mol Med. 1976;51(Pt s3):697s–698s.

Unger EC, Hersh E, Vannan M, McCreery T. Gene delivery using ultrasound contrast agents. Echocardiography. 2001;18(4):355–61.

Unger EC, Porter T, Culp W, Labell R, Matsunaga T, Zutshi R. Therapeutic applications of lipid-coated microbubbles. Adv Drug Deliv Rev. 2004;56(9):1291–314.

Valko M, Rhodes C, Moncol J, Izakovic M, Mazur M. Free radicals, metals and antioxidants in oxidative stress-induced cancer. Chem Biol Interact. 2006;160(1):1–40.

van Wamel A, Bouakaz A, de Jong N. Duration of ultrasound bubbles enhanced cell membrane permeability. In: Ultrasonics. Piscataway: IEEE; 2003. pp. 917–20.

Vivino AA, Boraker DK, Miller D, Nyborg W. Stable cavitation at low ultrasonic intensities induces cell death and inhibits 3H-TdR incorporation by Con-A-stimulated murine lymphocytes in vitro. Ultrasound Med Biol. 1985;11(5):751–9.

Waddington C, Perry MM. The ultrastructure of the developing urodele notochord. Proc R Soc Lond Ser B Biol Sci. 1962;156(965):459–82.

Wang JH-C, Thampatty BP. An introductory review of cell mechanobiology. Biomechan Model Mechanobiol. 2006;5(1):1–16.

Wang Y, Yuan F. Delivery of viral vectors to tumor cells: extracellular transport, systemic distribution, and strategies for improvement. Ann Biomed Eng. 2006;34(1):114–27.

Wang X, Liu Q, Wang Z, Wang P, Zhao P, Zhao X, Yang L, Li Y. Role of autophagy in sonodynamic therapy-induced cytotoxicity in S180 cells. Ultrasound Med Biol. 2010;36 (11):1933–46.

Wang P, Wingnang Leung A, Xu C. Low-intensity ultrasound-induced cellular destruction and autophagy of nasopharyngeal carcinoma cells. Exp Ther Med. 2011;2(5):849–52.

Ward M, Wu J, Chiu J-F. Ultrasound-induced cell lysis and sonoporation enhanced by contrast agents. J Acoust Soc Am. 1999;105(5):2951–7.

Ward M, Wu J, Chiu J-F. Experimental study of the effects of optison® concentration on sonoporation in vitro. Ultrasound Med Biol. 2000;26(7):1169–75.

Wu J, Ross JP, Chiu J-F. Reparable sonoporation generated by microstreaming. J Acoust Soc Am. 2002;111(3):1460–4.

Xu Z, Ludomirsky A, Eun LY, Hall TL, Tran BC, Fowlkes JB, Cain CA. Controlled ultrasound tissue erosion. IEEE Trans Ultrason Ferroelectr Freq Control. 2004;51(6):726–36.

Xu Z, Hall TL, Fowlkes JB, Cain CA. Effects of acoustic parameters on bubble cloud dynamics in ultrasound tissue erosion (histotripsy). J Acoust Soc Am. 2007;122(1):229–36.

Yan C, Resau JH, Hewetson J, West M, Rill WL, Kende M. Characterization and morphological analysis of protein-loaded poly (lactide-co-glycolide) microparticles prepared by water-in-oil-in-water emulsion technique. J Control Release. 1994;32(3):231–41.

Yock PG, Fitzgerald PJ. Catheter-based ultrasound thrombolysis shake, rattle, and reperfuse. Circulation. 1997;95(6):1360–2.

Zarnitsyn V, Rostad CA, Prausnitz MR. Modeling transmembrane transport through cell membrane wounds created by acoustic cavitation. Biophys J. 2008;95(9):4124–38.

Zhong W, Sit WH, Wan JM, Yu AC. Sonoporation induces apoptosis and cell cycle arrest in human promyelocytic leukemia cells. Ultrasound Med Biol. 2011;37(12):2149–59.

Zhong W, Chen X, Jiang P, Wan JM, Qin P, Yu AC. Induction of endoplasmic reticulum stress by sonoporation: linkage to mitochondria-mediated apoptosis initiation. Ultrasound Med Biol. 2013;39(12):2382–92.

MIX
Papier aus verantwortungsvollen Quellen
Paper from responsible sources
FSC® C105338

If you have any concerns about our products,
you can contact us on
ProductSafety@springernature.com

In case Publisher is established outside the EU,
the EU authorized representative is:
Springer Nature Customer Service Center GmbH
Europaplatz 3, 69115 Heidelberg, Germany

Printed by Libri Plureos GmbH
in Hamburg, Germany